DEEP EARTHQUAKES

This is the first book to present a comprehensive description and discussion of 'deep earthquakes' – earthquakes with origins deeper than 60 km, including some with depths as great as 650–700 km. Deep earthquakes are scientifically important because they account for approximately one-quarter of all earthquakes; because a few are very large and damaging; because they provide much of the data that constrain our knowledge of Earth structure and dynamics; and because scientists still don't fully understand the mechanical processes that allow them to occur.

The first three chapters explain what deep earthquakes are, why they are important, and how they were first discovered. The next three chapters describe the distributions of their depths and size, evaluate how they cluster in both time and space, and review observation about their source properties. There are then three chapters that discuss theories for the origin of deep earthquakes and the significance of deep earthquakes for contemporary science. The final chapter is a comprehensive geographic summary of the literature about deep seismicity for 27 individual terrestrial regions and the Earth's Moon.

Deep Earthquakes presents a comprehensive, topical, historical, and geographical summary of deep earthquakes and related phenomena. It will be of considerable interest to researchers and graduate students in the fields of earthquake seismology and deep Earth structure.

CLIFF FROHLICH received a Ph.D. in physics from Cornell University in 1976. Since 1978 he has been employed as a research scientist at the University of Texas at Austin where he is now Associate Director of the Institute for Geophysics. The majority of his many scholarly publications concern deep earthquakes, the statistical analysis of earthquakes catalogs, and regional tectonics. His first book, Texas Earthquakes, coauthored with Scott D. Davis, was published in 2002. Dr Frohlich also has a keen interest in the physics of sport and is the author of a paper on the physics of diving and gymnastics for Scientific American.

DEEP EARTHQUAKES

CLIFF FROHLICH

Institute for Geophysics, Jackson School of Geosciences
University of Texas at Austin

CAMBRIDGE
UNIVERSITY PRESS

CAMBRIDGE UNIVERSITY PRESS
Cambridge, New York, Melbourne, Madrid, Cape Town, Singapore,
São Paulo, Delhi, Dubai, Tokyo

Cambridge University Press
The Edinburgh Building, Cambridge CB2 8RU, UK

Published in the United States of America by Cambridge University Press, New York

www.cambridge.org
Information on this title: www.cambridge.org/9780521123969

First published 2006
This digitally printed version 2009

A catalogue record for this publication is available from the British Library

ISBN 978-0-521-82869-7 Hardback
ISBN 978-0-521-12396-9 Paperback

Contents

v

Preface

Why write a book on deep earthquakes?

On 21 January 1906 an earthquake with M_W of 7.4 occurred at a depth of 340 km beneath Japan. Fusakichi Omori, one of the world's finest seismologists of that time, located the event using (S-P) intervals, and concluded that at several Japanese seismograph stations the P waves were transverse and the S waves were longitudinal. This puzzled Omori, so he published a paper about this a year later (see Fig. 0.1 and Chapter 3). So began the story of deep earthquakes – Omori's (1907) paper is among the earliest publications in my collection of references on deep earthquakes.

Of course, for many reasons, Omori wouldn't have told the story of the beginning quite this way. In 1907 the magnitude scale hadn't yet been invented, and Omori's paper never uses the terms "P", "S", or "(S-P) interval"; rather, it mentions the "1st displacement," the "2nd displacement," and the "duration of 1st preliminary tremor." Omori also didn't know that the earthquake was deep; indeed, he didn't even know that "normal" earthquakes occurred at depths of 40 km or less. Our perspective on earthquakes has changed considerably since 1907; my collection of references on deep earthquakes now has more than 2000 entries. We have learned a great deal about earthquakes over the past 100 years. For example, one thing we have learned is that deep earthquakes are quite common – nearly a third of all earthquakes located by the International Seismological Centre have focal depths exceeding 60 km; of these, about one fifth exceed 300 km.

The purpose of this book is to summarize what a century of research has taught us about deep earthquakes, i.e., earthquakes with focal depths of 60 km and greater. As far as I can tell, no one has written a summary book about deep earthquakes – ever. Thus, I submit that the passage of 100 years, the appearance of 2000 publications, and the absence of any other book marks a good time to review what we now know about deep earthquakes.

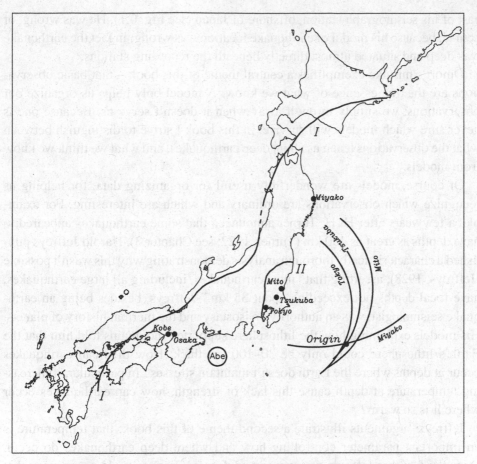

Fig. 0.1 Reproduction of Omori's 1907 figure summarizing his observations of the Japan earthquake of 21 January 1906. The arcs indicate event-station distances determined from (S-P) intervals recorded at Tokyo, Mt. Tsukuba, Mito, and Miyako. Omori assumed that the earthquake was shallow and occurred east of Japan where the arcs crossed; a modern location (Abe, 1985) finds that it was deep and 360 km further west (circle labeled "Abe"). Roman numerals I and II show the boundaries of the area of "slight motion" and "strong motion," respectively.

So, what do we know, and what don't we know? Even if you take the time to read all those 2000 publications, the answers aren't straightforward. This is because research papers don't report only established facts or observations; they also interpret these observations and attempt to explain them with models and speculation. Thus, much that the literature reports isn't well established – it often isn't even true. For example, for the 1906 earthquake Omori concluded that P waves were transverse and S waves were longitudinal because he believed all earthquakes occurred at shallow depths and thus the (S-P) intervals proved his event occurred east of his seismograph station, offshore of Japan (see Fig. 0.1). He was wrong, of course, because his model of earthquake location was wrong; in fact the earthquake was deep and situated almost directly beneath the recording stations.[1]

Omori's mistake exemplifies a central theme of this book – that basic observations are the true essence of what we know. A model only helps us organize our observations; we simply throw it away when it doesn't serve us. Because one is never sure which models will survive, in this book I strive to distinguish between what the observations teach us about deep earthquakes, and what we think we know from models.

Of course, models are wonderfully useful for organizing data, for helping us recognize which observations are ordinary and which are interesting. For example, a few years after H. H. Turner announced that some earthquakes appeared to have depths as great as 400 km (Turner, 1922; see Chapter 3), Harold Jeffreys published a characteristically thorough analysis demonstrating why this wasn't possible (Jeffreys, 1928), asserting that "most earthquakes, including all large earthquakes, have focal depths not exceeding about 35 km." Jeffreys, besides being an earthquake seismologist, was an authority on isostasy and the thermal history of planets. His models explaining how the lithosphere supported mountains told him that the Earth's lithosphere could only be 30–100 km thick. How then can earthquakes occur at depths where the Earth doesn't maintain stresses? In particular, if increasing temperature at depth cause this lack of strength, how can earthquakes occur where it is so warm?

Jeffreys' arguments illustrate a second theme of this book: that temperature is an important parameter controlling how and where deep earthquakes do occur. Does the fact that Jeffreys was wrong and deep quakes really do occur mean that temperatures where they occur are very low? Or do they represent a mechanical phenomenon that is fundamentally different from earthquakes that occur in the lithosphere?

This book is organized into five sections; the first four sections present a topical review of deep earthquake phenomena. The first section presents background information and consists of three chapters:

- Chapter 1 introduces deep earthquakes by presenting observations about several large and significant twentieth century deep earthquakes;
- Chapter 2 defines what "deep" is, presents a simple model to estimate temperature within subducted lithosphere, and discusses why deep earthquakes are worth studying;
- Chapter 3 describes the discovery of deep earthquakes and explores their early history, concentrating primarily on the period prior to 1970.

---

[1] We mustn't be too hard on Omori. His research was extraordinary enough that I found it worth reading almost 100 years later. If anybody bothers to read this book a century from now, please contact my descendants and tell them I said "Hello."

The second section, with two chapters, focuses on the properties of earthquake catalogs:

- Chapter 4 looks at the distributions of focal depths, magnitudes, and scalar moments;
- Chapter 5 evaluates spatial and temporal clustering of hypocenters.

The third section addresses the mechanism of deep earthquakes:

- Chapter 6 reviews observations of source properties;
- Chapter 7 considers various mechanical models that explain the occurrence of deep earthquakes.

The fourth section considers why deep earthquakes have been and still are worth studying:

- Chapter 8 shows how the analysis of deep earthquake signals has provided key information about a great many features of Earth structure and dynamics;
- Chapter 9 considers what we have learned over the past 100 years, what we have still to learn, and deep earthquake research's place in the history of the science.

Absent from these four sections is any sense of the character of deep seismicity in specific geographic regions; this is the topic of the fifth and last section – a single chapter – Chapter 10. This presents maps, cross sections, focal mechanisms, and a review of the literature for deep-focus and intermediate earthquakes in each of 28 geographic regions. The reader of Chapter 10 will learn that there are decided differences in the "flavor" of research in the various regions. For example, the majority of the literature concerning South America (Section 10.18) appears in well-known journals and concentrates on variations in the geometry of Wadati–Benioff zones, and also on the source properties of a number of extraordinarily large deep-focus earthquakes. In contrast, much of the literature on Romania (Section 10.24) appears in books, technical reports, and regional publications, and concerns models for evaluating hazard. Finally, many papers concerning Greek deep seismicity (Section 10.23) appear in European journals and focus on time-dependent trends in seismic activity or relations between deep seismicity and other phenomena, such as volcanism of geoelectric signals. Chapter 10 also includes descriptions of 58 individual deep earthquakes which are unusually large, important historically, or otherwise remarkable.

Each of the first nine chapters and each geographic summary in Chapter 10 is self-contained; i.e., mostly you can read them independently without reading preceding material. However, a few abbreviations that appear regularly throughout the book (see Table 0.1) won't be explained repeatedly in each chapter. Separate reference lists follow each of the first nine chapters and each geographic summary. Some of these lists are quite long; thus, for readers desiring further guidance I have

Table 0.1. *Abbreviations used throughout this book.*

Catalogs of hypocenters: See detailed explanation and references in Chapter 10.

Abe	Catalog published by Abe (1981) covering the period 1897–1980
CMT	Centroid moment tensor – and the catalog published by Harvard for the period 1977–present
CMT-Hist	Additional centroid moment tensors for the historical period 1906–1976 published by various sources
ISC	International Seismological Centre – and their catalog covering the years 1964–present
EHB	Engdahl, van der Hilst, and Buland (1998) – and their catalog comprised of selected relocations of ISC data 1964–2004
EV Centennial	Engdahl and Villaseñor (2002) – and their catalog covering the period 1900–1999
GR	Gutenberg and Richter (1954) – and their catalog covering the period 1904–1952

Magnitude scales:

M_W	Moment magnitude determined from scalar moment M_o by the formula $M_W[N\text{-}m] = (2/3)\log_{10}M_o - 6$
M_S	Surface wave magnitude; usually as reported in the GR catalog
m_b	Short period body-wave magnitude, usually as reported in the ISC catalog
m_B	Longer-period body-wave magnitude, usually as reported in the Abe catalog

Other abbreviations:

CLVD	Compensated linear vector dipole; see Section 6.2.1
h	Focal depth estimate for an earthquake
JHD	Joint hypocenter determination
NUVEL	Model describing the relative motions of tectonic plates; developed by DeMets *et al.* (1990; 1994)
pP depth	Focal depth determined from (pP–P) intervals; see Section 4.1.2
PREM	Preliminary reference earth model – a global model of velocity and elastic parameters proposed by Dziewonski and Anderson (1981)
USGS	United States Geological Survey
WWSSN	World Wide Standardized Seismograph Network

marked a few especially important, interesting, or comprehensive references with a "•".

Why was this book written, really?

In 1935 Konrad Lorenz published a classic paper on the learning behavior of young ducklings and goslings. He noted that at certain specific developmental periods some animals seemed instinctively primed to learn quite remarkable behaviors. As an example, he observed that soon after hatching, baby ducklings learned to follow the first conspicuous moving object they saw – usually the mother duck. But, they might also "imprint" on a bird of another species, a bright red ball, a scientist, or almost anything else, if they saw it at the appropriate time. Subsequently, the ducklings would continue to follow this imprint-parent, even if doing so didn't make much sense.

In my experience, we scholars aren't all that different from ducklings – at some key time in our university education – usually in graduate school – we open our eyes and imprint on some particular problem that seems interesting to us. Then we follow it blindly through our careers, steadfastly giving it the lion's share of our attention, even if doing so doesn't make much sense.

I here confess that, as a graduate student at Cornell University in about 1974, I imprinted on deep earthquakes. Actually, when I began graduate school I worked in a low-temperature laboratory operated by physicists John Reppy, Bob Richardson, and Dave Lee. The graduate student working on the apparatus next to mine was Doug Osheroff. Doug was having all kinds of trouble with a gauge that got stuck every time he tried to cool liquid helium mixtures below a certain temperature. It was clear to me that nothing would come out of this; Doug's gauge was obviously faulty and everyone knew that near absolute zero nothing much happened.[2] I became disillusioned with low-temperature physics and soon switched to geophysics and earthquake seismology. Within a year I was studying deep earthquakes.

In the three subsequent decades I have maintained a steadfast and passionate interest in deep earthquakes in spite of distractions provided by numerous wives, children, and research grants which should have encouraged me to focus on something more practical. But, I can't help myself; I just find deep earthquakes interesting; I am obsessed with deep earthquakes. Why I chose deep earthquakes and not something equally respectable, such as low-temperature physics, is a puzzle.

This book, then, is an attempt to do something useful with my obsession. If you share my interest in deep earthquakes, please use this book to move the subject forward in any way you can. Alternatively, if you are passionate about some other subject but for some reason need to know about deep earthquakes, this book may help you avoid sifting through what has become a considerable literature.

Who is not responsible?

Although I must accept full responsibility for the material in this book, numerous individuals helped and encouraged me during its preparation. First, I am indebted

---

[2] Time has demonstrated that it was my insight that was faulty, not the gauge. It turned out that Osheroff's gauge was working fine, but "sticking" as the helium mixture cooled through a previously unknown phase transition. Osheroff, Richardson, and Lee won the Nobel Prize for this experiment in 1996.

to Bryan Isacks and Muawia Barazangi, my earthquake mentors at Cornell when I imprinted on deep earthquakes. Second, I must thank Steve Kirby and Terry Wallace, who each almost coauthored this book with me. In 1995 Kirby and I proposed to write a book on deep earthquakes, but gave up when the book proposal was rejected. Later, Wallace approached me about coauthoring a book, but we both became overwhelmed with other projects and made little progress. The present manuscript came about largely due to encouragement from my wife, Jacqueline Henkel, and from a close friend, Frank Whigham. Both are professors of English, and they managed to convince me that writing a book might be a reasonable thing to do even if the scientific community generally values journal publications more than books.[3]

Finally, I am grateful to numerous colleagues who were supportive of this project. Those who reviewed various sections of this book and provided thoughtful criticism were: Geoff Abers, Muawia Barazangi, John Cassidy, George Choy, Mike Coffin, Patricia Cooper, Shamita Das, Diane Doser, Bob Engdahl, Xaq Frohlich, Alexei Gorbatov, Steve Grand, Jacqueline Henkel, James Jackson, Yan Kagan, Honn Kao, Junji Koyama, Rob McCaffrey, Yosio Nakamura, Jim Ni, Lani Oncescu, Martin Reyners, Ray Russo, David Rubie, Martha Savage, Cezar Trifu, Agustín Udías, Doug Wiens, and Ray Willemann. Scott Davis read a draft of the entire manuscript and caught numerous blunders, subtle and otherwise. I owe special thanks to Bob Engdahl, who sent me a recently updated version of the EHB catalog complete to December 2004. Inés Benlloch and Yosio Nakamura provided helpful assistance with translations. Geoff Abers, Michael Antolik, Shamita Das, Brad Hacker, and Shun-ichiro Karato were kind enough to provide electronic copies of figures. A book like this couldn't happen without access to a good library and professional support from a good librarian; for these I am indebted to Dennis Trombatore and The University of Texas Library.

References

Abe, K., 1981. Magnitudes of large shallow earthquakes from 1904 to 1980, *Phys. Earth Planet. Int.*, **27**, 72–92.

1985. Re-evaluation of the large deep earthquake of Jan. 21, 1906, *Phys. Earth Planet. Int.*, **39**, 157–166.

DeMets, C., R. G. Gordon, D. F. Argus, and S. Stein, 1990. Current plate motions, *Geophys. J. Int.*, **101**, 425–478.

1994. Effect of recent revisions to the geomagnetic reversal timescale on estimates of current plate motions, *Geophys. Res. Lett.*, **21**, doi:10.1029/94GL02118, 2191–2194.

---

[3] My wife and my friend may be wrong. Indeed, no less an authority on science than Thomas Kuhn, the celebrated inventor of the concept of scientific paradigms, has stated that "The scientist who writes [books] is more likely to find his professional reputation impaired than enhanced" (Kuhn, 1970).

Dziewonski, A. M. and D. L. Anderson, 1981. Preliminary reference earth model, *Phys. Earth Planet. Int.*, **25**, 297–356.

Engdahl, E. R., R. van der Hilst, and R. Buland, 1998. Global teleseismic earthquake relocation with improved travel times and procedures for depth determination, *Bull. Seismol. Soc. Amer.*, **88**, 722–743.

Engdahl, E. R. and A. Villaseñor, 2002. Global seismicity: 1900–1999, In *International Handbook of Earthquake and Engineering Seismology*, San Diego, CA, Academic Press for International Association of Seismology and Physics of the Earth's Interior, 665–690.

Gutenberg, B. and C. F. Richter, 1954. *Seismicity of the Earth and Associated Phenomena* (2nd edition), Princeton, N.J., Princeton University Press, 310 pp.

Jeffreys, H., 1928. The times of transmission and focal depths of large earthquakes, *Mon. Not. Roy. Astron. Soc., Geophys. Supp.*, **1**, 500–521.

Kuhn, T. S., 1970. The Structure of Scientific Revolutions, (2nd edition), Chicago, University Chicago Press, 210 pp.

• Omori, F., 1907. Seismograms showing no preliminary tremor, *Bull. Imperial Earthquake Investigation Committee*, **1**, No. 3, 145–154.

Turner, H. H., 1922. On the arrival of earthquake waves at the antipodes, and on the measurement of the focal depth of an earthquake, *Mon. Not. R. Astron. Soc., Geophys. Supp.*, **1**, 1–13.

Part I

Background and introductory material

1

The big, the bad, and the curious

If you asked most seismologists for a brief summary of what they know about intermediate- and deep-focus earthquakes, they might write something like:

Although nearly all of the world's earthquakes occur within the crust, there are a few with foci in the mantle having depths between 60 and 700 km. These so-called "deep earthquakes" occur near deep-ocean trenches within planar groups of hypocenters called Wadati–Benioff zones. In comparison with shallow-focus earthquakes, deep earthquakes tend to be smaller in size, have higher stress drops, more impulsive source-time functions, and few or no aftershocks. Generally, deep earthquakes aren't destructive because of their small size, distance from the surface, and tendency to occur in oceanic regions.

Unfortunately, most of the assertions in the above statement aren't strictly true; i.e., some are never true, some are true only sometimes, and some we aren't yet sure about. For example, in global earthquake catalogs about 25% of all earthquakes have reported focal depths exceeding 60 km (Table 1.1); the 75% that are shallower are hardly "nearly all." Moreover, even for very large earthquakes with M_W or m_B larger than 8.0, those with focal depths of 60 km and more make up about a quarter of all events.[1] Moreover, we shall see that not all deep earthquakes are associated with oceanic trenches; many important deep quakes occur as isolated events and not within planar groups. Whether deep quakes have higher stress drops and more impulsive source time functions is subject to debate. Finally, some deep earthquakes do have fairly numerous aftershocks and some deep earthquakes are highly destructive.

To illustrate some important characteristics of deep earthquakes, this chapter will present descriptions of several events. In different ways, each shows why the above statement is problematic. In addition, each of the quakes discussed is significant

[1] In the twentieth century, deep earthquakes are absent among very, very large events with M_W exceeding 8.5. However, until the Sumatra earthquakes of 26 December 2004 and 28 March 2005 there had been no earthquakes in the Harvard CMT catalog of any depth with M_W exceeding 8.5, and the largest m_B in the Abe catalog is only 8.3. Indeed, Abe assigned an m_B of 7.9 to both the 1960 Chile and 1964 Alaska earthquakes.

3

Table 1.1. *Fractions of shallow and deep earthquakes reported in the Harvard,*
Abe (1981), and ISC catalogs.

Catalog	Number; years	Shallow $h < 60\,km$	Deep $h > 60\,km$	Deep-focus $h > 300\,km$
Harvard; $M_W \geq 8.0$	13; 1977–2004	0.77	0.23	0.08
Abe; $m_B \geq 8.0$	13; 1897–1976	0.77	0.23	–
Harvard; $M_W \geq 7.0$	376; 1977–2004	0.73	0.27	0.09
Abe; $m_B \geq 7.0$	1110; 1897–1976	0.62	0.38	0.07
Harvard; $M_W \geq 5.6$	9403; 1977–2004	0.76	0.24	0.07
ISC; $m_b \geq 5.3$	18840; 1964–2000	0.70	0.30	0.05

because it is a celebrated representative of a particular category of deep earthquakes, or because it raises important questions about mantle dynamics.

1.1 The big: 9 June 1994 Bolivia

This earthquake was deep – Harvard assigned it a focal depth of 647 km – and really, really big, with a magnitude M_W of 8.2. Indeed, when it occurred it was the largest earthquake of any depth the world had experienced in the 17 years since the Sumbawa earthquake of 1977. It is also the largest earthquake ever recorded with a focal depth greater than 300 km.

A remarkable feature of South America is that it possesses a disproportionate fraction of the world's large very deep earthquakes (Table 1.2). For quakes with focal depths exceeding 300 km, South America has experienced the world's largest (1994 Bolivia), the second and third largest (Colombia, 1970; and northern Peru, 1922), and also the tenth, 13th, 15th, and 18th largest (Peru–Bolivia border: August 1963, November 1963, 1961, and 1958). Why such large earthquakes occur here and not elsewhere is a significant question. The answer isn't obvious, since the lithosphere subducting beneath South America is neither particularly old and fast (as in Tonga) nor particularly young and slow (as in the northwestern United States). What then, is special about South America?

Although the 1994 Bolivia earthquake was felt throughout much of South America, it caused only minor damage. It broke windows in tall buildings and caused some structural damage in La Paz, Cochabamba, and Oruro, all towns within about 500 km of the epicenter. It caused numerous landslides in southern Peru, which allegedly were responsible for numerous injuries and four deaths.

An unusual feature of the Bolivia earthquake was that it was felt in North American cities at distances of 50–80° from the epicenter (Fig. 1.1). For example,

Table 1.2. *Earthquakes with depths greater than 300 km and moments exceeding* 2.5×10^{20} *N-m occurring between 1906 and 2004. Source: Huang and Okal (1998) augmented by CMT catalog for events occurring since 1996.*

Rank	Date	Area	Depth (km)	Moment (N-m)	M_W	Environment
1	09 Jun 1994	Bolivia	647	2.6×10^{21}	8.3	bend
2	31 Jul 1970	Colombia	623	1.4×10^{21}	8.1	isolated
3	17 Jan 1922	North Peru	664	9.4×10^{20}	7.9	isolated
4	17 Jun 1996	Flores Sea	589	7.9×10^{20}	7.9	bend
5	29 Mar 1954	Spain	630	7.0×10^{20}	7.9	isolated
6	29 Sep 1973	North Korea	593	5.0×10^{20}	7.8	bend/edge
7	11 Jun 1972	Celebes Sea	332	4.7×10^{20}	7.7	
8	19 Aug 2002	Fiji	699	4.3×10^{20}	7.7	
9	26 May 1932	Fiji	560	4.0×10^{20}	7.7	bend
10	15 Aug 1963	Peru–Bolivia	573	3.9×10^{20}	7.7	bend
10	28 Feb 1950	Sea of Okhotsk	339	3.9×10^{20}	7.7	
12	25 May 1907	Sea of Okhotsk	548	3.7×10^{20}	7.7	
13	09 Nov 1963	Peru–Bolivia	573	3.5×10^{20}	7.7	
13	19 Aug 2002	Fiji	631	3.5×10^{20}	7.7	
15	19 Aug 1961	Peru–Bolivia	620	3.4×10^{20}	7.7	
16	09 Mar 1994	Fiji	563	2.7×10^{20}	7.6	
16	23 May 1956	Fiji	436	2.7×10^{20}	7.6	
18	26 Jul 1958	Peru–Bolivia	592	2.6×10^{20}	7.6	bend

the *Minneapolis Star Tribune* quoted a woman living on the 12th floor of a condominium who

. . . felt her bed rocking and thought the wind must be blowing something fierce. "Then I looked outside the building, no wind," she said. She went into the living room and asked her father whether he had felt the shock. He had, and the two of them hurried downstairs and into their car in the parking lot. They used a car phone to call the police. "The guy thought I was crazy," the woman said.

Many of the felt reports came from people situated in buildings having three to twelve stories, and correspond to intensities of MM-I or MM-II.[2] Steel and concrete buildings with N stories have natural periods of approximately $0.08N$ (Kanai, 1983), which for buildings of 3–12 stories corresponds to frequencies of 1–4 Hz.

The 1994 Bolivia earthquake is the only deep-focus event to generate felt reports at such great distances. After the earthquake Anderson *et al.* (1995) evaluated these

---

[2] Felt reports at teleseismic distances are most often caused by surface waves from very large shallow earthquakes. For example, after the Alaska earthquake of 28 March 1964 ($M_W = 9.2$), newspapers widely reported water sloshing over the sides of swimming pools, and waves up to 1.8 meters high overturned small boats and caused minor damage $45°$ away in several channels along the Texas and Louisiana coasts.

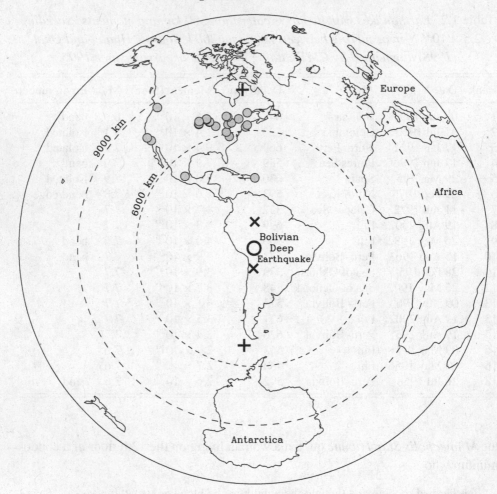

Fig. 1.1 Felt reports at great distances (filled circles) for the 9 June 1994 Bolivia deep-focus earthquake. Plus (+) and (×) symbols indicate locations at distances greater than 1000 km from the epicenter where P and S waves respectively have maximum amplitude as they leave the focal region and reach the Earth's surface. The timing of the distant felt reports and their location nearer the local P-maximum suggests that P, and not S, was responsible. Plotted felt report data are from Anderson *et al.* (1995).

reports, and concluded that the reported times indicated that people were experiencing the arrival of P or PcP waves, rather than S or surface waves (Table 1.3). Moreover, seismograms at North American cities indicated that the highest accelerations were attributable to P rather than S or surface waves (Fig. 1.2), with peak accelerations in the frequency range of 0.5–5.0 Hz. The surface waves for this quake had low amplitudes because of its great focal depth. And, the quake's

Table 1.3. *Reported times in relation to estimated P and S arrival times for felt effects in North American cities following the 1994 Bolivian earthquake. Source for locations and times of felt effects is compilation by Jim Dewey of the U.S. Geological Survey, augmented by additional newspaper reports.*

Time reported	Number of observations
One hour or half-hour before P arrival	2
One minute before P arrival	1
Between P arrival and three minutes after	5
Between S arrival and five minutes after	0
Range of times reported includes both P and S arrivals	2
Long after S arrival	1

focal mechanism had an orientation that produced maximum-amplitude P in north central North America (Fig. 1.1).

A second peculiar feature of the Bolivia earthquake was its location beneath a part of South America where few deep-focus earthquakes had occurred previously (Fig. 1.3). To the northwest, the rare large and isolated deep quakes that occur beneath Colombia lie approximately along the extension of a linear trend of deep-focus activity that extends along the Peru–Brazil border, and the southern extension of this trend meets the Peru–Bolivia border. Then, to the southeast, a second linear group of earthquakes extends from central Bolivia into central Argentina. Both the northwest and southeast groups are in accord with the classical model of deep earthquake occurrence (Fig. 1.4) – that seismic activity occurs along a planar region, the Wadati–Benioff zone, which delineates the cold core of subducting oceanic lithosphere. However, this model doesn't fit the 1994 Bolivia hypocenter. It lies within a distinct gap between the northwest and southeast groups. One possible explanation is that it occurred within a kink in subducted lithosphere that connects the two groups.

What actually occurred in the mantle beneath Bolivia when the 1994 earthquake took place? Ihmlé (1998) modeled the rupture as a continuous process, and concluded that the rupture consisted of six pulses lasting about 50 seconds that released moment along a subhorizontal planar region (Fig. 1.5). He found slip of 4 meters over a region with dimensions of about 20×30 km^2, and slip of at least a meter over a roughly circular area with a diameter of 120 km. He concluded that the rupture front traveled with a velocity of 2 km/s or less. Ihmlé's results are generally consistent with other investigations of the rupture process that modeled the individual pulses as point sources (Beck *et al.*, 1995; Estabrook and Bock, 1995; Goes and Ritsema, 1995; Silver *et al.*, 1995; Antolik *et al.*, 1996). However, because

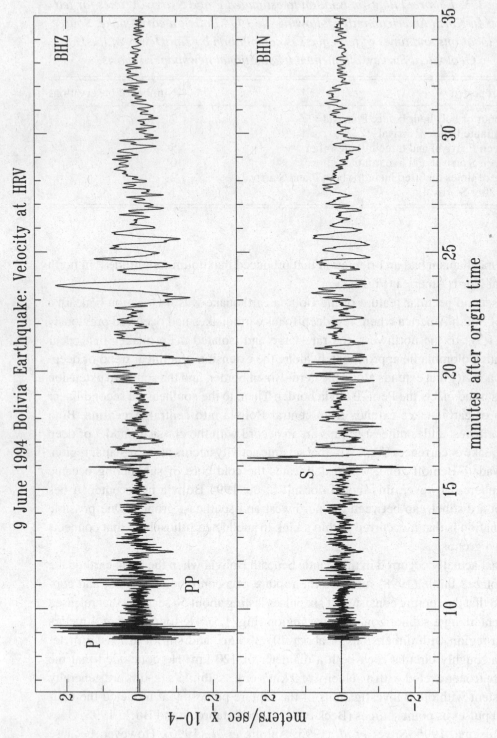

Fig. 1.2 Velocity and acceleration seismograms for the Bolivia earthquake of 9 June 1994 recorded at HRV (Harvard, MA; Δ = 56°, azimuth = 356°). While peak velocities for P and S are about the same, the highest peak accelerations accompany the higher-frequency P arrival.

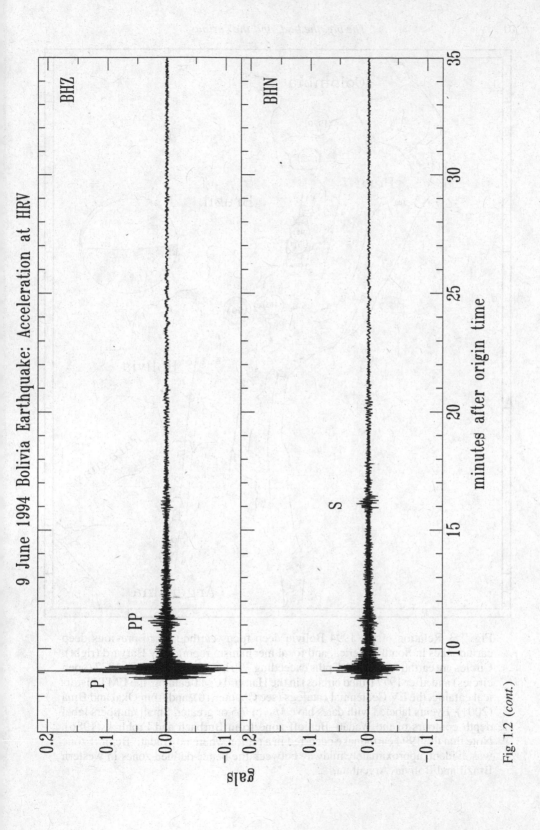

Fig. 1.2 (cont.)

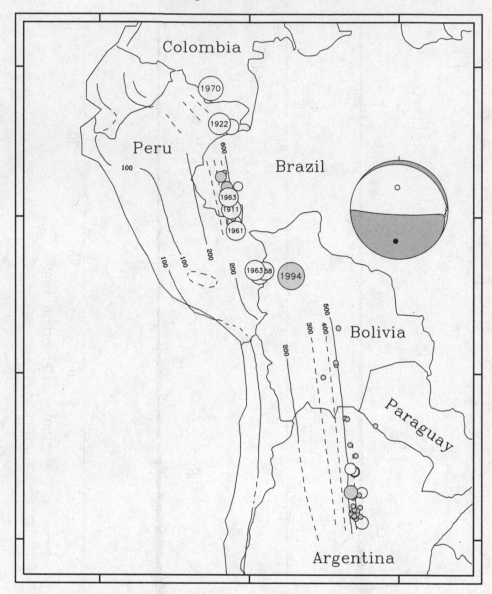

Fig. 1.3 Relation of the 1994 Bolivia deep-focus earthquake to previous deep earthquakes in South America, and focal mechanism reported by Harvard (right). Circles are earthquakes with depths exceeding 300 km occurring before 1977 (open circles) and after 1977 (filled circles) in the Harvard CMT catalog, the CMT historical catalog, the EV Centennial catalogs (see Chapter 10), and from Okal and Bina (2001). Events labeled with dates have M_W of 7.5 or greater. Small numbers label depth contours of the Wadati–Benioff zone from Burbach and Frohlich (1986). Note that the 1994 earthquake occurred in a region where no Wadati–Benioff zone was evident, approximately midway between the better-defined zones in western Brazil and Bolivia–Argentina.

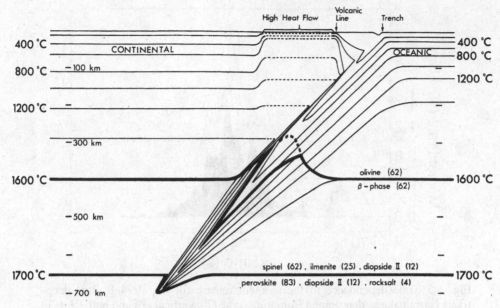

Fig. 1.4 A model of the thermal and petrological environment where subduction-zone deep earthquakes typically occur; thin lines denote isotherms, and the heavy lines represent the positions of the olivine–spinel and spinel-to-oxides phase transitions. As cold oceanic lithosphere penetrates the warmer mantle, its temperature changes because of conduction, because of frictional heating along its upper surface, and because of heat released or absorbed as subducting material undergoes phase transitions. Earthquakes occur within roughly planar zones where the stress is sufficiently high and the temperature sufficiently low. The model shown here is after Liu (1983).

he modeled rupture as occurring along a front rather than at a sequence of localized subevents, Ihmlé found that the subevent locations often did not coincide with the regions of maximum moment release (Fig. 1.6; and see Fig. 6.10).

Serendipitously, when the 1994 Bolivia earthquake occurred there were two temporary local networks operating in Peru, Bolivia, and Brazil, making it possible to detect 89 aftershocks occurring over a period of 20 days. Myers *et al.* [1995] located 45 of these and found them to occur within a slab-like region having a thickness of about 30 km, a lateral dimension of about 55–60 km, and dipping at approximately 45° to the northeast. Although the aftershocks' epicenters did correspond roughly to Ihmlé's zone of maximum slip, they did not delineate a plane coincident with either of the nodal planes of the mainshock. And the majority were situated beneath the mainshock hypocenter at depths up to 660 km. Tinker *et al.* [1995] determined focal mechanisms for 12 of the aftershocks and concluded that most had near-vertical P axes and all possessed mechanisms that were significantly different from that of the mainshock.

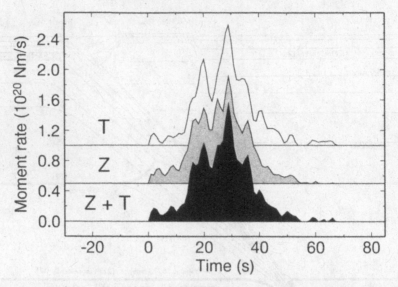

Fig. 1.5 Time dependence of the moment release for the 1994 Bolivia deep-focus earthquake as determined from transverse (T), vertical (Z) and both (Z + T) component data at 50 teleseismic stations. Figure reproduced from Ihmlé and Jordan (1995).

The geometries of both the rupture and the aftershock zones have received considerable scrutiny because they provide important constraints on the physical mechanism of deep earthquakes. One proposed mechanism (see Chapter 7) is that deep-focus earthquakes nucleate when numerous microscopic "anticracks" occur within a thin (~10 km thick) metastable olivine wedge that, because it is colder than normal mantle, survives at the core of subducted lithosphere even at depths exceeding 600 km. However, Silver *et al.* [1995] argue that the dimensions of both rupture and aftershock zones for the 1994 Bolivia earthquake are too large to fit within such a region. This has provided support for proponents of other physical mechanisms, such as shear-induced thermal instabilities which might lead to melting along the rupture surface (Kanamori *et al.*, 1998; Bouchon and Ihmlé, 1999; and Karato *et al.*, 2001). Of course, a difficulty for all interpretations is that the low rate of seismic activity in this region makes it difficult to determine the geometry of the subducting lithospheric slab with certainty.

Because of its size and because digital seismograms from around the world were available to the entire seismological community, more research papers have been published concerning the 1994 Bolivia earthquake than any other deep earthquake, ever. While some of the investigations appear to be redundant, they provide information about what features of earthquakes are well- and which are poorly-determined. For example, the investigations all generally found similar focal

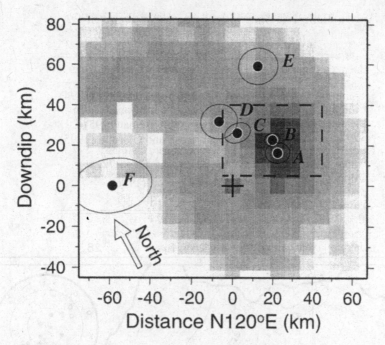

Fig. 1.6 Slip distribution (shading) for the 1994 Bolivia deep-focus earthquake, and subevent locations determined if rupture is modeled as six point sources. Data is plotted along a plane that strikes N120° E and dips 11° towards the north-northeast. The figure is from Ihmlé (1998), and the dashed line delineates the aftershock zone determined by Myers *et al.* (1995).

mechanisms for the mainshock (e.g., Fig. 1.3), they concluded that the rupture proceeded roughly northward along a near-horizontal plane, and found that it consisted of several pulses of moment release occurring over about 50 seconds. However, there was little agreement about the static stress drop; estimates varied from less than 500 bars to almost 3000 bars (Table 1.4). As we shall find in Chapter 6, lack of agreement between different studies is typical for stress drop estimates.

1.2 The bad: 4 March 1977 Romania and 25 January 1939 Chile

A few deep earthquakes kill people – lots of people. On the evening of 4 March 1977 an earthquake with $M_{W(CMT)}$ of 7.5 ruptured the mantle at a depth of 80–110 km beneath Vrancea County in Romania. This earthquake killed more than 1500 people, mostly in the city of Bucharest, a city of about two million people situated about 150 km south of the epicenter (Fig. 1.7). Fattal *et al.* (1977) report that it destroyed or seriously damaged some 33,000 housing units and left some 200,000 people homeless. It destroyed some 374 schools, 11 hospitals, and one orphanage.

Table 1.4. *Static stress drops* $\Delta\sigma$ *reported for the 9 June 1994 Bolivia earthquake.*

$\Delta\sigma$ (bars)	Reference
1100	Kikuchi and Kanamori (1994)
500	Estabrook and Bock (1995)
2830	Goes and Ritsema (1995)
1140	Lundgren and Giardini (1995)
170–310	Myers *et al.* (1995)
710	Goes *et al.* (1997)
1500	Antolik *et al.* (1996)
700–1900	Ihmlé (1998)

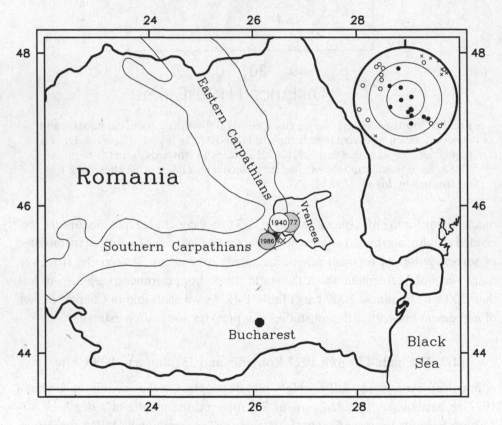

Fig. 1.7 Epicenters of intermediate-depth earthquake activity of Romania. Open circles, filled circles, and crosses are epicenters from the EV Centennial, CMT, and EHB catalogs, respectively (see Chapter 10). Solid lines indicate boundaries of Romania and Vrancea County, and the approximate extent of the Carpathian Mountains. Equal area plot at upper right summarizes focal mechanisms for well-determined Harvard CMT; solid circles are T axes, crosses are B axes, and open circles are P axes.

As in many deadly earthquakes, most of the deaths in Romania are attributable to poor construction. In Bucharest the earthquake caused the collapse of 33 buildings, mostly 8–12 story concrete frame apartment buildings constructed prior to 1940. Typically, the first story was higher than the others and almost devoid of walls so as to accommodate stores and other non-residential facilities. Within the upper stories masonry infill walls and partitions were used liberally to provide enclosures for apartment or office space and to function as bracing against wind action. Thus, the structures possessed laterally stiff upper stories resting on relatively flexible columns at the base. They weren't designed to withstand lateral forces, and the bottom stories pancaked when exposed to lateral forces in the earthquake (Fig. 1.8). Strong-motion instruments in Bucharest indicated that the highest accelerations were predominantly horizontal and arrived as a single strong pulse with an amplitude of 0.2 g and a period of about 1.4 s.

Except for the death and damage it caused, the 1977 earthquake was not particularly unusual for a large intermediate-depth event. The focal mechanism had a near-vertical T axis, which is typical of earthquakes in Vrancea and the most common type for intermediate-depth earthquakes elsewhere. The rupture consisted of at least four pulses of moment release occurring over about 20 seconds, all occurring between depths of 80 and 110 km and extending over a region with horizontal dimensions of about 70 km. Räkers and Müller (1982) found a stress drop of 52 bars, which is somewhat lower than reported for most large intermediate earthquakes.[3] Local stations recorded 140 aftershocks over a period of two months (Fuchs *et al.*, 1979). These included events with depths between 70 and 130 km in the region where the rupture occurred as well as crustal activity above a depth of 40 km.

From a tectonic perspective, perhaps the most notable feature of the 1977 Romanian earthquake was that it occurred in an intercontinental environment where there is no longer ongoing subduction. Although there was active subduction about 10 million years ago it has now ceased, leaving as remnants the Carpathian mountain chain and the Vrancea deep earthquakes zone. The intermediate-depth earthquakes in Romania have been called the Vrancea "nest" because activity is confined within a horizontal region having dimensions of about 15 km × 40 km (Fig. 1.9). However, since their focal depths range from 60 km to more than 160 km; they are not nearly as localized as the earthquakes in the classic nest beneath Bucaramanga, Columbia, where all the activity occurs within a region having vertical and horizontal dimensions of 10 km or less. In Vrancea the earthquakes only appear to form a nest because they occur within a nearly vertical zone.

In spite of being highly destructive and having a magnitude M_W of 7.5, the 1977 earthquake had numerous precedents in Romania. The historical record here

---

[3] However, as explained in Chapter 6, reported values for static stress drop are among the least reliable of all earthquake source parameters.

Fig. 1.8 Partial collapse of the Wilson apartment building in Bucharest caused by the Romanian earthquake of 4 March 1977. The ground floor of this 12-story structure failed because of inadequate reinforcement in the columns that support the upper portion of the building. Photo by Niculai Mândrescu of the Institute for Physics of the Earth in Bucharest.

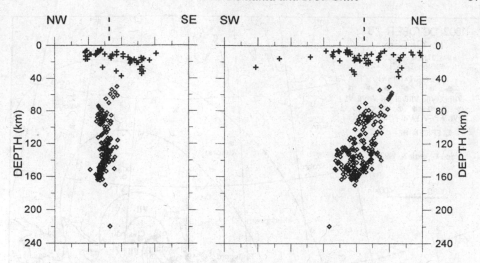

Fig. 1.9 Cross sections of hypocenters of 262 Vrancea microearthquakes recorded between 1982 and 1989 by a regional network. All hypocenters had at least seven P and three S arrivals; locations determined by the JHD method. The figure is reproduced from Oncescu and Bonjer (1997).

extends back to Roman times and includes similar intermediate-depth earthquakes with magnitudes of 7 or greater that occur about 2–3 times each century. On 10 November 1940, for example, a Romanian intermediate-depth earthquake with M_W of 7.7 killed about 1000 people, including 267 crushed by the collapse of the Carlton Hotel in Bucharest (see Section 10.24.3). Over the past several centuries earthquakes with probable magnitudes of 7 or greater caused damage in Bucharest in 1681, 1738, 1802, 1829, 1838, and (more recently) in 1940 and 1986.[4]

According to Mândrescu and Radulian (1999) the 26 October 1802 earthquake was the largest of these earthquakes (M_W of 7.9). It was known by contemporaries and in the memory of subsequent generations as the "big earthquake" and was "felt over a huge area, from St. Petersberg to the Greek islands and from Belgrad to Moscow" (see Fig. 1.10). Its felt area places it among the largest earthquakes known on the interior of the Eurasian continent. During the earthquake all the church towers in Bucharest fell down and many churches and houses collapsed. Nearly all descriptions of the earthquake mention that it collapsed the Tower of Coltea, a venerated local landmark.[5] Radu and Utale (1992) estimate a depth of

[4] Considering Romania's well-documented history of damaging earthquakes, it is remarkable that a *Science News Letter* [1940] article describing the 1940 earthquake had as its title "Rumania not known as seismically active region."

[5] The Tower of Coltea was a large stone fire observation tower in central Bucharest. My readings into the historical seismicity of Romania left me with the impression that the entire subject is recorded in the collapse of various towers and churches.

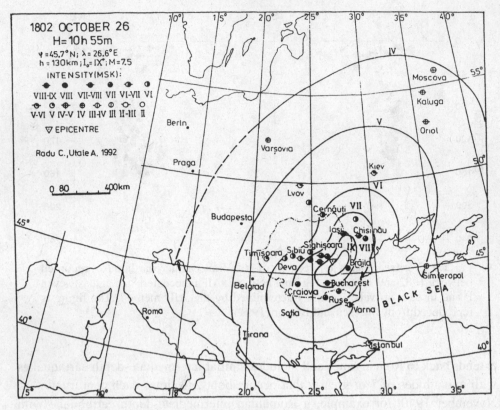

Fig. 1.10 Felt area map for the Romanian earthquake of 26 October 1802. Comparison of corresponding isoseismals for the 1802, 1940, and 1977 earthquakes suggests (see Radu *et al.*, 1979; Bălan *et al.*, 1982) that the 1802 earthquake may have been the largest of the three. Figure reproduced from Radu and Utale (1992).

130 km from an analysis of felt reports; Oncescu and Bonjer (1997) favor a depth of 110–130 km because present-day activity rates are low and these depths did not rupture in the 1940, 1977, or 1986 earthquakes.

Although most seismologists know that intermediate-depth earthquakes produced severe hazard in Romania, there is less awareness that those in South America may be even more deadly. For example, the second most deadly earthquake in South America in the twentieth century was the M_S 7.8 Chillan, Chile, earthquake of 25 January 1939 that killed 28,000 people.[6] The Abe catalog reports its depth as 60 km and Beck *et al.* (1998) as 80–100 km; this places it within the upper portion

---

[6] The most deadly twentieth century South American earthquake occurred near the coast of northern Peru on 31 May 1970; it reportedly killed 66,000 people, including about 20,000 by avalanches triggered by the quake. Although the ISC catalog reports a focal depth of 49 km as constrained by pP–P intervals, the EHB catalog depth is 73 km. Thus this earthquake, too, may have had an intermediate focal depth.

of an approximately straight Wadati–Benioff zone that has a dip of about 30° and extends to a depth of about 200 km beneath Chile and Argentina.

When the Chillan earthquake occurred the reports over the first two days focused on damage in the city of Concepcion, the provincial capital.[7] It soon became clear that Chillan, 80 km further inland, was subject to the most severe damage. A pilot who flew over Chillan reported that "144 blocks of dwellings and first-class buildings were razed. Only three buildings are left standing and can be inhabited." Most of the buildings were "of the old colonial type"; one of the collapsed buildings was the Municipal Theatre, where 1000 people were reportedly enjoying a United States film when the earthquake struck (Fig. 1.11).[8]

The weather compounded the misery in Chillan. In the days following the quake there were periods of both blistering sun and torrential rain. The *New York Times* (28 January 1939) stated that:

Despite the frantic efforts of heroic relief workers to bury the dead as quickly as possible, either in man-made mass burial pits or in the great fissures that were opened in the earth by the temblors, sanitary conditions, aggravated by the intense heat, became almost intolerable in numerous places and the danger of contagious disease was ever-present.

There were shortages of drinking water and medical facilities, especially as the earthquake had destroyed Chillan's hospitals. It took several days to get aid to the city because the quake had damaged both highways and railroads. Ultimately, the news reports make it clear that there were casualties and significant damage in numerous towns and cities within a region that extended about 200 km along the Chilean coast.

Considering the magnitude of the Chillan disaster, there is surprisingly little written about it. This may be partly because the Chilean government imposed strict censorship and apparently exercised control over casualty reports. Thus, there is absolutely no mention of casualties in the June 1939 government-authorized engineering report on the effects of the earthquake (del Canto *et al.*, 1940), even though its authors visited the affected area and censused damage "casa por casa y calle por calle" (house by house and street by street). Their census of 3500 structures in Chillan indicates that 86% were adobe or unreinforced masonry; of all structures only 3% were undamaged while 56% collapsed completely or had walls that fell or were beginning to fall. Similarly, two university professors (Komischke, 1939a; Bastiancig, 1939) who toured the damage zone three weeks after the quake say

---

[7] The morning following the earthquake, the Associated Press story in the *New York Times* stated that "an amateur radio operator . . . had reports of great damage at Concepcion and Terremoto." Obviously the writer was unaware that "terremoto" means "earthquake" in Spanish.

[8] The report is from the *New York Times*. My academic colleagues specializing in film history tell me that Spanish language versions of American films were very popular in South America in the late 1930s. I have been unable to confirm that the film was *Boys' Town* (see Fig. 1.11) although I have spent way too much time researching this.

EFECTOS DEL TERREMOTO EN CHILLAN. — 1. Vista exterior del teatro. — 2. La segunda torre de la iglesia del Carmen caída en el techo del templo. — 3. De estos techos cayeron todas las tejas. — 4. Vista interior del teatro. — 5. Estas murallas exteriores de ladrillos no reforzados cayeron, y las internas quedaron intactas por tener refuerzos de madera. — 6. Un edificio de murallas de adobes. Al derrumbarse estas murallas, cayó el techo que era de hierro y cristal. — 7. Este hombre se salvó milagrosamente saliendo a gatas de su casa semi-destruida. — 8. Murallas de adobes. — 9. La terraza del teatro caída en el suelo.

Fig. 1.11 Photos published in the journal *Scientia* (**5**, no. 19, p. 136) showing damage from the 25 January 1939 Chillan earthquake. Photos 1, 4, and 9 are of the Municipal Theatre where 1000 people died. In a grim and fateful twist, the text of the poster in photo 1 (Hijo sin Hogar = boy without a home) suggests the film they were watching may have been *Boys' Town*, a 1938 Spencer Tracy film about an orphanage in the United States.

nothing about casualties.[9] Instead they describe the thick sediments that underlie Chillan and present numerous sketches indicating that liquifaction and variations in the water table compounded the earthquake damage. Bastiancig does include a map of damaged areas in Chillan with notes describing the characteristics of buildings that survived the shaking.

Finally, there is the report of Saita (1940), a Japanese seismologist who toured the region six months after the earthquake. He notes that the Chilean government reports less than 7000 deaths, but his own estimates using information from the Japanese consulate in Chile suggests that in Chillan alone the earthquake flattened 60% of the houses and caused 25,000 deaths.

1.3 The curious: 29 March 1954 Spain

The Spanish deep earthquake of 29 March 1954 had a depth of 630 km and was extraordinarily large – larger even than the Romanian and Chillan earthquakes. Indeed, its magnitude M_W of 7.9 makes it one of the largest earthquakes ever to occur on the European continent. And, when it occurred it was the second largest earthquake anywhere with a depth exceeding 300 km (Table 1.2).

Yet even though it occurred beneath a populated, highly developed country like Spain, it went almost unmentioned in contemporary news accounts. In newspapers in New York, London, and Paris, the only front-page headlines that were remotely seismological concerned the fact that the U.S. was testing H-bombs in the South Pacific. Indeed, the sole mention of the earthquake in the *New York Times* appeared on page 14, where there was a single article. In its entirety it stated:

Spain Shaken by Quake
MADRID, March 29 (AP) – Panic-stricken Spaniards rushed from their homes in Malaga and Granada today as the strongest earthquake in years toppled walls in those southern cities.

Paris' *Le Monde* gave only slightly more information in an article buried on page 16, the very last page. The article presented details about how seismographs in Germany, Italy, Spain, and France recorded the quake, including the fact that one of the stations reported that the epicenter was in Greece or Turkey. The article also stated that the collapsed walls were in buildings under construction. The Madrid daily paper *ABC* didn't mention the earthquake at all, reporting instead on the celebration of the 15th anniversary of the Franco regime, some international news, and regional culture (soccer and bullfighting). How could such a large earthquake go

---

[9] In a later article focusing on the historical context of the Chillan event Komischke (1939b) states without comment that the government reports 6000 deaths. The Komischke and Bastiancig articles are devilishly difficult to obtain. I am indebted to Ildefonso Reyes Acevedo at the Universidad Técnica Federico Santa María in Valparaiso, Chile, for sending electronic copies.

virtually unnoticed by the press? Almost certainly because its extraordinary depth mitigated its effects at the surface; thus it killed no one and caused only limited damage.

Of course, its size and depth did capture the attention of scientists. Bonelli and Carrasco (1956) observed that this was the first earthquake with focal depth exceeding 300 km in this region. Hodgeson and Cock (1956) evaluated first motions and found a focal mechanism with one vertical and one horizontal nodal plane. In spite of its large size there were no recorded fore- or aftershocks. However, Chung and Kanamori (1976) performed the most thorough analysis and found that the rupture consisted of six or more subevents occurring over a period of about 40 seconds; they relocated two of the subevents and found they occurred well below the initial shock, suggesting that rupture occurred on the vertical nodal plane. Although Chung and Kanamori reported that the complex waveform was "very different from typical deep events" and that the earthquake "lasted much longer than other deep earthquakes," we now know that both features are typical for very large deep events.

Also remarkable about the Spanish deep earthquake is its location. Richter (1958) called it "one of the most important earthquakes ever recorded." He suggested that its occurrence directly below a mountainous region indicated that steeply dipping faults must extend to great depth. He also concluded that similar very deep earthquakes might occur beneath other convergent zones, such as the Aleutians and Mexico.

Although 50 years have passed and plate tectonic interpretations have modified our understanding of convergent margins, the Spanish deep earthquake continues to be a puzzle for several reasons. First, how can a deep-focus earthquake occur in the absence of the usual regional indicators of ongoing subduction? There is no deep ocean trench and no active volcanism. Does the very occurrence of a deep earthquake prove that there is ongoing subduction? If so, in Spain there is little else to confirm it.

Second, what is the significance of the fact that the Spanish deep earthquake does not occur within a Wadati–Benioff zone? Although there are a few tiny earthquakes beneath Spain and North Africa with depths as great as 160 km (see Section 10.21), there is a complete absence of seismic activity between 200 km and 600 km. Is there a seismic gap in the subducted lithosphere here? Or did the Spanish quake occur in a piece of detached lithosphere? Or does the Spanish quake demonstrate that some deep earthquakes have nothing to do with subduction?

Third, how can earthquakes like the 1954 Spanish and 1994 Bolivian earthquake be so large, considering that they occurred at such a great depth? Laboratory experiments demonstrate that above a critical temperature and pressure, rocks fail by ductile flow rather than sudden brittle rupture (see Fig. 1.12). Thus, thermal conduction will warm subducting lithosphere (see Chapter 2), reducing its effective

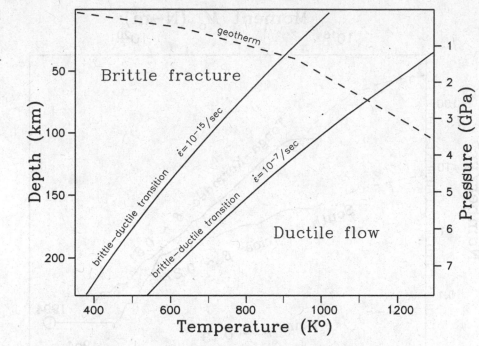

Fig. 1.12 Approximate relationship between pressure (depth), temperature, strain rate, and the mechanism of rock failure. At low temperatures, low pressures, and high strain rates rocks tend to fracture, whereas at high temperatures, high pressures, and low strain rates they fail by creep. The brittle–ductile transition depends on strain rate $\dot{\varepsilon}$. For temperature/pressure conditions corresponding to mantle depths exceeding about 40–70 km, brittle fracture does not occur, suggesting that some other mechanism must be responsible for deep earthquakes. Relationships for brittle–ductile transitions determined by equating eq. 6 and eq. 9 from Houseman and England (1986); the plotted geotherm is from Stacey (1992).

thickness as it reaches 600 km depth. How can such a physically limited region store enough energy to produce earthquakes with M_W of 7.9–8.2? For the Spanish earthquake, is it possible that the thermal conditions of a subducting slab have little or nothing to do with its occurrence?

Fourth, since such a large earthquake did occur beneath Spain, why don't numerous, smaller-magnitude earthquakes occur there as well? Between 1954 and 2000 only three other earthquakes have occurred in the hypocentral region of the 1954 shock (in 1973, 1990, and 1993, see Section 10.21), and all these have magnitudes of 4.5 or less. Thus the distribution of earthquake sizes is significantly different for Spain than for South America, which in turn is different than Tonga (Fig. 1.13). What is the significance of these regional differences?

Finally, as suggested by Richter, are there other regions like Spain where large deep earthquakes may occur but are so rare that they occur once a century or less?

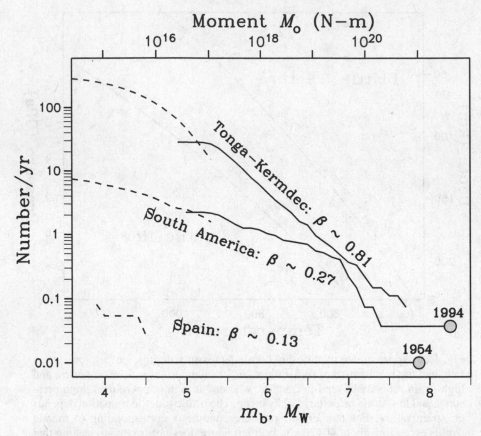

Fig. 1.13 Cumulative distributions of magnitude for earthquakes with focal depths exceeding 300 km in Tonga–Kermadec, South America; and Spain. Dashed lines indicate mean annual number N of events with magnitude exceeding m_b in the ISC catalog 1964–2001. Solid lines for Tonga–Kermadec and South America are mean annual numbers for Mw from the CMT catalog 1977–2003. For Spain the solid line is the 29 March 1954 earthquake; the annual number calculated assuming historical record spans 100 years. Circles labeled '1954' and '1994' are 29 March 1954 and 9 June 1994 Bolivia earthquakes. Note that slopes β of $\log(M_o)$ vs $\log(N)$ are very different for the three regions.

One wonders if we would believe the reported depths for the three small Spanish deep earthquakes mentioned above if the 1954 earthquake had never occurred. After all, small and poorly-recorded earthquakes with anomalously great depths are reported in numerous geographic regions, and seismologists seldom choose to take the locations seriously. For example, an earthquake occurred in Mexico on 26 January 1976, and the ISC located it at a depth of 704 km, using P arrivals at five stations. Although this depth is probably inaccurate, because of the Spanish deep earthquake of 1954 we can't dismiss it entirely.

Table 1.5. *Characteristics of earthquakes discussed in this chapter.*

Alleged characteristic	Bolivia 1994	Romania 1977	Chile 1939	Spain 1954
Small size	False	False	False	False
Associated with deep-ocean trench	True	False	True	False
Occurs in planar Wadati–Benioff zone	Uncertain	Uncertain	True	False
High stress drop	Uncertain	False	Unknown	Uncertain
Impulsive source-time function	False	False	Unknown	False
Few or no known aftershocks	False	False	False	True
Not highly destructive	True	False	False	True

1.4 What we know and what we don't know

The four earthquakes described in this chapter illustrate why the summary statement that opened this chapter isn't strictly true (Table 1.5). In particular:

- Some deep earthquakes are exceedingly large, with magnitudes of 7.5 or more. In some geographic regions, such as Europe, the largest deep earthquakes have magnitudes that are as large or nearly as large as any shallow earthquakes.
- A few deep earthquakes are highly destructive. In South America earthquakes with depths of about 60 km produced the highest numbers of fatalities of all earthquakes in the twentieth century.
- A few deep earthquakes occur in regions where there is no active, ongoing subduction. For deep earthquakes that are associated with active subduction zones, some are isolated events whose relationship to the subducted lithosphere is unclear.
- The ruptures associated with large deep earthquakes aren't simple and impulsive, but consist of several pulses of moment release, much like large shallow earthquakes.
- Some large deep earthquakes have numerous aftershocks; others have none.

Possibly the reason that seismologists make assertions like the statement that opens this chapter is because the vast majority of all deep earthquakes occur beneath the subduction zones that surround the Pacific (Fig. 1.14). Most of these aren't large, aren't destructive, but are clustered within Wadati–Benioff zones. Most have no reported aftershocks, although the significance of this is uncertain since most occur in remote areas where earthquakes with magnitudes less than about 5.0 aren't often detected. The deep quakes with magnitudes smaller than about 6.0 appear to be impulsive since they occur beneath the complex structures that make up the crust and uppermost mantle and which generate reverberations complicating the coda on records of shallow earthquakes. Thus the summary statement is true for the majority of all deep earthquakes, even though it isn't true in general.

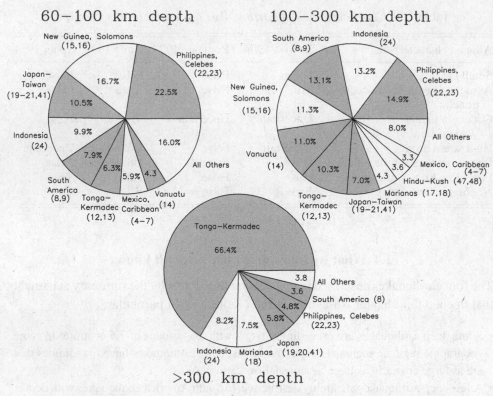

Fig. 1.14 Pie graphs of the geographic distribution of earthquakes with depths of 60–100 km, 100–300 km, and >300 km. Data are for all earthquakes in the ISC catalog 1964–2000 with $m_b \geq 5.0$; shaded sections are for five regions that include more than 50% of all events. The geographic distribution depends on cutoff magnitude because there are regional differences in both the size distribution (see Fig. 1.13) and the detection of smaller-magnitude events.

What isn't clear from our description of the four earthquakes in this chapter is how the physical properties of the mantle regulate deep seismic activity. Important questions include:

• How does mantle temperature affect the physical mechanism that generates deep earthquakes?
• What role do phase transitions in the mantle have in generating deep quakes?
• What is the significance of geographic differences in the properties of deep seismicity? Why are the distributions of earthquake size and depth different in different geographic regions?

These questions and their answers will occupy our attention for the remainder of this book.

1.5 References

Abe, K., 1981. Magnitudes of large shallow earthquakes from 1904 to 1980, *Phys. Earth Planet. Int.*, **27**, 72–92.

Anderson, J. G., M. Savage, and R. Quaas, 1995. 'Strong' ground motions in North America from the Bolivia earthquake of June 9, 1994 ($M_W = 8.3$), *Geophys. Res. Lett.*, **22**, doi:10.1029/95GL01808, 2293–2296.

Antolik, M., D. Dreger, and B. Romanowicz, 1996. Finite fault source study of the great 1994 deep Bolivian earthquake, *Geophys. Res. Lett.*, **23**, doi:10.1029/96GL00968, 1589–1592.

Bălan, S., V. Cristescu, and I. Cornea, eds, 1982. *Cutremurul de pămînt din România de la 4 martie 1977 (The March 4, 1977 Earthquake)*, Bucharest, Romania, Editura Academiei Republicii Socialiste România, 516 pp.

Bastiancig, A., 1939. El terremoto del 24 de enero de 1939 en Chile: Observaciones y consideraciones relacionadas con la edificación, *Scientia*, **5**(20), 178–188.

Beck, S. L., S. Barrientos, E. Kausel, and M. Reyes, 1998. Source characteristics of historic earthquakes along the central Chile subduction zone, *J. South Amer. Earth Sci.*, **11**, doi:10.1016/S0895-9811 (98) 00005-4, 115–129.

Beck, S. L., P. Silver, T. C. Wallace, and D. James, 1995. Directivity analysis of the deep Bolivian earthquake of June 9, 1994, *Geophys. Res. Lett.*, **22**, doi:10.1029/95GL01089, 2257–2260.

Bonelli, J. and L. E. Carrasco, 1956. El primer sismo europeo de foco profundo, *Revista de Geofisica*, **15**, 301–321.

Bouchon, M. and P. Ihmlé, 1999. Stress drop and frictional heating during the 1994 deep Bolivia earthquake, *Geophys. Res. Lett.*, **26**, doi:10.1029/1999GL005410, 3521–3524.

Burbach, G. V. and C. Frohlich, 1986. Intermediate and deep seismicity and lateral structure of subducted lithosphere in the circum-Pacific region, *Rev. Geophys.*, **24**, 833–874.

Chung, W.-Y. and H. Kanamori, 1976. Source process and tectonic implications of the Spanish deep-focus earthquake of March 29, 1954, *Phys. Earth Planet. Int.*, **13**, 85–96.

del Canto, H., P. Godoy P., E. Aguirre S., J. Muñoz-Christi, and J. Ibáñez V., 1940. Informe de la comisión gubernativa sobre los efectos producidos por el terremoto de enero 1939, *Anales del Instituto de Ingenieros de Chile*, Oct-Nov, 376–395.

Estabrook, C. H. and G. Bock, 1995. Rupture history of the great Bolivian earthquake: slab interaction with the 660-km discontinuity? *Geophys. Res. Lett.*, **22**, doi:10.1029/95GL02234, 2277–2280.

Fattal, G., E. Simiu, and G. Culver, 1977. *Observations on the Behavior of Buildings in the Romania Earthquake of March 4, 1977*, U.S. National Bureau of Standards Special Publication 490, 168 pp.

Fuchs, K., K.-P. Bonjer, G. Bock, I. Cornea, C. Radu, D. Enescu, D. Jianu, A. Nourescu, G. Merkler, T. Moldoveanu, and G. Tudorache, 1979. The Romanian earthquake of March 4, 1977. II. Aftershocks and migration of seismic activity, *Tectonophysics*, **53**, doi:10.1016/0040–1951(79)90068–4, 225–247.

Goes, S. and J. Ritsema, 1995. A broadband P wave analysis of the large deep Fiji Island and Bolivia earthquakes of 1994, *Geophys. Res. Lett.*, **22**, doi:10.1029/95GL02011, 2249–2252.

Goes, S., L. Ruff, and N. Winslow, 1997. The complex rupture process of the 1996 deep Flores, Indonesia earthquake (M_W 7.9) from teleseismic P-waves, *Geophys. Res. Lett.*, **24**, doi:10.1029/97GL01245, 1295–1298.

Hodgson, J. H. and J. I. Cock, 1956. Direction of faulting in the deep-focus Spanish earthquake of March 29, 1954, *Tellus*, **8**, 321–328.

Houseman, G. and P. England, 1986. A dynamical model of lithospheric extension and sedimentary basin formation, *J. Geophys. Res.*, **91**, 719–729.

Huang, W.-C. and E. A. Okal, 1998. Centroid moment tensor solutions for deep earthquakes predating the digital era: discussion and inferences, *Phys. Earth Planet. Int.*, **106**, doi:10.1016/S0031-9201(97)00111-8, 191–218.

Ihmlé, P. F., 1998. On the interpretation of subevents in teleseismic waveforms: the 1994 Bolivia deep earthquake revisited, *J. Geophys. Res.*, **103**, doi:10.1029/98JB00603, 17919–17932.

Ihmlé, P. F. and T. H. Jordan, 1995. Source time function of the great 1994 Bolivia deep earthquake by waveform and spectral inversions, *Geophys. Res. Lett.*, **22**, doi:10.1029/95GL01437, 2253–2256.

Kanai, K., 1983. *Engineering Seismology*, Tokyo, University of Tokyo Press, 251 pp.

Kanamori, H., D. L. Anderson, and T. H. Heaton, 1998. Frictional melting during the rupture of the 1994 Bolivian earthquake, *Science*, **279**, doi:10.1126/science.279.5352,839, 839–842.

Karato, S.-I., M. R. Riedel, and D. A. Yuen, 2001. Rheological structure and deformation of subducted slabs in the mantle transition zone: Implications for mantle circulation and deep earthquakes, *Phys. Earth Planet. Int.*, **127**, doi:1016/S0031-9201(01)00223-0, 83–108.

Kikuchi, M. and H. Kanamori, 1994. The mechanism of the deep Bolivia earthquake of June 9, 1994, *Geophys. Res. Lett.*, **21**, doi:10.1029/94GL02483, 2341–2344.

Komischke, A., 1939a. Observaciones sobre el terremoto del 24 de enero de 1939 en Chile central, *Scientia*, **5**(20), 163–175.

1939b. Observaciones sobre el terremoto del 24 de enero de 1939 en Chile central, *Scientia*, **6**(21), 2–7, 1939b.

Liu, L., 1983. Phase transformations, earthquakes, and the descending lithosphere, *Phys. Earth Planet. Int.*, **32**, 226–240.

Lomnitz, C., 1970. Major earthquakes and tsunamis in Chile during the period 1535 to 1955, *Geol. Rundshau*, **59**, 938–960.

Lundgren, P. and D. Giardini, 1995. The June 9 Bolivia and March 9 Fiji deep earthquakes of 1994: I. Source processes, *Geophys. Res. Lett.*, **22**, doi:10.1029/95GL02233, 2241–2244.

Mândrescu, N. and M. Radulian, 1999. Macroseismic field of the Romanian intermediate-depth earthquakes. In *Vrancea Earthquakes: Tectonics, Hazard and Risk Mitigation*, eds. F. Wenzel, D. Lungu and O. Novak, The Netherlands, Kluwer, 374 pp., 163–174.

Myers, S. C., T. C. Wallace, S. L. Beck, P. G. Silver, G. Zandt, J. Vandecar, and E. Minaya, 1995. Implications of spatial and temporal development of the aftershock sequence for the M_W 8.3 June 9, 1994 deep Bolivian earthquake, *Geophys. Res. Lett.*, **22**, doi:10.1029/95GL01600, 2269–2272.

Okal, E. A. and C. R. Bina, 2001. The deep earthquakes of 1997 in Western Brazil, *Bull. Seismol. Soc. Amer.*, **91**, 161–164.

Oncescu, M. C. and K. P. Bonjer, 1997. A note on the depth recurrence and strain release of large Vrancea earthquakes, *Tectonophysics*, **272**, doi:10.1016/0040-1951(96)00263-6, 291–302.

Radu, C. and A. Utale, 1992. The Vrancea (Romania) earthquake of October 26, 1802, *Proc. XXIII ESC General Assembly, Prague*, 110–113.

Radu, C., V. Kárník, G. Polonic, D. Prochazkova and Z. Schenkova, 1979. Isoseismal map of the Vrancea earthquake of March 4, 1977, *Tectonophysics*, **53**, doi:10.1016/0040–1951(79)90063–5, 187–193.

Räkers, E. and G. Müller, 1982. The Romanian earthquake of March 4, 1977. III. Improved focal model and moment determination, *J. Geophys.*, **50**, 143–150.

Richter, C. F., 1958. *Elementary Seismology*, San Francisco, W. H. Freeman, 768 pp.

Saita, T., 1940. The great Chilean earthquake of January 24, 1939 (in Japanese), *Bull. Earthquake Res. Inst., Tokyo Univ.*, **18**, 446–459.

Science News Letter, 1940. Rumania not known as seismically active region. *Science News Letter*, **38**, 324.

Silver, P. G., S. L. Beck, T. C. Wallace, C. Meade, S. C. Myers, D. E. James, and R. Kuehnel, 1995. Rupture characteristics of the deep Bolivian earthquake of 9 June 1994 and the mechanism of deep-focus earthquakes, *Science*, **268**, 69–73.

Stacey, F. D., 1992. *Physics of the Earth*, (3rd edn), Brisbane, Australia, Brookfield Press, 513 pp.

Tinker, M. A., T. C. Wallace, S. L. Beck, P. G. Silver, and G. Zandt, 1995. Aftershock source mechanisms from the June 9, 1994, deep Bolivian earthquake, *Geophys. Res. Lett.*, **22**, doi:10.1029/95GL01090, 2273–2276.

2

What, where, how, and why?

2.1 What are deep-focus and intermediate-focus earthquakes?

When we call an earthquake "deep," what do we mean? As we shall see, there is surprisingly little agreement about this among seismologists. However, in this book I use these definitions:

- A *shallow* or *shallow-focus earthquake* is any earthquake with a believable reported focal depth less than 60 km;
- An *intermediate-focus, intermediate-depth*, or *intermediate earthquake* is any earthquake with a believable reported focal depth equal to or greater than 60 km but less than 300 km;
- A *deep-focus earthquake* is any earthquake with a believable reported focal depth of 300 km or more;
- A *deep earthquake* is any deep-focus or intermediate-focus earthquake, i.e., a quake with a believable reported focal depth of 60 km or more.

Although it is confusing for "deep-focus" and "deep" to specify distinctly different depth categories, it is often desirable (as with the title of this book) to specify intermediate- and deep-focus earthquakes together. Since about 1970 a few scientists have called them "mantle" earthquakes instead of "deep" earthquakes (e.g., Isacks and Molnar, 1971). This is misleading since not all shallow-focus earthquakes occur within the crust; e.g., nearly all oceanic intraplate earthquakes are shallow but occur in the mantle. Finally, there are now already hundreds of published papers which used "deep earthquake" to specify both deep-focus and intermediate-focus events. Since there is no way to undo this, and since merely suggesting a new name is unlikely to change language in future publications, I will use the terms "deep" and "deep-focus," and be careful throughout this book to apply them precisely.

I thank Doug Wiens for reviewing an earlier draft of this chapter.

Essentially, the definitions above correspond to those proposed by the principal discoverer of deep earthquakes, Kiyoo Wadati (see Section 3.3.1). In his second paper on deep earthquakes, Wadati (1929) stated that:

The name of shallow is given to such earthquakes which . . .generally situate at the depth of within about 60 km. The name of deep is generally given to such earthquakes which are considered to take place at the deeper part than the case of shallow ones. Among deep earthquakes, such ones of more than 300 km deep are comparatively frequent and they are called 'deep earthquakes' in a narrow sense and 'very deep earthquakes' if it is required to avoid the confusion. On the other hand 'intermediate earthquakes' whose depths are about 100–200 km are not so numerous as one expected.

Most subsequent investigators have adhered to Wadati's definitions only rather loosely; the resulting confusion is seldom serious but it does create problems when one tries to count the quakes in different categories. For example, Honda (1962) states:

Earthquakes are classed here as shallow when the focal depth does not exceed 100 km, intermediate when it is from 100 km to 250 km, and deep when it exceeds 250 km.

Duda (1965) states:

The earthquakes investigated are classified according to their focal depth. The limiting depth between shallow and intermediate shocks is taken as 65 km, as usual. The limiting depth between intermediate and deep earthquakes is taken as 450 km. This is different from what has been used hitherto but strongly recommended from an investigation of the strain release versus focal depth.

In Gutenberg and Richter's (1949) classic book *Seismicity of the Earth*, the authors state:

Shocks are here classified as shallow when the depth does not exceed 60 km; intermediate when the depth is from 70 to 300 km; deep when it exceeds 300 km.

This not only begs the question of how to categorize events with focal depths of 65 km, it also differs from their Geological Society of America Special Paper (Gutenberg and Richter, 1941) where they include quakes with focal depths of 60 km in their tables of intermediate shocks.

Some of this confusion seems to have come about because seismologists expected that natural depth categories would emerge upon an improved understanding of earth structure and seismicity. For example, Richter (1958) describes Gutenberg and Richter's (1949) classification scheme, and then states:

Present evidence indicates that the Moho is a level of discontinuity in all parts of the Earth, oceanic as well as continental, although its depth varies widely. In some future time, when we have reasonably good information for all parts of the Earth, it should be possible to divide all earthquakes into shallow, originating above the Moho, and deep-focus, originating below it.

However, this definition has never become popular, in part because it isn't always easy to delineate the Moho in subduction zones where deep earthquakes are most common. Recent global surveys indicate that the maximum depth of most sub-duction zone thrust events is about 50 km (Bilek *et al.*, 2004), and beneath this depth there is also an observable change in the frequency of aftershocks (see Section 4.2.5). Thus, we would probably choose 50 km as the shallow/intermediate boundary if we were defining it today.

In any case, recent authorities still differ slightly among themselves concerning the shallow/intermediate boundary; time has apparently not encouraged definitions to converge. While the *Regional Catalogue of the International Seismological Centre* defines "shallow focus" as $0 \leq h \leq 60$ km; intermediate focus as $60 < h \leq 300$ km, and deep focus as $h > 300$ km, Rothé (1969) defines "normal" earthquake focal depths to be 0–69 km. The third edition of the *Dictionary of Geological Terms* (Bates and Jackson, 1984) has entries for deep-focus and intermediate-focus earthquakes, defined as earthquakes with foci "between depths of 300–700 km" and "between depths of about (sic) 60 km and 300 km". However, there is still confusion because it defines:

shallow-focus earthquake – an earthquake with a focus at a depth of less than 70 km.

Finally, Frohlich (1987) (who should have known better) states incorrectly that the nomenclature used by both Wadati and the International Seismological Centre is that "earthquakes with focal depths exceeding 300 km are deep earthquakes, and those with depths between 70 and 300 km are intermediate earthquakes."

Nowadays, some authors occasionally use "deep earthquake" to describe a shallow-focus earthquake that is deeper than, say, 15 km, especially when it is associated with volcanic activity (e.g., Weaver *et al.* 1983; Klein *et al.*, 1987; Castellano *et al.*, 1997; Nakamichi *et al.*, 2003; Wolfe *et al.*, 2003). Moreover, a few authors apply "intermediate" to earthquakes with focal depths shallower than 60 km (e.g., Coco *et al.*, 1997; Quintanar *et al.*, 1999). However, this is rare enough

--→

Fig. 2.1 Geographic distribution of large intermediate-depth earthquakes. Symbols are hypocenters with magnitudes of 7.0 or greater; open circles are events occurring between 1900 and 1976 in the Abe or EV Centennial catalogs; filled circles are events between 1977 and 2004 in the Harvard CMT catalog. The labeled regions are the boundaries of 27 geographic regions discussed in Chapter 10. In some, like Italy (see Section 10.22), intermediate-depth earthquakes certainly occur but none have magnitudes exceeding 7.0; in others, like the South Shetland Islands (see Chapter 10.19), deep events have been reported but their existence hasn't been unequivocally confirmed. The text discusses earthquakes labeled 1914, 1925, 1947, 1950, 1964, and 1970, which are examples of earthquakes whose reported depth or location is almost certainly erroneous.

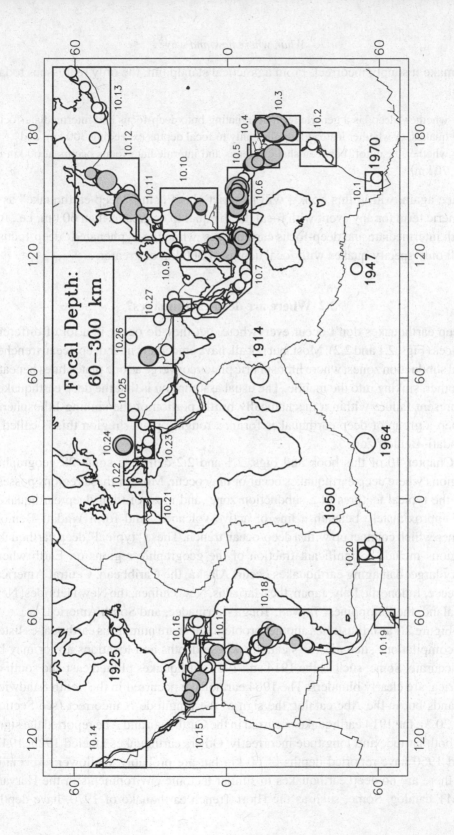

Focal Depth: 60–300 km

to make it simply incorrect. From a practical standpoint, the only real issues today are:

(1) whether "deep" is a generic term designating both deep-focus and intermediate-focus quakes, or whether instead it applies only to focal depths exceeding 300 km; and
(2) whether the cutoff between shallow-focus and intermediate-focus occurs at 60 km or 70 km.

Once again, within this book I will be strict about using "deep earthquake" as a generic term for any event with focal depth equal to or greater than 60 km, i.e., for both intermediate and deep-focus earthquakes, while the hyphenated "deep-focus" will only specify quakes with focal depths of 300 km or greater.

2.2 Where are deep earthquakes?

Deep earthquakes don't occur everywhere, but they do occur in a lot of different places (Figs. 2.1 and 2.2). Most, but not all, have epicenters near deep ocean trenches and subduction zones, where lithospheric plates converge as one plate slides beneath another, sinking into the mantle. The usual assumption is that the deep earthquakes represent failure within a mechanically brittle portion of the sinking lithosphere; when a group of deep earthquakes forms a roughly planar region this is called a Wadati–Benioff zone.

Chapter 10 of this book and Figs. 2.1 and 2.2 delineate some 28 geographic regions where deep earthquakes occur or may occur. Most of these regions possess all the typical features of a subduction zone, and most of their deep earthquakes lie approximately beneath a line of active volcanoes and form Wadati–Benioff zones which connect up with a deep-ocean trench. These "typical" deep earthquake regions include a significant fraction of the geographic regions on Earth where very large, damaging earthquakes occur: Alaska, the Caribbean, Central America, Greece, Indonesia, Italy, Japan, the Marianas, New Guinea, the New Hebrides, New Zealand, the Philippines, Taiwan, Tonga–Kermadec, and South America.

Figure 2.1 also illustrates another problem: there are numerous earthquakes listed in compilations such as Abe's catalog whose depths and locations are or may be inaccurate. Some, such as the 1914 and 1964 earthquakes plotted east and south of Africa, are clearly blunders. The 1964 earthquake occurred in the South Sandwich Islands but in the Abe catalog the sign of the longitude is incorrect (see Section 10.20.3); the 1914 earthquake occurred in the Caribbean and Abe reported the signs of both latitude and longitude incorrectly. Others earthquakes labeled 1925, 1947, and 1950 have reported depths of 60 km but are probably shallower, especially as there are no deep earthquakes in similar tectonic environments in the Harvard CMT catalog. Some, such as the Hjort Trench earthquake of 1970, have depths

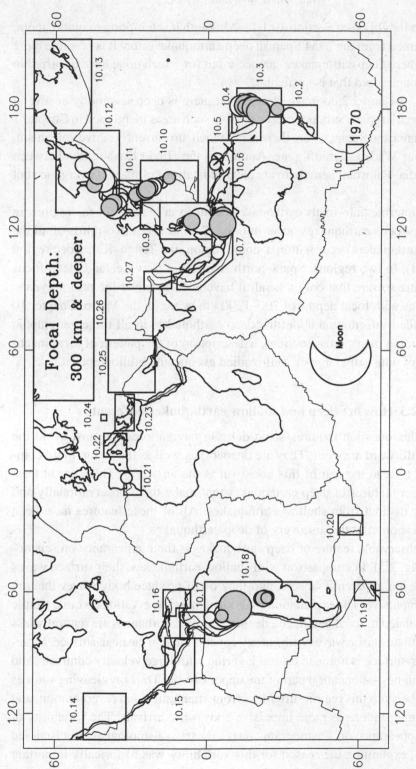

Fig. 2.2 Geographic distribution of large deep-focus earthquakes. Symbols are as in Fig. 2.1. The Moon icon at bottom center indicates that there is seismic activity with depths of 700–1200 km on the Moon (see Section 10.29).

of uncertain reliability (see Section 10.1.1). Although it is tempting to ignore them, the lesson learned from the 1954 Spanish deep earthquake is that it is possible there are regions where deep earthquakes can occur but for which none have occurred in the century-long record that is available.

What Figs. 2.1 and 2.2 do make clear is that there is deep seismicity in several regions that aren't typical subduction zones. Two such areas mentioned in Chapter 1 include Romania and Spain, where there is no trench, no currently active volcanism, and no obvious Wadati–Benioff zone. Another is the Hindu Kush region between China and India–Pakistan; here there are clear indications of plate convergence but no ocean.

Although intermediate-focus earthquakes occur in at least 25 of the regions in Fig. 2.1, deep-focus earthquakes occur in only 12. And about two-thirds of all the deep-focus earthquakes occur within a single region, the Tonga–Kermadec region (see Fig. 1.14). In two regions, Spain–north Africa and New Zealand, deep-focus earthquakes are so rare that only a handful have occurred over the past 50 years. Finally, quakes with focal depths of 700–1200 km occur on the Moon. Chapter 10 presents detailed information about the deep earthquakes in all these geographic regions, including maps, cross sections, a description of the pattern of focal mechanisms, lists of unusually large or well-studied events, and bibliographies.

2.3 How are deep and shallow earthquakes different?

Addressing this question requires some delicate navigation: one must avoid the terse, straightforward answer ("They are deeper") as well as the smug, more complete answer ("Read the rest of this book – it is the answer"). There are at least four significant features of deep earthquakes that make them observationally and mechanically distinct from shallow earthquakes. All of these features have been evident since soon after the discovery of deep earthquakes.

The first observable feature of deep earthquakes is their appearance on seismograms (see Fig. 2.3). In comparison with shallow earthquakes, their surface waves are weak or even apparently absent, and they often produce body waves that are much more impulsive. The explanation for both of these observations is quite simple and closely related to focal depth, i.e., deeper elastic disturbances are naturally less likely to stimulate elastic waves with energy concentrated in the near-surface. Moreover, the near-surface is home to crustal layering with large velocity contrasts, and overlies the highest-attenuation part of the upper mantle. Thus rays leaving sources which occur beneath this region suffer less from attenuation and reverberation, and arrive at seismic stations as more impulsive body-wave arrivals. The variability of earthquake appearance on seismograms puzzled early seismologists; as discussed in Chapter 3, explaining the reason for this variability was historically important

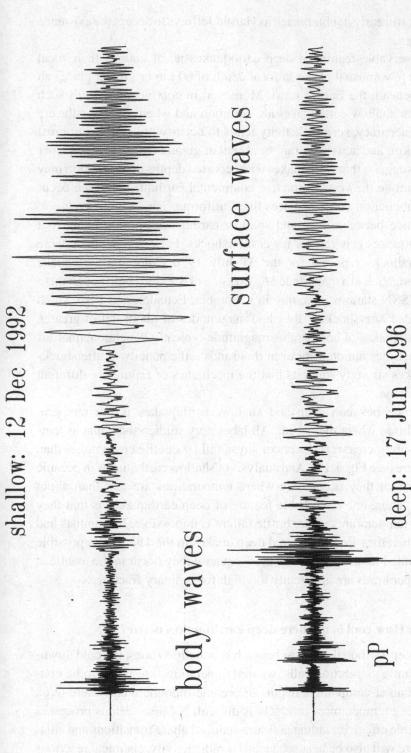

Fig. 2.3 Seismograms from shallow and deep earthquakes look distinctly different. For deep earthquakes, surface waves are often weak or absent, but they often produce surface-reflected phases such as pP that aren't generated by shallow foci. The figure compares 50 minute segments of vertical-component records for two large earthquakes in Java recorded at a distance of 82° at station SPA (South Pole). Earthquakes shown are: 12 December 1992, $M_{\mathrm{W(CMT)}} = 7.8$, $h_{\mathrm{CMT}} = 20$ km; 17 June 1996, $M_{\mathrm{W(CMT)}} = 7.9$, $h_{\mathrm{CMT}} = 580$ km.

as it led some powerful early doubters such as Harold Jeffreys to accept the existence of deep earthquakes.

The second observable feature of deep earthquakes is, of course, their focal depth. Except in a few unusual places, a focal depth of 60 km or greater places an earthquake well beneath the Earth's crust. Moreover, in continental regions such as California where shallow earthquakes are common and where focal depths are known with great accuracy, seismic activity tends to occur within the upper crust at depths of 5–15 km, and activity is rare or absent at greater depths (e.g., Bonner *et al.*, 2003). This suggests that earthquakes with reported depths of 60–650 km may be mechanically unlike the very destructive continental earthquakes which occur well away from subduction zones in places like California, Turkey, China, etc.

A third difference between deep and shallow earthquakes is that many (but not all) deep earthquakes generate far fewer aftershocks. For example, there were exactly zero aftershocks reported for the 31 July 1970 Colombia earthquake (Fig. 2.4), even though it had a magnitude $M_{W(CMT-Hist)}$ of 8.1, and even though there were several WWSSN stations operating in Colombia, Ecuador, and Peru which would have recorded aftershocks if they had magnitudes of about 4.0 of greater. Since shallow earthquakes of comparable magnitude typically generate numerous aftershocks – sometimes hundreds or even thousands – the paucity of aftershocks for deep earthquakes strongly suggests that the mechanics of failure are different for the two phenomena.

A final difference between deep and shallow earthquakes is the temperature/pressure condition where they occur. All laboratory studies show that as temperature and pressure increase, rocks under stress fail by ductile creep rather than sudden brittle failure (see Fig. 1.12). And analysis of shallow earthquakes in oceanic lithosphere shows that they occur only where temperatures are less than about 700 °C (Fig. 2.5). Thus, one remarkable feature of deep earthquakes is that they appear to occur at temperatures where brittle failure is impossible. If scientists had understood this when they first discovered deep quakes in the 1920s, it is possible they might have called them "hot earthquakes" because they occur in the mantle at depths where temperatures are apparently too high for ordinary fracture.

2.4 How cool is it where deep earthquakes occur?

Of course, most deep earthquakes occur beneath subduction zones, the cold downgoing limbs of mantle convection cells, where temperatures are likely to be considerably cooler than at comparable depths in oceanic lithophere elsewhere (e.g., Fig. 2.5). Estimating temperatures precisely is difficult because various processes have a non-negligible effect, including pressure-induced phase transitions and adiabatic heating. There will also be heat produced by radioactivity, chemical reactions associated with dewatering, and friction. However, at least since the appearance of

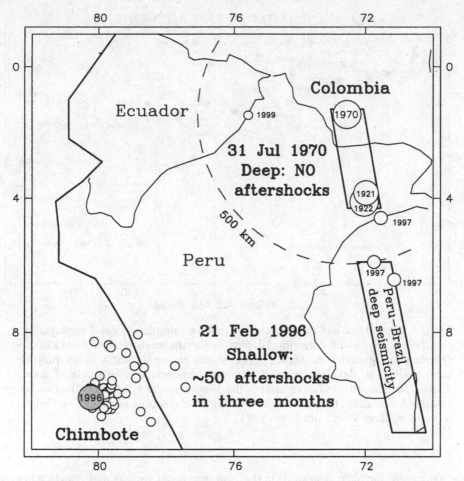

Fig. 2.4 Some large deep earthquakes have few or no aftershocks. The 31 July 1970 ($h_{CMT-Hist} = 623$ km; $M_{W(CMT-Hist)} = 8.1$) earthquake in Colombia was remarkably isolated, as the only cataloged hypocenters within a sphere of radius 500 km of the focus are two large deep-focus earthquakes occurring more than 200 km to the south in 1921–22, and three small earthquakes with m_b of 4.5 or less occurring in 1997 and 1999. In contrast, the ISC catalog includes 47 aftershocks occurring within three months after the shallow focus Chimbote earthquake of 21 February 1996 ($M_{W(CMT)}=7.5$).

McKenzie's (1969) paper, geoscientists realized that at depths where earthquakes occur, conduction through the upper surface of the subducting lithosphere was the most important process affecting temperature. This section presents a simple analysis to assess the approximate temperatures within subducting lithosphere under various conditions.

For conduction, the analysis shows (Box 2.1 and Fig. 2.6) that the temperature perturbation in a slab subducted to depth h depends primarily on three parameters:

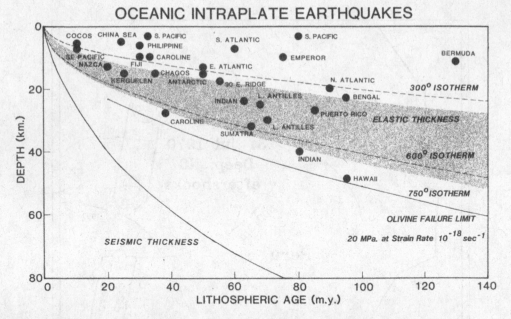

Fig. 2.5 Crustal age and focal depths for shallow intraplate oceanic earthquakes (solid circles) with well-determined depths. Isotherms are determined from a lithospheric cooling model; the stipled region shows elastic thickness as determined from analysis of flexure; seismic thickness is determined from Rayleigh wave dispersion. The olivine failure limit is the lower limit at which 20 MPa can be sustained, calculated for a dry olivine rheology and a strain rate of 10^{-18}/s. Figure reproduced from Wiens and Stein (1983).

- the age t_{Age} of the plate at subduction (i.e., the time spent with its upper surface held at the ocean temperature);
- the time $t_{sub \to h}$ after subduction that the plate spends in the mantle before it arrives at depth h;
- the temperature difference ΔT_{Mantle} between the unperturbed mantle and the ocean.

Indeed, if the temperature perturbation is normalized by ΔT_{Mantle}, the rather remarkable result is that the minimum temperature in the slab depends only on the ratio $t_{Age}/t_{sub \to h}$ (Fig. 2.7).

Since Kirby *et al.* (1991) most analyses have expressed the temperature influence using the so-called thermal parameter Φ, defined as the product of t_{Age}, the age, and $V_{\perp}$, the vertical component of the subduction velocity. There is abundant evidence that the existence and character of deep seismicity correlates with Φ; e.g., deep-focus earthquakes only occur in regions where Φ exceeds about 5000 km (see Fig. 2.8, and Chapters 4 and 7).

Box 2.1
The thermal parameter Φ and the approximate temperature within the interior of subducting lithosphere

There are compelling reasons to believe that thermal conditions within subducting lithosphere affect whether and how deep earthquakes occur. This box describes a simple model for approximating the thermal structure within subducted lithosphere. It also explains why a useful statistic to assess temperature is the so-called thermal parameter Φ, defined as the product of t_{Age}, the age of the lithosphere when subduction begins, and $V_{\perp}$, the vertical component of the subduction velocity.

First, imagine a "lithosphere factory" at a mid-ocean ridge that manufactures blocks of lithosphere having a constant temperature T_{Mantle} (Fig. 2.6). After manufacture they move along a plate tectonic "conveyor belt" where they cool by conduction through their upper surface, which is in contact with the ocean with constant temperature T_{Ocean}. Some time later the blocks arrive at a trench and subsequently their upper surfaces are again held at the mantle temperature T_{Mantle}. In this simple model, the only process that affects temperature within the lithospheric blocks is conduction. Moreover, the temperature a distance Z beneath the surface can be found analytically, and the coldest temperature within the block depends only on the temperature difference and the ratio of the times the surface of the block is in contact with T_{Ocean} and T_{Mantle}.

In particular, before the blocks reach the trench the temperature at depth Z beneath the ocean floor is (e.g., Turcotte and Schubert, 2002; Carslaw and Jaeger, 1959):

$$T(t, Z) = T_{Ocean} + \Delta T_{Mantle} \mathrm{erf}[Z/2\sqrt{\kappa t}], \qquad (2.1)$$

where ΔT_{Mantle} is the difference in temperature between the ocean and the mantle, κ is the thermal diffusivity, t is the time after manufacture, and erf is the error function:

$$\mathrm{erf}(x) = \frac{2}{\sqrt{\pi}} \int_0^x e^{-\xi^2} d\xi, \qquad (2.2)$$

If the lithospheric blocks reach the trench at time t_{Age}, subsequently the temperature distribution a distance z beneath the upper surface is (Carslaw and Jaeger, 1959):

$$T(t, Z) = T_{Mantle} + \Delta T_{Mantle}(\mathrm{erf}[Z/2\sqrt{\kappa(t_{sub \to h} + t_{Age})}] - \mathrm{erf}[Z/2\sqrt{\kappa t_{sub \to h}}]),$$
$$(2.3)$$

where $t_{sub \to h}$ is the time that elapses after subduction begins, i.e., $t - t_{Age}$. We who study deep earthquakes seldom have direct knowledge of $t_{sub \to h}$; however, we usually know the earthquake depth h and the vertical component of subduction $V_{\perp}$, and so $t_{sub \to h} = h/V_{\perp}$. Thus eq. 2.3 allows us to calculate the temperature distribution within the subducting slab.

For any depth h we can manipulate eq. 2.3 to determine where within the subducting lithosphere the temperature is coldest; this will occur where the partial derivative with respect to Z is zero. Some algebra shows that this occurs when:

$$Z^2 = 2\kappa t_{\text{Age}} \frac{h}{\Phi}\left(1 + \frac{h}{\Phi}\right)\ln\left(1 + \frac{\Phi}{h}\right), \tag{2.4}$$

where we have used the fact that $\Phi = t_{\text{Age}}V_\perp$. If we substitute this into eq. 2.3 we find that the temperature contrast depends only on the initial temperature difference and on the ratio of Φ and h:

$$\frac{T - T_{\text{Mantle}}}{\Delta T_{\text{Mantle}}} = \text{erf}\left(\sqrt{\frac{1}{2}\frac{h}{\Phi}\ln\left(1 + \frac{\Phi}{h}\right)}\right) - \text{erf}\left(\sqrt{\frac{1}{2}\left(1 + \frac{h}{\Phi}\right)\ln\left(1 + \frac{\Phi}{h}\right)}\right). \tag{2.5}$$

This has a value of 0.0 where the temperature is at the mantle temperature, and 1.0 where it is at the ocean temperature. Note that h/Φ is just the ratio of $t_{\text{sub}\to h}$ and t_{Age}, the times the block surface spends in contact with the mantle and the ocean, respectively. The appeal of eqs. 2.3, 2.4, and 2.5 is that they provide estimates of the temperature distribution in a closed form (e.g., see Fig. 2.7).

What is the physical meaning of Φ and Φ/h? Since $t_{\text{sub}\to h} = h/V_\perp$:

$$\frac{t_{\text{Age}}}{t_{\text{sub}\to h}} = \frac{t_{\text{Age}}V_\perp}{h} = \frac{\Phi}{h}. \tag{2.6}$$

Thus Φ/h is actually just the ratio of ages, useful for expressing temperature because the temperature effects of conduction are expressable in terms of the ratio (see eqs. 2.3 and 2.5). For calculations, $V_\perp$ is usually expressed in mm/yr and t_{Age} in millions of years so their product Φ will have units of km. Thus very old, fast subduction where t_{Age} is 200 million years and $V_\perp$ is 75 mm/yr produces Φ of 15,000 km. Hot, slow subduction with t_{Age} of 10 million years and $V_\perp$ of 10 mm/yr produces Φ of 100 km.

So how cold are the interiors of subducting lithosphere? According to eq. 2.5, Φ/h of 10 corresponds to temperature contrasts $\Delta T/\Delta T_{\text{Mantle}}$ of 0.52, i.e., the minimum slab temperature is 52% colder than normal mantle. If the ocean floor and mantle are at $0\,^\circ$C and $1300\,^\circ$C respectively, then 0.52 corresponds to a temperature contrast of $675\,^\circ$C, and an interior temperature of $625\,^\circ$C. In Tonga estimates of Φ are typically 9,000–15,000 km, so at 600 km depth Φ/h is 15–25, the temperature contrast is 0.58–0.65, and the interior temperature would be 450°–550°. These temperatures should be comparable to those at 80–120 km depth in central Aleutians, where Φ/h is also about 15–25 since Φ is about 2000 km.

Of course, these results are only approximate. Adiabatic heating will increase T by about $280\,^\circ$C between the surface and depths of 600 km.[1] However, this

[1] Some publications discussing the mantle's thermal properties refer to the so-called *potential temperature*. This is the temperature a material would have if brought to zero pressure with no exchange of heat, i.e., adiabatically. Since the asthenosphere is conducting vigorously its temperature is close to the adiabat. Thus, the potential

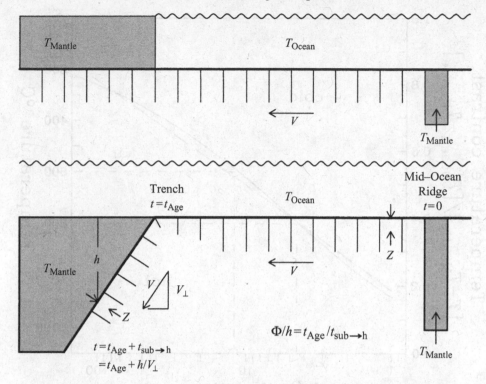

Fig. 2.6 (Top) Consider blocks of lithosphere which form at the right with constant temperature T_{Mantle} and then move towards the left while losing heat by conduction through their upper surface, which is held first at temperature T_{Ocean}, and then at temperature T_{Mantle}. The temperature within the blocks can be expressed analytically (eq. 2.3); and the coldest temperature within the blocks depends only on the temperature difference and the ratio of the times spent at each temperature (eq. 2.5).

(Bottom) If lithosphere arrives at the trench with age t_{Age} and subducts to depth h at a vertical rate $V_\perp$, the ratio of times spent at the surface and in the mantle is Φ/h, where $\Phi = t_{Age}V_\perp$ is the thermal parameter. After subduction, eqs. 2.3, 2.4, and 2.5 express the temperature contrast produced by conduction through the surface of the lithosphere.

occurs roughly linearly with depth and affects both normal mantle and subducted lithosphere; thus it doesn't influence the temperature contrast (Molnar *et al.*, 1979). Turcotte and Schubert (2002) estimate that the olivine–spinel phase change contributes about 133 °C. However, because of uncertainty about the kinetics of the phase change it isn't clear how this is distributed over depth. Finally, to obtain a

temperature of the asthenosphere is approximately constant, whereas it increases markedly downward within the lithosphere, where heat is exchanged by conduction.

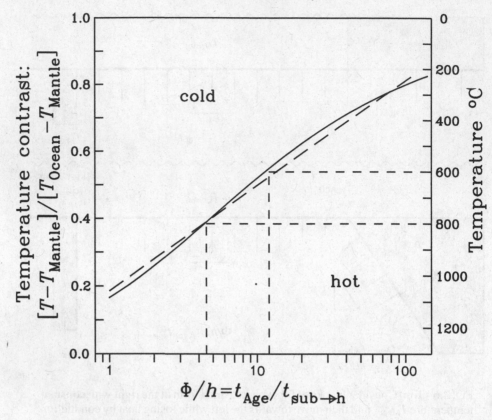

Fig. 2.7 Dependence of temperature contrast within subducted lithosphere on the ratio Φ/h (eq. 2.5). Φ is the thermal parameter and h is depth of the subducted lithosphere's upper surface. On the left scale, a value of 0.0 implies interior temperature has warmed to the temperature of normal mantle; 1.0 corresponds to temperature of surface before subduction. The right scale indicates the temperature mantle for a model where T_{Ocean} and T_{Mantle} are 0°C and 1300°C. Horizontal and vertical dashed lines indicate values of Φ/h corresponding to temperatures of 600–800 °C, approximately the maximum temperature where shallow oceanic earthquakes occur (see Fig. 2.5). The slanted dashed line indicates a linear approximation to eq. 2.5:

$$\frac{I - I_{Mantle}}{I_{Ocean} - I_{Mantle}} \cong 0.18 + 0.31 \log_{10} \frac{\Phi}{h}.$$

Fig. 2.8 Variation in thermal parameter Φ and depth of earthquake activity along the Indonesian subduction zone. Map (top) and cross section (center) present locations and depths for selected and relocated earthquakes with depths exceeding 150 km, while the bottom plot shows there is a systematic increase in thermal parameter Φ as we move eastward along the Indonesian arc. Note that deep-focus earthquakes occur only east of 105°E, where Φ exceeds 5000 km. On the lower plot distances where there are two values of Φ are areas where the age of the subducting lithosphere is uncertain. Figure reproduced from Kirby *et al.* (1996).

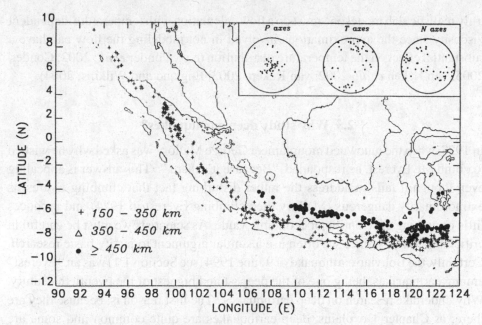

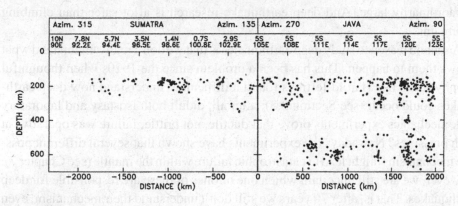

Along-Strike Distance vs Thermal Parameter, Indonesia

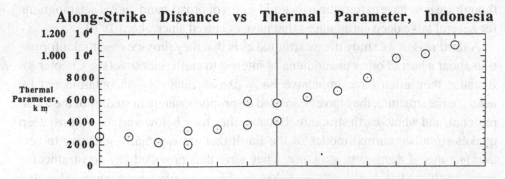

truly realistic slab model one must do a flow calculation with temperature-dependent viscosity since the approximations involved in not modeling the flow can have a substantial effect on the temperature distribution (e.g., Conder *et al.*, 2002; Conder, 2005; van Keken *et al.*, 2002; van Keken, 2003; England and Wilkins, 2004).

2.5 Why study deep earthquakes?

In 1923, when the renowned mountaineer George Mallory was asked why he wished to climb Mt. Everest, he responded, "Because it's there."[2] This answer is appealing even though it fails to address the rather disturbing fact that climbing Everest is extraordinarily dangerous (Mallory died climbing Everest in 1924) and produces little or nothing of scientific or economic value. As scientists we must be careful in criticizing Mallory since we often use a similar argument to justify basic research. Certainly, the Bolivian earthquake of 9 June 1994 (see Section 1.1) was an "Everest" among earthquakes, being one of the deepest and biggest of the twentieth century. Why undertake research on deep earthquakes? The first reason is, because they are there; as Chapter 1 explains, deep earthquakes are quite common and some are extraordinarily large. And deep earthquake research is a lot safer than climbing mountains.

A second reason to study deep earthquakes is that we don't really understand what allows them to happen. This has been a problem since the 1920s when thoughtful geophysicists like Harold Jeffreys found it difficult to understand how deep earthquakes could occur (see Section 3.3); after all, didn't both isostasy and laboratory rock mechanics experiments prove that ductile, not brittle, failure was operative at such pressures? Nowadays the experiments have shown that several different possible mechanisms might lead to catastrophic failure within the mantle (see Chapter 7). However, we are still uncertain which one or ones of these are responsible for deep earthquakes. That is, after 70 years we still don't understand their mechanism, even though we now have extraordinarily good suites of broad-band digital seismograms for several large deep earthquakes that have occurred since about 1970.

A third reason we study deep earthquakes is that they provide essential information about a host of other phenomena of interest to earth scientists (see Chapter 8). Because they often have impulsive body phases relatively uncontaminated by near-source structure, they have been used disproportionately in studies of the local, regional, and whole-earth structure. Because they have below-surface origins, deep quakes stimulate normal modes of the Earth that are especially sensitive to certain features of deep-earth structure. They have thus provided key constraints for gross earth models of velocity and density. Most people now acknowledge that earth scientists' acceptance of plate tectonics between 1965–1975 was one of the

---

[2] See *New York Times* 18 March 1923.

great paradigm shifts of the past two hundred years. But, deep earthquakes played a critical role in this acceptance, since Wadati–Benioff zones delineated the path of downgoing material near convergent margins. If subduction had been aseismic, how long would it have taken to convince the world that mantle convection occurred, and that beneath deep ocean trenches there was colder, higher-velocity mantle material? Just as most people don't realize that deep earthquakes are common, they also fail to recognize that deep earthquakes have taught us much of what we have learned about the Earth from analyzing earthquake records.

Finally, in several places intermediate-depth earthquakes pose a severe hazard to mankind. Chapter 1 already has described two of these, the 1939 Chillan and 1977 Romanian earthquakes. Damaging intermediate-depth earthquakes are certain to occur in the future in both regions. Moreover, it is possible one could even occur in the continental United States: Abe (1981) assigned a focal depth of 60 km to the 13 April 1949 Seattle earthquake with m_B of 6.9. Although some other investigators have assigned a slightly shallower depth, the Seattle–Tacoma area experienced damage from earthquakes with focal depths of 40 km or more in 1965 and 2001, and events with greater depths seem possible.

2.6 Previous reviews of deep earthquake literature

In spite of their ubiquity, utility, and hazard, many highly respected seismology and geophysics texts devote little or no space to deep earthquakes. They are barely mentioned in Dahlen and Tromp (1998), Shearer (1999), Udías (1999), Aki and Richards (2002), Turcotte and Schubert (2002), and Bolt (2004). Scholz (2002) devotes only a couple of pages to them. Even texts like Lay and Wallace (1996) and Fowler (2004) mention them only on a few pages where they discuss seismicity in subduction zones.

Surprisingly, the present book is apparently the first ever book-length monograph devoted exclusively to deep earthquakes. However, there have been a few review papers that strived to be comprehensive. Three appeared about a decade after the discovery of deep earthquakes; these were by Leith and Sharpe (1936), Gutenberg and Richter (1938), and Jeffreys (1939). Although neither Leith nor Sharpe ever published much else about deep earthquakes, their paper is by far the best from a modern perspective (see Chapter 3). Gutenberg and Richter (1938) focus mostly on summarizing then-available knowledge on the geographical distribution of deep earthquakes, while Jeffreys (1939) writes in his characteristically rambling, self-focused style in which he buries gems of extraordinary geophysical insight between crotchety arguments about the best way to interpolate tables.[3] For reasons that

---

[3] Let me be perfectly clear: I think Harold Jeffreys may well be the finest geophysicist of the twentieth century. As such, his publications thus prove that one doesn't have to be a great prose writer to attain this status.

are unclear there were no other comprehensive reviews published until Frohlich (1989).

There have been several reviews concerning the geographical distribution of deep earthquakes and their focal mechanisms. Many of these reviews appeared before it was possible for ordinary seismologists to routinely manipulate earthquake catalogs graphically. In addition to the catalogs of Gutenberg and Richter (1941; 1949; 1954) and Rothé (1969), these include a series of papers by Koning (1952), a series of papers by *Santo* [1969a; 1969b; 1969c; 1970a; 1970b; 1970c; 1970d], and a set of maps published by Barazangi and Dorman (1969). Subsequently, several investigators have interpreted catalog data to construct contours of or fit smooth surfaces to Wadati–Benioff zones (Bevis and Isacks, 1984; Burbach and Frohlich, 1986; Yamaoka *et al.*, 1986; Chiu *et al.*, 1991). Isacks and Molnar (1971) and Apperson and Frohlich (1987) reviewed how the focal mechanism orientations depend on the geometry of the Wadati–Benioff zone (see also Chapters 6 and 10). Astiz *et al.* (1988) performed a global geographic survey of large intermediate-depth earthquakes, including a complete catalog of focal mechanisms for earthquakes occurring after 1960 and having magnitudes exceeding 6.5.

Since 1990 there have been authoritative review papers concerning various closely related topics. One of the most valuable is Lay's (1997) comprehensive review on the nature of subducting slabs. Lay thoroughly discusses Wadati–Benioff zone structure, the influence that subducted material has on seismic waves, results of tomographic studies, and what these tell us about mantle convection. Lay's review is far more detailed than the corresponding material in this book in Chapter 8.

There have also been two important reviews discussing the nature of deep earthquake source mechanisms. Green and Houston (1995) discuss current knowledge of slab temperature distribution, focal mechanism patterns, deep earthquake stress drops, and source-time functions. They then argue that intermediate- and deep-focus earthquakes have different mechanical origins. At intermediate depths the dehydration of subducted, hydrated minerals allows brittle failure to occur at pressures where microcracks wouldn't ordinarily survive. And, at greater depths metastable olivine transforms to spinel along so-called anticracks (see Section 7.3.3), which coalesce and allow deep-focus earthquakes to occur. Although Kirby *et al.* (1996) organize their review around their interest in the mechanism problem, they present the most thorough summary presently available concerning the characteristics of deep-focus earthquakes. They also conclude that transformational faulting of olivine to spinel is responsible for deep-focus earthquakes. However, both of these reviews are quite forceful in their arguments supporting particular mechanisms, and thus don't provide a broadly comprehensive summary of the various competing mechanisms. Because the mechanics of deep rupture depends ultimately on the properties of mantle materials, Poli and Schmidt's (2002) review of the petrology of subducted slabs is an invaluable complement to the other reviews.

Finally, there have been several collections of papers on various topics related to deep seismicity, most produced after conferences. The *Journal of Geophysical Research* published a special section with 16 contributed papers in 1987 entitled "deep and intermediate focus earthquakes, phase transitions, and the mechanics of deep subduction" (see Frohlich, 1987). A number of significant papers on deep earthquakes and related processes appeared in an AGU special volume published in 1996 (Bebout *et al.*, 1996). And in 2001 the *Physics of the Earth and Planetary Interiors* published 16 papers on the "processes and consequences of deep subduction" (see Rubie and van der Hilst, 2001). Among these is a paper by Wiens (2001), which is perhaps the best summary available concerning the effect of temperature on the observed characteristics of deep earthquakes.

2.6 References

Abe, K., 1981. Magnitudes of large shallow earthquakes from 1904 to 1980, *Phys. Earth Planet. Int.*, **27**, 72–92.

Aki, K. and P. G. Richards, 2002. *Quantitative Seismology*, (2nd edn), Sausalito, CA, University Science Books, 700 pp.

Apperson, K. D. and C. Frohlich, 1987. The relationship between Wadati–Benioff zone geometry and P, T and B axes of intermediate and deep focus earthquakes, *J. Geophys. Res.*, **92**, 13821–13831.

Astiz, L., T. Lay, and H. Kanamori, 1988. Large intermediate-depth earthquakes and the subduction process, *Phys. Earth Planet. Int.*, **53**, 80–166.

Barazangi, M. and J. Dorman, 1969. World seismicity map of ESSA Coast and Geodetic survey data for 1961–1967, *Bull. Seismol. Soc. Amer.*, **59**, 369–380.

Bates, R. L. and J. A. Jackson, 1984. *Dictionary of Geological Terms*, (3rd edn), New York, Doubleday, 571 pp.

Bebout, G. E., D. W. Scholl, S. H. Kirby, and J. P. Platt (eds.), 1996. *Subduction, Top to Bottom*, Geophys. Mon. 96, Washington, D.C., American Geophysical Union, 384 pp.

Bevis, M. and B. L. Isacks, 1984. Hypocentral trend surface analysis: probing the geometry of Benioff zones, *J. Geophys. Res.*, **89**, 6153–6170.

Bilek, S. L., T. Lay, and L. J. Ruff, 2004. Radiated seismic energy and earthquake source duration variations from teleseismic source time functions for shallow subduction zone earthquakes, *J. Geophys. Res.*, **109**, doi:10.1029/2004JB003039, B00308.

Bolt, B. A., 2004. *Earthquakes*, (5th edn), New York, W. H. Freeman, 378 pp.

Bonner, J. L., D. B. Blackwell, and E. T. Herrin, 2003. Thermal constraints on earthquake depths in California, *Bull. Seismol. Soc. Amer.*, **93**, 2333–2354.

Burbach, G. V. and C. Frohlich, 1986. Intermediate and deep seismicity and lateral structure of subducted lithosphere in the circum-Pacific region, *Rev. Geophys.*, **24**, 833–874.

Carslaw, H. S., and J. C. Jaeger, 1959. *Conduction of Heat in Solids*, Oxford, Oxford University Press, 510 pp.

Castellano, M., F. Bianco, S. Imposa, G. Milano, S. Menza, and G. Vilardo, 1997. Recent deep earthquake occurrence at Mt. Etna (Sicily, Italy), *Phys. Earth Planet. Int.*, **102**, doi:10.1016/S0031–9201(96)03214–1, 277–289.

Chiu, J.-M., B. L. Isacks, and R. K. Cardwell, 1991. 3-D configuration of subducted lithosphere in the western Pacific, *Geophys. J. Int.*, **106**, 99–111.

Coco, M., J. Pacheco, S. K. Singh, and F. Courboulex, 1997. The Zihuatanejo, Mexico, earthquake of 1994 December 10 ($M = 6.6$): source characteristics and tectonic implications, *Geophys. J. Int.*, **131**, 135–145.

Conder, J. A., 2005. A case for hot slab surface temperatures in numerical viscous flow models of subduction zones with an improved fault zone parameterization, *Phys. Earth Planet. Int.*, **149**, doi:10.1016/j.pepi.2004.08.018, 155–164.

Conder, J. A., D. A. Wiens, and J. Morris, 2002. On the decompression melting structure at volcanic arcs and back-arc spreading centers, *Geophys. Res. Lett.*, **29**, doi:10.1029/2002GL015390.

Dahlen, F. A. and J. Tromp, 1998. *Theoretical Global Seismology*, Princeton, NJ, Princeton University Press, 1025 pp.

Duda, S. J., 1965. Secular seismic energy release in the circum-Pacific belt, *Tectonophysics*, **2**, doi:10.1016/0040–1951(65)90035–1, 409–452.

England, P. and C. Wilkins, 2004. A simple analytical approximation to the temperature structure in subduction zones, *Geophys. J. Int.*, **159**, doi:10.1111/j.1365–246X.2004.02419.x, 1138.

Fowler, C. M. R., 2005. *The Solid Earth: An Introduction to Global Geophysics*, (2nd edn.), Cambridge, U.K., Cambridge University Press, 685 pp.

Frohlich, C., 1987. Kiyoo Wadati and early research on deep focus earthquakes: Introduction to the special section on deep and intermediate focus earthquakes, *J. Geophys. Res.*, **92**, 13777–13788. Reprinted in *History of Geophysics*, **4**, 166–177, 1990.

  1989. The nature of deep-focus earthquakes, *Ann. Rev. Earth Planet. Sci.*, **17**, doi:10.1146/annurev.ea.17.050189.001303, 227–254.

•Green, H. W. and H. Houston, 1995. The mechanics of deep earthquakes, *Ann. Rev. Earth Planet. Sci.*, **23**, doi:10.1146/annurev.ea.23.050195.001125, 169–213.

Gutenberg, B., and C. F. Richter, 1938. Depth and geographical distribution of deep-focus earthquakes, *Geol. Soc. Amer. Bull.*, **49**, 249–288.

  1941. Seismicity of the Earth, *Geol. Soc. Amer. Spec. Pap.*, **34**, 1–130.

  1949. *Seismicity of the Earth and Associated Phenomena*, Princeton, N.J., Princeton University Press, 273 pp.

  1954. *Seismicity of the Earth and Associated Phenomena*, (2nd edn), Princeton, N.J., Princeton University Press, 310 pp.

•Honda, H., 1962. Earthquake mechanism and seismic waves, *Geophys. Notes, Tokyo Univ.*, **15**, Supp., 1–97.

Isacks, B. and P. Molnar, 1971. Distribution of stresses in the descending lithosphere from a global survey of focal-mechanism solutions to mantle earthquakes, *Rev. Geophys. Space Phys.*, **9**, 103–174.

•Jeffreys, H., 1939. Deep-focus earthquakes, *Ergebnisse kosmischen. Physik*, **4**, 75–105.

Kirby, S. H., W. B. Durham, and L. A. Stern, 1991. Mantle phase transitions and deep-earthquake faulting in subducting lithosphere, *Science*, **252**, 216–225.

•Kirby, S. H., S. Stein, E. A. Okal, and D.C. Rubie, 1996. Metastable mantle phase transformations and deep earthquakes in subducting oceanic lithosphere, *Rev. Geophys.*, **34**, doi:10.1029/96RG01050, 261–306.

Klein, F. W., R. Y. Koyanagi, J. S. Nakata, and W. R. Tanigawa, 1987. The seismicity of Kilauea's magma system. In *Volcanism in Hawaii*, eds. R. W. Decker, T. L. Wright, and P. H. Stauffer, U.S. Geol. Surv. Prof. Pap., 1350, 1019–1185.

Koning, L. P. G., 1952. Earthquakes in relation to their geographical distribution, depth and magnitude, *Proc. Koninklijke Nederlandse Akad. v. Wetenschappen*, **B55**, 60–77, 174–193, 194–206, 263–292.

•Lay, T., 1994. *Structure and Fate of Subducting Slabs*, San Diego, CA., Academic Press, 185 pp.; reprinted from volume **35** of *Advances in Geophysics: Seismological Structure of Slabs*, eds. R. Dmowska and B. Saltzman, 1–185, 1994.

Lay, T. and T. C. Wallace, 1996. *Modern Global Seismology*, San Diego, Academic Press, 521 pp.

•Leith, A. and J. A. Sharpe, 1936. Deep focus earthquakes and their geological significance, *J. Geol.*, **44**, 877–917.

•McKenzie, D. P., 1969. Speculation on the consequences and causes of plate motions, *Geophys. J. Roy. Astron. Soc.*, **18**, 1–32.

Molnar, P., D. Freedman, and J. S. F. Shih, 1979. Lengths of intermediate and deep seismic zones and temperatures in downgoing slabs of lithosphere, *Geophys. J. Roy. Astron. Soc.*, **56**, 41–54.

Nakamichi, H., H. Hamaguchi, S. Tanaka, S. Ueki, T. Nishimura, and A. Hasegawa, 2003. Source mechanisms of deep and intermediate-depth low-frequency earthquakes beneath Iwate volcano, northeastern Japan, *Geophys. J. Int.*, **154**, doi:10.1046/j.1365-246X.2003.01991.x, 811–828.

Poli, S. and M. W. Schmidt, 2002. Petrology of subducted slabs, *Ann. Rev. Earth Planet. Sci.*, **30**, doi:10.1146/annurev.earth.30.091.201.140550, 207–235.

Quintanar, L., J. Yamamoto, and Z. Jiménez, 1999. Source mechanism of two 1994 intermediate-depth-focus earthquakes in Guerrero, Mexico, *Bull. Seismol. Soc. Amer.*, **89**, 1004–1018.

Richter, C. F., 1958. *Elementary Seismology*, San Francisco, W. H. Freeman, 768 pp.

Rothé, J. P., 1969. *The Seismicity of the Earth, 1953–1965*, Paris, United Nations Educational, Scientific and Cultural Organization (UNESCO), 336 pp.

Rubie, D. C. and R. D. van der Hilst, 2001. Processes and consequences of deep subduction: introduction, *Phys. Earth Planet. Int.*, **127**, doi:10.1016/S0031-9201(01)00217-5, 1–7.

Santo, T., 1969a. Characteristics of seismicity in South America, *Bull. Earthquake Res. Inst., Tokyo Univ.*, **47**, 635–672.

Santo, T., 1969b. Regional study on the characteristic seismicity of the world. Part I. Hindu Kush region, *Bull. Earthquake Res. Inst., Tokyo Univ.*, **47**, 1035–1048.

Santo, T., 1969c. Regional study on the characteristic seismicity of the world. Part II. From Burma down to Java, *Bull. Earthquake Res. Inst., Tokyo Univ.*, **47**, 1049–1061.

Santo, T., 1970a. Regional study on the characteristic seismicity of the world. Part III. New Hebrides Islands region, *Bull. Earthquake Res. Inst., Tokyo Univ.*, **48**, 1–18.

Santo, T., 1970b. Regional study on the characteristic seismicity of the world. Part IV. New Britain Island region, *Bull. Earthquake Res. Inst., Tokyo Univ.*, **48**, 127–144.

Santo, T., 1970c. Regional study on the characteristic seismicity of the world. Part V. Bonin–Mariana Islands region – Comparative study on the reliability of USCGS seismic data since 1963, *Bull. Earthquake Res. Inst., Tokyo Univ.*, **48**, 363–380.

Santo, T., 1970d. Regional study on the characteristic seismicity of the world. Part VI. Colombia, Rumania and South Sandwich Islands, *Bull. Earthquake Res. Inst., Tokyo Univ.*, **48**, 1089–1105.

Scholz, C. H., 2002. *The Mechanics of Earthquakes and Faulting*, (2nd edn), Cambridge, U.K., Cambridge University Press, 471 pp.

Shearer, P. M., 1999. *Introduction to Seismology*, Cambridge, U.K., Cambridge University Press, 260 pp.

Turcotte, D. L. and G. Schubert, 2002. *Geodynamics*, (2nd edn), Cambridge, U.K., Cambridge University Press, 456 pp.

Udías, A., 1999. *Principles of Seismology*, Cambridge, U.K., Cambridge University Press, 475 pp.

Van Keken, P. E., 2003. The structure and dynamics of the mantle wedge, *Earth Planet. Sci. Lett.*, **215**, doi:10.1016/S0012-821X(03)00460-6, 323–338.

Van Keken, P. E., B. Kiefer, and S. M. Peacock, 2002. High-resolution models of subduction zones: implications for mineral dehydration reactions and the transport of water into the deep mantle, *Geochem., Geophys., Geosyst.*, **3**, doi:10.1029/2001GC00256, 1056.

Wadati, K., 1929. Shallow and deep earthquakes (2nd paper), *Geophys. Mag.*, **2**, 1–36.

Weaver, C. S., J. E. Zollweg, and S. D. Malone, 1983. Deep earthquakes beneath Mount St. Helens: evidence for magmatic gas transport, *Science*, **221**, 1391–1394.

• Wiens, D. A., 2001. Seismological constraints on the mechanism of deep earthquakes: temperature dependence of deep earthquake source properties, *Phys. Earth Planet. Int.*, **127**, doi:10.1016/S0031-9201(01)00225-4, 145–163.

Wiens, D. A. and S. Stein, 1983. Age dependence of oceanic intraplate seismicity and implications for lithospheric evolution, *J. Geophys. Res.*, **88**, 6455–6468.

Wolfe, C. J., P. G. Okubo, and P. M. Shearer, 2003. Mantle fault zone beneath Kilauea Volcano, Hawaii, *Science*, **300**, doi:10.1126/science.1082205, 478–480.

Yamaoka, K., Y. Fukao, and M. Kumazawa, 1986. Spherical shell tectonics: effects of sphericity and inextensibility on the geometry of the descending lithosphere, *Rev. Geophys.*, **24**, 27–53.

3

The history of deep earthquakes

3.1 Solid earth geophysics in 1900

Nowadays, we often forget how recently we came to our present understandings about the structure and dynamics of the Earth (e.g., see Brush, 1980). In the year 1900, Emil Wiechert had just proposed that the Earth had a core, which he said was solid and made of iron (Wiechert, 1897); Clarence Dutton had just coined the word "isostasy" (Dutton, 1889; reprinted 1925); one of the first international geophysics journals – *Gerlands Beiträge zur Geophysik* – had just published its first issue in 1887; and nobody had yet rejected Alfred Wegener's theory of continental drift, because it hadn't yet been proposed (Wegener, 1915).

In seismology, there was much confusion about the meaning of the various seismic phases observed on the few crude seismographs available at the time (e.g., Dewey and Byerly, 1969). While Poisson had established the existence of P and S waves earlier in the nineteenth century, it wasn't clear to all that they would be observable. Knott (1888), for example, performed a theoretical analysis of the reflection and refraction of waves at discontinuities. He concluded that P and S would be mixed up by conversions and reflections on passing through heterogeneous materials in the earth's crust, so it would be futile to look for a separation of the two kinds of waves, or to infer anything about the origin of the shock from the nature of the surface motion. Rayleigh (1885) had predicted the presence of surface waves; however, when scientists did obtain seismograms they often incorrectly identified the prominent surface wave arrivals as S phases. Only in 1900 did Oldham correctly identify P, S, and surface waves on seismograms (Oldham, 1900). Even then there was considerable uncertainty about why surface waves often lasted so long, especially after Lamb's (1904) analysis for a half-space showed that they should be just a simple pulse instead of a dispersed wave train. And, when

I thank Diane Doser for reviewing an earlier draft of this chapter.

Omori (1907) located a large 1906 Japanese deep earthquake by using time differences between P and S at several stations to determine event-to-station distances, he presumed wrongly that it was shallow and thus concluded that at the Tokyo station, the initial P phase was transversely polarized, while the larger S phase was longitudinally polarized (see the Preface).

So, how and when did we come to accept the existence of deep earthquakes? In this chapter I explore something of this history, focusing on their discovery and the role they played in earth science up to the 1960s when plate tectonics became established.

3.2 The depth of normal earthquakes

3.2.1 Normal is shallow; normal is deep

At first, the question wasn't: "How deep are the deepest earthquakes?" Rather, it was: "What is the depth of normal earthquakes?" Today it seems obvious that many large, destructive earthquakes must be shallow because they sometimes produce fault scarps, offset curbs and fences, and cause fissures in the ground. However, a century ago it was quite logical to think that these fault-associated phenomena might simply be a product of some deeper, more essential earthquake process. Thus the scientific literature before about 1928 demonstrates a considerable range of opinion on the matter of earthquake depth; then, as now, geologists and geophysicists seemed impressed by different and apparently contradictory features of the observations.

On the one hand, the presence of surface faulting and the relatively small area of intensive damage for some earthquakes strongly suggested that their focal depth must be quite shallow. For example, one chapter of Milne's (1891) book on earthquakes is entitled "The depth of the earthquake centrum"; in a section entitled "Greatest depth of an earthquake origin" he reviews the literature of the time and tentatively concludes that "the greatest possible depth of any earthquake impulse occurring in our planet is limited to . . . 30.64 geographical miles." For the San Francisco, California earthquake of 1906, Reid (1910) interpreted travel times and concluded that the focal depth was "not more than 40 km or so"; moreover, from the falloff of shaking with distance he concluded that the fault "would hardly extend to a greater depth than 20 km, and probably not so deep." There were various methods proposed to determine focal depths including using the variation with distance of travel times for P waves, as well as P waves refracted or reflected at the crust-mantle boundary (De Quervain, 1914; Mohorovičić, 1914; Gutenberg, 1923); all obtained depths of about 20–60 km. However, because these methods often relied on the availability of numerous accurately known phase arrival times to determine inflection points and other features of the travel time curves, they were not widely applicable and not entirely convincing.

Nevertheless, other reported geophysical observations seemed to support considerably greater depths. For example, for the Owen's Valley, California earthquake of 26 March 1872, Whitney (1872) concluded that the geographical extent of the macroseismal area implied that the "depth was very great – probably not less than 50 miles."[1] Pilgrim (1913) calculated focal depths from apparent velocities for ten events including the 1906 San Francisco earthquake, which he found to be at 140 km depth. Mainka (1915) applied Pilgrim's method to available European teleseismic observations of 95 earthquakes, and concluded that only 37% had focal depths in the 0–100 km range, while the remainder had greater depths. Walker (1921) analyzed emergence angles for P phases and concluded that typical focal depths were about 1250 km. Oldham (1923) interpreted intensity data for an 1895 Italian quake and concluded its depth must be "on the order of 100 miles or so."

How was such confusion possible? In the case of Walker, it now appears he did not properly understand the effect of the free surface and crustal layering on phase amplitudes. Furthermore, some reputable scientists believed that there was no connection between faulting and the origin of earthquakes. For example, for the 1906 San Francisco quake, Oldham (1909) stated "the growth of our knowledge of earthquakes is making it continuously more evident that, whether great or small, they have little or no connexion with the faults which reach the surface of the earth." He then specifically mentions the San Andreas fault, and says "the earthquake, therefore, cannot be regarded as an incident in the growth of the fault, nor the fault as the cause of the earthquake."

3.2.2 Turner's 1922 paper – high-focus hocus pocus

So it was, in 1922, that Turner (1922) published the first paper in the first volume of the *Monthly Notices of the Royal Astronomical Society, Geophysical Supplement*, entitled "On the arrival of earthquake waves at the antipodes, and on the measurement of the focal depth of an earthquake." This paper was significant for a number of reasons; the *Geophysical Supplement* of the *Monthly Notices* was destined to become one of the world's most respected geophysical journals – *Geophysical Journal International*. And H. H. Turner was director of the organization

---

[1] Whatney also comments:

That there ever can be any hope of our being able to predict the time of occurrence of an earthquake shock, is in the highest degree improbable. We can say that such and such regions are more liable to be visited by these catastrophes than others are; but when they will happen, there is no possibility of ascertaining in advance.

Modern journal publishers should also note that Whitney's two full-length scientific articles on the Owen's Valley earthquake appeared in issues of the *Overland Monthly* dated only six months after the earthquake's occurrence.

that prepared the *International Seismological Summary*, which cataloged earthquake phases reported by seismological observatories around the world. Because of the 1922 paper, Turner has widely been credited as the "discoverer" of deep earthquakes.

. But, reading the paper today, one is not so sure. For earthquakes occurring in South America between 1913 and 1916, Turner analyzed available P travel time residuals for European stations at epicentral distances beyond 140°.[2] He noted that for some earthquakes the residuals at these stations were 20 to 30 s, larger than the average of residuals at nearer stations, and large even considering the uncertain timing of seismograph clocks of the time. He also noted that negative residuals, corresponding to antipodal phases that arrived earlier than expected, were more common than positive residuals. After some analysis of the best available travel time tables (Knott, 1919), he concluded that:

There is clearly a systematic error for each particular earthquake which is ascribed to a particular depth of focus . . . the two extreme cases of systematic error, viz. +16 sec and −48 sec, can be satisfactorily explained as due to depth of focus −0.021 and +0.061 of the earth's radius, the negative sign indicating that the focus must be *above* the depth (*d*) corresponding to the tables in use. Hence *d* must be > 0.021, and perhaps has some such value as 0.04.

Thus, the implication is that "deep-focus" earthquakes exist which have depths as much as 400 km deeper than normal earthquakes; but "high-focus" earthquakes also exist which may have depths as much as 130–250 km shallower than normal earthquakes. Thus, normal earthquakes must occur at depths of at least 130 km.

3.2.3 Turning on Turner

Turner's paper immediately elicited considerable interest from some of the best geophysicists of the day (e.g., see Wrinch and Jeffreys, 1922; Gutenberg, 1923). Perhaps the most thoughtful response was from S. K. Banerji, the director of the Bombay Observatory in India (Banerji, 1925); he noted that both theoretical and observational studies indicated that earthquakes produced both body waves and surface waves. However, while theory implied that the amplitudes of body waves for shallow-focus and deep-focus earthquakes should be about the same, it showed clearly that Rayleigh waves should be much, much smaller for deep-focus events. He supported these assertions using Lamb's (1904) equations for surface waves in an elastic solid. After some analysis, he concluded that for Rayleigh waves with periods of 12–20 s on a flat earth:

---

[2] For the antipodal point at 180° he suggested the name "hypocentre." Obviously, this never became widely accepted.

If, now, *A* denotes the amplitude of the Rayleigh waves recorded at any station, produced by an earthquake the focus of which is situated on the surface, then, when the focus is

(1) 10 km. deep, the amplitude $= Ae^{-1}$,
(2) 100 km. deep, the amplitude $= Ae^{-10}$,
(3) 1000 km. deep, the amplitude $= Ae^{-100}$,

the sources being supposed to be of the same intensity and all in the same vertical line.

On a round Earth he found that the diminution of amplitude with depth is even more rapid, and thus concluded that, since "all records of distant earthquakes . . . show that the amplitude of the long (surface) wave phase is generally much larger than the amplitudes of the other phases" that "a depth of even 100 km is too much for an earthquake focus." He further concluded that Turner's (1922) "cases of 'high' focus will not be inconsistent with the results obtained in this paper (but) the values obtained for the depths of the 'deep' focus are inappropriate with the conclusions of this paper."

But, was it true that surface waves were always larger than body waves? Almost immediately, Byerly (1925) noted examples of earthquakes with body wave amplitudes that did exceed surface wave amplitudes. Somewhat later Stoneley (1931) made a more thorough study of about a dozen alleged deep-focus earthquakes reported by the *International Seismological Summary* that occurred between 1922 and 1927, and concluded that surface waves were indeed absent or greatly reduced in amplitude in comparison with normal events.[3] As often happens in science, still later Gutenberg (1933) pointed out that Zoeppritz (1912) had noticed this phenomenon previously; indeed, Zoeppritz's paper, published after his death, clearly describes "two different types of earthquakes." The first type produced seismograms with distinct and and impulsive phases, but little or no "Hauptwellen," or surface waves; as examples of this type Zoeppritz mentioned the Japanese earthquake of 21 January 1906 and the Tonga earthquake of 31 March 1907, now thought to have had depths of 340 km and 400 km, respectively. The second type produced seismograms with weaker, less impulsive preliminary phases and strong surface waves; examples of these included the 1906 San Francisco earthquake, as well as two shallow Tongan earthquakes.[4]

[3] He also noted that the terms "deep-focus earthquake" and "normal-focus earthquake" were cumbersome, and suggested that instead "bathyseism" and "orthoseism" be used. Earlier, Oldham (1926) had suggested the term "bathyseism" and "episeism". Indeed, he states:

The word bathyseism has met with the approval of sound classical scholars of my acquaintance. The word episeism is as unanimously condemned; but they have been unable to suggest a correctly formal word which would express the concept of an earthquake originated at a shallow depth, or at a higher level than the bathyseism. The word episeism is, at any rate, no more barbarous than many other scientific terms in common use.

[4] Zoeppritz's explanation for these two types of earthquakes is very wrong but very novel. He notes that during an earthquake potential energy stored in the earth changes to kinetic energy $\frac{1}{2}mv^2$, as rock of mass *m* is moved

Perhaps a more typical contemporary response to Turner's paper was that of Oldham (1926), who suggested that the approximate focal depth of an earthquake could be determined from an analysis of felt reports or what we would now call intensity maps.[5] Oldham was clearly someone worth listening to; much earlier it was he who had first clearly identified P, S, and surface waves (Oldham, 1900) as well as seismic phases which had penetrated the earth's core (Oldham, 1906). There was considerable intensity information available; e.g., Oldham possessed a catalog of intensities for more than 5000 Italian earthquakes felt between 1897 and 1910. He suggested that since the energy of an earthquake wave varies inversely as the square of the distance from the source, for an earthquake at depth h the ratio of the energy E_d a distance d away to E_e at the exact epicenter will be:

$$E_d/E_e = h^2/(h^2 + d^2). \tag{3.1}$$

Thus, h can be determined from observations of E_e, E_d, and d. Oldham thus undertook a reasonably thorough review of available estimates of what value of E_d corresponds to particular Mercalli and Rossi-Forel intensities; he then made some tortuous corrections to account for the effects of attenuation. For the quakes in the Italian catalog Oldham noted that "it may be taken as certain that over 90 per cent, and possibly 95 per cent, of the whole group originated at depths of under, and mostly well under, 5 miles from the surface."

However, in spite of his considerable professional achievements, Oldham seemed to lack confidence in the general applicability of this result, and apparently had considerable respect for the work of Turner. He goes on to conclude:

With the exception of a small proportion, the 'earthquakes' fall into two main classes: the largest of these . . . is formed by the local earthquakes, which can be felt, but which do not give rise to long-distance records; the smaller is composed of disturbances which can be registered, by suitable instruments, at long distances from the origin, and are known to be associated, in some cases with surface shocks that can be felt. The first-named class originates almost exclusively at depths of less than 30 km from the surface and, prevailingly, at depths of about 5 km. The second class originates at depths between 50 and 600 km, but, as regards the vast majority, mainly at about 200 km below the surface.

Oldham also seemed unwilling to disagree with either Pilgrim's 140 km estimate or with Reid's 20 km estimate concerning the depth of the 1906 California earthquake; he states:

The bathyseism (i.e., an event of the deeper class) may give rise not only to a long-distance record, but also to a local episeism of greater or less degree of violence . . . The direct

---

at velocity v. He then posits that if the mass m is relatively large but v is small, one has an earthquake with large, long-period surface waves; while if v is relatively large and m is small, the earlier phases are stronger.

[5] Oldham, however, used the term "intensity" for the energy in an arriving earthquake wave, and not, as we do today, to connote descriptions of felt reports such as those of Mercalli or Rossi-Forel.

effect of the disturbance, propagated from the bathyseism, might be insufficient to cause any damage at the surface, but, reinforced by the strains set up, and the fractures produced, by molar displacements of the surface-crust, a very high degree of violence may be induced in the epicentral region.

Oldham suggests that "near earthquakes" – which were shallow and damaging – might be a complete different phenomenon than teleseismically recorded earthquakes. This idea was plausible to many of Oldham's contemporaries, and only refuted after the Montana earthquake of 28 June 1925 ($M_S = 6.75$) occurred; it was felt locally and well-recorded at both regional and teleseismic distances. Byerly (1926) published extensive travel times for regional distances and concluded they were consistent with a shallow focus ($\sim$25 km). Jeffreys (1928) evaluated the times of teleseismic phase arrivals, and found them to be no different than arrivals from other teleseisms.

3.3 The discovery of deep earthquakes

3.3.1 Wadati's 1928 and 1929 papers

Meanwhile, in 1925 a 22 year old Japanese physics student named Kiyoo Wadati had just graduated from Tokyo University and started his first real job, working for the seismological section of the Japan Central Meteorological Observatory in Tokyo. According to Wadati (1989), "my routine job at that time was mainly to reduce the seismological reports sent from local observing stations and determine the positions of the foci of the respective earthquakes." At that time the Japan Meteorological Agency operated the densest network of seismograph stations in the world, and there was considerable practical interest in the possibility of predicting earthquakes; indeed, while at the university Wadati had experienced the great Kwanto earthquake and fire of 1 September 1923, which killed approximately 143,000 people.

On 23 May 1925, about a month after Wadati began work, a strong earthquake occurred on the western coast of central Japan, killing 40 people. This earthquake had many aftershocks, but the largest aftershock interested Wadati because "this earthquake showed a special wave propagation feature, areas of high seismic intensity extending far from the epicenter." Thus Wadati's career began.

Wadati's (1928) paper is a classic, and still worth reading today. In the introduction, he presents a thorough review of the literature, and makes it very clear that determining the "normal" depth of earthquakes is very much the problem of interest. Because the clocks available for use at the time were highly inaccurate, Wadati could not rely on absolute travel times as the primary information for locating earthquakes; moreover, since the Earth's velocity structure was then poorly known, he could not simply multiply a velocity factor times an S–P interval to estimate station-to-quake distances and thence, focal depths.

Instead, he devised a clever geometrical scheme using ratios of S–P intervals observed at four or more stations to determine an earthquake's epicenter and focal depth; for a constant-velocity Earth this method is entirely independent of velocity (Fig. 3.1). Using this method, he determined the focal depths for 15 earthquakes in the Kwanto district (east-central Japan), and found them all to be between 30 and 46 km. Since these quakes were typical of at least one class of "normal" (and sometimes deadly and highly destructive) earthquakes in Japan, this clearly demonstrated that some "normal" earthquakes were shallow, and not at depths of 130 km or more, as suggested by Turner.

However, Wadati then proceeded to plot maps of S–P intervals for numerous other large, well-recorded earthquakes. These maps of "iso-(P-S) lines" – as he called them – exhibited distinctly different patterns for some events (Fig. 3.2). The minimum observed S–P interval at the epicenter was 10 s or less for most "ordinary" earthquakes; but for others it was 40 s or more. Since the distribution of velocity with depth for P and S was then unknown, Wadati made a series of calculations for earth models where $1/v$ varied linearly with depth; for all reasonable models both the minimum (S–P) times and the inflection points of the travel time curves (Fig. 3.3) implied that some of the earthquakes had focal depths of 300 km or more.

Moreover, the larger of these deep earthquakes were exactly those which possessed "abnormal distribution of seismic intensity," i.e., no regions of extreme damage or intensity, as observed for "normal" (shallow) earthquakes, but instead zones of relatively high intensity separated by zones where the event was not felt (Fig. 3.4). At some stations these "abnormal" earthquakes also produced peculiar seismograms which showed no "preliminary tremor," a phenomenon which had been noted by numerous Japanese seismologists beginning with Omori (1907) (see the Preface). Wadati correctly concluded that the so-called "preliminary tremor" was simply the P wave, which in most cases was much smaller than the S wave; when at some stations the P wave was very large relative to the S wave the "preliminary tremor" seemed to be "absent." This was undoubtedly a result of the radiation pattern of P and S over the focal sphere. Wadati was unsure about the cause of the abnormal distribution of intensity, stating only that it might be due to "convergence of the seismic rays, induced secondary earthquakes, or existence of absorbing media in the earth crust."

Wadati's next paper in English (Wadati, 1929) demonstrated that the occurrence of deep earthquakes was not limited to a particular range of focal depths; instead he found earthquakes occurring at all depths down to about 500 km. In this paper Wadati also proposed the scheme for classifying earthquakes as shallow, intermediate, or deep which is still in general use today (see Section 2.1).

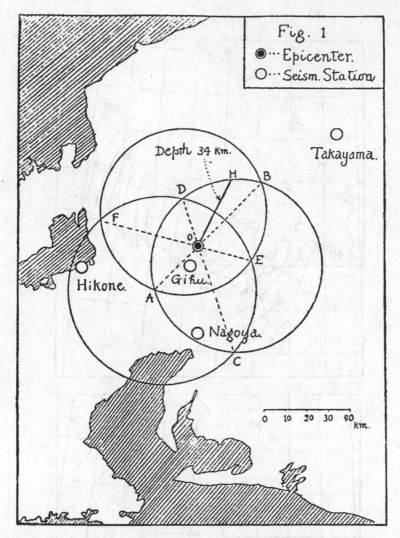

Fig. 3.1 Wadati's (1928) clever method for locating the earthquake of 8 July, 1925, and demonstrating that some earthquakes are, indeed, shallow. For any station such as Hikone, the quake–station distance Δ_H equals

$$\frac{V_P V_S}{V_P - V_S}(t_S^{[H]} - t_P^{[H]}),$$

where V_P, V_S, $(t_P^{[H]})$, and $t_S^{[H]}$ are the P and S velocities and arrival times. For any two stations, e.g. Hikone and Gihu, the ratio of distances Δ_H/Δ_G equals $(t_S^{[H]} - t_P^{[H]})/(t_S^{[G]} - t_P^{[G]})$, independent of the velocities. By a well-known but seldom-remembered theorem of elementary geometry, the locus of points having a constant ratio of distances to two fixed points is a sphere. Thus, Wadati could find the epicenter as the point O of intersection of three straight lines (AB, CD, EF) which connect the points of intersection of any two spheres and the Earth's surface; moreover, the depth of the epicenter (in this case, 34 km) is the half-length of the shortest chord (OH) passing through the epicenter in any of the circles. While Wadati's method assumes that V_P and V_S are constant, it does not require knowledge of their actual value; this was important because at the time there were no reliable regional velocity models. Figure reproduced from Wadati (1928).

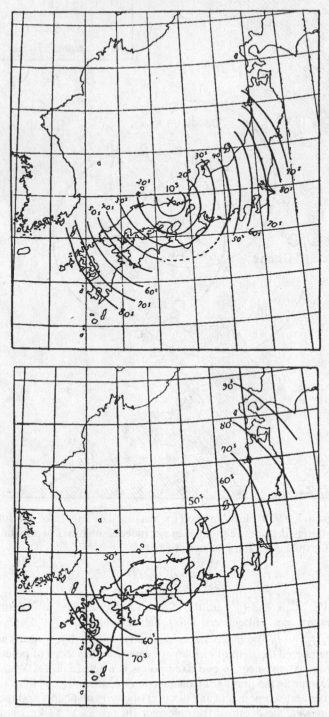

Fig. 3.2 S–P isochrons for (top) the shallow North Tazima earthquake of 23 May 1925 and (bottom) the deep earthquake of 15 January, 1927 (focal depth = 420 km). Figure reproduced from Wadati (1928).

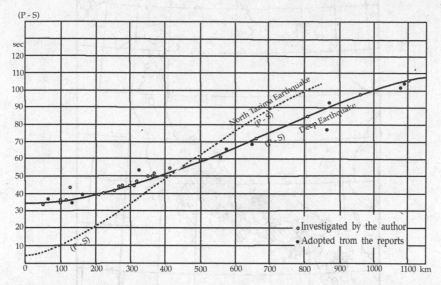

Fig. 3.3 Comparison of S–P intervals for the deep earthquake of 26 July 1926 (focal depth = 360 km) and the shallow North Tazima earthquake. For any such travel-time curve the minimum corresponds to rays leaving the focus vertically, while the inflection point corresponds to phase arrivals from rays leaving the focus horizontally; while in theory this provides a good way to determine focal depth, in practice it is subject to systematic errors. Figure reproduced from Wadati (1928).

3.3.2 *The impact of Wadati's papers*

Who discovered deep earthquakes? The question of who deserves credit for important scientific discoveries can be complex (see Brush, 1980); the individual who first
proposes an idea may support it with flawed observations or an incorrect theory;
for earthquake depth this criticism applies to Pilgrim (1913), Walker (1921), and
Turner (1922). There also may be individuals who are correct about both data and
theory, but who never communicate effectively, who never manage to convince the
scientific community about their ideas. Perhaps the most famous example of this is
Gregor Mendel, who in 1865 announced and subsequently published the essential
elements of modern genetics (Mendel, 1866); however, his work went unnoticed
until 1900 when it was discovered by others who had independently reached the
same conclusions (e.g., Correns, 1900).

For deep earthquakes, the closest parallel to Mendel's case would be Toshi Shida,
a geophysicist at Kyoto University.[6] At the dedication of the Beppu Geophysical
Laboratory in October, 1926, Shida presented a speech proposing the existence of

---

[6] Unlike Mendel, Shida was highly regarded during his lifetime for his scientific accomplishments. He was among
the first to publish observations of the solid earth tide; today the three dimensionless parameters describing Earth
tides are properly called Love–Shida numbers.

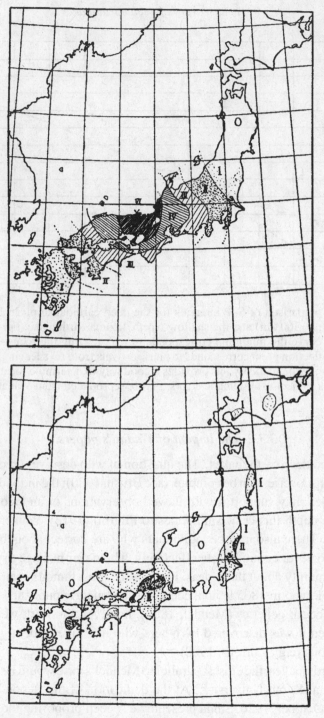

Fig. 3.4 Distribution of seismic intensity for (top) the North Tazima earthquake of 23 May 1925, and (bottom) the deep earthquake of 26 July 1926 (focal depth = 360 km). The intensity scale is that used by Japanese investigators of the time, where VI corresponds to a modified Mercalli intensity of X or XI. Figure reproduced from Wadati (1928).

deep earthquakes. He observed that since P arrivals are longitudinal waves, then the first motions of some Japanese earthquakes showed they must have focal depths near 300 km; he presented specific data for an especially well-recorded earthquake that had occurred on 26 July 1926, and also for the earthquake of 21 January 1906, which was especially large. However, Shida did not publish these results, and the text of his speech was not published until more than 10 years later, just after his death (Shida, 1937). However, Wadati was certainly aware of Shida's work, as he states, "Prof. T. Shida already considered that (the 27 July 1926 earthquake)[7] must be an earthquake of deep orgin, but his paper seems to appear not yet" (Wadati, 1928).

Yet by Brush's (1980) criteria, Kiyoo Wadati clearly deserves full credit for the discovery of deep earthquakes. Wadati's observations, published in both Japanese (Wadati, 1927) and in English (Wadati, 1928), demonstrate quite convincingly that there exist earthquakes with focal depths of 300 km as well as 30 km. And, there is little doubt that these publications convinced the scientific world. Before they appeared, the best papers such as those of Banerji (1925) and Oldham (1926) focus on demonstrating that earthquake focal depths are probably shallow.

Indeed, Byerly (1924) studied seismograms for the 17 January 1922 earthquake, now known to be a single, large ($M_{w(CMT-Hist)} = 7.9$) deep-focus earthquake occurring at approximately 660 km beneath Peru (see Okal and Bina, 1994; and Section 10.18.5). The reliable stations available to Byerly were roughly clustered in Europe, eastern North America, and western North America; thus, when Byerly constructed S–P arcs he found that they formed a triangle. Although Byerly stated that "the most obvious solution would be a single earthquake" he was apparently unwilling to consider the possibility of deep focus; thus he concluded:

The seismograms have been found to be the superimposed records of three distinct South American earthquakes. The first of these occurred at $O = 3^h\ 50^m\ 20^s$, M.G.T., with an epicenter in Brazil ... The second occurred at $O = 3^h\ 50^m\ 22^s$, with an epicenter in Venezuela ... The third occurred at $O = 3^h\ 50^m\ 24^s$, with an epicenter just off the coast of Ecuador.

Even Harold Jeffreys, who initially believed Turner's (1922) observations (see Wrinch and Jeffreys, 1922), upon further reflection had determined that an isostatic analysis of mountain elevations implied that the Earth had no strength at depth. Thus he concluded that the Earth could not support the stresses necessary for earthquakes with focal depths exceeding about 35 km (Jeffreys, 1928).

In contrast, immediately after Wadati's papers appeared, subsequent work focused on confirming the existence of deep earthquakes (e.g., Stoneley, 1931), or on developing techniques for their analysis. Scrase (1931; 1933), for example, noted that for stations at appropriate distances, deep earthquakes should produce

---

[7] This earthquake occurred at 18.54 GMT on 26 July, which was 27 July local time in Japan.

additional reflected phases besides PP, SP, PS, and SS; he proposed calling the upgoing reflected phases pP, sP, pS, and sS,[8] and reported observations of them for deep earthquakes. Stechschulte (1932) performed a detailed analysis of teleseismic recordings of the Japanese earthquake of 29 March 1928, which Wadati (1929) had called "the largest deep earthquake in recent years," and confirmed Wadati's conclusion that it must have focal depth of about 400 km. And, somewhat later, Visser (1936a; 1936b) and Tillotson (1938) looked carefully at Turner's (1922) "high-focus" earthquakes and concluded that all were apparently ordinary shallow-focus events.

Moreover, the existence of deep earthquakes explained many previous discrepancies in phase arrival times and uncertainties in phase identification. Thus, in the ten years following Wadati's initial paper, scientists at many of the world's geophysical laboratories undertook research efforts to develop accurate travel time tables, and thence to infer global earth structure. The most notable efforts were in Great Britain (Jeffreys, 1932; 1935; 1939b; 1939c; Jeffreys and Bullen, 1940); in the United States (Gutenberg and Richter, 1935a; 1935b; 1936a; 1936b, 1937); and in Japan (Honda, 1931; Wadati *et al.*, 1932; Wadati, 1932; Wadati and Oki, 1932; 1933; Wadati and Masuda, 1933; 1934). As noted by Frohlich (1987), the mantle velocity models determined during this period differ surprisingly little from those in use today.

3.4 The discovery of Wadati–Benioff zones

3.4.1 Earthquake catalogs and the plasticity of the mantle

Wadati's papers and rapid improvements in both seismography and travel-time tables encouraged other scientists to look for deep earthquakes elsewhere besides Japan. Early global maps were published by Turner (1930), Conrad (1933), Visser (1936a; 1936b), Hayes (1936), Leith and Sharpe (1936), and Yamaguti (1936); together these indicate the presence of intermediate- and deep-focus events in nearly all the geographic regions where they are known today as well as a few places such as Africa and the northeastern United States where we no longer believe they occur. Several of the early investigators also noticed the association between deep earthquakes, deep ocean trenches, and volcanoes.

However, Beno Gutenberg and Charles Richter undoubtedly prepared the most comprehensive and reliable earthquake catalog. Basically, they evaluated locations using data from the *International Seismological Summary*, augmented with seismograms available to them in Pasadena, California. In a series of papers concentrating

---

[8] While Scrase gets credit for inventing the now-generally-accepted notation for pP, etc., in his paper he uses the notations PcPcP, ScPcS, and ScPcP for the phases we now call PKP, SKS, and SKP.

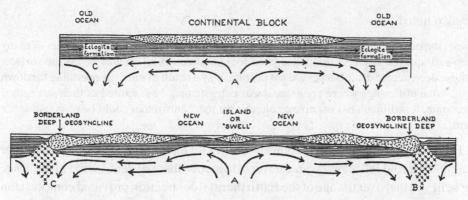

Fig. 3.5 Holmes' (1933) sketch describing how gross features of surficial geology might be controlled by convection in the Earth's mantle. He suggests that there is "distension of the continental block on each side of *A* (and) formation of new ocean floor from rising basaltic magma. The front parts of advancing continental blocks are thickened into mountainous borderlands with oceanic deeps in the adjoining ocean floor due to the accumulation of eclogite at *B* and *C*." Reproduced with permission from the Washington Academy of Sciences.

on deep earthquakes, they produced travel time tables for various focal depths and phases, and compiled catalogs of hypocenters (Gutenberg and Richter, 1934; 1936a; 1937; 1938; 1939a). Subsequently, these investigations formed the basis for their influential reference book, *Seismicity of the Earth*, which summarized global earthquake activity (Gutenberg and Richter, 1941; 1945; 1949; 1954). Seismologists still use this catalog today as it is one of the better sources of information about global seismicity in the first half of the twentieth century.

Meanwhile, as these data accumulated other scientists engaged in speculation about the nature of physical processes active in the Earth's mantle and their possible relationship to seismicity and surface geological features. For example, Holmes (1931; 1933) argued that the mantle, although solid, over long enough times ought to convect if its viscosity were lower than about 10^{25} Pa-s;[9] this was significant since Jeffreys (1926) had estimated that the mantle's dynamic viscosity was actually about 1.2×10^{19} Pa-s. A more thorough analysis by Haskell (1935) subsequently placed the value at 10^{21} Pa-s (see historical discussion by Mitrovica, 1996). Moreover, Holmes suggested that mantle convection cells might explain the presence of oceanic islands, and also mountain ranges at the margins of continents; his maps and sketches supporting these assertions (e.g., Fig. 3.5) look almost contemporary.

[9] Readers who wish to compare values for mantle viscosity reported in the literature should keep in mind that many papers specify dynamic viscosity η, while others specify kinematic viscosity ν (which is just η divided by density ρ). Moreover, both cgs and mks units are commonly used. Thus, Haskell's (1935) value of 10^{21} Pa-s for the mantle's dynamic viscosity corresponds to a kinematic viscosity of about 2.9×10^{17} m²/s. In cgs units these values are 10^{22} poise and 2.9×10^{21} cm²/s.

He also noted that:

(Near) the oceanic deeps of the western margins of the Pacific . . . lie the epicenters of many deep earthquakes that have originated from foci hundreds of kilometres below the surface. If these deep-seated disturbances are not in some way a result of currents operating far down in the zone of flowage where great masses of eclogite may be fractured as their strength is overcome, it is difficult to conceive an alternative mechanism that could be responsible for them.

However, the convection hypothesis was by no means universally accepted. For example, Harold Jeffreys was an able proponent of a thermal contraction hypothesis. He believed that over the age of the Earth thermal contraction provided compression of about the right order to account for mountain-building; moreover, in the first edition of his book, *The Earth* (Jeffreys, 1924), he had stated:

It has been seen that the strength of the earth's crust is finite at the surface, increases to a maximum at a depth probably of the order of 100 km, and then gradually decreases with depth, probably becoming insignificant at a depth of about 400 km. The greatest earthquakes should take place where the greatest stress is relieved by fracture, in other words at the level of greatest strength, some 100 km below the surface. Earthquakes below that level should increase in frequency and decrease in violence with depth, while other earthquakes of moderate intensity should originate at depths between zero and 100 km. We should expect, therefore, that the foci of the greatest earthquakes would be at depths of the order of 100 km, though they would have a considerable range about this depth.

3.4.2 Wadati discovers Benioff zones

Meanwhile, Wadati (1935) published another observational paper in which he presented the first contour map ever published showing an inclined zone of hypocenters (Fig. 3.6). In this paper he states:

. . . contour lines and the distribution of active and dormant volcanoes have a close connection with one another . . . Generally speaking, the possibility of drawing contour lines of the focal depth suggests that there exists in the crust something like a weak surface at where the earthquake outburst is liable to occur. This surface extends slopewise in the crust near the Japanese Islands . . . Contour lines corresponding to larger depths lie on one side of the active volcanic chains and those of small depth on the other side. It is interesting that the volcanic activity seems to have no connection with the occurrence of shallow-focus earthquakes, but is closely connected with that of earthquakes having rather deep foci . . . Generally speaking, volcanoes are found in a chain running nearly parallel to the continental margin and earthquakes occur also in a zone along a volcanic chain. But deep-focus earthquakes are apt to take place on one side nearer the continent and shallow focus ones on the other side where it is in most cases bordered on a very deep sea. This tendency seems to be observable in many volcanic regions in the world. Of course, we cannot say decisively but if the theory of continental drift suggested by A. Wegener be true, we may perhaps be able to see its traces of the continental displacement in the neighbourhood of Japan consulting the figures of contour lines as well as the distributions of volcanoes . . .

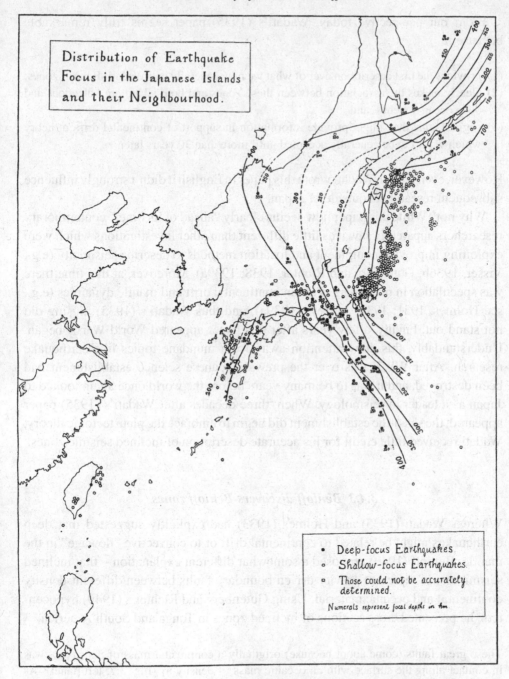

Fig. 3.6 Wadati's (1935) map showing contour lines of equal focal depth for deep earthquakes near Japan.

From our perspective today, Wadati's (1935) paper seems truly remarkable because it:

(1) provides the first description ever of what we now generally call Wadati–Benioff zones;
(2) clearly makes the association between these zones and both island arc volcanoes and deep ocean trenches; and
(3) suggests that they might provide information in support of continental drift, a theory which wasn't to be generally accepted until more than 30 years later.

However, even though Wadati wrote this paper in English it didn't strongly influence subsequent researchers outside of Japan.

Why not? While this paper now seems clearly "ahead of its time," contemporary researchers apparently saw it as little different than other investigations which were exploiting improving equipment and location methods to describe seismicity (e.g., Visser, 1936b; Gutenberg and Richter, 1938; 1939a). Moreover, at this time there was speculation in many papers about continental drift and mantle dynamics (e.g., see Holmes, 1931; 1933; Jeffreys, 1939a), and thus Wadati's (1935) musing did not stand out. Finally, a few years after this paper appeared World War II began. Understandably, this drew attention away from mundane topics like earthquake research. After WWII was over the prewar Japanese science establishment had been destroyed, and it was to be many years before the world once again looked to Japan as a leader in seismology. When, three decades after Wadati's (1935) paper appeared, the scientific establishment did begin to embrace the plate tectonic theory, Wadati received little credit for his accurate description of inclined seismic zones.

3.4.3 Benioff discovers Benioff zones

Whereas Wadati (1935) and Holmes (1933) had explicitly suggested that deep earthquakes might be related to continental drift or to convective "flowage" in the mantle, Benioff (1949) proposed a somewhat different explanation – that inclined seismic zones represent gravity-driven boundary faults between different-density continental and oceanic material. Using Gutenberg and Richter's (1949) hypocenters, he presented cross sections of inclined zones in Tonga and South America.

These great faults (come about because) originally a continental mass of density δ_2 was in contact along the surface with an oceanic mass of density δ_1 (Fig. 3.7, left panel). As a result of secular flow in the masses a hydrostatic pressure developed . . . (it) produced a differential stress, acting perpendicular to the boundary. . . Acting over a sufficiently long time this differential stress pattern produced a distortion of the two masses which resulted in a configuration shown in Fig. 3.7 (right panel). . . The downwarping of the oceanic block thus produced the depressions which form the oceanic deeps.

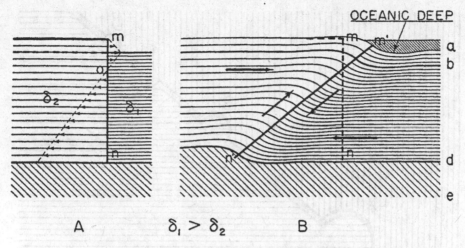

Fig. 3.7 Benioff's (1949) sketch "illustrating the proposed origin of the great Tonga–Kermadec fault." He suggests that over time gravitational stresses cause thinner, denser oceanic material to flow beneath thicker, lighter continental; thus an initially vertical boundary (left) deforms into a dipping planar surface (right) along which earthquakes occur.

In addition, to explain the change in dip of the inclined seismic zone in South America beneath about 300 km, he suggested that there may be layers of material with two different densities on the continental side. Subsequently he developed these ideas further (Benioff, 1954) and presented measurements at both intermediate and deep depths of inclined seismic zone dip angles, determined from about a half dozen cross sections for various geographic regions around the Pacific.

While Benioff's (1949, 1954) papers exercised a strong influence on the developers of plate tectonics – convincing them of the importance of inclined seismic zones – in retrospect they seem far less insightful than Wadati's. Instead, more impressive by far is Benioff's effort to quantify the relationship between individual earthquakes and fault motion. While his approach isn't quite modern his ideas have since developed into the concepts of scalar moment and seismic asperities. In particular, he observes that

Thus on a given fault the square root of the energy of an earthquake is proportional to the elastic strain rebound increment which generates it . . .

i.e., that $E^{1/2}$ is proportional to the 'strain rebound increment'. He utilizes Gutenberg and Richter's (1942) relationship to determine $E^{1/2}$ in terms of magnitude M:[10]

$$\log_{10} E^{1/2} = 6.0 + 0.9M \tag{3.2}$$

[10] Here, E is in cgs units, or ergs. Note the similarity of this equation to the Hanks and Kanamori (1979) relationship which defines moment magnitude $M\mathrm{w}$ in terms of scalar moment M_0 : $\log_{10} M_0 = 16.0 + 1.5 M_\mathrm{W}$.

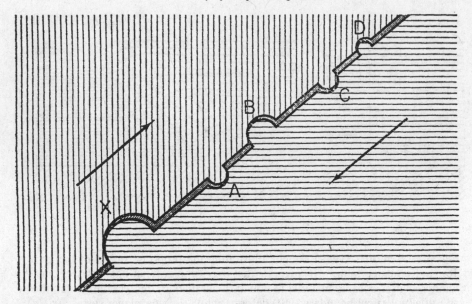

Fig. 3.8   Benioff's (1949) sketch illustrating how individual earthquakes might
represent the breaking of "indentations and elevations . . . over which fault surfaces
are locked."

Finally, using Gutenberg and Richter's (1949) data he constructs plots of the accumulated strain rebound increments over 30 years for the Tonga–Kermadec region,
commenting that a significant fraction – about 7% – of the total strain rebound increment is attributable to the largest earthquake – the $M_W = 7.7$ deep-focus quake of 26
May 1932. He further suggests that individual earthquakes represent the breaking
of "indentations and elevations (which) represent points or areas over which fault
surfaces were locked" (Fig. 3.8).

In spite of the fact that Benioff's contribution was one of synthesis – analyzing
global geophysical observations and bringing together information from geology,
seismology, and physics – his style apparently was not to reference his sources
in detail; in his 1949 and 1954 papers he referenced only eight papers and ten
papers respectively; these include three by himself, and none by Wadati. Instead,
for information on seismicity Benioff seems to have relied heavily on Gutenberg
and Richter, his colleagues on the faculty at CalTech. Nevertheless his density-
difference model to explain the origin of inclined seismic zones seems to be at odds
with Gutenberg and Richter's (1945) paper in which they state:

It should not be overlooked that, in spite of the occurrence of deep-focus earthquakes, the
data of seismology indicate no difference in the accessible physical properties between
rocks underlying oceanic and continental areas at depths in excess of about 70 km.[11]

---

[11] Incidentally, Gutenberg and Richter did refer to Wadati's (1935) paper in their 1938 paper and in the 1949
and 1954 editions of their book, *Seismicity of the Earth*, but not in the 1939, 1941 or 1945 papers which were
essentially preliminary versions of the book.

Moreover, in at least one matter Benioff seems to have glossed over important details; in Benioff (1954) he states:

In a recent paper, Honda and Masatsuka (1952) have studied the direction of motion of some 145 intermediate and deep earthquakes of Japan using observations of a large number of Japanese stations. Since these stations are all in or near the epicentral region of the shocks, the reliability of the first-motion determinations is relatively high. Their results indicate clearly that for these shocks the dip of the faults and the direction of slip are in substantial agreement with the assumptions and findings of the present investigation.

In fact, to a modern observer this appears to be untrue, as the orientations of nodal planes for deep earthquakes beneath Japan are highly variable, and generally do not coincide with the plane of the Wadati–Benioff zone (see Section 10.11).

Hugo Benioff died in 1968; subsequently in western literature the inclined zones that he described have often been called Benioff zones. Meanwhile, in Japan they are called Wadati zones. Why has Benioff's name been applied to these zones, considering that Wadati's descriptions were earlier, closer to the mark, and published in English?

Dickinson and Hatherton (1967) were apparently the first to call them Benioff zones. William Dickinson and Trevor Hatherton (personal communication) both make it clear that they were strongly influenced by publications of Benioff and Gutenberg and were largely unfamiliar with Wadati's publications. Hatherton notes that he was influenced by his association with Benioff while spending a year at CalTech; Dickinson states that since 1969 he has referred to them only as "inclined seismic zones," partly because he now feels it is inappropriate to give them any man's name and partly "because the term Benioff zone has been distorted by some into a catchy synonym for subduction zone." In any case, it is clear that Benioff was an effective spokesman for the dynamic importance of the inclined zones, and that ultimately his work strongly influenced his contemporaries. However, it is also clear that Wadati was their true discoverer.

Today, considering that both terms are in wide use, the only fair and practical solution is to call them Wadati–Benioff zones, as suggested by Uyeda (1971). I now follow this practice in all my publications.

3.5 The deep earthquake source

3.5.1 *Phase transitions in the upper mantle*

Almost as soon as scientists established the existence of deep earthquakes, they began speculating about the mechanical nature of the failure process that generated them. From the very beginning, people seemed attracted to the idea that phase transitions might have something to do with the process, even though deep earthquake seismograms seemed similar in many ways to those of shallow quakes. For example, Stechschulte (1932) said:

The apparent predominance of shear waves must be taken into account in any hypothesis that one might put forth in regard to the mode of origin of a shock at so great a depth as 410 km. The records would seem to preclude anything in the way of a mere explosive activity. The question arises whether possibly changes in volume by crystalization may take place and at such a rate that the resulting stress differences cannot be quickly enough realized by plastic flow and relief must be had in the earthquake shock.

Meanwhile, Jeffreys (1936) had interpreted earthquake travel-time curves and concluded that there must be a velocity discontinuity in the uppermost mantle. Jeffreys had noted that there was a sharp increase in the slope of the travel-time curve at quake-station distances of about 20°. Although the available travel-time compilations did not report multiple arrivals as might occur at a cusp (Fig. 3.9), he concluded that the change in slope implied there was a distinct increase in seismic velocity within the mantle at a depth of approximately 480±20 km.

Nevertheless, when Wadati and Oki (1933), Gutenberg and Richter (1939b), and Jeffreys (1939c) finished inversion of travel time data and published global earth models, they included no first-order velocity discontinuity in the 400–700 km depth range.[12] Apparently not everybody was convinced of the presence of mantle velocity discontinuities. Leith and Sharpe's (1936) review stated:

It has . . . been pointed out that the mantle of the earth, to a depth of 1000 km, is characterized by a lack of definite structural discontinuities.

They were probably strongly influenced by Gutenberg and Ricther (1936b), who said:

We find no evidence for any first-order discontinuity within the mantle, but a discontinuity of the second order appears to exist at about 100 km, and two others near 2000 km.

In fact, it was about thirty years before it was confirmed that such discontinuities or strong gradients did exist in the mantle, that there were actually two of them – near depths of about 400 km and 650 km (Fig. 3.10), and that these were explainable in terms of phase transitions in olivine. There were several reasons that it took this long. Initially, World War II severely curtailed the pertinent areas of geophysical research. Subsequently, Cold War pressures helped encourage investigations of mantle structure. Ultimately this gave seismologists a much improved collection of short-period regional seismic data, which demonstrated the existence of the mantle discontinuities (e.g., Golenetskii and Medvedeva, 1965; Archambeau *et al.*, 1969).

---

[12] In Jeffreys' (1939c) model the velocity does increase fast enough in the upper mantle to cause a small triplication in the travel time near about 20°.

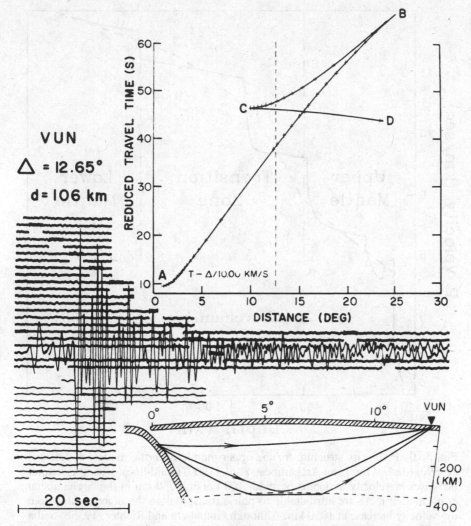

Fig. 3.9 A zone of rapidly increasing seismic velocity in the upper mantle causes a
triplication in the travel-time curve (top); the rays with the shallowest penetration
make up the *AB* branch; rays that turn within the zone of increasing velocity
(dotted line in lower sketch) form the *BC* branch; and rays that penetrate below the
zone make up the *CD* branch, which often has low amplitudes. The short-period
seismogram (center) for a Tonga earthquake with a focal depth of 106 km recorded
in Fiji at a distance of 12° clearly exhibits arrivals corresponding to the *AB* and
BC branch. Figure reproduced from Frohlich *et al.* (1977) with permission from
Blackwell Publishers, Ltd.

Meanwhile, Birch (1952) made a detailed theoretical study of the petrological
nature of the upper mantle using an equation of state based on finite strain theory. He
concluded that upper mantle properties were consistent with an olivine–pyroxene–
garnet composition, but that the elastic properties of the lower mantle were significantly different and resembled those of the relatively close-packed oxides such as

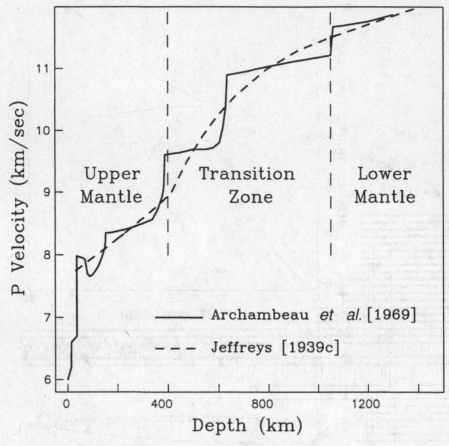

Fig. 3.10 P velocity structure for the upper mantle from the models of Jeffreys (1939c) (dashed line) and Archambeau *et al.* (1969) (solid line). Note the distinct increases in velocity at depths of about 400 km and 650 km in the Archambeau *et al.* model; these are attributable to phase transitions in the mantle. Also note the velocity increase at 1050 km. Although Gutenberg and Richter (1936b) and a few other investigators have reported a velocity increase at about this depth, it is absent in most subsequent global models.

corundum, periclase, rutile and spinel. Thus, Birch proposed that a series of major phase transformations characterized the upper 1000 km of the mantle.[13]

---

[13] Although Birch's (1952) paper is a classic for its serious content, it is perhaps most remembered for its arch disclaimer:

Unwary readers should take warning that ordinary language undergoes modification to high-pressure form when applied to the interior of the Earth; a few examples of equivalents follow:

High-pressure form	*Ordinary meaning:*
certain	dubious
undoubtedly	perhaps
positive proof	vague suggestion
unanswerable argument	trivial objection
pure iron	uncertain mixture of all the elements.

However, high-pressure laboratory apparatus had to improve significantly (see Fig. 3.11) before it was possible to experiment at the pressures and temperatures necessary to confirm a suggestion by Bernal in 1936 (see Section 7.2.2) that phase changes in olivine might, indeed, cause velocity discontinuities. Indeed, it was 1963 before scientists achieved static pressures in the laboratory corresponding to depths greater than 300 km (see Ringwood, 1970, 1996,[14] for reviews), and somewhat later still before experiments confirmed that olivine-to-spinel structure (e.g., Ringwood and Major, 1970) and spinel-to-oxide (Ming and Bassett, 1975) phase transformations could explain the observed velocity discontinuities at 400 and 650 km.

3.5.2 Deep and shallow sources are different

Despite the fact that deep and shallow earthquake radiation patterns were apparently quite similar, even when deep quakes were first discovered there were strong reasons to believe that the failure processes responsible must be fundamentally different. Laboratory studies of brittle failure clearly indicated that rock strength increased markedly as confining pressure – and thus, depth – increased; this implies that fracture at the depths of deep-focus earthquakes requires truly extraordinary shear stresses. For example, while summarizing an experimental study of rock deformation, Griggs (1936) stated:

May it be emphasized that these results should be accepted not as principles which may be immediately applied to geology, but as a small step in our groping for the physical principles underlying structural geology. For example, the character of the relation between strength and confining pressure is such as to suggest that rocks are excessively strong at depths in the earth, and hence the forces necessary to produce deformation (such as deep focus earthquakes) are colossal. This, however, cannot be safely inferred until the effect of time and temperature have been measured.

Numerous subsequent laboratory observations have since confirmed Griggs' observation that rock strength increases with pressure. Indeed, empirical studies show that, for a wide range of brittle materials, shear strength τ and confining pressure σ_n vary approximately[15] as $\tau = a + b\sigma_n$.

However, other experiments and observations such as isostasy demonstrate that, as temperature increased towards melting, rock strength decreased and failure occurred as ductile flow, rather than fracture. For example, observations show that

---

[14] Ringwood's (1996) paper, published three years after his death, gives a nice description of the historical development of knowledge about mantle phase transitions among petrologists, and the general skepticism that pioneers such as Birch and Ringwood faced from their peers.

[15] This relationship is known as Byerlee's law, after James D. Byerlee.

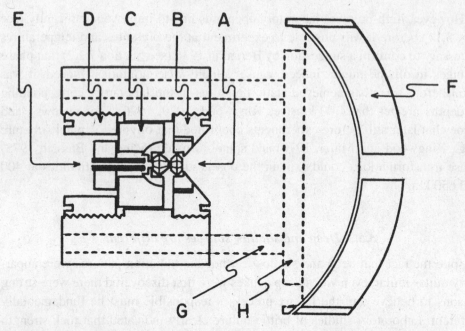

Fig. 3.11 The invention of the so-called diamond anvil apparatus made it possible to routinely achieve mantle pressures and temperatures, and thus extend experimental rock mechanics to regimes where deep earthquakes occur. In diamond anvil devices, millimeter-sized samples of material are placed between the facets of two large gem-quality diamonds (*A*) that are set within a fist-sized device that uses driver screws (*D*) to apply pressure. The mechanical strength of the diamonds makes it possible to achieve exceptionally high pressures; however, since diamonds transmit light and are poor conductors they allow one to heat the samples with laser beams and photograph the samples (note film cassette *F*) as pressure and temperature change. Figure reproduced from Bassett *et al.* (1967) with permission from the American Institute of Physics.

under such conditions the shear stress τ and the strain rate $\dot{\varepsilon}$ depend on temperature T as $\tau = A\dot{\varepsilon}^{1/3}B^{1/T}$. If we equate this relation and Byerlee's law, using reasonable values for the constants a, b, A, and B, we obtain an empirical relation showing how the brittle–ductile transition depends on temperature, pressure, and flow rate (Fig. 1.12). For reasonable flow rates and geotherms, this demonstrates clearly that ordinary brittle fracture cannot occur at depths exceeding 100 km or so.

Thus, the deep earthquake paradox was: how could deep earthquakes occur at all, if the rocks at their source must be either too strong or too weak to fracture? Thus, Bridgman (1936) stated:

There is a theoretical possibility that must be kept in mind. There is no thermodynamic reason why transitions should not occur under shearing stress which would never occur under any combination of temperature and hydrostatic pressure . . . It seems to me that the recognition of the possibility of simultaneous existence of two phenomena, rupture and flow, must react on our pictures of how yielding takes place deep in the earth. In particular it would seem that this sort of rupture must be a factor in deep-seated earthquakes.

Although Stechschulte (1932), Leith and Sharpe (1936) and others all recognized that the radiation pattern of deep earthquakes showed no apparent isotropic component, the idea that deep earthquakes might somehow be related to phase transitions is so appealing that it keeps being revisited. For example, Bridgman (1945) stated:

Conditions are conceivable in which a transition might run catastrophically because of a change in equilibrium conditions associated with crustal movements, thus giving rise to deep seated earthquakes, but the chance for this sort of thing appears more remote than the chances for a catastrophic release of stress by a fracture phenomenon after a critical stress has been built up slowly by progress of a transition.

Similarly, when Benioff (1963; 1964) investigated strainmeter records from two deep-focus Peruvian earthquakes, he proposed that a 3% volume change at the source was the most probable cause of deep-focus (600 km) earthquakes, stating that the collapse "may be accompanied by faulting at its periphery."

The search for implosive components of deep earthquakes has continued right up to the present. For example, Gilbert and Dziewonski (1975) suggested that the Colombia earthquake of 31 July 1970 ($M_{\text{W(CMT−Hist.}} = 8.1$; $h_{\text{CMT−Hist}} = 623 \, \text{km}$) possessed an isotropic precursor and a substantial non-double-couple component. However, Okal and Geller (1979) and Russakoff *et al.* (1997) have disputed this result. Since broadband digital data have become widely available there have been several additional efforts to evaluate isotropic components of deep quakes; so far, none have been found (e.g., Kawakatsu, 1991; 1996; Okal, 1996).

Finally, in addition to the search for exotic source mechanisms for deep events, papers regularly appear which propose that various exotic mechanisms are responsible for many different kinds of quakes, shallow as well as deep. For example, Evison (1963; 1967) suggested that "phase transitions may be the true cause of earthquakes" and "sudden faulting may be essentially a surface phenomenon . . . (essentially) a form of earthquake damage." Robson *et al.* (1968) proposed that the upward rise of magmas may produce "fracture by extension failure, and . . . give rise to a motion . . . consisting of displacements in three mutually perpendicular directions." They state that such quakes are "well known in Japan," citing as examples two deep quakes occurring more than 35 years previously on 3 June 1929 and 2 June 1931. They then conclude that the magma process "provides a plausible mechanism for the generation of deep earthquakes."

3.5.3 Deep and shallow sources are alike

Not surprisingly, Japanese seismologists performed the most careful early studies of the radiation pattern of earthquake sources. In Japan the world's best regional station network overlayed the deep seismic zone and provided relatively good azimuthal coverage. For comparison there were also abundant shallow earthquakes, some with surface faulting. Because some of these shallow quakes were extraordinarily destructive seismological research was a respectable, fully developed scientific discipline. For example, Honda (1932) noted that:

The depth of focus of the North Idu earthquake which occurred on Nov. 25, 1930 was very shallow and there appeared two faults, the greater one of which ran north and south, and the eastern part of it was displaced northward and the western part southward . . . After the results of the triangulations made before and after the earthquake, the earth crust at the hypocentral region was contracted in the direction NW–SE and elongated in the direction NE and SW. The deformation of the earth crust as above may be surely caused by the pressure and tension acting in each direction respectively. The initial motion of the seismic motion observed on the earth's surface was thought to be the direct effect of the deformation of the earth crust at the hypocentral region.

Honda (1932, 1934a) also carefully measured the amplitudes and first motions of both P and S waves for 45 Japanese deep earthquakes, concluding:

. . . by analogy from the results acquired in the investigation of shallow earthquakes, we may safely conclude that the deep earthquakes also are generated as the effect of the sudden deformation of the earth crust due to the pressure acting in the direction of the rarefaction waves of the initial motion, or the tension in the direction of the condensational waves.

Furthermore, he proposed that "the displacements are well expressed by the formula $\sin 2\theta \cos \varphi$, where θ is colatitude and φ is the azimuth in spherical coordinates" in a system with an appropriately oriented polar axis. Finally, he published maps showing the quakes' nodal planes and pressure and tension axes (Figs. 3.12 and 3.13), stating:

One can see at once there has been acting the pressure in the deep layer of about two or three hundred km depth, in the direction nearly east and west and mostly inclined by half a right angle from the horizontal plane, the eastern side of which being directed upward . . . The tension is perpendicular to the former, that is directed east and west, and mostly inclined by half a right angle from the horizontal plane, the western part of which being directed upward.

Studies such as these suggested that deep and shallow quakes might be quite similar; however, for various reasons there was confusion. For shallow quakes it wasn't possible to obtain information about seismic radiation over very much

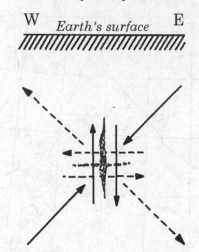

Fig. 3.12 Honda's (1934b) cartoon showing the orientation of stresses, equivalent force couples, and nodal planes "in the deep layer beneath the southern part of the 'principal deep earthquake zone' " in Japan.

of the focal sphere since upgoing and laterally directed rays were confined to the near field. And, seismologists were well aware of Reid's (1910) interpretation of the geodetic studies before and after the 1906 San Francisco earthquake (Fig. 3.14); the data indicated clearly that the 1906 quake had relieved shear strains that had existed for at least fifty years along regions adjacent to the San Andreas Fault.

One apparently plausible interpretation of Reid's (1910) data was that the expected strain release was equivalent to that from sudden application of two opposing forces directed along the fault trace. Subsequently this has been known as a single-couple or a type I source; for such a source no shear-wave energy travels along the direction of the fault trace, and a maximum travels in the direction normal to the fault plane. However, Honda's (1932; 1934b) data for deep quakes indicated strong shear waves traveling both perpendicular to and parallel to the nodal planes as indicated by P radiation; theoretical analysis showed this could arise from strain release equivalent to two perpendicular pairs of opposing forces – known subsequently as a double-couple or type II source.

Between 1930 and about 1965 there were numerous published investigations trying to determine whether ordinary earthquakes possessed single-couple or double-couple sources, or both (e.g., see the fine review by Honda, 1957). Double couples were the ultimate winner; the most careful observations generally favored them, and theorists finally concluded that single couples were impossible since they would violate the conservation of angular momentum around the strained

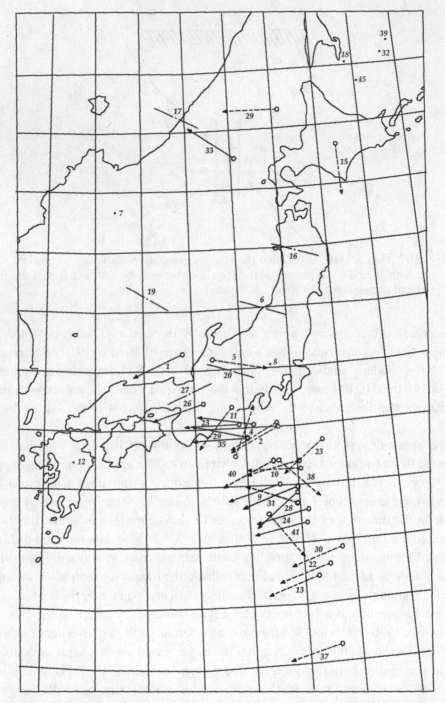

Fig. 3.13 Honda's (1934b) map showing the directions of "the maximum pressure" for deep earthquakes that occurred near Japan between 1927 and 1932. Numbers refer to specific earthquakes in a table in his paper; dotted arrows indicate directions that are "somewhat ambiguous."

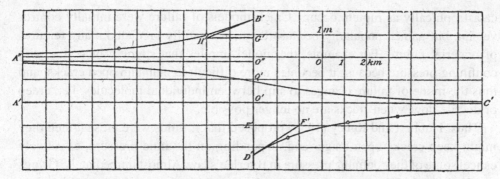

Fig. 3.14 Reid's (1910) figure summarizing the results of geodetic surveys prior to and after the 1906 San Francisco earthquake. The surveys, performed in the intervals I, 1851–1865, II, 1874–1892, and III, 1906–1907, indicated that "the line which, at the time of the II survey, was straight, as $A'O'C'$, had, at the time of the III survey, been broken at the fault and curved into the form $A''B'$, $D'C'$." Furthermore, the surveys results showed the movement was generally parallel to the fault, without significant compression or extension perpendicular to the fault. These results formed the basis for the elastic rebound model of faulting, and were widely referenced as justification for the single-couple or type I models of earthquake source radiation. Figure reproduced courtesy of the Carnegie Institution of Washington.

region.[16] In retrospect, the confusion is attributable to the difficulty of separating true S from P-to-S conversions and to the before-mentioned paucity of observations over most of the focal sphere.

Nevertheless, over the years there have been numerous seismologists who would agree with Gutenberg and Richter (1937), who stated:

The problem of the causative mechanism of these shocks has not yet been solved completely. The complex patterns of initial observed compressions and dilatations, and the frequent observations of large shear waves (S), make explanations in terms of an explosive or collapse type of origin improbable . . . The most probable conclusion, in the present state of our knowledge, is that normal and deep-focus earthquakes are brought about by the same forces, which may act either at the surface or at great depth.

How is this possible, considering the weight of laboratory-based evidence (Fig. 1.12) indicating that deep events must not be ordinary brittle fracture?

The most successful, classical explanations for brittle fracture began with Griffith (1921), who proposed that shear stress opens numerous microscopic tensile cracks. Ultimately these cracks become linked together as stress increases, and coelesce

---

[16] If this confuses you today, you can take comfort that you would have been in the company of some fine seismologists in 1964. In Benioff's (1964) paper, for example, he notes that while it is established that most quakes have double couple sources, K. Aki suggests that shallow dip slip quakes may be single couples.

catastrophically as rupture occurs. Crack models of failure were initially controversial, but gained popularity because they successfully predicted many features
of material failure. For example, the model implies that strength increases with
confining pressure because it becomes more difficult to maintain open cracks, and
thus the mode of failure changes to slip between individual molecules, i.e., creep
or ductile flow, when cracks are no longer possible.

Thus, Hubbert and Rubey (1959) proposed that if fluids were present then they
might keep cracks open longer and, depending on the fluid pressure, effectively
cancel some of the confining pressure in Byerlee's law. Almost immediately, Griggs
and Handin (1960) suggested then that the presence of trapped pore fluids might
explain why deep quakes occurred at greater depths than indicated by Fig. 1.12; the
working fluid could be water or, as suggested by Griggs and Baker (1969), partially
melted rock generated by accelerating creep processes in the fault zone. Raleigh
(1967) suggested an even more plausible hypothesis – that the natural dewatering
of serpentine at depth generated the trapped fluids. Serpentine is essentially olivine
metamorphosed by water. As it subducts beneath deep ocean trenches it reaches
conditions of temperature and pressure where the water is expelled; Raleigh and
Paterson (1965) showed that this was likely to occur at depths corresponding to
intermediate-focus earthquakes.

3.6 The deep earthquakes of history

3.6.1 Philosophically deep

A friend of the author, who is a Fellow of the American Geophysical Union but
who shall go unnamed, presented a paper at a 1996 meeting in St. Louis arguing
that the Lisbon, Portugal, earthquake of 1755 was a deep-focus earthquake. Indeed,
the paper's abstract states:

We now think that the Lisbon earthquake occurred nearly 400 km beneath the surface of
the Earth, and it is a dramatic example of what is known as a 'deep' earthquake.

In support of this hypothesis, he cited its similarities to the Spanish deep earthquake
of 1954 – both possessed very large felt areas even though there were no reports of
an epicentral region with very high seismic intensities.

Now, the Lisbon earthquake is of special interest because it is undoubtedly the
single most influential earthquake in the history of western civilization. It occurred
at approximately the time of High Mass on All Saints' Day ($\sim$10 AM, 1 Nov 1755),
and a significant number of its victims died in church; it was directly responsible for
an enormous body of philosophical and scientific literature attempting to explain
how such an earthquake could happen, and how it could happen on All Saints'

Day. For example, the Lisbon earthquake afflicts the hero of Voltaire's 1758 work *Candide*, the still-famous satire spoofing the philosopher Leibniz' contention that – since that God is perfect and all powerful – our world is the best of all possible worlds.[17]

Nevertheless, only a little library work is required to demonstrate that the Lisbon earthquake could not have been a deep earthquake. There are exceptionally well-documented reports of tsunamis which killed more than 1000 people along the Atlantic coast of the Iberian Peninsula and northwest Africa (e.g., Martínez Solares and López Arroyo, 2004); these could only have been produced by a shallow event occurring several hundred km offshore of the Straits of Gibralter.

3.6.2 The earliest deep earthquakes

What, then, are the earliest known deep earthquakes? Among the earliest with focal depths constrained by teleseismic records are those in Abe's (1981) catalog. Abe reports several events with depths between 100 km and 450 km occurring near Japan between 1904 and 1906. Of these, the earthquake of 21 January 1906 has the most historical significance, as Zoeppritz (1912), Omori (1907) and Wadati (1928) all studied it and puzzled over its non-ordinary properties (see Preface and Section 10.11.3).

Is it possible to use historical records to identify deep earthquakes that occurred prior to the invention of the seismograph? Several twentieth century intermediate-depth earthquakes have been very damaging, e.g., the earthquakes of 25 January 1939 in Chile, 4 March 1977 in Romania, and 23 November 1979 in Colombia (see Sections 1.2, 10.17.3, 10.18.5 and 10.24.3). Similarly, in the twentieth century in Greece and the surrounding islands there have been ten earthquakes with reported focal depths between 60 and 100 km that were large or damaging enough to appear in the Dunbar *et al.* (1992) catalog of significant historical earthquakes. In these and other areas where there is vigorous modern intermediate activity and a long historical record that includes the description and location of earthquake damage, it is logical to identify certain pre-twentieth century events as probable or possible intermediate earthquakes.

For example, Dunbar *et al.* (1992) report focal depths of 100 km for earthquakes occurring in Greece, Crete, or the Dodecanese Islands in 368 BC, 222 BC,

---

[17] Another earthquake that generated a remarkable literary trail is the Santiago, Chile, earthquake of 13 May 1647 that reportedly killed 2000 people. I am disappointed to report that this earthquake apparently wasn't deep (Ramirez, 1990). However, it played a central role in Heinrich von Kleist's 1806 novella, *Das Erdbeben in Chile*. Kleist is himself interesting: as an early architect of the short story, for his ultra-long embedded Germanic sentences, for his focus on the grotesque, for the confused sexuality in his works and in his own life, and for his death in a murder-suicide at age 34 in 1811. His writing went almost unnoticed until the twentieth century; subsequently *Das Erdbeben in Chile* has become the subject of a 1975 movie written and directed by Helma Sanders, a opera by the Czech composer Jan Cikker, and a 2003 opera by the Armenian composer Avet Terterian.

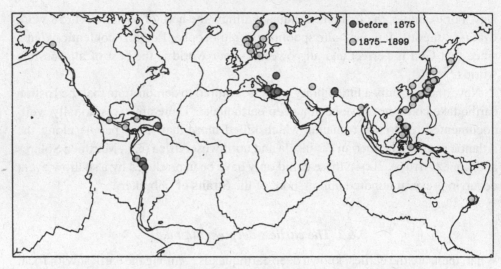

Fig. 3.15   Reported intermediate- and deep-focus earthquakes occurring prior
to 1900, as listed in the Dunbar *et al.* (1992) catalog of significant historical
earthquakes and the USGS nineteenth century earthquake catalog. Prior to 1875
nearly all such earthquakes occurred in Greece, Romania, and South America. The
reported depths for the earthquakes in northern Europe are probably erroneous, as
deep earthquakes are unknown there today.

183 BC, AD 155, AD 365, AD 796, AD 1304, and AD 1508; these are the eight
earliest intermediate-depth earthquakes in the catalog.[18] The catalog also reports
earthquakes in Peru with focal depths between 60 and 120 km occurring in 1590,
1604, 1650, 1715, and 1784 (Fig. 3.15).

However, the best-documented examples of "historical deep earthquakes" are
in Romania. Romania experiences two to five intermediate earthquakes with M_W
greater than 7 each century, with the most recent occurring in 1908, 1940, 1977
and 1986. The Romanian catalog is thought to be complete at this level back to
1411 (Oncescu *et al.*, 1999). Mândrescu and Radulian (1999a; 1999b) indicate that
earthquakes in 1471, 1620, 1738, 1802, 1829, 1838, 1893 and 1894 caused great
damage. Of these, the 26 October 1802 earthquake was the largest ($M_W = 7.9$); it
collapsed all the church towers in Bucharest and was felt from St. Petersburg to

---

[18] Thus, the answer to the trivia question: "What is the earliest reported intermediate- or deep-focus earthquake"
would be "The earthquake that struck Crete in 368 BC." The earliest such earthquake having a known date
also occurred in Crete, on 21 July AD 365. Over its long history Crete has experienced numerous earthquakes,
including one which may have been responsible for severely damaging the palace at Knossos in the seventeenth
or eighteenth century BC, and others associated with an enormous volcanic eruption on the island of Thera
about 1500 BC. The dates and even the existence of each of these early earthquakes is open to debate; e.g.,
Guidoboni *et al.* (1994), the most authoritative source on ancient Mediteranncan earthquakes, presents different
dates than Dunbar for all of them.

the Greek islands and from Belgrad to Moscow (see Fig. 1.10). Because the 1940, 1977, 1986 and 1990 Romanian earthquakes all had focal depths between 80 and 160 km, it is plausible that the previous large earthquakes also had intermediate focal depths.

Although I have not made a thorough search I have been able to find only a few other regions where there are reports of historical intermediate earthquakes occurring prior to 1890. Leeds' (1974) historical catalog of Nicaraguan earthquakes lists earthquake depths as "normal" or "intermediate" (70 to 200 km); he specifies twelve nineteenth century earthquakes as intermediate, including one as early as 1809. Similarly, Dorel (1981) determined focal depth for historical earthquakes in the Lesser Antilles by assessing how the seismic intensity decreased with distance away from the maximum as in Fig. 3.4. He lists two nineteenth century earthquakes with intermediate focal depths; the largest, with a magnitude M_S of 7.5, had an estimated depth of 150 km and occurred on 9 January 1888. Dunbar *et al.* (1992) report a depth of 100 km for an earthquake that occurred on 27 August 1886 in Turkey.

Clearly, there is considerable uncertainty concerning the focal depths of all these "historical deep earthquakes." We should be dubious about all such reportedly deep foci unless they are accompanied by accurately reconstructed intensity maps, and associated with modern deep earthquakes in the region which show a similar pattern of damage. Some reported "historical deep earthquakes" are almost certainly spurious. For example, a catalog of nineteenth century earthquakes available from the USGS includes 18 earthquakes in northern Europe with reported depths between 63 and 500 km (Fig. 3.15). As there are no contemporary deep earthquakes known there, and no structural features there such as an island arc suggesting that they might occur, these "historical deep earthquakes" probably had shallow depths.

A presently unanswered question is: can we identify any large ($M_W > 7.5$) geographically isolated deep-focus earthquakes that occurred prior to the invention of seismometers? In the past 50 years three extraordinarily large deep-focus earthquakes have occurred in regions where deep-focus earthquake activity is sparse; these are the Spanish deep earthquake of 1954, the Colombian deep earthquake of 1970, and the Bolivian deep earthquake of 1994 (see Sections 1.1, 1.3 and 5.1.4). Since 1900, these events represent three of the five largest earthquakes known having focal depths greater than 300 km. Is it possible to identify any similar events in past centuries? Have large isolated deep-focus earthquakes occurred in other geographic locations? Since such earthquakes presumably occur in every century and are widely felt and written about; the author of this book challenges any individuals with appropriate historical information to propose candidate events.

3.7 References

Abe, K., 1981. Magnitudes of large shallow earthquakes from 1904 to 1980, *Phys. Earth Planet. Int.*, **27**, 72–92.

Archambeau, C. B., E. A. Flinn, and D. G. Lambert, 1969. Fine structure of the upper mantle, *J. Geophys. Res.*, **74**, 5825–5865.

• Banerji, S. K., 1925. On the depth of earthquake focus, *Philosophical Mag.*, *Ser. 6*, **49**, 65–80.

Bassett, W. A., T. Takahashi and P. W. Stook, 1967. X-ray diffraction and optical observation on crystalline solids up to 300 kb, *Rev. Sci. Instr.*, **38**, 37–42.

Benioff, H., 1949. Seismic evidence for the fault origin of oceanic deeps, *Geol. Soc. Amer. Bull.*, **60**, 1837–1856.

1954. Orogenesis and deep crustal structure – additional evidence from seismology, *Geol. Soc. Amer. Bull.*, **65**, 385–400.

1963. Source wave forms of three earthquakes, *Bull. Seismol. Soc. Amer.*, **53**, 893–903.

1964. Earthquake source mechanisms, *Science*, **143**, 1399–1406.

Birch, F., 1952. Elasticity and constitution of the earth's interior, *J. Geophys. Res.*, **57**, 227–286.

• Bridgman, P. W., 1936. Shearing phenomena at high pressure of possible importance for geology, *J. Geol.*, **44**, 653–669.

1945. Polymorphic phase transitions and geological phenomena, *Amer. J. Sci*, **243A**, 90–97.

• Brush, S. G., 1980. Discovery of the Earth's core, *Amer. J. Phys.*, **48**, 705–724.

• Byerly, P., 1924. The South American earthquakes of January 17, 1922, *Bull. Seismograph Stations, (Berkeley)*, **2**, 50–54.

1925. Review: 'On the depth of earthquake focus,' by S. K. Banerji, *Bull. Seismol. Soc. Amer.*, **15**, 148–152.

1926. The Montana earthquake of June 28, 1925, GMCT, *Bull. Seismol. Soc. Amer.*, **16**, 209–265.

Conrad, V., 1933. Die zeitliche Folge von Beben mit tiefem Herd, *Gerlands Beitr. z. Geophysik*, **40**, 113–133.

Correns, C. G., 1900. Mendels Regel über das Verhalten der Nachkommenschaft der Rassenbastarde, *Berichten der Deutschen Botanischen Gesellschaft*, **18**, 158–168.

De Quervain, A., 1914. Über die Herdtiefenberechnung aus einer oder zwei herdnahen Stationen und die hierzu erdforderliche Zeitgenauigkeit, *Gerlands Beitr. z. Geophysik*, **13**, 148–162.

Dewey, J. and P. Byerly, 1969. The early history of seismometry (to 1900), *Bull. Seismol. Soc. Amer.*, **59**, 183–227.

Dickinson, W. R. and T. R. Hatherton, 1967. Andesitic volcanism and seismicity around the Pacific, *Science*, **157**, 801–803.

Dorel, J., 1981. Seismicity and seismic gap in the Lesser Antilles arc and earthquake hazard in Guadeloupe, *Geophys. J. Roy. Astron. Soc.*, **67**, 679–695.

Dunbar, P. K., P. A. Lockridge, and L. S. Whiteside, 1992. *Catalog of Significant Earthquakes 2150 BC – A D 1991*, Rept. SE 49, World Data Center A, U.S. Department of Commerce, Boulder, CO.

Dutton, C., 1889. On some of the greater problems of physical geology, *Bull. Phil. Soc. Washington*, **11**, 51–64; reprinted in *J. Wash. Acad. Sci.*, **15**, 359–369, 1925.

Evison, F. F., 1963. Earthquakes and faults, *Bull. Seismol. Soc. Amer.*, **53**, 873–891.

1967. On the occurrence of volume change at the earthquake source, *Bull. Seismol. Soc. Amer.*, **57**, 9–25.

Frohlich, C., 1987. Kiyoo Wadati and early research on deep focus earthquakes: introduction to the special section on deep and intermediate focus earthquakes, *J. Geophys. Res.*, **92**, 13777–13788. Reprinted in *History of Geophysics*, **4**, 166–177, 1990.

Frohlich, C., M. Barazangi, and B. L. Isacks, 1977. Upper mantle structure beneath the Fiji Plateau: seismic observations of second P arrivals from the olivine–spinel phase transition zone, *Geophys. J. Roy. Astron. Soc.*, **50**, 185–213.

Gilbert, F. and A. M. Dziewonski, 1975. An application of normal mode theory to the retrieval of structural parameters and source mechanisms from seismic spectra, *Phil. Trans. Roy. Soc., London*, **A278**, 187–269.

Golenetskii, S. I. and G. Y. Medvedeva, 1965. On discontinuities of the first kind in the earth's upper mantle, Izv. Akad. Nauk SSSR, *Phys. Solid Earth*, **5**, 318–322.

Griffith, A. A., 1921. The phenomenon of rupture and flow in solids, *Phil. Trans. Roy. Soc., London*, A221, 153–198.

Griggs, D. T., 1936. Deformation of rocks under high confining pressures, I. Experiments at room temperature, *J. Geol.*, **44**, 541–577.

Griggs, D. T. and D. W. Baker, 1969. The origin of deep-focus earthquakes. In *Properties of Matter Under Unusual Conditions*, eds. H. Mark and S. Fernbach, New York, NY, Interscience, 389 pp., 23–42.

Griggs, D. and J. Handin, 1960. Observations on fracture and a hypothesis of earthquakes, *Geol. Soc. Amer. Mem.*, **79**, 347–373.

Guidoboni, E., A. Comastri, and G. Traina, 1994. *Catalogue of Ancient Earthquakes in the Mediterranean Area Up to the 10th Century*, Rome, Italy, Istituto Nazionale di Geofisica, 504 pp.

Gutenberg, B., 1923. Neue Methoden zur Bestimmung der Herdtiefe von Erdbeben aus Aufzeichnungen an herdnahe gelegenen Stationen, *Zeitschrift für angewandte Geophysik*, **1**, 65–75.

1933. Über Erdbeben mit Herdtiefen von mehreren hundert Kilometern, *Geologische Rundschau*, **24**, 229–239.

Gutenberg, B. and C. F. Richter, 1934. Contributions to the study of deep-focus earthquakes, *Gerlands Beitr. z. Geophysik*, **41**, 160–169.

1935a. On seismic waves (first paper), *Gerlands Beitr. z. Geophysik*, **43**, 56–133.

1935b. On seismic waves (second paper), *Gerlands Beitr. z. Geophysik*, **45**, 280–360.

1936a. Materials for the study of deep-focus earthquakes, *Bull. Seismol. Soc. Amer.*, **26**, 341–390.

1936b. On seismic waves (third paper), *Gerlands Beitr. z. Geophysik*, **47**, 73–131.

1937. Materials for the study of deep-focus earthquakes (second paper), *Bull. Seismol. Soc. Amer.*, **27**, 157–183.

1938. Depth and geographical distribution of deep-focus earthquakes, *Geol. Soc. Amer. Bull.*, **49**, 249–288.

1939a. Depth and geographical distribution of deep-focus earthquakes (second paper), *Geol. Soc. Amer. Bull.*, **50**, 1511–1528.

1939b. On seismic waves (fourth paper), *Gerlands Beitr. z. Geophysik*, **54**, 94–136.

1941. Seismicity of the Earth, *Geol. Soc. Spec. Pap.*, **34**, 1–130.

1942. Earthquake magnitude, intensity, energy and acceleration, *Bull. Seismol. Soc. Amer.*, **32**, 143–191.

1945. Seismicity of the Earth (Supplementary paper), *Geol. Soc. Amer. Bull.*, **56**, 603–668.

1949. *Seismicity of the Earth and Associated Phenomena*, Princeton, NJ, Princeton University Press, 273 pp.

• 1954. *Seismicity of the Earth and Associated Phenomena*, (2nd edn.), Princeton, NJ, Princeton University Press, 310 pp.

Hanks, T. C. and H. Kanamori, 1979. A moment magnitude scale, *J. Geophys. Res.*, **84**, 2348–2350.

Haskell, N. A., 1935. The motion of a fluid under a surface load, 1, *Physics*, **6**, 265–269.

Hayes, R. C., 1936. Normal and deep earthquakes in the south-west Pacific, *New Zealand J. Sci. Tech.*, **17**, 691–701.

Holmes, A., 1931. Radioactivity and Earth movements, *Trans. Geol. Soc. Glasgow*, **18**, 559–606.

• 1933. The thermal history of the Earth, *J. Wash. Acad. Sci.*, **23**, 169–195.

Honda, H., 1931. The velocity of the P-wave in the surface layer of the Earth's crust, *Geophys. Mag.*, **4**, 28–35.

1932. On the types of the seismograms and the mechanism of deep earthquakes, *Geophys. Mag.*, **5**, 301–326.

• 1934a. On the amplitude of the P and the S waves of deep earthquakes, *Geophys. Mag.*, **8**, 153–164.

1934b. On the mechanism of deep earthquakes and the stress in the deep layer of the Earth crust, *Geophys. Mag.*, **8**, 179–185.

1957. The mechanism of the earthquakes, *Sci. Rept. Tohoku Univ., Ser. 5*, **9**, 1–46.

Honda, H. and A. Masatsuka, 1952. On the mechanisms of the earthquakes and the stresses producing them in Japan and its vicinity, *Sci. Rept. Tohoku Univ., Ser. 5, Geophys.*, **4**, 42–60.

Hubbert, M. K. and W. W. Rubey, 1959. Role of fluid pressure in overthrust faulting, *Geol. Soc. Amer. Bull.*, **70**, 115–206.

Jeffreys, H., 1924. *The Earth, Its Origin, History and Physical Constitution*, Cambridge, Cambridge University Press, 278 p.

1926. The viscosity of the Earth (fourth paper), *Mon. Not. Roy. Astron. Soc., Geophys. Supp.*, **1**, 412–424.

1928. The times of transmission and focal depths of large earthquakes, *Mon. Not. Roy. Astron. Soc., Geophys. Supp.*, **1**, 500–521.

1932. *Tables of the Times of Transmission of the P and S Waves of Earthquakes*, London, British Association for the Advancement of Science.

1935. Some deep-focus earthquakes, *Mon. Not. Roy. Astron. Soc., Geophys. Supp.*, **3**, 310–343.

1936. The structure of the Earth down to the 20° discontinuity, *Mon. Not. Roy. Astron. Soc., Geophys. Supp.*, **3**, 401–422.

• 1939a. Deep-focus earthquakes, *Ergebnisse kosmischen Physik*, **4**, 75–105.

1939b. Some Japanese deep-focus earthquakes, *Mon. Not. Roy. Astron. Soc., Geophys. Supp.*, **4**, 424–460.

1939c. The times of P, S and SKS and the velocities of P and S, *Mon. Not. Roy. Astron. Soc., Geophys. Supp.*, **4**, 498–533.

Jeffreys, H. and K. E. Bullen, 1940. *Seismological Tables*, London, Gray-Milne Trust.

Kawakatsu, H., 1991. Insignificant isotropic component in the moment tensor of deep earthquakes, *Nature*, **351**, doi:10.1038/351050a0, 50–53.

1996. Observability of the isotropic component of a moment tensor, *Geophys. J. Int.*, **126**, 525–544.

Knott, C. G., 1888. Earthquakes and earthquake sounds as illustrations of the general theory of elastic vibrations, *Seismological Society of Japan, Trans.*, **12**, 115–136.

1899. Reflexion and refraction of elastic waves, with seismological applications, *Philosophical Mag., Ser. 5,* **48**, 64–82, reprinted in *Proc. Roy. Soc. Edinburgh.,* **39**, 159–208, 1919.

Lamb, H., 1904. On the propagation of tremors over the surface of an elastic solid, *Phil. Trans. Roy. Soc., London,* **A203**, 1–42.

Leeds, D. J., 1974. Catalog of Nicaraguan earthquakes, *Bull. Seismol. Soc. Amer.,* **64**, 1135–1158.

• Leith, A and J. A. Sharpe, 1936. Deep-focus earthquakes and their geological significance, *J. Geol.,* **44**, 877–917.

Mainka, C., 1915. Über Zeitdifferenzen aufallender Einsätze in einem Seismogramm gegen den ersten, *Gerlands Beitr. z. Geophysik,* **14**, 39–84.

Mândrescu, N. and M. Radulian, 1999a. Seismic microzoning of Bucharest (Romania): a critical review. In *Vrancea Earthquakes: Tectonics, Hazard and Risk Mitigation,* eds. F. Wenzel, D. Lungu and O. Novak, The Netherlands, Kluwer, 374 pp., 109–121.

1999b. Macroseismic field of the Romanian intermediate-depth earthquakes. In *Vrancea Earthquakes: Tectonics, Hazard and Risk Mitigation,*.eds. F. Wenzel, D. Lungu and O. Novak, The Netherlands, Kluwer, 374 pp., 163–174.

Martínez Solares, J. M. and A. López Arroyo, 2004. The great historical 1755 earthquake. Effects and damage in Spain, *J. Seismology,* **8**, doi:10.1023/B:JOSE.0000021365.94606.03, 275–294.

Mendel, J. G., 1866. Versuche über Pflanzenhybriden, *Verhandlungen des naturforschenden Vereins in Brünn,* **4**, 3–47.

Milne, J., 1891. *Earthquakes and Other Earth Movements,* New York, Appleton.

Ming, L. and W. A. Bassett, 1975. The postspinel phases in the Mg_2SiO_4–Fe_2SiO_4 system, *Science,* **187**, 66–68.

Mitrovica, J. X., 1996. Haskell (1935) revisited, *J. Geophys. Res.,* **101**, doi:10.1029/95JB03208, 555–570.

Mohorovičić, S., 1914. Die reduzierte Laufzeitkurve und die Abhangigkeit der Herdtiefe eines Bebens von der Entfernung des Inflexionspunktes der primaren Laufzeitkurve, *Gerlands Beitr. z. Geophysik,* **13**, 217–240.

Okal, E. A., 1996. Radial modes from the great 1994 Bolivian earthquake: no evidence for an isotropic component to the source, *Geophys. Res. Lett.,* **23**, doi:10.1029/96GL00375, 431–434.

Okal, E. A. and C. R. Bina, 1994. The deep earthquakes of 1921–1922 in northern Peru, *Phys. Earth Planet. Int.,* **87**, 33–54.

Okal, E. A. and R. J. Geller, 1979. On the observability of isotropic seismic sources: the July 31, 1970 Colombian earthquake, *Phys. Earth Planet. Int.,* **18**, 176–196.

Oldham, R. D., 1900. On the propagation of earthquake motion to great distances, *Phil. Trans. Roy. Soc., London,* **A194**, 135–174.

1906. The constitution of the interior of the Earth, as revealed by earthquakes, *Quart. J. Geol. Soc. London,* **62**, 456–473.

1909. The geological interpretation of the earth-movements associated with the Californian earthquake of April 18, 1906, *Quart. J. Geol. Soc. London,* **45**, 1–16.

1923. The earthquake of 7th August, 1895, in northern Italy, *Quart. J. Geol. Soc. London,* **79**, 231–236.

1926. The depth of origin of earthquakes, *Quart. J. Geol. Soc. London,* **82**, 67–93.

Omori, F., 1907. Seismograms showing no preliminary tremor, *Bull. Imperial Earthquake Investigation Committee,* **1**, No. 3, 145–154.

Oncescu, M. C., V. I. Marza, M. Rizescu and M. Popa, 1999. The Romanian earthquake catalogue between 984 – 1997. In *Vrancea Earthquakes: Tectonics, Hazard and Risk*

Mitigation, eds. F. Wenzel, D. Lungu and O. Novak, The Netherlands, Kluwer, 374 pp., 43–47.

Pilgrim, L., 1913. Die Berechnung der Laufzeiten eines Erdstosses mit Berücksichtigung der Herdtiefen, getstützt auf neurere Beobachtungen, *Gerlands Beitr. z. Geophysik*, **12**, 363–483.

Raleigh, C. B., 1967. Tectonic implications of serpentinite weakening, *Geophys. J. Roy. Astron. Soc.*, **14**, 113–118.

Raleigh, C. B. and M. S. Paterson, 1965. Experimental deformation of serpentinite and its tectonic implications, *J. Geophys. Res.*, **70**, 3965–3985.

Ramirez, L. D., 1990. Estimation of some focal parameters of an historical Chilean earthquake. In *Symposium International, Géodynamique andine, resumés des communications*, Paris, ORSTOM, 7–12.

Rayleigh, Lord (R. J. Strutt), 1885. On waves propagated along the plane surface of an elastic solid, *Proc. London Math. Soc.*, **17**, 4–11.

Reid, H. F., 1910. *The California Earthquake of April 18, 1906, Report of the State Earthquake Investigation Commission*. II. *The Mechanics of the Earthquakes*, Washington, DC, Carnegie Institution of Washington, 192 pp.

Ringwood, A. E., 1970. Phase transformations and the constitution of the mantle, *Phys. Earth Planet. Int.*, **3**, 109–155.

     1996. Phase transformations in the earth's mantle, *Phys. Earth Planet. Int.*, **96**, doi:10.1016/S0031-9201(96)90021-7, 79–84.

Ringwood, A. E. and A. Major, 1970. The system Mg_2SiO_4–Fe_2SiO_4 at high pressures and temperatures, *Phys. Earth Planet. Int.*, **3**, 89–108.

Robson, G. R., K. G. Barr, and L. C. Luna, 1968. Extension failure: an earthquake mechanism, *Nature*, **218**, 28–32.

Russakoff, D., G. Ekström and J. Tromp, 1997. A new analysis of the great 1970 Colombia earthquake and its isotropic component, *J. Geophys. Res.*, **102**, doi:10.1029/97JB01645, 20423–20434.

Scrase, F. J., 1931. The reflected waves from deep-focus earthquakes, *Proc. Roy. Soc. London*, **A132**, 213–235.

   • 1933. The characteristics of a deep focus earthquake: a study of the disturbance of February 20, 1931, *Phil. Trans. Roy. Soc., London*, **A231**, 207–234.

Shida, T., 1937. Thank-you address at the dedication ceremony of Beppu Geophysical Laboratory (Proposition for the existence of deep focus earthquakes) (in Japanese), *Chikyu Butsuri Geophysics*, **1**, 1–5.

Stechschulte, V. C., 1932. The Japanese earthquake of March 29, 1928, and the problem of depth of focus, *Bull. Seismol. Soc. Amer.*, **22**, 81–137.

Stoneley, R., 1931. On deep-focus earthquakes, *Gerlands Beitr. z. Geophysik*, **29**, 417–435.

Tillotson, E., 1938. The "high-focus" earthquakes of the International Seismological Summary, *Gerlands Beitr. z. Geophysik*, **52**, 377–407.

• Turner, H. H., 1922. On the arrival of earthquake waves at the antipodes, and on the measurement of the focal depth of an earthquake, *Mon. Not. R. Astron. Soc., Geophys.* **Supp.**, **1**, 1–13.

     1930. Deep focus, *International Seismological Summary for 1927 Jan., Feb., Mar.*, pp. 1–4, Berkshire, U. K., International Seismological Centre.

Uyeda, S., 1971. *The New View of the Earth*, San Francisco, CA, W. H. Freeman

Visser, S. W., 1936a. Some remarks on the deep-focus earthquakes in the International Seismological Summary, *Gerlands Beitr. z. Geophysik*, **47**, 321–332.

1936b. Some remarks on the deep-focus earthquakes in the International Seismological Summary. Second paper, *Gerlands Beitr. z. Geophysik*, **48**, 254–267.

Wadati, K., 1927. Existence and study of deep earthquakes (in Japanese), *J. Meteorol. Soc. Jpn., Ser. 2*, **5**, 119–145.

• 1928. Shallow and deep earthquakes, *Geophys. Mag.*, **1**, 161–202.

1929. Shallow and deep earthquakes (2nd paper), *Geophys. Mag.*, **2**, 1–36.

1932. On the travel time of earthquake waves, II, *Geophys. Mag.*, **7**, 101–111.

• 1935. On the activity of deep-focus earthquakes in the Japan Islands and neighbourhoods, *Geophys. Mag.*, **8**, 305–325.

1989. Born in a country of earthquakes, *Ann. Rev. Earth Planet. Sci.*, **17**, doi:10.1146/annurev.ea.17.050189.000245, 1–12.

Wadati, K. and K. Masuda, 1933. On the travel time of earthquake waves, V, *Geophys. Mag.*, **7**, 269–290.

1934. On the travel time of earthquake waves, VI, *Geophys. Mag.*, **8**, 187–194.

Wadati, K. and S. Oki, 1932. On the travel time of earthquake waves, III, *Geophys. Mag.*, **7**, 113–137.

1933. On the travel time of earthquake waves, IV, *Geophys. Mag.*, **7**, 139–153.

Wadati, K., K. Sagisaka, and K. Masuda, 1932. On the travel time of earthquake waves, I, *Geophys. Mag.*, **7**, 87–99.

Walker, G. W., 1921. The problem of finite focal depth revealed by seismometers, *Phil. Trans. Roy. Soc, London*, **A122**, 45–51.

Wegener, A., 1915. *Die Enstehung der Kontinente und Ozeane*, No. 23, 1–94, Brunswick, Sammlung Vieweg.

Whitney, J. D., 1872. The Owen's Valley earthquake: II. General conclusions, *Overland Monthly*, **9**, 266–278.

Wiechert, E., 1897. Über die Massenverteilung im Innern der Erde, *Königliche Gesellschaft der Wissenschaften zu Göttingen Nachtrichten, Mathematisch-physicalische Klasse*, **3**, 221–243

Wrinch, D. and H. Jeffreys, 1922. The variable depth of earthquake foci, *Nature*, **110**, 310.

Yamaguti, S., 1936. A discussion on the results of a statistical investigation of earthquake phenomena, *Bull. Earthquake Res. Inst., Tokyo Univ.*, **14**, 399–414.

• Zoeppritz, K., 1912. Über Erdbebenwellen V.A.5, zwei verschiedene Arten von Erdbeben, *Nachr. Akad. Wiss., Göttingen Math.-Phys. Klasse*, **K1**, 132–134.

Part II

Properties of intermediate- and deep-focus earthquakes

4

The distributions of depth and size

Distributions of parameters describing individuals within a population generally contain information about the processes that affect the individuals; e.g., the age distribution of an animal species will depend on whether individuals most often succumb to episodes of bad weather, to random predation of adults, or to "old age." In this chapter, the individuals are earthquakes, the populations reside in earthquake catalogs, and the parameters of interest are focal depth and "size," i.e., magnitude or moment. What can the distributions of focal depth and size tell us about the physical processes that control deep seismicity?

4.1 What is focal depth?

Although Wadati originally established that deep earthquakes existed by evaluating (S-P) isochrons and felt intensity data, nowadays most catalogs report depths determined using one of three types of data. These are: (1) P arrival times; (2) (pP–P) intervals; and (3) body and surface waveforms.

4.1.1 Depths from P arrival times

The most common method to locate earthquakes – the method used to construct the catalog of the International Seismological Centre (ISC) – is to compare observed P arrival times with arrivals predicted by a reference earth model. The location is then the latitude, longitude, depth, and origin time that produces the best fit between model and observations; in principle this is unique if there are P times recorded at four or more stations. The advantage of this method is that it is widely applicable since P arrivals from numerous stations are often available for earthquakes with magnitudes as small as 4 or less and as large as 8 or more.

I thank Bob Engdahl and RayWillemann for reviewing an earlier draft of this chapter.

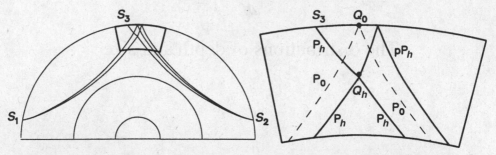

Fig. 4.1 Focal depth and origin time are inherently uncertain for locations determined solely from P arrivals. For the earthquake Q_h with focal depth h and origin time T_h, P arrival times at teleseismic stations S_1 and S_2 will be nearly identical to those for earthquake Q_0 occurring at the surface with origin time $T_h - t_{0toh}$, where t_{0toh} is the time it takes a P wave to travel from the surface to depth h. To remove the uncertainty about depth, arrival times must be available for rays that leave the source and travel upward, such as P arrivals at local station S_3 or pP arrivals at a distant station such as S_2.

However, this method suffers from at least three serious problems. First, there may be errors in the reported P arrival times. When seismologists carefully reread P arrivals and compare them to bulletin data it is not uncommon to find errors or serious uncertainties in about one quarter to one half of the arrivals (e.g., Frohlich *et al.*, 1997). This is likely to cause serious inaccuracy if there are only a few stations (say, 10) providing data, or if there are only one or two stations in certain key azimuths.

Second, even when data are numerous and error-free, the focal depth and origin time determined from P arrivals are inherently uncertain unless data are available from stations near the epicenter (Fig. 4.1) which record upward-traveling rays leaving the earthquake source. For example, Longshot was a large ($m_b = 5.8$), extremely well-recorded nuclear explosion set off 660 meters beneath the Earth's surface on 29 October 1965; yet from P arrivals at 221 stations the ISC determined a focal depth of 46 km. Undoubtedly similar location errors afflict ISC locations for many earthquakes.

Finally, the reference earth model may be incorrect – this is especially true for earthquakes along convergent margins, where seismic velocities can differ by 10% or more from global average models. The ISC regularly reports focal depths exceeding 60 km along the Zagros in Iran and beneath the Tibetan plateau. However, in both regions waveform modeling and local seismic observations indicate that focal depths are shallow (see sections 10.25 and 10.26); the discrepancies occur because the ISC's reference earth model is inappropriate here.

4.1.2 Depths from (pP−P) intervals

In the absence of local station data, a useful approach for determining accurate depths utilizes the difference in the arrival times of P and pP. This works because pP energy must initially travel upward from the hypocenter to the surface and back and before following a path similar to that of the P phase (Fig. 4.1). Thus the (pP−P) interval depends strongly on focal depth. Generally catalog-makes compute these "pP depths" or "(pP−P) depths" precisely using traveltime tables; but, a rule of thumb that yields the focal depth within about 20% of the true depth is:

$$h = \frac{1}{2} V_{\text{Mantle}} t_{\text{pP}-\text{P}} \cong (5\,\text{km/s}) t_{\text{pP}-\text{P}}. \tag{4.1}$$

An advantage of this method is that one can determine depth if there are reliable P and pP observations at a single station, even if there are clock errors and absolute arrival times are unknown.

There are some disadvantages, e.g., short-period pP is not always visible for low-magnitude earthquakes and is sometimes not easily identifiable for very large events. Also, it is sometimes difficult to identify pP, to decide whether it corresponds to a reflection from the ocean surface or the ocean bottom, or to distinguish it from other reflected phases such as sP. Yet when several clear (pP−P) observations are available it provides unequivocal confirmation of focal depth; because it is independent of clock errors it was usually the method of choice for all earthquakes occurring before about 1960.

Because the inclusion of times from up- and downgoing rays improves earthquake locations, some catalogs systematically incorporate them into the location process. In the so-called EHB catalog Engdahl *et al.* (1998) use phase arrivals from the ISC catalog to determine locations using a variety of phases in addition to P, including pP, sP, S, and various core phases. The ISC doesn't routinely use (pP−P) observations to determine focal depth; however, when several pP arrivals are available it reports both the P-only and (pP−P) depths.

4.1.3 Depths from waveform matching

The third method for determining focal depth is to compare recorded waveforms with synthetic seismograms generated for a particular earth model, with the assigned depth, epicenter, and focal mechanism corresponding to that of the best-matching synthetic. To be routinely applicable this method requires digital seismic data and a fast, accurate way to calculate synthetics; both have been available since about 1980. Usually only long-period data are used ($T > 45$ s) so that the synthetics need not take into account the details of near-surface crustal structure, such as the presence of oceans and continents. Presently, the Harvard group determines mechanisms for about 1000 earthquakes each year (e.g., see Dziewonski and Woodhouse, 1983); there have also been efforts to apply this method to historical data (e.g., see Huang *et al.*, 1998).

Waveform matching seems to provide quite accurate focal depths; this is because the incorporation of S waveforms fixes the origin time; moreover, shallow and deep earthquake seismograms really are quite different (see Fig. 2.3) and waveform matches must properly sum the relative contributions of many different up- and downgoing waves (pP, sP, sS, PP, PcP, ScP, etc.). The main difficulty with the method is that it requires earthquakes with relatively well-recorded long-period body waveforms. At teleseismic distances this usually means events with M_W greater than about 5.4; to match both body and surface waveforms earthquakes must be even larger ($M_W > 6.0$).

Also, it is worth remembering that waveform matching methods determine a centroid depth corresponding to an average location for the entire seismic rupture, while traveltime-based methods find the point of initiation of the rupture. Especially for large earthquakes, these two depths can differ by 20 km or more.

4.1.4 Comparison of depths determined from different methods

How do depths determined by different methods compare? For the 1977–2002 period there are 3796 deep earthquakes common to the Harvard, ISC, and EHB catalogs for which the ISC also reported depths determined from (pP−P) intervals; these represent about 95% of all deep events in the Harvard catalog. There are systematic differences in the depths determined by the various methods (Table 4.1 and Fig. 4.2); the Harvard CMT depths average about 4–5 km deeper than the others, which generally are within about 2 km of one another. The standard deviation σ of the differences indicates that the EHB and pP depths are the most consistent ($\sigma \sim$ 11 km), while ISC and CMT depths are the least ($\sigma \sim 22$ km).

But, how reliable are depths reported in the various catalogs? When we consider depth differences between two catalogs, variability in both catalogs contributes to the standard deviation; we wish to know how much the variation is within each of the catalogs. Box 4.1 proposes a method to determine this when the differences are statistically independent. This is a reasonable assumption for the CMT, pP, and ISC depths since they are determined using different data; it is a poor assumption for the EHB catalog since it relies heavily on the ISC's P and pP arrival times. The results (Table 4.2) indicate that the pP depths suffer the least variability ($\sigma_{pP} \sim 6$ km) while the CMT and ISC depths are the most variable ($\sigma_{CMT} \sim 15$ km; $\sigma_{ISC} \sim 17$ km). If we assume the distributions of differences are roughly Gaussian (Fig. 4.2) then about 95% of reported depths are within 2σ of the "true" depth, i.e.,

Table 4.1. *Means and standard deviations for differences in depth, latitude, and longitude in four global catalogs 1977–2002. Included events all have Harvard CMT depths of 60 km or greater and non-zero depths reported in all four catalogs. To avoid blunders the statistics exclude events with depth differences exceeding 100 km. Units for all entries are km.*

$60 < h_{CMT} < 700$	$N = 3796$	CMT-pP	pP-ISC	ISC-CMT	CMT-EHB	EHB-ISC	pP-EHB
Depth	std	15.7	17.7	22.4	14.1	16.9	11.1
	mean	4.0	0.8	−4.8	4.7	0.1	0.7
Latitude	std			27.5	27.6	5.3	
	mean			0.5	−0.3	−0.2	
Longitude	std			27.6	28.4	6.8	
	mean			2.0	−3.5	1.6	

$60 < h_{CMT} < 100$	$N = 1035$	CMT-pP	pP-ISC	ISC-CMT	CMT-EHB	EHB-ISC	pP-EHB
Depth	std	18.0	17.2	21.3	15.3	16.0	12.7
	mean	3.9	−2.5	−1.4	3.0	−1.6	−0.8
Latitude	std			29.6	29.9	5.5	
	mean			3.8	−4.4	0.6	
Longitude	std			30.4	31.0	6.5	
	mean			1.0	−0.7	0.2	

$100 < h_{CMT} < 700$	$N = 2761$	CMT-pP	pP-ISC	ISC-CMT	CMT-EHB	EHB-ISC	pP-EHB
Depth	std	14.8	17.9	22.8	13.6	17.2	10.4
	mean	4.0	2.0	−6.1	5.4	0.8	1.3
Latitude	std			26.6	26.7	5.2	
	mean			−0.7	1.3	−0.5	
Longitude	std			26.5	27.4	6.9	
	mean			2.3	−4.5	2.2	

within 12 km for the pP depths and within about 30–35 km for the Harvard and ISC catalogs. Of course, this applies only to those earthquakes common to all the catalogs, generally the larger events.

4.2 The global distribution of earthquake focal depths

Some general features of the global distribution of earthquake depth are clear regardless of how one chooses to determine focal depth: about two thirds of all quakes are shallow, having depths less than 60 km; then, between 60 km and

Table 4.2. *Depth uncertainties for the CMT, pP and ISC catalogs. Entries are as determined from the statistics in Table 4.1 using the method described in Box 4.1.*

	σ_{CMT}	σ_{pP}	σ_{ISC}
All: $h_{CMT} \geq 60$ km	14.8 km	5.4	16.9
60 km $\leq h_{CMT} \leq 100$ km	15.5	9.1	14.6
All: $h_{CMT} \geq 100$ km	14.5	3.1	17.6

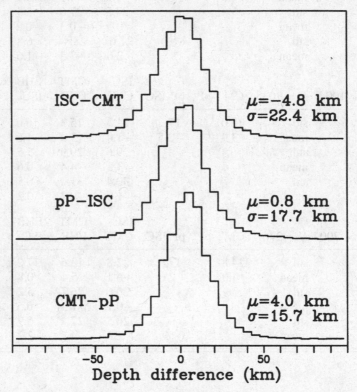

Fig. 4.2 Histograms of focal depth differences as reported in the Harvard CMT, ISC, and pP catalogs for the 3796 earthquakes with focal depths 60–700 km analyzed in Table 4.1.

about 300 km there is an approximately exponential decrease in occurrence rate (Fig. 4.3). The rate increases beneath about 400 km and reaches a maximum near about 600 km; then, it decreases quite abruptly and ceases completely at about 690 km. Since there are major changes in phase and/or composition in the Earth's mantle at depths of about 400 km and 650 km, most scientists believe that these features

Box 4.1
How reliable are depths reported in catalogs?

A simple model of statistical variance allows us to estimate the depth uncertainties for earthquakes reported in three or more catalogs. In particular, suppose that h_{CMT}, h_{pP}, and h_{ISC} are the depths in the Harvard CMT, ISC, and pP catalogs, and suppose each differs from a "true" depth h_{true} by a systematic term μ and a randomly distributed error term ε:

$$h_{CMT} = h_{true} + \mu_{CMT} + \varepsilon_{CMT}$$
$$h_{pP} = h_{true} + \mu_{pP} + \varepsilon_{pP}$$
$$h_{ISC} = h_{true} + \mu_{ISC} + \varepsilon_{ISC}. \qquad (4.2)$$

The errors ε_{CMT}, ε_{pP}, and ε_{ISC} have (unknown) variances σ^2_{CMT}, σ^2_{pP}, and σ^2_{ISC}. However, if the depth differences are independent their variances should add; e.g. the variance V_{CMT-pP} for $(h_{CMT} - h_{pP})$ will be $\sigma^2_{CMT} + \sigma^2_{pP}$; the variance for $(h_{pP} - h_{ISC})$ will be $\sigma^2_{pP} + \sigma^2_{ISC}$, etc. For the three catalogs together this gives three equations in three unknowns:

$$\sigma CMT^2 + \quad \sigma_{pP^2} \qquad\qquad = V_{CMT-pP}$$
$$\sigma_{pP^2} + \sigma ISC^2 = V_{pP-ISC}$$
$$\sigma CMT^2 \qquad\qquad + \sigma ISC^2 = V_{ISC-CMT}. \qquad (4.3)$$

Thus, we can solve these equations with least squares to find the unknown depth uncertainties σ_{CMT}, σ_{pP}, and σ_{ISC}.[1] (see Table 4.1).

are somehow responsible for the rate increase beneath 400 km and the cessation at 690 km.[2]

4.2.1 The exponential falloff with depth

Is there significance in the apparent exponential decrease in rate? Sykes (1966) and Isacks *et al.* (1968) were perhaps the first to notice this, observing that between the surface and 200 km the decrease was a factor of ten every 100 km. One theory of the origin of intermediate seismicity is that water released by dehydration reactions in

---

[1] Equation 4.3 is a linear system of the form $A\vec{x} = \vec{v}$ with solution $\vec{x} = A^{-1}\vec{v}$. In principle it is possible to extend this method and include depth differences with the EHB catalog; with four catalogs eq. 4.3 becomes six equations in four unknowns, and one can use least squares to determine a best-fitting solution of the form $\vec{x} = (A^tA)^{-1}A^t\vec{v}$. However, EHB depths rely heavily on pP observations, which thus violates the assumption that depth differences are independent.

[2] Alternatively, Yan Kagan (personal communication) suggests that "It is quite possible that the present distribution [of depths] is an accident of nature. A few millions of years later or before the present, the distribution may have been very different: a subducting slab may behave like a crumpled piece of paper below 200–300 km and its spatial distribution may be quite random."

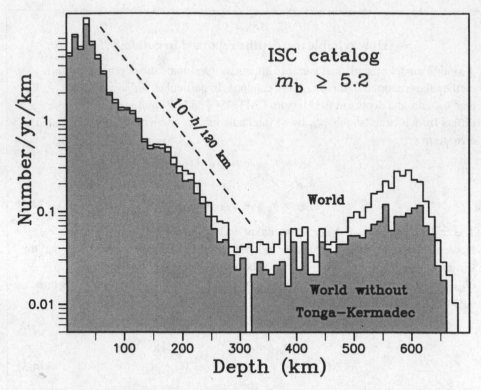

Fig. 4.3 Annual occurrence rates vs focal depth for earthquakes with $m_b \geq 5.2$ in the ISC catalog 1964–2001. The upper line shows rates for the entire world, while the lower line excludes earthquakes in the Tonga–Kermadec region. Dashed line shows that the rate is approximately exponential between 60 and 300 km, with a factor of 10 decrease approximately every 120 km.

subducting lithosphere makes brittle fracture possible at depth (see Section 7.3.1); thus numerous investigators including Isacks *et al.* (1968) have suggested the exponential decrease reflects the progression of the dehydration reaction in response to changes in pressure and temperature. Alternatively, Vassiliou and Hager (1988) suggested that both the exponential decrease and the increase beneath 400 km might come about if seismicity rates were proportional to shear stress; they performed calculations to reproduce this stress pattern for slabs sinking under gravitational forces that met a barrier at 670 km. In any case, the data indicates the fit is fairly good; for the ISC catalog (Fig. 4.3) between 60 km and 300 km the rate decreases by a factor of ten about every 120 km.

However, there are several reasons indicating these explanations are true only in the most general terms. The main problem is that the depth distribution doesn't fit an exponential very well for individual geographic regions (Fig. 4.4). Cross sections

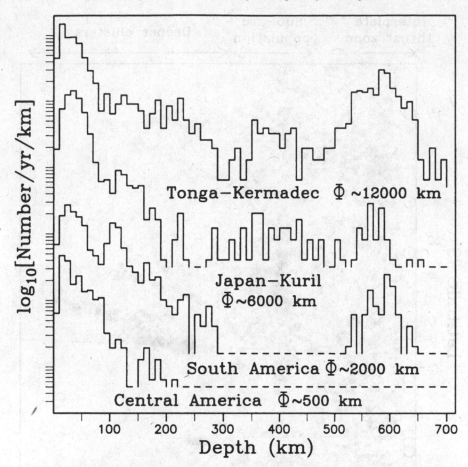

Fig. 4.4 Annual occurrence rates vs depth for four geographic regions; earthquakes in the EHB catalog 1964–2002 with $m_b \geq 5.4$. Data are graphed on a semilog scale as in Fig. 4.3, but offset vertically for clarity; dashed lines indicate no activity. Boundaries of regions defined as in Chapter 10. Labels indicate approximate thermal parameter Φ for each region.

normal to the arc in individual regions often indicate that intermediate-depth earthquakes are highly clustered rather than being distributed nearly uniformly (e.g., see Tonga–Kermadec and South America, Figs. 10.7 and 4.5). And the depth distributions are quite different for different geographic regions. For example, about two thirds of all the world's earthquakes deeper than 300 km occur in the Tonga–Kermadec region (see Fig. 1.14). A few regions such as South America possess both intermediate- and deep-focus earthquakes but have no activity between about 225 km and 500 km. And, several regions such as Central America, the Aleutians and the Hindu Kush have earthquakes down to depths of perhaps 250 km but none deeper.

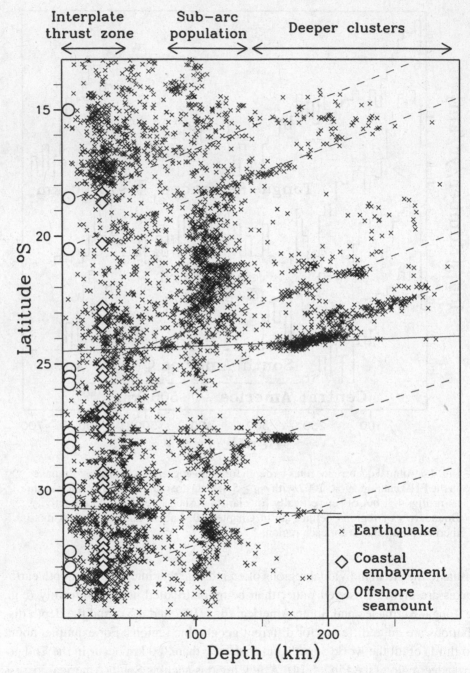

Fig. 4.5 North–south cross section showing intermediate-depth seismicity beneath South America occurring in response to subduction of the Nazca plate. The figure is redrawn from Kirby *et al.* (1996a) with seismicity updated using well-located 1964–2002 events reported in the EHB catalog. Kirby *et al.* (1996a) observed that many earthquakes occur along roughly linear features that connect at the surface to offshore seamount chains and coastal embayments (dashed lines). Other linear trends (solid lines) may be indicative of subducted features on the ocean floor.

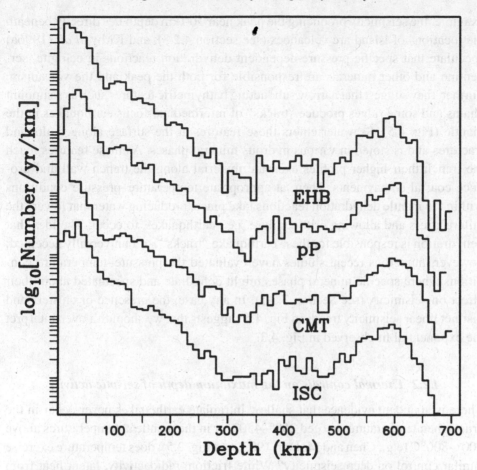

Fig. 4.6 Annual occurrence rates vs depth for four global catalogs. Data are graphed on a semilog scale as in Fig. 4.3, but offset vertically for clarity. Rates are for events with m_b or M_W of 5.4 or greater; for EHB catalog events with poorly determined hypocenters are excluded.

Moreover, the fit to an exponential isn't especially good even for the global distribution if we limit consideration to catalogs with very well-determined focal depths. For example, in the CMT, pP and EHB catalogs (Fig. 4.6) there is a local minimum at about 80 km depth and a distinct peak at 100 km, and possible indications of other features near about 250 km.[3]

Several scientists have suggested that the dehydration of subducting lithosphere may be responsible for the local peaks and minima in Figs. 4.4 and 4.6. For

[3] These features may be real or they may be location artifacts. For example, it is well-known that focal depths often cluster near layer boundaries for travel time-based regional location algorithms that utilize layered flat-earth crustal models. It is possible that similar systematic errors afflict teleseismic location algorithms.

example, the seismicity producing the peak near 100 km depth lies directly beneath the locations of island arc volcanoes (see section 4.2.3), and Kirby *et al.* (1996a) speculate that specific pressure-dependent dehydration reactions in eclogite, serpentine and other minerals are responsible for both the peak and the volcanism. Further, they suggest that narrow subducting bathymetric features such as seamount chains and some ridges produce "tracks" of intermediate-focus earthquakes in the mantle (Fig. 4.5). Seawater enters these features at the surface along faults and fractures and is stored in various hydrous mineral phases. As these features reach the trench their higher profiles excavate material along the trench wall and produce coastal embayments. Then, at appropriate temperature–pressure conditions within the mantle dehydration reactions take place, producing water that lowers the failure stress and allowing brittle failure, i.e., earthquakes, to occur. The idea that dehydration is responsible for these earthquake "tracks" isn't universally accepted; however, numerous recent studies have evaluated the pressure–temperature conditions where specific mineral phases might dehydrate and speculated about their effect on seismicity (see Section 7.3.1). In any case, the presence of clusters and distinct linear seismicity trends in Fig. 4.5 suggests that we shouldn't over-interpret the exponential fit observed in Fig. 4.3.

4.2.2 Thermal controls on the maximum depth of seismic activity

There is abundant evidence that shallow intraplate earthquakes never occur in the crust when temperatures exceed $250°$–$450°$, or in the mantle at temperatures above $600°$–$800°C$ (e.g., Chen and Molnar, 1983, and Fig. 2.5); does temperature exercise similar control on deep seismicity? While friction, radioactivity, latent heat from chemical and phase transitions, etc., all affect the interior temperature within subducting lithosphere, conduction of heat away from the slab's upper surface exerts the strongest influence (see Section 2.4).

Although various investigators have used different approaches to evaluate the effects of temperature on deep seismicity, all reach similar conclusions. Among the earliest was McKenzie (1969), who stated:

The temperature structure of the sinking lithosphere is determined by the time constant of the slab and by the spreading rate, and in turn governs the distribution of intermediate and deep focus earthquakes.

Similarly, several empirical analyses of the length L of Wadati–Benioff zones have concluded that it approximately satisfies:

$$L = V t_{Age}/10, \tag{4.4}$$

where V is the convergence rate at the trench and t_{Age} is the age of the ocean floor when subduction commences (e.g. Molnar *et al.*, 1979; Jarrard, 1986). However, if we multiply both sides by sin θ, where θ is the dip angle of the Wadati–Benioff zone, then $L\sin\theta$ is depth h and $V\sin\theta$ is the vertical component of subduction $V_\perp$. Since $V_\perp t_{Age}$ is the thermal parameter Φ, eq. 4.4 is equivalent to saying that earthquakes stop when the ratio Φ/h exceeds 10. For the model in Box 2.1 and Figs. 2.6 and 2.7 this suggests that earthquakes don't occur when interior temperatures exceed about 625 °C.

However, more detailed analysis shows that the relationship between Φ and the maximum depth of seismicity D_{max} isn't a simple linear relationship as indicated by eq. 4.4. Kirby *et al.* (1991; 1996b) evaluated Φ and D_{max} in 24 subduction zones; their plot indicated a roughly linear relationship for Φ less than 2000 km. Except for a few anomalous regions D_{max} only exceeded 300 km when Φ exceeded about 4000–5000 km.

Gorbatov and Kostoglodov (1997) have performed the most thorough study of the relationship between D_{max} and Φ (Fig. 4.7). Rather than determine a single value for Φ for each subduction zone, they evaluated only selected regions and within adjacent cross sections having widths of approximately 100 km (Gorbatov *et al.*, 1996; 1997). They were also careful to estimate lithospheric age at the time of subduction, rather than simply using the age of presently subducting material. Finally, they considered separately areas where the age of subducting lithosphere is uncertain or where plate reconstructions suggest there may be age discontinuities within the subducted material.

It is noteworthy that different investigators often obtain quite different results when they determine Φ. For example, for Tonga, Kirby *et al.* (1996b) find Φ is 17,000 km, Wiens and Gilbert (1996) get 11,800 km, and Gorbatov and Kostoglodov (1997) obtain 4540 km to 8970 km. This reflects that there is often considerable uncertainty about both the age of the subducted lithosphere and the vertical descent rate, especially in arcs such as Tonga and the Marianas where there is back-arc spreading. Also, analysis of ocean-floor bathymetry suggests that lithospheric thickness remains approximately constant after its age reaches 80 Ma, and thus one could argue that this should be the maximum value used for calculating Φ.

Nevertheless Gorbatov and Kostoglodov's (1997) results (Fig. 4.7) are generally consistent with those of Kirby *et al.* (1996b) but with somewhat less scatter.[4] Gorbatov and Kostoglodov find a reasonably consistent and straightforward relationship when Φ is less than about 3500 km and D_{max} is less than 300 km. Then,

---

[4] In their paper, Gorbatov and Kostoglodov (1997) fit their results to a fourth-order polynomial; however, the coefficients they present are incorrect; the correct coefficients are:

$$D_{max} = 54 + 0.10\Phi + 0.44 \times 10^{-4}\Phi^2 - 0.37 \times 10^{-7}\Phi^3 + 0.63 \times 10^{-11}\Phi^4.$$

It is unclear why they used a fourth order polynomial since there are only two inflection points in the data, and a third order polynomial fits nearly as well (see Fig. 4.7).

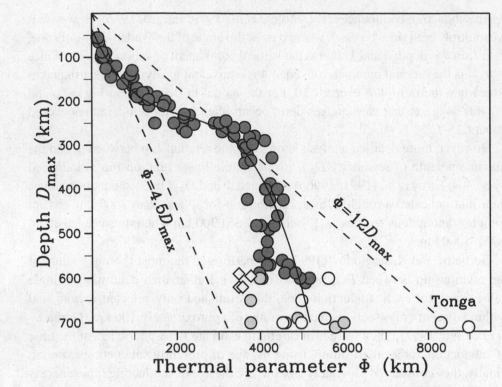

Fig. 4.7 Maximum earthquake depth D_{max} vs thermal parameter Φ. Data are from Gorbatov and Kostoglodov (1997) for selected portions of the Aleutians, Japan, Kamchatka, Kurils, Marianas, Mexico, New Hebrides, South America, Suamatra, and Tonga subduction zones. Light gray and white circles are data from Tonga and Marianas where back-arc spreading may affect Φ; light gray circles assume no spreading, and white circles include spreading. White diamonds are from Argentina, where a deep slab may be detached. The solid line is a cubic equation providing an approximate fit to the data above 550 km:

$$D_{max} = 35 + 0.22\Phi - 0.89 \times 10^{-4}\Phi^2 + 1.37 \times 10^{-8}\Phi^3.$$

Dashed lines show that most data points fall in the region where $4.5\,D_{max} < \Phi < 12\,D_{max}$, corresponding to minimum slab temperatures of approximately $600\,°-800\,°C$ (see Fig. 2.7).

for Φ of 3500–5000 km the range of maximum depths extends from 300–700 km. And, in those few regions where Φ exceeds 5000 km the maximum depths always are 600 km or more.

What can we conclude from Fig. 4.7 about the processes that control deep seismicity? First, these results suggest that temperature does place some overall constraints on seismic activity. Nearly all the data points fall between the lines $\Phi = 4.5\,D_{max}$ and $\Phi = 12\,D_{max}$, which, for the simple-minded conduction model of Box 2.1, correspond to interior temperatures of about 600°–800°C. Second, at depths shallower than 300 km temperature may be the primary parameter controlling the process that allows earthquakes to occur. The fact that the curve isn't precisely linear may well be a limitation of the simple-minded model implied by the use of Φ as a proxy for temperature. Indeed, for depths less than about 250 km the curve is nearly identical to Spencer's (1994) calculated estimates for the depths of the olivine brittle–ductile transition within subducted lithosphere.

However, when Φ reaches about 3500 km there is a broad range of maximum depths; this suggests that something in addition to temperature controls seismicity beneath 300 km. An obvious possibility is the olivine–spinel phase change; it is exothermic but metastable at lower temperatures, and thus Fig. 4.7 may simply indicate that it depends strongly on both temperature and pressure. However, this is highly speculative; one implication of Fig. 4.7 is that intermediate- and deep-focus earthquakes may have different failure mechanisms.

For regions where earthquakes occur at depths exceeding 300 km, there has been some effort to evaluate regional differences in the depth distribution of seismicity. For example, Helffrich and Brodholt (1991) determined the depth of the local minimum in moment release for eight regions where there are earthquakes below 300 km. They concluded that "whatever the cause of the increase in the numbers of deep earthquakes [at depths·exceeding the moment minimum], it has a thermal dependence similar to that of the phase transformations in olivine." Similarly, Bina (1997), Bina *et al.* (2001), and Guest *et al.* (2003) modeled buoyancy stresses attributable to mantle phase transitions and observed that they peaked between 400 km and 650 km, thus possibly explaining the high activity rates observed at these depths in Tonga–Kermadec. And Furukawa (1994) observed that intense peaks in activity rates occur between 400 and 600 km in Tonga and Izu–Bonin, while the rate was nearly constant in the Marianas and Japan. Both he and Wortel and Vlaar (1988) attribute the higher rates of seismic activity at these depths to thermal stresses surrounding the core of the subducting lithosphere. However, Fischer and Jordan (1991) offered a somewhat different explanation, concluding that the slab's resistance to penetration as it passes the 650 discontinuity is responsible for the stress and high activity.

4.2.3 Island arc volcanism and focal depth

Wadati (1935) noted an apparent relationship between intermediate-depth seismicity and island-arc volcanism; shortly thereafter Berlage (1937) stated that oceanic arc islands coincided with the 100 km seismicity contour. Subsequent data summarized in Gill (1981) and Tatsami (1986) indicated that Wadati–Benioff zones lay at 125 ± 38 km and 112 ± 19 km depth, respectively, beneath the volcanic front, defined as the trenchward limit of stratovolcanoes in subduction zones.

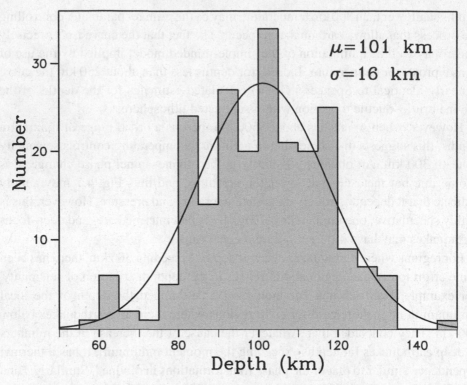

Fig. 4.8 Depths beneath volcanoes on top of the Wadati–Benioff zone for 223 volcanoes investigated by England *et al.* (2004). The solid line is the best-fitting normal distribution; μ and σ are mean and standard deviation of the depths.

Recently England *et al.* (2004) reevaluated this relationship (Fig. 4.8) using relocated hypocenters from the EHB catalog. The mean depth to the upper surface of the Wadati–Benioff zone was 101 km, almost exactly in agreement with Berlage. Although individual depths ranged from 65 km to 130 km, within individual arc segments the depth variation was only a few km. Surprisingly, the depth variation did not correlate with either the age of the descending ocean floor or the thermal parameter Φ. Instead, they found an inverse correlation with the vertical descent speed of the subducting plate (Fig. 4.9), i.e., Wadati–Benioff zones were at shallower depths between faster-descending plates.

This and the absence of correlation with Φ contradict the hypothesis that the depth of seismicity underlying volcanoes depends on the interior temperature of the subducting lithosphere. However, calculations by England and Wilkins (2004) indicate that there is an inverse correlation between vertical descent speed and the temperature of the uppermost surface of the descending slab; faster-descending plates more effectively entrain hot material from the overlying wedge, raising the temperature at shallow depths. Thus the results of Fig. 4.9 are consistent with the idea that pressure–temperature conditions at the slab–mantle interface

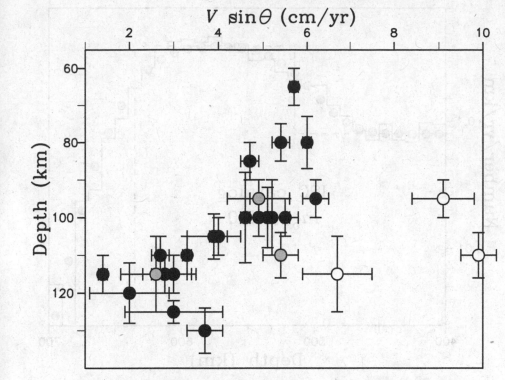

Fig. 4.9 Comparison of volcano-to-Wadati–Benioff zone depth and vertical descent rate $V \sin \theta$ for 24 island arcs investigated by England *et al.* (2004). Here V is the convergence velocity normal to the arc, and θ is the dip angle of the Wadati–Benioff zone. Grey symbols are Tonga, Kermadec, and Marianas which have significant backarc spreading; white symbols are same arcs with backarc spreading rate included.

control the generation of magmas responsible for island arc volcanism (e.g., see Davies and Stevenson, 1992; Davies, 1999).

It is worth emphasizing that the results in Fig. 4.9 aren't in contradiction with the Φ vs D_{max} relationship in Fig. 4.7. The seismicity immediately below volcanic zones generally occurs 100 km or more above the deepest earthquakes in a Wadati–Benioff zone. Thus, while the deepest earthquakes are likely to occur in the cold interior of subducting lithosphere, the processes that control the formation of magmas act at the upper surface.

4.2.4 *How deep are the deepest deep earthquakes?*

The vast majority of all earthquakes with depths exceeding 300 km occur in the western Pacific, the Philippines, and Indonesia (see Fig. 1.14). Elsewhere, between 300 and 500 km there are only a handful of well-determined hypocenters in South America and the Mediterranean. At depths exceeding 500 km, they

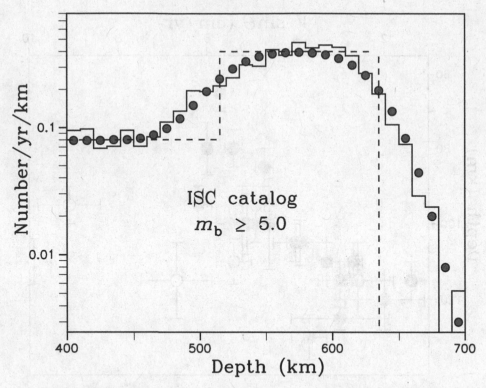

Fig. 4.10 Annual occurrence rate vs focal depth for deep-focus earthquakes with
$m_b \geq 5.0$ in the ISC catalog 1964–2001 (solid line). The observation that the rate
tails off below 600 km is consistent with several models, including one where the
"true" rate is constant between 515 km and 635 km and zero at greater depths
(dashed line). If statistical errors in determining focal depth are normally distributed with a standard deviation σ of 25 km, the observed hypothetical distribution (filled circles) is nearly identical to the actual observed distribution.

occur regularly in Tonga–Kermadec–Vitiaz, Indonesia, the Philippines, the Marianas, and the Kurils. Elsewhere, they are common in South America but otherwise
there are none other than the remarkable Spanish earthquakes (see Sections 8.1.3
and 10.21) and a few small events beneath New Zealand (see Section 10.2).

What is nature of the distribution of very deep events? Stark and Frohlich (1985)
noted that if the standard deviation in determining focal depth were as large as
25 km, it would be possible to explain the distribution of all earthquakes beneath
635 km simply as a statistical "tail" caused by location errors (Fig. 4.10). However,
it is clear from Fig. 4.2 and Table 4.2 that uncertainties in depth determination
are somewhat less than 25 km. Willemann (1991) showed that the shape of the
distribution below 400 km – including both the increasing rate to about 600 km
and then the "tail" beneath – is consistent with a simple temperature-controlled

stress–strain model. However, his model failed to explain why seismicity terminates so abruptly at depths exceeding 635 km.

Yet some earthquakes clearly do occur with depths between 635 km and 690 km; what is especially remarkable is these very deep earthquakes occur in several different geographic regions and some are quite large. Indeed, earthquakes with magnitude M_W of 6.9 or greater and depths of 635 km or more have occurred beneath Tonga–Kermadec, Indonesia, the Philippines, the Marianas, the Kurils, South America, and Spain. This abrupt size–depth cutoff is significant because it argues against the simplest thermal models (e.g., see Willemann, 1991) in which warming causes the subducting slab to become progressively thinner and less capable of producing large events.

But, what is the depth of the very deepest earthquakes?[5] Three fairly careful studies of focal depth all concur that a few quakes have well-constrained depths as great as 680 km to 695 km (Stark and Frohlich, 1985; Rees and Okal, 1987; Okal and Bina, 1998). For example, the earthquake of 17 June 1977 was especially well recorded; it had an ISC location constrained at 201 stations, including 5 stations within $10°$, and a depth of 684 km as constrained by 14 (pP−P) intervals; when it occurred it was the deepest earthquake in the entire Harvard CMT catalog with an assigned depth of 693.9 km. Rees and Okal (1987) relocated numerous historical earthquakes and concluded that 690 km was the approximate depth for a Tonga earthquake occurring on 10 June 1954 and a Banda Sea earthquake occurring on 11 May 1955. While there are depths between 700 km and 848 km reported in the ISC catalog and other generally reliable sources, in every case studied the above-referenced authors concluded that these depths were either reported improperly, poorly determined, or determined using insufficient data. Thus, Okal and Bina (1998) concluded: "there are no earthquakes reliably located at depths greater than 690 km, with a precision estimated at 10 km." They also argue that the absence between 1964 and 1995 of observed earthquakes with depths exceeding 700 km indicates that none with moment exceeding 10^{17} N-m ($M_W = 5.3$) did occur in that time period; and, if such earthquakes do occur, activity rates must be 400–8000 times less than between depths of 400–650 km.

Undoubtedly it will be necessary to revisit this question from time to time, especially since there is often contradictory evidence about depth for even relatively large earthquakes. For example, the ISC's depth from (pP−P) intervals for the 9 June 1994 Bolivia earthquake ($M_{W(CMT)} = 8.2$) was 713 km, although virtually all other organizations assigned it depths of 630–650 km. Since the Okal and Bina study, the Harvard CMT catalog has reported two extraordinarily deep earthquakes in Tonga (2 September 2000, $h_{CMT} = 698$ km, $M_{W(CMT)} = 6.0$; 19 August 2002,

---

[5] Sadly, the *Guinness Book of Records* makes no statements whatsoever about the maximum depth of earthquake activity, focusing instead on such mundane details such as the costliest, deadliest, and biggest earthquakes.

$h_{CMT} = 699$ km, $M_{W(CMT)} = 7.7$). The EHB catalog assigned these events depths of 693 km and 679 km, respectively. For the second event Tibi *et al.* (2003) obtained 664 km using data from both teleseismic and local stations.

4.2.5 *How deep are the shallowest deep earthquakes?*

How deep are the shallowest "deep" earthquakes? Soon after the discovery of deep earthquakes, various proposed classification schemes arbitrarily set the shallow–intermediate cutoff depth at 60 km, a depth used since by the ISC. This cutoff depth is in such wide use that it has become fixed by tradition (see Section 2.1); however, we might ask what depth we would choose today, with another half-century of data at our disposal. The perfect depth would be one where there was both a clear change in the observable properties of individual earthquakes, and where physical arguments suggested that the mechanism causing earthquakes changed in a fundamental, physical way. Unfortunately, shallow and deep earthquakes exhibit considerable overlap in most of their observable properties, such as focal mechanism, stress drop, etc.

Perhaps our best option would be to choose the shallow–deep cutoff at 50 km. This is about the maximum depth for most subduction zone thrust earthquakes (e.g., Bilek *et al.*, 2004) and also coincides with a distinct change in the frequency of aftershocks. The observations indicate that earthquakes with focal depths shallower than 50 km are much more likely to have numerous aftershocks than are deeper quakes (Fig. 4.11). There is considerable variation; i.e., some shallow earthquakes have few or no aftershocks, while a few rare deep events (see Section 5.2.3) have well-developed aftershock sequences. But, if we group earthquakes and take averages, it appears that some distinct mechanical change affecting aftershock behavior occurs at about 50 km depth. Also, Houston (2001) evaluated moment-scaled source time functions and observed that at depths exceeding 40 km they exhibited a distinct decrease in duration and complexity. It is noteworthy that a depth of 40–50 km is approximately where the brittle/ductile failure transition occurs (see Fig. 1.12).

4.3 What is earthquake size?

4.3.1 *Prologue: properties of a good statistic*

A statistic is simply a number; a statistic is most useful if it passes three tests:

(1) it is easy to measure;
(2) its value depends on a phenomenon of interest;
(3) other phenomena don't influence the value very much.

If we are very fortunate a statistic will also pass two other tests –

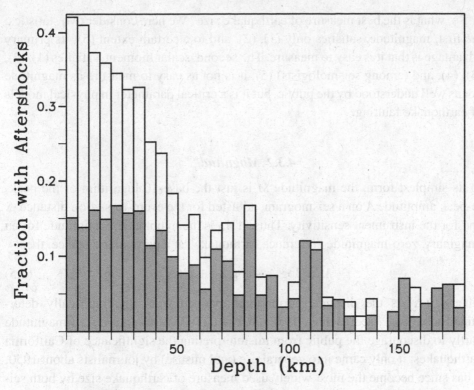

Fig. 4.11 Depth dependence of aftershock occurrence. The vertical axis is the fraction of earthquake sequences having exactly one aftershock (dark bars) or more
than one aftershock (white bars). For these data, mainshocks are earthquakes in
the EHB catalog 1964–2002 that have $m_b \geq 5.2$, well-determined focal depths,
and which are not aftershocks of previous events. Aftershocks are found by applying a ratios test (Frohlich and Davis, 1985) to relative times of all ISC catalog
events occurring within two years and 100 km of each mainshock; these are aftershocks if they occur within a period following the mainshock where the chance
of any earthquake occurring is 1 per cent or less. Note that mainshocks with
focal depths shallower than 45–50 km have aftershocks more often than deeper
earthquakes.

(4) it is an important parameter in a physical model explaining the phenomenon of interest;
and
(5) almost everyone understands the statistic.

Of course, most statistics don't pass all these tests. For example, for American
football players a commonly reported statistic is weight; this satisfies (1), (2),
(5), and possibly (4), but certainly not (3). Popular discussions of human native
intelligence often focus on IQ; this satisfies (2) only, as it is hard to measure,
strongly dependent on all kinds of social and cultural influences, and misunderstood
by almost everybody.

So, what is the best measure of earthquake size? We here consider two statistics: the first, magnitude, satisfies only (1), (2), and to a certain extent (5); its primary advantage is that it is easy to measure. The second, scalar moment, satisfies (1), (2), (3), (4), and (among seismologists) (5); it is not as easy to measure as magnitude nor as well understood by the public, but it is a critical parameter in physical models of earthquake faulting.

4.3.2 Magnitude

In its simplest form, the magnitude M is just the base-10 logarithm of the peak-to-peak amplitude A on a seismogram, adjusted for the event-to-station distance Δ and for the instrument sensitivity. Thus, if A_0 is the hypothetical amplitude for an imaginary, zero-magnitude "reference earthquake" at a particular distance, then:

$$M = \log A - \log A_0 + C(\Delta), \tag{4.5}$$

where $C(\Delta)$ is the distance correction, obtained from an empirically determined table (e.g., see Richter, 1958). Richter (1935) first introduced magnitude partly to discourage the public from misinterpreting the significance of California earthquakes; it only came into general use (and misuse) by journalists about 1950; it has since become the most widely used measure of earthquake size by both scientists and the general public. It has great utility because of its wide acceptance and because it is very easy to use; a simple ruler quite literally becomes the measuring "scale" of the Richter scale.

But, there are problems.[6] Some are quite fixable; e.g., it is possible to construct tables which correct for the effects of focal depth as well as distance (e.g., see Gutenberg, 1945; Gutenberg and Richter, 1956; Båth, 1985; Murphy and Barker, 2003), or to permit station corrections accounting for regional difference in earth structure and attenuation (e.g., Rezapour, 2003). Also, different seismographs have their peak sensitivity at different frequencies. For example, the classical WWSSN station had seismographs that peaked at both short ($T_{\text{instrument}} \sim 1$ s) and long periods ($T_{\text{instrument}} \sim 30$ s); the largest phases at short periods are usually P or S, while at long periods the largest (for shallow earthquakes) are surface waves.

This led to defining different magnitude scales, with each measuring particular phases and each indicated by a different notation. The most important:

- M_S – surface-wave magnitude – determined from surface waves with periods near 20 s; unfortunately, M_S is unreliable for deep earthquakes since surface-wave amplitude is strongly dependent on focal depth;

[6] Because reported magnitudes are so variable even though they rely on the logarithm of an observation, this book's author once swore he would never publish a paper about magnitudes. However, this, like so many of my principles, I have since compromised (see Frohlich and Davis, 1993; Frohlich, 1998).

- m_b – body-wave magnitude – determined from short-period seismographs using phases such as P or S with periods near about 1 s; and
- m_B – a longer-period body-wave magnitude determined from phases with periods between about 0.5 and 12 s; for deep earthquakes m_B provides a reasonable alternative to M_S;
- M_m – mantle magnitude – determined from the spectral amplitude of Rayleigh waves and adjusted for depth, developed especially to determine meaningful sizes for historical earthquakes (Okal, 1990);
- M_W – moment magnitude – defined (by Hanks and Kanamori, 1979) from scalar moment M_o (in units of N-m) using the equation:

$$M_W = (2/3)\log_{10} M_o - 6.0. \qquad (4.6)$$

In addition, there is a veritable alphabet soup of other scales, each defined in terms of the amplitude, duration, or period of some specific phase or phases as recorded on some particular seismograph system. Of course, all this causes confusion because any earthquake has not one magnitude, but three or more at each seismograph station. Creators of new magnitude scales generally try to make them agree with established scales (usually M_S or M_W) over a specific range. However, these efforts are only partially successful, and there is a morass of literature attempting to establish linear and non-linear relationships between the scales (see Båth, 1981; Kanamori, 1983; Utsu, 2002).

A more fundamental problem occurs because some magnitude scales saturate, i.e., they seldom determine a magnitude larger than some particular value M_{max}. For m_b this is about 6.7; for M_S this is about 8.5. Why does saturation occur? If a fault of length L_{fault} slips and the rupture proceeds at speed $V_{rupture}$, it radiates seismic waves for a duration $T_{dur} \sim L_{fault}/V_{rupture}$. If a seismograph has peak amplification at period $T_{instrument} = T_{dur}$, slip from each portion of the fault will increase the motion of the seismometer and the instrument will measure a meaningful magnitude. But, if $T_{instrument} \ll T_{dur}$, the instrument only responds to motion along portions of the fault of dimension $\sim V_{rupture}T_{instrument}$, and the remainder of the rupture simply increases the duration of the measured signal, rather than its amplitude (see Fig. 4.12). Thus the magnitude saturates at M_{max} for all ruptures with dimensions exceeding $\sim V_{rupture}T_{instrument}$.

Finally, magnitudes vary significantly over time because of many "human" factors. These include changes in policy or of personnel at institutions that routinely read records or determine magnitudes, and the addition/deletion of "key" stations in regional or global networks (Habermann, 1987; 1991; Eneva *et al.*, 1994).

4.3.3 Moment

Nowadays, most seismologists agree that the most useful measure of earthquake size is scalar moment M_o, usually just called "moment". Although its formal definition

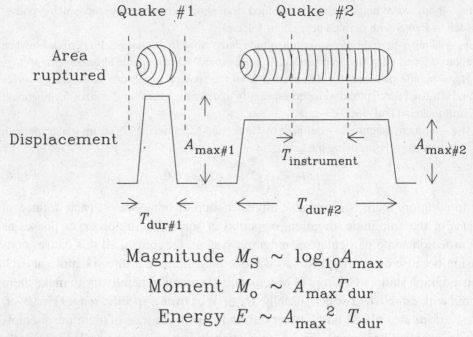

$$\text{Magnitude } M_S \sim \log_{10} A_{max}$$
$$\text{Moment } M_o \sim A_{max} T_{dur}$$
$$\text{Energy } E \sim A_{max}^2 \, T_{dur}$$

Fig. 4.12 Schematic explanation of how magnitude M_S, scalar moment M_o, and radiated energy E depend on the area and duration of rupture on a planar fault. Magnitude M_S depends only on the maximum displacement A_{max}. For Quake #1 (upper left) the rupture duration T_{dur} is shorter than the period $T_{instrument}$ where the seismograph has peak sensitivity, and thus displacement A_{max} incorporates information about the area and slip along the entire fault; for Quake #2 (upper right) rupture duration T_{dur} exceeds $T_{instrument}$, and thus A_{max} depends only on part of the rupture. However, scalar moment M_o depends on the time integral of the displacement pulse, i.e., on both A_{max} and T_{dur}. Since the radiated energy of an elastic wave (like the energy in a spring) is proportional to the square of A_{max}, a small fault which ruptures with a high stress drop (left) can have a larger magnitude than a quake with a much longer duration and a larger scalar moment (right).

is in terms of a tensor describing earthquake radiation patterns (see Chapter 6, Figs. 6.17 and 6.19), for this chapter four features of moment are important:

- Scalar moment is determined by comparing observed and synthetic low-pass-filtered seismograms, with M_o being the multiplicative normalization factor that provides the best fit. If the filtering is performed at long enough periods (e.g., 300 s), saturation is not a problem even for huge earthquakes (Fig. 4.12).
- Scalar moment has units of energy – dyne-cm or newton-m. Note, however, *an earthquake's scalar moment is not the same as its energy release or its radiated energy*. In

fact, earthquake moment bears no simple relationship to energy; it is a different physical quantity.[7]

- Unlike magnitude, moment is measured on a linear scale, not a log scale. Typically, for small earthquakes ($m_b \sim 2$), M_o is 10^{12} newton-m; for moderate earthquakes ($M_W \sim 6$), M_o is 10^{18} newton-m; and the very largest earthquakes have M_o of 10^{23} newton-m.
- Moment is directly interpretable in terms of physical parameters describing the earthquake source; for slip-on-a-fault motion:

$$M_o = \mu A S, \tag{4.7}$$

where A is the area of rupture surface, S is the average slip over this area, and μ is the elastic rigidity of the surrounding rocks prior to the rupture.

This last relationship (eq. 4.7) makes moment extraordinarily useful for evaluating questions concerning the physical significance of various-sized earthquakes. For example, one can address questions like: why does M_S saturate at about 8.5? When earthquakes occur the velocity of rupture is comparable to the shear velocity V_S; thus, a magnitude scale defined with peak sensitivity at period $T_{instrument}$ will record motions generated along fault patches with maximum dimensions $L_{fault} = V_S T_{instrument}$ (see Fig. 4.12). Empirical studies (Kanamori and Anderson, 1975) find that mean fault slips S are typically about $10^{-4} L_{fault}$; thus since area $A \sim L_{fault}^2$ the moment corresponding to saturation will be:

$$M_{oSat} = \mu A S = 10^{-4} \mu (V_S T_{instrument})^3. \tag{4.8}$$

If μ is 65×10^9 Pa and V_S is 4.5 km/sec, appropriate values for the mantle at a depth of about 100 km (Masters and Shearer, 1995), then for $T_{instrument}$ of 30 sec, M_{oSat} is 1.6×10^{22} newton-m. If we convert M_o to magnitude using eq. 4.6, the calculated saturation magnitude is 8.8, in good agreement with the observed value.

Moreover, summing moments along a Wadati–Benioff zone provides information about the average slip released by earthquakes (Fig. 4.13). In particular, suppose a Wadati–Benioff zone is inclined with dip angle θ and we sum moments over depth bins of size Δh for earthquakes within an along-strike distance D_{A-S}. Then, the Wadati–Benioff zone's surface area within this depth range is $D_{A-S} \Delta h / \sin \theta$, and we can define an "effective slip" $S_{effective}$ along this surface by dividing the moment sum by the product of this area and μ:

$$S_{effective} = \Sigma M_o \sin \theta / D_{A-S} \Delta h. \tag{4.9}$$

In all three zones considered in Fig. 4.13, $S_{effective}$ is approximately 0.01–0.10 cm/yr for depths exceeding 100 km. Of course, we mustn't interpret $S_{effective}$ literally

---

[7] A recently published geology text (which shall here go unnamed) defines seismic moment as "a measurement of the total amount of energy released during an earthquake." Do not be fooled; this is incorrect. Seismic moment and radiated energy have the same units but are fundamentally different.

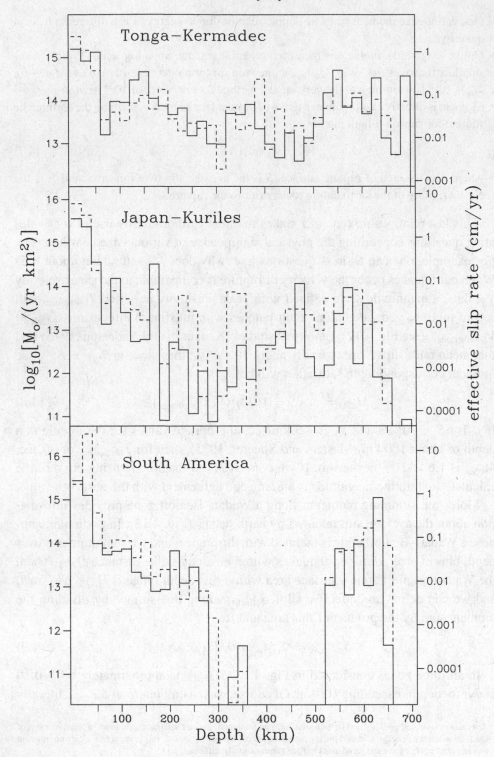

as the slip along the planar surface of the subducting lithosphere since neither the focal mechanisms nor locations of deep quakes indicate that they occur at the surface. And, since we have only about a century of data it is possible the plotted moment rates are somewhat less than the true long-term rates. Nevertheless, it is remarkable that $S_{effective}$ is only about one per cent of the NUVEL convergence rate, that its magnitude is about the same in regions as dissimilar as Tonga and South America, and that it doesn't vary much with focal depth.

4.4 The global distribution of deep earthquake sizes

At first glance, certain general features of the distribution of earthquake sizes (Fig. 4.14) are similar regardless of which magnitude scale you consider or whether you consider magnitude or scalar moment. In particular, there is a range of magnitudes between $M_{uniform}$ and M_{max} over which plots of $\log N$ vs magnitude M or $\log M_0$ are roughly linear; the occurrence rate N approximately satisfies:

$$\log_{10} N = a - b M,$$
$$\log_{10} N = A - \beta \log_{10} M_0. \tag{4.10}$$

Then, for earthquakes with sizes both below and above this range there are fewer earthquakes than predicted by this linear relationship.

Scientists who interpret earthquake size distributions generally attempt to fit the observations to determine parameters for one of several hypothetical distributions. The most common are:

- Gutenberg–Richter[8] – the linear relationship of eq. 4.10 holds for all M above $M_{uniform}$;
- truncated Gutenberg–Richter – the linear relationship holds between $M_{uniform}$ and M_{max}, and there are no larger earthquakes; thus, M_{max} is a "hard" limit;

Fig. 4.13 Depth dependence of rate of scalar moment released for earthquakes in three geographic regions. Data for solid lines are for 1977–2003 period from Harvard CMT catalog; for dashed line period is extended back to 1900 by augmenting with data from the EV Centennial catalog (Engdahl and Villaseñor, 2002) and estimating moment using eq. 4.6. For the left scale rate is determined by dividing the sum of moments by the time period, bin thickness (20 km), and along-strike length of the seismic zone on the Earth's surface. For the right hand scale the resulting rate is converted to effective slip using eq. 4.7. For Tonga–Kermadec, data are from 14.8 °–38 °S and along-strike length is 2170 km; for Japan–Kuril: 35 °–55 °N and 3720 km; for South America: 0 °–40 °S and 4450 km. To construct the right scale for all three regions we assume μ is 10^{11} Pa and θ is 45 °.

---

[8] Although the relationship described by eq. 4.10 is often called the "Gutenberg–Richter law", Japanese scientists Mishio Ishimoto and Kumizi Iida (Ishimoto and Iida, 1939) were actually the first to formulate it.

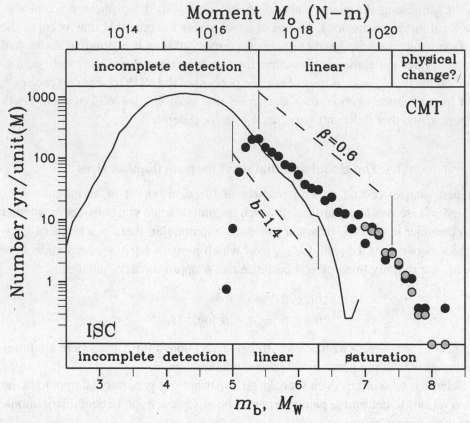

Fig. 4.14 Distribution of magnitudes for earthquakes with focal depth exceeding 100 km in the ISC catalog 1964–2001 (solid line, magnitude m_b), Harvard CMT catalog 1977–2003 (dark circles, magnitude M_w), and the EV centennial catalog 1900–1999 (lighter circles, magnitudes of 7 or greater only). For both the ISC and CMT catalogs, the distribution is roughly linear over a limited range of magnitudes indicated by vertical lines. Occurrence rates are lower outside of this range; for lower magnitudes this is almost certainly due to incomplete detection of events; for higher magnitudes it may be an artifact caused by saturation of the magnitude scale or a physical change in the mechanism that generates earthquakes.

- modified Gutenberg–Richter – above $M_{uniform}$, there are two or more linear regions, each with different values of b or β;
- Gamma – the probability density function for the distribution of M_o that leads to eq. 4.10 is multiplied by an exponential $\exp(-M_o/M_{max})$. Thus, the distribution is nearly identical to Gutenberg–Richter at small magnitudes but rolls off at large magnitudes; here M_{max} is a "soft" limit since magnitudes exceeding it are possible.

What is the physical significance of M_{max} and $M_{uniform}$? Earthquakes larger than M_{max} are rare or absent, either because a different physical process generates large earthquakes (see below) or simply because saturation afflicts the measuring scale (see above). Research papers occasionally appear which attribute some deep physical significance to $M_{uniform}$, such as a size-dependent change in the earthquake process for very small earthquakes (e.g., Taylor *et al.*, 1990; Trifu and Radulian, 1991; Spiedel and Mattson, 1993). This is possible but unproven, since in most cases $M_{uniform}$ is simply the magnitude below which seismograph networks and the people who operate them begin to "miss" some earthquakes. Although various regional networks report b values for small earthquakes that differ from the b values of Fig. 4.14, this is largely because there are systematic differences in magnitude scales. Thus the significance of $M_{uniform}$ is cultural rather than physical; if networks were more sensitive and measuring scales defined more carefully the straight portion of the distribution would probably extend to smaller magnitudes.

4.4.1 What are b and β values telling us?

Both b and β simply measure the relative proportions of large and small earthquakes. Convection proceeds in the Earth's mantle, the tectonic plates move, lithosphere subducts, and earthquakes accompany some of this activity. But, do very large or very small earthquakes accommodate most of the motion? This is analogous to the question we ask ourselves when we are short of money; are we short because we are overwhelmed by small daily expenses such as food, gas, utilities, insurance? Or is the problem a few big-ticket items, like a new house, a new car, or kids at Yale?

For earthquakes, is most of the moment released by large or small events? Equation 4.10 implies that the total number of quakes N is proportional to $M_o^{-\beta}$; i.e., the total moment released by all the quakes scales as $M_o^{(1-\beta)}$. Thus, if β is exactly 1.0, the rate of moment production is the same for quakes of all sizes; when β exceeds 1.0 the smallest quakes produce most of the moment; when β is less than 1.0 the very largest quakes produce most of the moment. Figure 4.14 indicates that β is about 0.60; thus globally the large quakes dominate the moment budget. For example, one deep-focus earthquake, the Bolivia earthquake of 9 June 1994 with M_W of 8.25 and a moment of 2.64×10^{21} newton-m possessed almost exactly one-third of all the moment released between 1977 and 2003 by all deep-focus earthquakes in the Harvard catalog.

Aki (1981) went a step further, arguing that b and β are statistical indicators of the geometry of the region of faulting, with the fractal dimension corresponding to 3β. That is, $\beta = 0.33, 0.67$, and 1.0 (i.e., b of 0.5, 1.0, and 1.5) correspond respectively to earthquakes generated along faults with linear, planar, and volume-filling geometry. Subsequent investigators have extended this analysis to evaluate the significance of the decrease in activity above M_{max}, which they attribute to the fact that the finite thickness of the lithosphere confines the dimensions of very large ruptures (see

Table 4.3. *Model explaining dependence of β on the relationship between fault length L, width W, and slip S. The model assumes the number N of earthquakes is proportional to the total available fault area A_{tot}, so that $N \sim A_{tot}/LW$. Also, it assumes that slip S equals a constant c times L or W. Thus β depends on how L and W vary as the scalar moment $M_0 = \mu LWS$ changes. Small earthquakes are roughly equidimensional; however, for larger earthquakes the thickness of subducting lithosphere limits the fault width W. How this changes β depends on whether the slip is proportional to length L or width W. In the table the underlined parameters are fixed, thus only the remaining parameters determine β as they must accommodate changes in moment and area.*

dimensions	slip S	area A	moment M_0	$N \sim A_{tot}/A$	β value
$L \times L$ (small)	$\underline{c}L$	L^2	μcL^3	$\sim 1/M_0^{2/3}$	2/3
$L \times W$ (large)	$\underline{c}L$	$\underline{W}L$	μcWL^2	$\sim 1/M_0^{1/2}$	1/2
$L \times W$ (large)	$\underline{c}W$	$\underline{W}L$	μcW^2L	$\sim 1/M_0$	1

Table 4.3). They thus consider a modified Gutenberg–Richter model and search for changes in β at large magnitudes. Unfortunately, there are spirited disagreements among them about how to interpret these changes (e.g., Rundle, 1989; Pacheco *et al.*, 1992; Romanowicz and Rundle, 1993; Scholz, 1994). For deep-focus earthquakes Giardini (1988) and Okal and Kirby (1995) have applied a similar analysis. However, at present these analyses are highly speculative, especially since only a handful of large-magnitude earthquakes are available to constrain β.

Analyses of acoustic emissions recorded during laboratory rock fracture experiments provide alternative interpretations for the significance of b and β. For example, Mogi (1962) reported that b values depend on the heterogeneity of rock samples, with b decreasing as heterogeneity decreased. Scholz (1968) observed that b decreased as confining pressures increased. Amitrano (2003) found that the critical variable was the mechanical mode of failure, as b values decreased when temperature–pressure conditions increased towards the brittle–ductile transition. He further suggested that this may explain both Mogi's and Scholz's results, as homogeneity and pressure generally both increase as a material becomes increasingly ductile.

4.4.2 Measurements of b and β

One problem that afflicts all these investigations is that measuring b or β and M_{max} is really quite tricky. The problem is not a mathematical one; indeed, there is a nice family of papers explaining how to determine the slope and confidence limits for b or β from observations over a finite range of sizes, even accounting for systematic errors that arise when sizes are reported in "binned" groups (Aki, 1965; Page, 1968; and Bender, 1983), or when mainshocks are separated from aftershocks (Lombardi, 2003). For the variants of the Gutenberg–Richter distribution, a more serious problem is that the distribution between $M_{uniform}$ and M_{max} isn't really very straight; thus, for any particular data set one obtains quite different answers depending on the values chosen for the limits. Especially for the gamma distribution, the value of M_{max} depends strongly on only one or two of the largest earthquakes (Kagan, 1991a; 1991b; 1999), which may be historical and have poorly determined sizes. Finally, as described above, the determination of magnitude is subject to all kinds of size- and time-dependent systematic errors (e.g., see Frohlich and Davis, 1993).

So, approximately what are the values for b and β for the global distribution of deep- and intermediate-focus earthquakes? For the ISC catalog, Frohlich and Davis (1993) found that the answer varies significantly depending on the magnitude range chosen. They found b_{ISC} to be 1.33 for m_b between 5.0 and 6.0 and depths 70–300 km, and 1.34 for depths 300–700 km. Giardini (1988) determined b_{ISC} to be 1.39 for m_b between 5.1 and 5.8 for depths exceeding 350 km. Since β is approximately 2/3 of b (see eq. 4.6), these correspond to β values of 0.89–0.93. Analysis of the CMT catalog produces significantly smaller β values. Frohlich and Davis (1993) obtained β of 0.54 and 0.60 for intermediate- and deep-focus earthquakes, respectively; Okal and Kirby (1995) reported β between 0.58 to 0.70 as determined in 100-km depth ranges.

By far the most thoughtful statistical analysis of β is by Kagan (1999; 2002), who fits moments in the CMT catalog with a gamma distribution. He found that β was close to 0.60 for both intermediate- and deep-focus events. Moreover, he evaluated the effect of removing aftershocks from the catalog, which affects β for shallow earthquakes but has little effect for intermediate and deep events. For the declustered catalog, he found that a single value for β (0.60 ± 0.02) was satisfactory for shallow, intermediate, and deep-focus earthquakes. Kagan also evaluated M_{max} for deep-focus earthquakes, concluding only that M_{max} must be greater than about 8.0. The occurrence of the 1994 Bolivia and Tonga deep-focus earthquakes strongly affected the value assessed for M_{max}; since 1994 it is no longer possible to determine an upper limit.

4.4.3 Regional differences in the size distribution

Even though Kagan (1999) concluded that the global distributions of shallow, intermediate, and deep-focus earthquakes have identical β values, there is evidence that β is not the same in all geographic regions, especially for deep-focus events. For example, several investigators report unusually high β values for Tonga–Kermadec deep-focus earthquakes (see Fig. 1.13). For the CMT catalog Giardini (1988), Frohlich and Davis (1993) and Kagan (1999) all obtained 0.81 for β,[9] and Okal and Kirby (1995) report values of 0.83–0.92. This is higher than in all other geographic regions with significant activity (Indonesia, Philippines, Marianas). At the other extreme, the lowest b and β values are in South America; there, Giardini (1988) obtained β of 0.27 and Kagan (1999) found 0.26, nearly identical to the "composite" value of 0.28 determined by Frohlich (1989) using a combination of ISC, Harvard, and historical data. Notwithstanding the uncertainties about what b- or β-value is "best," the differences between Tonga–Kermadec and South America signify a huge difference in the proportions of large and small events. If β is 0.80, for every earthquake with M_W of 7 there will be 250 with M_W of 5; however, if β is 0.28, for each M_W 7 earthquake there will be only seven with M_W of 5.

Furthermore, in certain restricted geographic areas the predominance of large-magnitude earthquakes may be even more extreme than in South America. In Spain, for example, there have been only four deep-focus earthquakes recorded; one on 29 March 1954 ($h_{EVC} = 627$ km; $M_{W(CMT\text{-}Hist)} = 7.9$) had the fifth largest moment observed for deep-focus earthquakes in this century (see Table 1.2). Yet in the half-century since there have been exactly three "aftershocks"– two with m_b of 3.9 and one with m_b of 4.5 (see Section 10.21); this is consistent with a β of about 0.13 (Fig. 1.13). Similarly, for the second largest deep earthquake known, the Colombia deep earthquake of 31 July 1970 ($h_{CMT\text{-}Hist} = 623$ km; $M_{w(CMT\text{-}Hist)} = 8.1$), there are no other earthquakes in the ISC catalog within 200 km of its hypocenter (see Fig. 2.4). Finally, the world's largest deep earthquake, the Bolivia earthquake of 9 June 1994, also occurred where there had been little previous activity, although – unlike the Spanish or Colombian earthquakes – it did possess a number of smaller aftershocks (Myers *et al.*, 1995; Tinker *et al.*, 1995). All these examples suggest that the physical conditions near the source of very large, isolated deep earthquakes may be very different from those in typical deep earthquake zones.

However, for deep-focus earthquakes Wiens and Gilbert (1996) and Wiens (2001) have asserted that regional variations in β correlate with the thermal parameter Φ. If this is true (see Fig. 4.15), the relatively higher values of β in Tonga–Kermadec may reflect how the relatively lower temperatures there influence the mechanical environment at the source. For example, a colder, faster-subducting slab may have higher β because material within it has had less opportunity to approach equilibrium, and thus may be more heterogeneous. As noted above, from laboratory experiments

---

[9] Actually, both papers report b values, here converted to equivalent β using eq. 4.6.

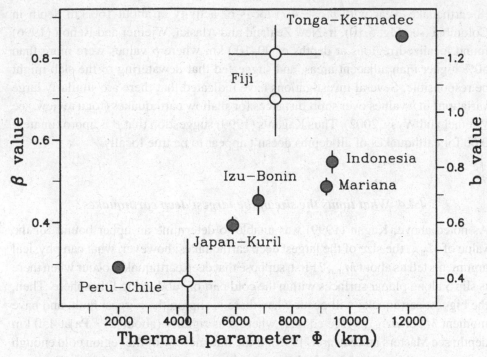

Fig. 4.15 Relationship between β or b values and thermal parameter Φ for deep-focus earthquakes. For the labeled regions, Wiens and Gilbert (1996) presented measurements (dark circles) suggesting there is strong correlation between Φ and b. However, there is considerable uncertainty about both variables; white circles and vertical lines are Kagan's (1999) values and standard deviations for β; horizontal lines indicate the range of values for Φ reported by Gorbatov and Kostoglodov (1997).

Mogi (1962) and Amitrano (2003) suggest that this produces higher β values. At the other extreme, activity in Andean South America and the very isolated 1954 Spanish and 1970 Colombian earthquakes may occur within warmer, older, more homogeneous material.

For intermediate-depth earthquakes the available research suggests that regional variations in b or β are smaller than for deep-focus events (Frohlich and Davis, 1993; Kagan, 1999). Moreover, where analysis suggests that b or β have relatively low or high values, there is little agreement among different magnitude scales concerning which geographic regions possessed the most extreme values.

However, there is evidence that geographic differences in b and β do occur within certain restricted environments. For example, Frohlich et al. (1995) found extraordinarily high b and β values of about 2.0 and 0.9 for earthquakes in the Bucaramanga nest, a very compact (dimensions of $\sim$10 km), very active

(8 earthquakes/year with $m_b \geq 4.7$) locus of activity at about 160 km depth in Colombia (see Fig. 5.10). In New Zealand and Alaska, Wiemer and Benoit (1996) found localized regions at depths of 90–100 km where b values were more than 50% higher than adjacent areas, and suggested that dewatering of the slab might be responsible. Several investigations have indicated that there are similarly large variations in b values over short distances for shallow earthquakes (for a review, see Wiemer and Wyss, 2002). Thus Kagan's (1999) suggestion that β is approximately 0.60 for earthquakes of all depths doesn't appear to be true locally.

4.4.4 What limits the size of the largest deep earthquakes?

As noted above, Kagan (1999) was unable to determine an upper bound for the value of M_{max}, the size of the largest deep earthquakes; however, what can physical arguments tell us about M_{max}? First, suppose that deep earthquakes occur when there is slip S along planar surfaces within the cold part of subducting lithosphere. Then, the biggest earthquake will occur when there is slip on the biggest fault, and have moment M_o of $\mu A_{max} S$ (see eq. 4.7) where μ is rigidity (about 10^{11} Pa at 450 km depth; see Masters and Shearer, 1995). If the thickness W of the region cold enough to support sudden rupture is 20 km, and if it ruptures a distance L laterally along a Wadati–Benioff zone – say L is 100 km – then the area A_{max} will be 2×10^9 m^2. Finally, assume that S is about $10^{-4}L$, or 10 meters as in eq. 4.8. Then M_o is 2×10^{21} N-m, which corresponds to M_W of 8.2 – identical to the magnitude of the 1994 Bolivian earthquake.

Next, what is M_{max} for the (unlikely) case where phase transitions in the mantle cause deep earthquakes, and the seismic radiation reflects an instantaneous accompanying volume change? Müller (2001) shows that a volume change ΔV will produce a seismic source with moment M_o that will be between:

$$(\lambda + 2\mu/3)\Delta V \leq M_o \leq (\lambda + 2\mu)\Delta V, \qquad (4.11)$$

where λ and μ are the Lame parameters; here the upper value corresponds to a spherically symmetric source, and the lower value to a planar geometry. Thus, suppose a spherical region with diameter D instantaneously undergoes a transition and deforms to a smaller sphere with 10% less volume. If D is 10 km and λ and μ both equal 10^{11} Pa, the moment M_o will be about 10^{22} N-m, which corresponds to M_W of 8.7. Finally, if substantial chunks of the cold core of subducting lithosphere can undergo an instantaneous phase transition, one can imagine that even larger events might occur. For example, a 10% volume collapse within a 10 km$\times$100 km$\times$100 km region would correspond to M_W of about 10.

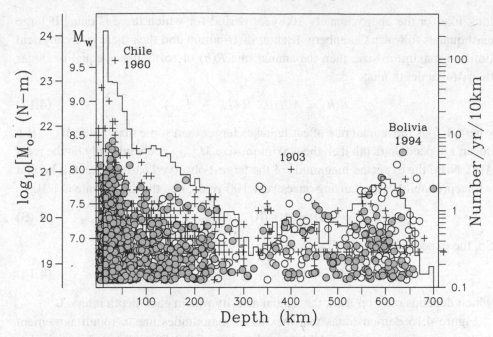

Fig. 4.16 Earthquake size and annual occurrence rates vs focal depth for large earthquakes. The sizes and depths of individual earthquakes (plotted symbols and left-hand scales) are as reported in three catalogs: Harvard CMT 1977–2003 (filled circles), CMT-Historical 1907–1976 (open circles) and EV Centennial 1900–1999 (+ symbols). Annual occurrence rate (solid line and right-hand scale) is for earthquakes in the Harvard CMT catalog with M_w of 5.4 or greater. Left and right axes are scaled using eq. 4.14 assuming $b = 0.9$ ($\beta = 0.6$) and $M_{min} = 5.4$; thus, the solid line is also the maximum expected size for the model explained in the text. Among the most egregious outliers are the 1960 Chile and 1994 Bolivia earthquakes with well-established depths and sizes, and the 1903 Tonga earthquake with a relatively poorly determined depth and size.

4.4.5 Is maximum size depth-dependent?

Subducting lithosphere meets two phase transitions and undergoes various thermal, mechanical, and possibly chemical changes as it travels from the surface into the lower mantle; thus, it is certainly plausible that maximum earthquake size M_{max} would depend on depth. Does the data support this idea? Fig. 4.16 suggests that it well might, since most earthquakes with magnitudes M_W exceeding about 7.8 seem to occur at depths shallower than 200 km or greater than 525 km.

However, Frohlich (1998) investigated the possibility that M_{max} was independent of depth, and that the decrease in maximum size observed between 300–500 km occurred only because events in this depth range are relatively rare. If M_{max} doesn't depend on depth, what would the observed distribution of maximum sizes M_{maxObs}

look like for the approximately 100-year period for which there is data? If large earthquakes follow a Gutenberg–Richter distribution and thus there is no physical limit on maximum size, then the annual rate $R(h)$ of earthquakes equal or larger than M_W at depth h is:

$$R(h) = A(h)10^{-b}(M_W - M_{min}),   (4.12)$$

where $A(h)$ is the annual rate of earthquakes larger than some magnitude M_{min}. If b doesn't depend on depth then the maximum size M_{maxObs} depends only on the term $A(h)$. Now, the expected magnitude of the largest observed earthquake will be that with a probability of occurring once every 100 years, i.e., the annual rate is 1/100:

$$R_{100} = 0.01 = A(h)10^{-b}(M_{maxObs} - M_{min}).   (4.13)$$

So, the expected value for M_{maxObs} is:

$$M_{maxObs} = M_{min} + [\log_{10} A(h) - \log_{10} R_{100}]/b,   (4.14)$$

which depends only on $A(h)$, the annual activity rate in each depth interval.

Figure 4.16 demonstrates that maximum magnitudes are in rough agreement with the predictions of eq. 4.14 for a b value of 0.9 ($\beta = 0.60$). In particular, the approximately exponential decrease between 60 and 300 km in the rate of occurrence $A(h)$ coincides with an approximately linear decrease in M_{maxObs}; and, the peak in rate near 600 km occurs approximately where M_{maxObs} reaches a local maximum. There are a few outlier events; however, a certain number of these are expected. Indeed, with 100 years of observations one tenth of the depth intervals should contain a 1000-year earthquake; the model predicts that its magnitude would be $1/b$ larger than M_{maxObs}, i.e., about 1.1 magnitude units.

However, most of the discrepancies with the predicted M_{maxObs} cluster within three depth intervals. The largest earthquakes observed exceed the predictions of eq. 4.14 between depths of 300 and 500 km, and again for depths exceeding about 625 km. A possible explanation is that some earthquakes in these intervals are from a population with a lower b or β value; indeed, the outliers beneath 625 km include the anomalous Spanish 1954 and the 1970 and 1994 South American earthquakes discussed previously. The third anomalous interval is between about 100–160 km, where the largest earthquakes are smaller than predicted. Earthquakes from this depth range are numerous, and thus the absence of earthquakes as large as predicted by eq. 4.14 may well suggest that M_{max} is about 7.6. However, Okal (1992) suggests significantly larger earthquakes have occurred at these depths, citing the events of 21 December 1939 ($h_{Abe} = 150$ km; $M_m = 8.4$) and 9 December 1950 ($h_{Okal} = 128$ km; $M_m = 8.3$).[10]

---

[10] M_m is the "mantle magnitude" as defined and assigned by Okal (1992).

In conclusion, as yet there is no compelling evidence that the maximum size of deep earthquakes depends on focal depth. Beneath 60 km nearly all the data in Fig. 4.16 are consistent with a depth-independent maximum possible size as long as it exceeds about 8, just as suggested by Kagan (1999).

4.5 References

Aki, K., 1965. Maximum likelihood estimate of b in the formula Log $N = a - bM$ and its confidence limits, *Bull. Earthquake Res. Inst., Tokyo Univ.*, **43**, 237–239.

1981. A probabilistic synthesis of precursory phenomena. In *Earthquake Prediction: an International Review*, Maurice Ewing Series, Vol. **4**, eds. D. W. Simpson and P. G. Richards, Washington, D.C., AGU, 566–574.

Amitrano, D., 2003. Brittle-ductile transition and associated seismicity: experimental and numerical studies and relationship with the b value, *J. Geophys. Res.*, **108**, No. 2044, doi:10.1029/2001JB000680.

Båth, M., 1981. Earthquake magnitude – recent research and current trends, *Earth Sci. Rev.*, **17**, 315–398.

1985. Surface-wave magnitude corrections for intermediate and deep earthquakes, *Phys. Earth Planet. Int.*, **37**, 228–234.

Bender, B., 1983. Maximum likelihood estimation of b values for magnitude grouped data, *Bull. Seismol. Soc. Amer.*, **73**, 831–851.

Berlage, H. P., 1937. A provisional catalogue of deep-focus earthquakes in the Netherlands East Indies 1918–1936, *Gerlands Beitr. z. Geophysik*, **50**, 7–17.

Bilek, S. L., T. Lay, and L. J. Ruff, 2004. Radiated seismic energy and earthquake source duration variations from teleseismic source time functions for shallow subduction zone earthquakes, *J. Geophys. Res.*, **109**, doi:10.1029/2004JB003039, B00308.

Bina, C. R., 1997. Patterns of deep seismicity reflect buoyancy stresses due to phase transitions, *Geophys. Res. Lett.*, **24**, doi:10.1029/97GL53189, 3301–3301.

Bina, C. R., S. Stein, C. Marton, and E. M. Van Ark, 2001. Implications of slab mineralogy for subduction dynamics, *Phys. Earth Planet. Int.*, **127**, doi:10.1016/S0031-9201(01)00221-7, 51–66.

Chen, W.-P. and P. Molnar, 1983. Focal depths of intracontinental and intraplate earthquakes and their implications for the thermal and mechanical properties of the lithosphere, *J. Geophys. Res.*, **88**, 4183–4214.

Davies, J. H., 1999. Simple analytic model for subduction zone thermal structure, *Geophys. J. Int.*, **139**, doi:10.1046/j.1365-246x.1999.00991.x, 823.

Davies, J. H. and D. J. Stevenson, 1992. Physical model of source region of subduction zone volcanics, *J. Geophys. Res.*, **97**, 2037–2070.

• Dziewonski, A. M. and J. H. Woodhouse, 1983. Studies of the seismic source using normal-mode theory. In *Earthquakes: Observation, Theory and Interpretation*, eds. H. Kanamori and E. Boschi, pp. 45–137, New York, North-Holland.

Eneva, M., R. E. Habermann, and M. W. Hamburger, 1994. Artificial and natural changes in the rates of seismic activity; a case study of the Garm Region, Tadjikistan (CIS), *Geophys. J. Int.*, **116**, 157–172.

Engdahl, E. R. and A. Villaseñor, 2002. Global seismicity: 1900–1999. In *International Handbook of Earthquake and Engineering Seismology*, San Diego, CA, Academic Press for International Association of Seismology and Physics of the Earth's Interior, 665–690.

Engdahl, E. R., R. van der Hilst, and R. Buland, 1998. Global teleseismic earthquake relocation with improved travel times and procedures for depth determination, *Bull. Seismol. Soc. Amer.*, **88**, 722–743.

England, P. and C. Wilkins, 2004. A simple analytical approximation to the temperature structure in subduction zones, *Geophys. J. Int.*, **159**, doi:10.1111/j.1365-246X.2004.02419.x, 1138.

•England, P., R. Engdahl, and W. Thatcher, 2004. Systematic variation in the depths of slabs beneath arc volcanoes, *Geophys. J. Int.*, **156**, 377–408, doi:10.1111/j.1365-246x.2003.02132.x.

Fischer, K. M. and T. H. Jordan, 1991. Seismic strain rate and deep slab deformation in Tonga, *J. Geophys. Res.*, **96**, 14429–14444.

Frohlich, C., 1989. The nature of deep-focus earthquakes, *Ann. Rev. Earth Planet. Sci.*, **17**, doi:10.1146/annurev.ea.17.050189.001303, 227–254.

   1998. Does maximum earthquake size depend on focal depth? *Bull. Seismol. Soc. Amer.*, **88**, 329–336.

Frohlich, C. and S. Davis, 1985. Identification of aftershocks of deep earthquakes by a new ratios method, *Geophys. Res. Lett.*, **12**, 713–716.

   1993. Teleseismic *b* values: or, much ado about 1.0, *J. Geophys. Res.*, **98**, 631–644.

Frohlich, C., K. Kadinsky-Cade and S. D. Davis, 1995. A reexamination of the Bucaramanga, Colombia, earthquake nest, *Bull. Seismol. Soc. Amer*, **85**, 1622–1634.

Frohlich, C., M. F. Coffin, C. Massell, P. Mann, C. L. Schuur, S. D. Davis, T. Jones, and G. Karner, 1997. Constraints on Macquarie Ridge tectonics provided by Harvard focal mechanisms and teleseismic earthquake locations, *J. Geophys. Res.*, **102**, doi:10.1029/96JB03408, 5029–5041.

Furukawa, Y., 1994. Two types of deep seismicity in subducting slabs, *Geophys. Res. Lett.*, **21**, doi:10.1029/94GL01083, 1181–1184.

Giardini, D., 1988. Frequency distribution and quantification of deep earthquakes, *J. Geophys. Res.*, **93**, 2095–2105.

Gill, J. B., 1981. *Orogenic Andesites and Plate Tectonics*, Berlin, Germany, Springer-Verlag, 390 pp.

•Gorbatov, A. and V. Kostoglodov, 1997. Maximum depth of seismicity and thermal parameter of the subducting slab: general empirical relation and its application, *Tectonophysics*, **277**, doi:10.1016/0040-1951(97)00084-X, 165–187.

Gorbatov, A., V. Kostoglodov, and E. Burov, 1996. Maximum seismic depth versus thermal parameter of subducted slab: application to deep earthquakes in Chile and Bolivia, *Geofísica Internacional*, **35**, 41–50.

Gorbatov, A., V. Kostoglodov, G. Suárez, and E. Gordeev, 1997. Seismicity and structure of the Kamchatka subduction zone, *J. Geophys. Res.*, **102**, doi:10.1029/96JB03491, 17883–17898.

Guest, A., C. Schubert, and C. W. Gable, 2003. Stress field in the subducting lithosphere and comparison with deep earthquakes in Tonga, *J. Geophys. Res.*, **108**, No. 2288, doi:10.1029/2002JB002161.

Gutenberg, B., 1945. Magnitude determination for deep-focus earthquakes, *Bull. Seismol. Soc. Amer.*, **35**, 117–130.

Gutenberg, B. and C. F. Richter, 1956. Magnitude and energy of earthquakes, *Ann. Geof*, **9**, 1–15.

Habermann, R. E., 1987. Man-made changes of seismicity rates, *Bull. Seismol. Soc. Amer.*, **77**, 141–159.

   1991. Seismicity rate variations and systematic changes in magnitudes in teleseismic catalogs, *Tectonophysics*, **193**, doi:10.1016/0040-1951(91)90337-R, 277–290.

Hanks, T. C. and H. Kanamori, 1979. A moment magnitude scale, *J. Geophys. Res.*, **84**, 2348–2350.

Helffrich, G. and J. Brodholt, 1991. Relationship of deep seismicity to the thermal structure of subducted lithosphere, *Nature*, **353**, doi:10.1038/353252a0, 252–255.

Houston, H., 2001. Influence of depth, focal mechanism, and tectonic setting on the shape and duration of earthquake source time functions, *J. Geophys. Res.*, **106**, doi:10.1029/2000JB900468, 11137–11150.

Huang, W.-C., E. A. Okal, G. Ekström, and M. P. Salganik, 1998. Centroid moment tensor solutions for deep earthquakes predating the digital era: the historical dataset (1907–1961), *Phys. Earth Planet. Int.*, **106**, doi:10.1016/S0031–9201(98)00081–8, 181–190.

Isacks, B., J. Oliver, and L. R. Sykes, 1968. Seismology and the new global tectonics, *J. Geophys. Res.*, **73**, 5855–5899.

Ishimoto, M. and K. Iida, 1939. Observations of earthquakes registered with microseismographs constructed recently (in Japanese), *Bull. Earthquake Res. Inst., Tokyo Univ.*, **17**, 443–478.

Jarrard, R. D., 1986. Relations among subduction parameters, *Rev. Geophys.*, **24**, 217–284.

Kagan, Y. Y., 1991a. Seismic moment distribution, *Geophys. J. Int.*, **106**, 123–134.

1991b. Likelihood analysis of earthquake catalogs, *Geophys. J. Int.*, **106**, 135–148.

•1999. Universality of the seismic moment-frequency relation, *Pure Appl. Geophys.*, **155**, doi:10.1007/s000240050277, 537–573.

2002. Seismic moment distribution revisited: I. Statistical results, *Geophys. J. Int.*, **148**, doi:10.1046/j.1365-246X.2002.01594.x.

Kanamori, H., 1983. Magnitude scale and quantification of earthquakes, *Tectonophysics*, **93**, doi:10.1016/0040-1951(83)90273-1, 185–199.

Kanamori, H. and D. L. Anderson, 1975. Theoretical basis of some empirical relations in seismology, *Bull. Seismol. Soc. Amer.*, **65**, 1073–1095.

Kirby, S., W. B. Durham, and L. A. Stern, 1991. Mantle phase changes and deep-earthquake faulting in subducting lithosphere, *Science*, **252**, 216–225.

Kirby, S., E. R. Engdahl, and R. Denlinger, 1996a. Intermediate-depth intraslab earthquakes and arc volcanism as physical expressions of crustal and uppermost mantle metamorphism in subducting slabs. In *Subduction, Top to Bottom*, eds. G. E. Bebout, D. W. Scholl, S. H. Kirby, and J. P. Platt, Geophys. Mon. 96, Washington, D.C., American Geophysical Union, 195–214.

Kirby, S. H., S. Stein, E. A. Okal, and D.C. Rubie, 1996b. Metastable mantle phase transformations and deep earthquakes in subducting oceanic lithosphere, *Rev. Geophys*, **34**, doi:10.1029/96RG01050, 261–306.

Lombardi, A. M., 2003. The maximum likelihood estimator of *b*-value for mainshocks, *Bull. Seismol. Soc. Amer.*, **93**, 2082–2088.

McKenzie, D. P., 1969. Speculation on the consequences and causes of plate motions, *Geophys. J. Roy. Astron. Soc.*, **18**, 1–32.

Masters, T. G. and P. M. Shearer, 1995. Seismic models of the Earth: elastic and anelastic. In *Global Earth Physics: a Handbook of Physical Constants*, ed. T. J. Ahrens, Washington, D.C., American Geophysical Union, 88–103.

Mogi, K., 1962. Magnitude frequency relations for elastic shocks accompanying fractures of various materials and some related problems in earthquakes, *Bull. Earthquake Res. Inst., Tokyo Univ.*, **40**, 831–853.

•Molnar, P., D. Freedman, and J. S. F. Shih, 1979. Lengths of intermediate and deep seismic zones and temperatures in downgoing slabs of lithosphere, *Geophys. J. Roy. Astron. Soc.*, **56**, 41–54.

Müller, G., 2001. Volume change of seismic sources from moment tensors, *Bull. Seismol. Soc. Amer.*, **91**, 880–884.

Murphy, J. R. and B. W. Barker, 2003. Revised distance and depth corrections for use in the estimation of short-period P-wave magnitudes, *Bull. Seismol. Soc. Amer.*, **93**, 1746–1764.

Myers, S. C., T. C. Wallace, S. L. Beck, P. G. Silver, G. Zandt, J. Vandecar, and E. Minaya, 1995. Implications of spatial and temporal development of the aftershock sequence for the M_W 8.3 June 9, 1994 deep Bolivian earthquake, *Geophys. Res. Lett.*, **22**, doi:10.1029/95GL01600, 2269–2272.

Okal, E. A., 1990. M_m: a mantle wave magnitude for intermediate and deep earthquakes, *Pure Appl. Geophys.*, **134**, 333–354.

• 1992. Use of the mantle magnitude M_m for the reassessment of the moment of historical earthquakes. II: Intermediate and deep events, *Pure Appl. Geophys.*, **139**, 59–85.

• Okal, E. A. and C. R. Bina, 1998. On the cessation of seismicity at the base of the transition zone, *J. Seismology*, **2**, doi:10.1023/A:1009789222914, 65–86.

Okal, E. A. and S. H. Kirby, 1995. Frequency–moment distribution of deep earthquakes; implications for the seismogenic zone at the bottom of slabs, *Phys. Earth Planet. Int.*, **92**, doi:10.1016/0031-9201(95)03037-8, 169–187.

Pacheco, J. F., C. H. Scholz, and L. R. Sykes, 1992. Changes in the frequency–size relationship from small to large earthquakes, *Nature*, **355**, 71–73.

Page, R., 1968. Aftershocks and microaftershocks of the great Alaska earthquake of 1964, *Bull. Seismol. Soc. Amer.*, **58**, 1131–1168.

Rees, B. A. and E. A. Okal, 1987. The depth of the deepest historical earthquakes, *Pure Appl. Geophys.*, **125**, 699–715.

Rezapour, M., 2003. Empirical global depth–distance correction terms for m_b determination based on seismic moment, *Bull. Seismol. Soc. Amer.*, **93**, 172–189.

Richter, C. F., 1935. An instrumental earthquake magnitude scale, *Bull. Seismol. Soc. Amer.*, **25**, 1–32.

1958. *Elementary Seismology*, San Francisco, W. H. Freeman, 768 pp.

Romanowicz, B. A. and J. B. Rundle, 1993. On scaling relations for large earthquakes, *Bull. Seismol. Soc. Amer.*, **83**, 1294–1297.

Rundle, J., 1989. Derivation of the complete Gutenberg–Richter magnitude–frequency relation using the principal of scale invariance, *J. Geophys. Res.*, **94**, 12337–12342.

Scholz, C. H., 1968. The frequency–magnitude relation of microfracturing in rock and its relation to earthquakes, *Bull. Seismol. Soc. Amer.*, **58**, 399–415.

Scholz, C. H., 1994. A reappraisal of large earthquake scaling, *Bull. Seismol. Soc. Amer.*, **84**, 215–218.

Spencer, J. E., 1994. A numerical assessment of slab strength during high- and low-angle subduction and implications for the Laramide orogenesis, *J. Geophys. Res.*, **99**, 9227–9236.

Spiedel, D. H. and P. H. Mattson, 1993. The polymodal frequency–magnitude relationship of earthquakes, *Bull. Seismol. Soc. Amer.*, **83**, 1893–1901.

Stark, P. B. and C. Frohlich, 1985. The depths of the deepest deep earthquakes, *J. Geophys. Res.*, **90**, 1859–1869.

Sykes, L. R., 1966. The seismicity and deep structure of island arcs, *J. Geophys. Res.*, **71**, 2931–3006.

Tatsami, Y., 1986. Formation of the volcanic front in subduction zones, *Geophys. Res. Lett.*, **13**, 717–720.

Taylor, D. W. A., J. A. Snoke, I. S. Sacks, and T. Takanami, 1990. Nonlinear frequency–magnitude relationships for the Hokkaido corner, Japan, *Bull. Seismol. Soc. Amer.*, **80**, 340–353.

Tibi, R., D. A. Wiens, and H. Inoue, 2003. Remote triggering of deep earthquakes in the 2002 Tonga sequences, *Nature*, **424**, doi:10.1038/nature01903, 921–925.

Tinker, M. A., T. C. Wallace, S. L. Beck, P. G. Silver, and G. Zandt, 1995. Aftershock source mechanisms from the June 9, 1994, deep Bolivian earthquake, *Geophys. Res. Lett.*, **22**, doi:10.1029/95GL01090, 2273–2276.

Trifu, C.-I. and M. Radulian, 1991. Frequency–magnitude distribution of earthquakes in Vrancea: relevance for a discrete model, *J. Geophys. Res.*, **96**, 4301–4311.

Utsu, T., 2002. Relationships between magnitude scales. In *International Handbook of Earthquake and Engineering Seismology*, San Diego, CA, Academic Press for International Association of Seismology and Physics of the Earth's Interior, 733–746.

Vassiliou, M. S. and B. H. Hager, 1988. Subduction zone earthquakes and stress in slabs, *Pure Appl. Geophys.*, **128**, 547–624.

Wadati, K., 1935. On the activity of deep-focus earthquakes in the Japan Islands and neighbourhoods, *Geophys. Mag.*, **8**, 305–325.

Wiemer, S. and J. P. Benoit, 1996. Mapping the *b*-value anomaly at 100 km depth in the Alaska and New Zealand subduction zones, *Geophys. Res. Lett.*, **23**, doi:10.1029/96GL01233, 1557–1560.

Wiemer, S. and M. Wyss, 2002. Mapping spatial variability of the frequency–magnitude distribution of earthquakes, *Adv. Geophys.*, **45**, 259–302.

• Wiens, D. A., 2001. Seismological constraints on the mechanism of deep earthquakes: temperature dependence of deep earthquake source properties, *Phys. Earth Planet. Int.*, **127**, doi:10.1016/S0031-9201(01)00225-4, 145–163.

Wiens, D. A. and H. J. Gilbert, 1996. Effect of slab temperature on deep-earthquake aftershock productivity and magnitude–frequency relations, *Nature*, **384**, doi:10.1028/384153a01, 53-156.

Willemann, R. J., 1991. A simple explanation for the depth distribution of deep earthquakes, *Geophys. Res. Lett.*, **18**, 1123–1126.

Wortel, M. J. R. and N. J. Vlaar, 1988. Subduction zone seismicity and the thermo-mechanical evolution of downgoing lithosphere, *Pure Appl. Geophys.*, **128**, 625–659.

5

Spatial and temporal clustering

Do deep earthquakes occur randomly in time and in space? Or does their occurrence exhibit regular patterns – perhaps correlated with other phenomena? If you consider the big picture and don't focus too hard on the details the answers are simple: deep earthquakes are almost random with respect to time; and they are very non-random in space. But what about those details? In this chapter I use a descriptive approach to characterize the most obvious features of deep earthquake clustering. Then I apply some simple statistical methods to evaluate various more subtle features.

5.1 Spatial clustering

5.1.1 Earthquakes and subduction zones

Geographically, the world's deep earthquakes are definitely not random; the great majority lie near deep-ocean trenches beneath well-defined subduction zones (see Figs. 2.1 and 2.2). Moreover, of those with depths exceeding 300 km, about two-thirds occur in the Tonga–Kermadec region (see Fig. 1.14). When plotted in cross sections oriented normal to the trench axis, intermediate- and deep-focus quakes commonly form approximately planar groups called Wadati–Benioff zones (Figs. 5.1 and 5.2).

What is the thickness of these planar zones? For intermediate-depth seismicity there are a few places where travel times from regional network data are available and there have been careful relative relocations of deep seismicity. At a depth of about 100–125 km beneath western Argentina, Smalley and Isacks (1987) determined that all quakes occurred within a zone approximately 20 km thick, with 90% confined to a 12 km thick region (Fig. 5.3). For southwest Japan and Costa Rica, Hacker *et al.* (2003) present data indicating that hypocenters fit within a 10–15 km

I thank Yan Kagan for reviewing an earlier draft of this chapter.

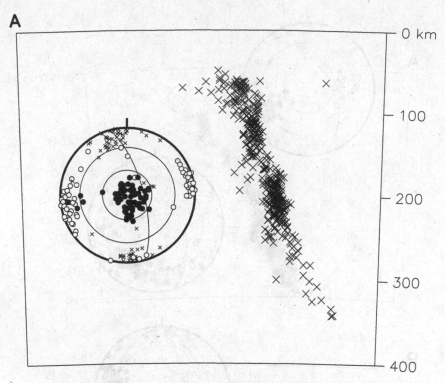

Fig. 5.1 Trench-perpendicular cross sections showing Wadati–Benioff zone in the northern New Hebrides; the section is centered on 13.5°S with a width of 220 km. Crosses are earthquakes occurring in 1964–2002 selected from the EHB catalog (Engdahl *et al.*, 1998). Equal area plot (map view, not back hemisphere projection) indicates orientations of T axes (filled circles), P axes (open circles), and B axes (crosses) from selected Harvard CMT; the curved line indicates the plane of the Wadati–Benioff zone. Note that seismic activity occurs at all depths down to about 300 km, and that nearly all focal mechanisms are of the downdip T, along-strike B, and slab-normal P type.

thick zone (Fig. 5.4); by analyzing location uncertainties they conclude that seismicity occurs within zones having thicknesses of about 7 km. Inspection of published cross sections suggests thicknesses are comparable to those in Figs. 5.3 and 5.4 for the upper zone in northern Japan (Hasegawa *et al.*, 1978a; 1978b; Zhao *et al.*, 1992; Sato *et al.*, 1996) and in Alaska (Reyners and Coles, 1982; Taber *et al.*, 1991; Abers, 1996; Ratchkovski and Hansen, 2002).

It is more difficult to determine the thicknesses of Wadati–Benioff zones for deep-focus earthquakes. For depths exceeding 300 km there is no place where activity rates are high and there is a regional network situated favorably for evaluating Wadati–Benioff zone thickness; thus, available estimates of thickness rely heavily

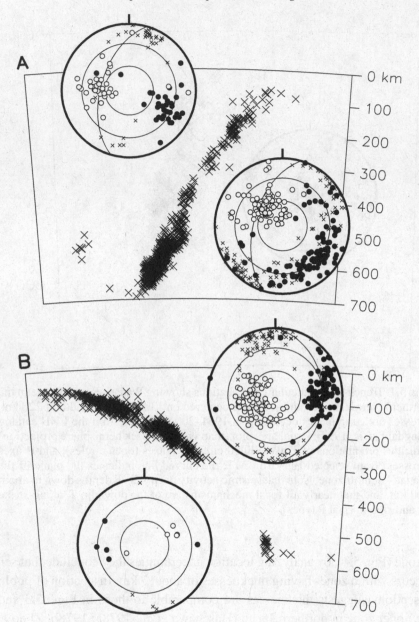

Fig. 5.2 Trench-perpendicular cross sections showing Wadati–Benioff zones in (A) Tonga and (B) South America. The Tonga section is centered on 22.5°S with a width of 130 km; the South America section is at 23.5°S with width 330 km. Circles are map-view equal area plots for intermediate (upper circle) and deep-focus (lower circle) earthquakes; curved lines indicate the approximate plane of the Wadati–Benioff zone. Data and symbols are as in Fig. 5.1. Note that focal mechanisms have downdip P axes in Tonga but downdip T in South America; and, that for deep-focus earthquakes in Tonga, T and B don't cluster about along-strike or slab-normal axes. Also note the gap in seismic activity between 300 and 500 km in South America, and the presence of rogue or outboard deep-focus events in both regions.

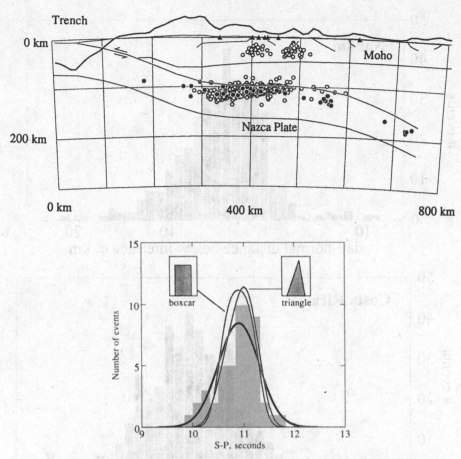

Fig. 5.3 (Top) East–west cross section of seismicity in South America at 31.25°S; open circles are quakes located by a local network (triangles); solid circles are selected ISC hypocenters. The east–west topographic profile shows the average elevation of a 100-km-wide swath centered on 31°S, with a vertical exaggeration of 4:1. (Bottom) Histogram of moveout-corrected S-P times for 51 selected quakes recorded digitally and occurring within epicentral distances of 30 km or less from station RTLL. The best-fitting Gaussian distribution is shown with a heavy line, while light lines indicate Gaussian convolved with a boxcar and a triangular distribution. 90% of the data have S-P times between 10.22 and 11.33s, which corresponds to a Wadati–Benioff zone thickness of 12 km. Figure reproduced from Smalley and Isacks (1987).

on teleseismic data. However, these studies also generally find that the thickness is in the range of 10–25 km (e.g., Wiens *et al.*, 1993; 1994).

The classical plate tectonic model is that Wadati–Benioff zones demarcate zones of high stress and delineate the cold, dense portions of subducting lithosphere as gravity causes it to sink into the mantle. In many regions the focal mechanisms

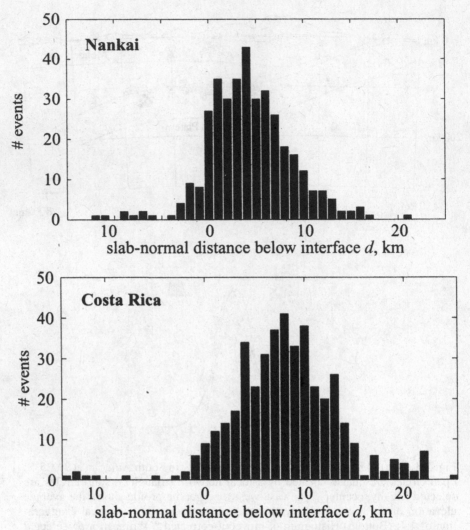

Fig. 5.4 Histograms showing distance d from the inferred surface of subducting lithosphere for hypocenters in southwest Japan (top) and Costa Rica (bottom). Data are from Protti *et al.* (1995) and Cummins *et al.*, (2002); figure reproduced from Hacker *et al.* (2003).

support this model (e.g., Figs. 5.1 and 5.2); often intermediate-depth quakes have either tensional (T) or compressional (P) axes that cluster along a downdip direction (Fig. 5.5). For deep quakes downdip T axes are rare and downdip P or near-vertical P clustering is common. The classical explanation for this is that the subducting slab experiences increasing resistance as it penetrates 650 km because mantle viscosity increases there. However, Fig. 5.5 is a bit misleading because in a majority of Wadati–Benioff zones the clustering of P or T isn't as simple as in Figs. 5.1 and 5.2

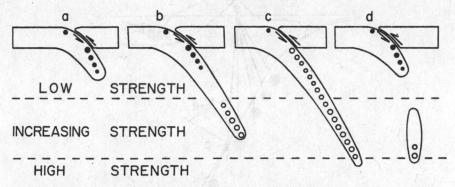

Fig. 5.5 Isacks and Molnar's (1971) cartoon explaining why intermediate- and deep-focus earthquake mechanisms have downdip T axes (filled circles) in some Wadati–Benioff zones and downdip P axes (open circles) in others.

(see Section 6.3). And where axial clustering does occur the scatter is typically ±10° or more, even in zones with relatively simple geometry and even when we exclude poorly determined mechanisms.

In several regions cross sections of Wadati–Benioff zones exhibit distinct bends or gaps (Fig. 5.2B); however, not all of these are real. In locations utilizing regional network data there can be spurious bends (Fig. 5.6) which are artifacts produced when the location procedure inadequately accounts for the velocity contrast between the subducting lithosphere and the adjacent mantle; this contrast can be 10% or more (for a review, see Lay, 1997). These spurious bends are most likely in regions such as the central Aleutians (e.g., see Taber *et al.*, 1991), where the aperture of the regional network is small because stations are situated only on volcanic islands.

More often, however, genuine Wadati–Benioff bends occur because subducted lithosphere is deformed; and, presumably "stagnant" lithosphere occupies many Wadati–Benioff zone gaps. Thus, numerous investigations use seismicity to determine contour lines and "map" subducted lithosphere (Isacks and Barazangi, 1977; Bevis and Isacks, 1984; Burbach and Frohlich, 1986; Yamaoka *et al.*, 1986; Chiu *et al.*, 1991). Where there are vertical gaps in seismicity there is always uncertainty; in the absence of tomographic evidence or reports of high-frequency phases traveling across the gap, it is always possible that the slab isn't continuous. In several regions where deep-focus earthquakes occur there are distinct gaps separating intermediate- and deep-focus activity, as in Fig. 5.2B; these include Indonesia, Izu–Bonin, New Zealand, parts of the Philippines, Spain, South America, and Vanuatu. Lay (1997) and Okal (2001) conclude that in Indonesia, Japan, Izu–Bonin, South America, and Tonga the slab is probably continuous, and in the New Hebrides (Vitiaz) and New Zealand[1] the slab is probably detached.

---

[1] Not all investigators concur that the slab is detached in New Zealand (see Section 10.2.5).

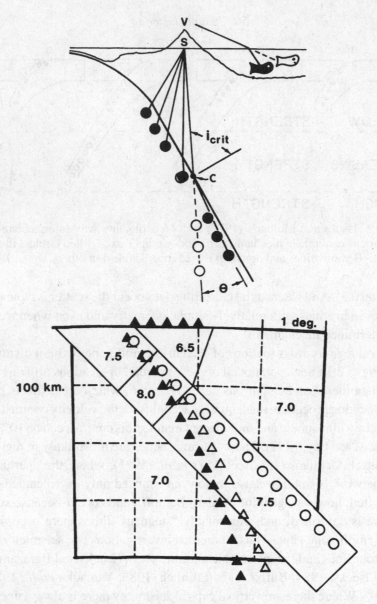

Fig. 5.6 (Top) Just as refraction at a water-air interface causes a viewer at V to see a fish at an apparent position that differs from its real position, the apparent position (open circles) of some earthquakes observed by a station on volcanic island V differs from the real position (solid circles). Figure reproduced from Frohlich *et al.* (1982). (Bottom) Locations determined using island-arc network data may determine an incorrect Wadati–Benioff geometry, even when S and P arrivals are used. In this simulated example, earthquakes (circles) situated in a subducting slab are located using travel times determined by tracing rays to stations (triangles at the surface) corresponding to a typical island-arc network. If the locating program uses a flat-layered velocity model the best-fitting locations exhibit a spurious bend or change in dip. Filled triangles correspond to locations determined using the structure shown (numbers are velocity in km/s); the open triangles are locations determined when the mantle–slab contrast is 7%. Figure reproduced from McLaren and Frohlich (1985).

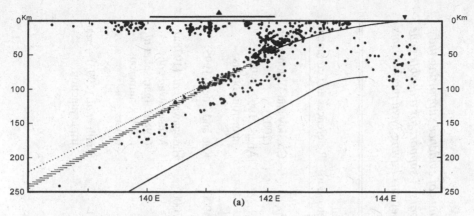

Fig. 5.7 Double seismic zone beneath the Tohoku district, Japan, as evident in an E–W cross section of microearthquakes located by a local network, assuming a flat-layered velocity model. The upper seismic zone is approximately coincident with the supposed upper surface from about 50–150 km; the lower seismic zone lies about 35–40 km below the upper zone and extends to about 160 km. The hatched zone indicates the position of the upper surface of subducted lithosphere if the velocity beneath is 6% higher than in the surrounding mantle. Figure reproduced from Hasegawa *et al.* (1978b) with permission from Blackwell Publishers, Ltd.

5.1.2 Double seismic zones

In several regions the Wadati–Benioff zone seems to be a "double zone," i.e., the intermediate-depth hypocenters fall along two roughly parallel planes that separate a region with no seismic activity (e.g., Fig. 5.7). While Sykes (1966) and Veith (1974) analyzed teleseismic locations and found preliminary evidence for a double zone in the Kuril arc, the most convincing examples have come from areas where regional seismic networks overlie the Wadati–Benioff zone and permit precise locations. For example, beneath Tohoku in northern Japan, Hasegawa *et al.* (1978a; 1978b; 1994) demonstrated that there are two zones about 40 km apart beginning at about 60–70 km depth and extending to a depth of about 160 km, where they merge into a single zone. Focal mechanisms of quakes in the upper zone tend to exhibit downdip compression, while mechanisms from the lower zone possess downdip tension.

There are reasons to be wary about reported double zones (Fig. 5.8 and Table 5.1), as there are at least four kinds of systematic errors that cause apparent, but spurious, double zones. First, if subducting lithosphere is segmented, a cross section containing hypocenters in adjacent along-strike segments can appear to be a double zone (e.g., Engdahl and Scholz, 1977; Kao and Chen, 1994). Second, the evidence for some double zones relies primarily on the existence of a few anomalous focal mechanisms (e.g., Kawakatsu, 1986b; Samowitz and Forsyth, 1981); this is worrisome

Table 5.1. *Reported intermediate-depth double seismic zones. There are varying amounts of evidence supporting the existence and properties of these double zones; where the table indicates evidence is "sparse" support for a double zone depends primarily on teleseismic locations and/or on the presence of a few lower plane events having focal mechanisms differing from the majority in the region.*

Location	Depth range (km)	Maximum separation (km)	Predominant focal mechanism		Thermal parameter (km)	References and comments
			Upper plane	Lower plane		
Subduction regular and local network observations available						
Cascadia 48.7° N, 123° W	40–60	5–10	??	Downdip T?	70–125	Cassidy and Waldhauser (2003)
Alaska peninsula (Shumagin) 55° N, 161.5° W	70–150	25	Downdip T	Downdip T	2500	Abers (1992; 1996); numerous others
Kuril–Kamchatka 46–52° N; 150–159° E	75–180	30–40	Downdip P	Downdip T	4000–6000	Kao and Chen (1995)
Northern Japan 39.5° N, 141° E	60–160	40	Downdip P	Downdip T	4000–7000	Hasegawa et al. (1978a); Matsuzawa et al. (1986); Igarashi et al. (2001); numerous others
Plate geometry complex or subduction highly oblique						
Mendocino 40.5° N, 124.5° W	13–30	10	Downdip T	?	80–150	Smith et al. (1993) – near triple junction

Northeast Taiwan 25° N, 122° E	40–130	Along-strike P	Downdip T?	1500	Kao and Rau (1999) – near bend in subduction zone, direction oblique
Northern Chile 18.5° S, 70° W	100–140	Downdip T?	Downdip P?	2200	Comte et al. (1999) – near bend in subduction zone
New Zealand 41° S, 174° E	50–80	Downdip T	Downdip T	2500	Reyners et al. (1997) – subduction oblique
New Britain 5.2° S, 151.3° E	60–160	Along-strike P	Downdip T	?	McGuire and Wiens (1995) – evidence sparse; near bend in subduction zone
Evidence for double zone is sparse					
Aleutians 52° N, 177.5° W	110–170	Downdip P	Downdip T	2200–3400	Engdahl and Scholz (1977) – double zone or segmented slab?
Central Chile 24.9° S, 68.7° W	150	Downdip T	Downdip P	4000	Araujo and Suárez (1994) – evidence based on one earthquake
Marianas 17.5° N, 145° E	80–120	Downdip P	Downdip T	4000–6000	Samowitz and Forsyth (1981)
Tonga 18.4–23.8° S, 174.5–177.5° W	60–185	Downdip P	Downdip T	5000–9000	Kawakatsu (1985; 1986b)

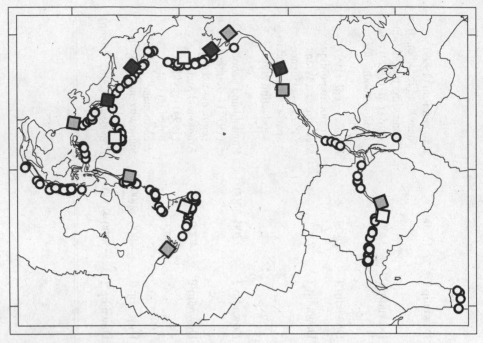

Fig. 5.8 Distribution of reported double seismic zones (rectangles) and outer rise earthquakes (circles). Dark rectangles indicate regions where subduction geometry is straightforward and evidence for double seismic zone is supported by local network observations; light gray rectangles are regions where subduction is oblique or the double zone occurs near a fold or bend in the subducting lithosphere; white rectangles are regions where supporting evidence is weaker. Outer rise earthquakes are from Christensen and Ruff (1988).

since peculiar or highly variable focal mechanisms occur in most geographic regions. Third, unless the same stations record all hypocenters, unmodeled lateral velocity variations can produce spurious location offsets in some hypocenters (e.g., Fig. 5.6) that make events appear to form two zones. And finally, misidentification of phases can lead to spurious zones; e.g., Cahill and Isacks (1986) alleged that misidentification of sP as pP was responsible for a double zone beneath northern Chile reported by Kono *et al.* (1985).

Nevertheless, double seismic zones clearly are present in several geographic regions, and differ with respect to depth, separation, and focal mechanism type. For example, the Cascadia and Mendocino double zones occur at shallow depths (< 60 m) and are separated by distances of 10 km or less; the Cook Inlet, Alaska, double zone is at depths of 50–90 km with a separation of about 25 km; while the Kuril–Kamchatka and northern Japan double zones are at 60–180 km depth with a separation of 30–40 km. In the Shumagin, Alaska, double zone, focal mechanisms indicate both upper and lower planes are in tension; in northern Japan and

Kuril–Kamchatka the upper zone is under compression and the lower zone is in compression.

What causes double seismic zones? The first explanation proposed was slab unbending; i.e. that upper and lower planes were zones of compressional and tensional stress in the top and bottom of the subducting lithosphere, respectively, caused because the curvature of subducting lithosphere changes after it passed the trench and straightens out on its path down into the mantle (e.g., Isacks and Barazangi, 1977; Engdahl and Scholz, 1977; Sleep, 1979; Tsukahara, 1980; Kawakatsu, 1986a). This is appealing because it explains the pattern of downdip compressional and tensional focal mechanisms observed in Japan and Kuril–Kamchatka. However, it fails to explain the different stress patterns observed in regions like the Alaska peninsula, and also doesn't explain why the double zones appear to persist to depths far greater than where the slab curvature actually decreases. A second proposed explanation was that the two planes are due to thermoelastic stresses produced as the slab warmed on its path into the mantle (Fujita and Kanamori, 1981; House and Jacob, 1982; Hamaguchi *et al.*, 1983; Goto *et al.*, 1985). Kao and Liu (1995) review various mechanisms, but for the lower plane favor a third explanation, concluding that seismicity is due to metastable phase transitions within the downgoing slab.

However, the most successful explanation for double seismic zones is that the two planes occur where temperature–pressure conditions permit the dehydration of minerals in the subducting oceanic crust and mantle (Nishiyama, 1992; Seno and Yamanaka, 1996). What is compelling about this explanation is that it appears to explain many of the variations between different regions. For example, by modeling temperature–pressure conditions, Hacker *et al.* (2003) showed that in both Cascadia and northern Japan . . .

Upper seismic zones . . . correspond spatially to dehydration of the crust. Lower seismic zones . . . correspond spatially to dehydration of the mantle. Seismic gaps between double zones correspond spatially to the thermally cold core of the slab. Aseismic mantle below the zone(s) of active seismicity is anhydrous.

This observation is appealing since Cascadia and northern Japan may well represent the warmest and coldest subducting lithosphere on the planet. Similar modeling studies are able to explain most of the properties of double seismic zones in the Alaska peninsula, Mendocino, northern Chile, and northeast Taiwan, and show why there is apparently only one zone in regions like southwest Japan and Costa Rica (e.g., Peacock and Wang, 1999; Seno *et al.*, 2001; van Keken *et al.*, 2002; Yamasaki and Seno, 2003).

These thermal modeling studies presume, of course, that the physical mechanism responsible for intermediate-depth seismicity is dehydration embrittlement (Kirby *et al.*, 1996; also see Section 7.3.1). Moreover, while it is plausible that hydrated oceanic crust provides the sources of hydrated minerals in the upper planes of double zones, an important unanswered question is how the mantle where lower planes occur becomes hydrated. One possible answer is that the hydration may occur along deep faults associated with outer rise earthquakes (see Fig. 5.8; also Savage, 1969; Peacock, 2001; Ranero et al., 2003).

At present the thermal models do not explain all the regional variations in focal mechanisms within double seismic zones (Table 5.1). Several of the reported double zones occur adjacent to a triple junction (Mendocino) or where there is a distinct change in the orientation of the trench (northern Chile, northeast Taiwan, Cook Inlet, New Britain). Thus it now seems likely that lateral bending of the subducting slab controls the stress pattern, while temperature and pressure conditions control the depth of mineral dehydration that permits seismicity to occur.

There have been two reports of double seismic zones for deep-focus earthquakes, one in Tonga at depths of 350–460 km (Wiens et al., 1994) and one in Izu–Bonin at 300–400 km (Iidaka and Furukawa, 1994; Iidaka and Obara, 1997); both have separations of 20–40 km. If these are confirmed they are apparently a phenomenon mechanically unrelated to intermediate double zones (see Guest et al., 2004).

5.1.3 Intermediate-depth earthquake nests

So far we have focused on clustering visible in trench-normal cross sections; however, obvious gaps and clusters are also visible in trench-parallel cross sections. For example, Fig. 4.5 reveals that in South America between 100–130 km depth there are several pockets of high activity surrounded by zones of lower activity at shallower and greater depth. Moreover, at depths of 150–275 km seismicity seems concentrated along several "fingers" that have along-trench horizontal dimension of 75 km or less and are well-separated from one another. These demonstrate clearly that much of the seismicity occurs within distinct clusters, even though the finger-like appearance in Fig. 4.5 is largely an artifact of the exaggerated depth scale. How common are these clusters, and what is their nature?

Beneath the industrial town of Bucaramanga, Colombia, there exists a remarkable cluster of intermediate-depth earthquakes known as the Bucaramanga nest (Fig. 5.9). Since 1964 the ISC has reported an average of eight earthquakes per year with m_b of 4.7 or greater from within a 30-km cube centered on 6.8° N, 73° W, depth 160 km. Tryggvason (1968) apparently was the first to recognize the unusual intensity of this clustering; Santo (1969a) was apparently the first to refer to this zone as a "nest" to distinguish its consistently high activity and small source volume from a swarm or aftershock sequence. Because there are only two reported earthquakes from this region prior to 1950, Lawson (1967) suggested that the nest had only recently become active. However, this is almost certainly untrue; this apparent

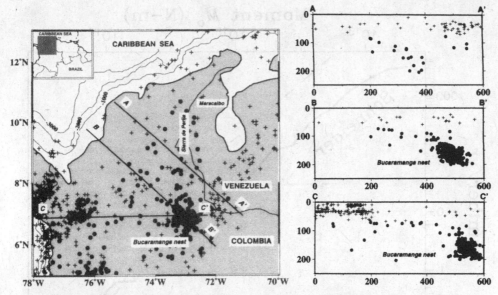

Fig. 5.9 The Bucaramanga earthquake nest, which is the intense cluster of seismic
activity near the border of northern Colombia and western Venezuela. Crosses
indicate quakes reported by the National Earthquake Information Center since
1963 with focal depths $h \leq 70$ km; solid circles indicate those with $h > 70$ km;
thick solid lines indicate the location of the three cross sections at right. Figure
reproduced from Malavé and Suárez (1995).

change is attributable to the fact that the largest quakes in the nest have m_b of about
5.9, and thus they were not routinely catalogued before 1960 (see Section 10.17).
A study of local newspapers and other historical sources demonstrated that while
catastrophic earthquakes are unknown in Bucaramanga, since at least 1930 there
have been occasional earthquakes strong enough to be felt by local residents
(Hurtado and de Jesus, 1999).

Subsequently scientists have investigated the Bucaramanga nest quite thoroughly. Tryggvason and Lawson (1970) and Dewey (1972) concluded that the
maximum dimension of the nest is on the order of 10 km, although with available
teleseismic observations they were unable to distinguish it from a point source.
However, in 1976 and 1979 there were two regional microearthquake surveys which
recorded about 15 nest microearthquakes per day; from these data Pennington
et al. (1979) and Schneider *et al.* (1987) concluded that the nest had a finite volume
with dimensions about $5 \times 4 \times 4$ km^3. Subsequently Frohlich *et al.* (1995) demonstrated that the Schneider (1984) hypocenters fit within a polyhedron of volume
11 km^3. Ojeda and Havskov (2001) have analyzed data from the Colombia national
seismic network and compared the nest's location to other features of the regional
seismicity.

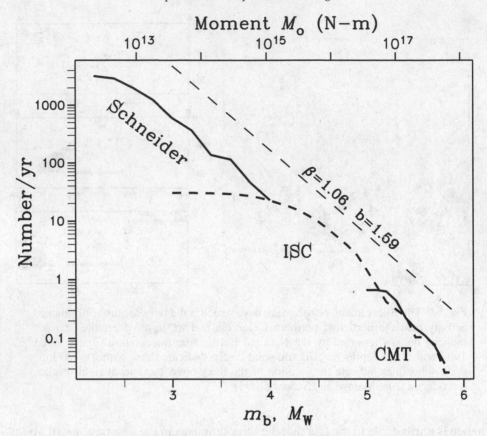

Fig. 5.10 Magnitude–frequency and moment–frequency distribution for earthquakes in the Bucaramanga nest. Lines indicate cumulative annual rates with magnitudes m_b or M_W exceeding the plotted value. Schneider (1984) data are moments M_o from 140 earthquakes recorded over a 16-day period by a temporary local network; ISC data are reported m_b for 1170 earthquakes occurring between 1964 and June 2002; Harvard CMT data are from 18 earthquakes recorded between 1977 and 2003. The straight dashed line shows that b- and β-values of 1.59 and 1.06 fit these data approximately.

Furthermore, since seismic moment M_o equals the product of rigidity μ, fault area A, and average slip S, Frohlich *et al.* (1995) calculated that a fault with aerial dimensions of $\sim 4 \times 4 \, \text{km}^2$ is adequate to produce the largest observed nest quakes, which have a scalar moment of about 5×10^{17} N-m. Measured β values for the nest exceed 1.0 (Fig. 5.10); this high value suggests that a significant fraction of the nest's moment release comes from small-magnitude events (see Section 4.4.1). They also found no evidence for aftershock behavior among nest events; the sequence of event occurrence times was indistinguishable from a Poisson process.

The origin of the Bucaramanga nest has been the subject of much speculation. Schneider *et al.* (1987) favor an explanation related to the generation and migration of fluids, possibly related to incipient subduction-zone volcanism, or to dehydration reactions accompanying phase changes and, perhaps, the subduction of a buoyant feature such as an oceanic ridge or island arc.[2] While several investigators have looked carefully at the geometry of regional earthquake locations (Pennington, 1981; Malavé and Suárez, 1995), there is little agreement as to whether the nest activity corresponds to subduction of the Nazca or the Caribbean plate. Van der Hilst and Mann (1994) suggest that the nest may simply represent the interaction between two colliding subducting lithospheric slabs, and Frohlich *et al.* (1995) note that the nest's annual moment release rate corresponds quite well to expected relative slip rate between the plates.

How common or how unusual is the Bucaramanga nest? Santo (1969a; 1969b; 1970) used the term "nest" to describe nine clusters in South America, the Hindu Kush, Vrancea, and the South Sandwich arc; he further stated that he had adopted the term "nest" from C. Tsuboi who had described similar clusters in Japan. Tryggvason and Lawson (1970) suggested that there might be nests beneath Burma and Italy.[3] Other investigators have also made similar comparisons between earthquakes in the Bucaramanga nest and those in the Hindu Kush (Chatelain *et al.*, 1980); Vrancea, Romania (Oncescu, 1984; Oncescu and Bonjer, 1997); the "nest of Socampa" in Peru (Sacks *et al.*, 1967); and the "Iliamna cluster" beneath Cook Inlet, Alaska (Pulpan and Frohlich, 1985; Ratchkovsky *et al.*, 1997a; 1997b). Zarifi and Havskov (2003) review the available literature on earthquake nests, and suggest that Tongan deep-focus earthquakes at approximately 22° S may form a nest.

Nevertheless, the Bucaramanga nest is different in significant respects from all others; all appear to have a larger volume, a lower activity rate, or less isolation from nearby activity. The Vrancea nest (Figs. 1.7 and 1.9) seems to have a much smaller *b* value, about 0.5 (Trifu and Radulian, 1991) in comparison to 1.5 or more for Bucaramanga (Fig. 5.10). However, it is difficult to make quantitative comparisons, since most of the nest investigations are from analysis of regional network data and involve earthquakes too small to appear in global catalogs.

5.1.4 Isolated and exceptional earthquakes

So, do intermediate- and deep-focus earthquakes occur *only* within Wadati–Benioff zones? And, do they only occur within actively subducting lithosphere? The answer to the first question is clearly "No," as some well-located deep earthquakes lie

---

[2] Activity in the Bucaramanga nest almost certainly explains the pronounced peak at about 160 km in a plot of global seismicity rates vs depth presented by Kirby *et al.* (1996).

[3] There is no evidence for nests in either Burma or Italy from teleseismically determined hypocenters.

well-separated from neighboring seismicity and not within anything resembling a planar group of hypocenters.[4] Kirby *et al.* (1991) have called these isolated events "rogue earthquakes";[5] in other papers they are called "isolated," "detached," or "outboard" earthquakes (Lundgren and Giardini, 1994; Chen and Brudzinski, 2001; Okal, 2001). I will return to the second question at the end of this section.

Several of these rogues are among the largest and most unusual deep-focus earthquakes known. The Spanish deep earthquake of 29 March 1954 ($M_{W(CMT-Hist)}$ = 7.9; $h_{CMT-hist}$ = 630 km; see Sections 1.3 and 10.21) is 500 km distant from all other events in the ISC catalog except for three small "aftershocks" which occurred in 1973, 1990 and 1993. Nearby there are no island arcs, trenches, active volcanoes, etc. to suggest that subduction zone is ongoing. Similarly, the Colombia earthquake of 31 July 1970 ($M_{W(CMT-Hist,}$ = 8.1; $h_{CMT-Hist}$ = 623 km) possesses no foreshock or aftershocks (Fig. 2.4; see also Section 10.18.4). Although it is situated about 250 km from two earthquakes with M_W of 7.4 and 7.9 that occurred in 1921 and 1922 (Okal and Bina, 1994; 2001), its nearest neighbor in the ISC catalog is about 500 km distant. Finally, the 9 June 1994 Bolivia earthquake ($M_{W(CMT)}$ = 8.2; h_{CMT} = 647 km; see Fig. 1.3, Sections 1.1 and 10.18.4) occurred in a small group of earthquakes that lies between and approximately 200 km distant from much more active deep-earthquake zones in Peru–Brazil and Argentina. Incidentally, Kirby *et al.* (1996) state that deep-focus earthquakes with moments exceeding 1.5×10^{19} N-m (M_W = 6.75) tend either to be isolated, to occur near the bottoms of Wadati–Benioff zones, or near the lateral edges. They offer no statistical verification of any of these assertions. However, the 1922, 1954, 1970, and 1994 earthquakes are four of the five largest deep-focus earthquakes known (Table 1.2).

We can classify most isolated deep-focus earthquakes into four categories, although these categories aren't completely distinct. First, some occur within gaps between or along extensions of active Wadati–Benioff zone segments. Examples would include the 1994 Bolivia earthquake mentioned above, the 12 May 1990 Sakhalin Island earthquake ($M_{W(CMT)}$ = 7.2; h_{CMT} = 613 km; see Section 10.12) and several New Zealand earthquakes that occur at about 610 km depth along the extension of the intermediate seismic zone but more than 200 km beneath other New Zealand seismicity (Adams, 1963; Adams and Ferris, 1976; and see Section 10.2.5 and Fig. 10.5).

Second, some appear to occur within subhorizontal zones that extend beyond the deepest Wadati–Benioff zone activity. Earthquakes of this type occur near the

---

[4] Many reports of poorly-recorded isolated intermediate- or deep-focus earthquakes are probably mislocations. For example, using data from only six stations, the ISC reported a depth of 717 km for a 19 June 1979 Iran earthquake (m_b = 4.4). Similarly, they used only five stations to determine a depth of 705 km for a Guerrero, Mexico, earthquake that occurred on 26 January 1976.

[5] This name is appealing, since according to *Webster's New World Dictionary*, a rogue is "an animal, as an elephant, that wanders apart from the herd and is fierce and wild."

Banda, Chile (Fig. 5.2B), Kuril, Izu–Bonin, and Tonga–Fiji (Fig. 5.2A) subduction zones (Okino *et al.*, 1989; Giardini, 1992; Glennon and Chen, 1993; Lundgren and Giardini, 1994; Okal and Kirby, 1998). Tomography studies (Fukao *et al.*, 1992) and analysis of phases such as T waves (Okal, 2001) suggests that these earthquakes may occur within "stagnant" lithosphere that is still connected to the Wadati–Benioff zone, but has changed direction as it met increased resistance to subduction near the 650 km discontinuity.

Third, some isolated earthquakes aren't clearly associable with any currently active subduction zone. These include the 1954 Spanish earthquake discussed above, and possibly the Vitiaz deep-focus earthquakes that occur between about 580 and 680 km depth and lie between the Tonga–Fiji and Vanuatu subduction zones (see Section 10.4 and Okal and Kirby, 1998). One interpretation is that these earthquakes occur within stagnant lithosphere generated along subduction zones that are no longer active; i.e., along the Africa–Eurasia plate boundary for the Spanish earthquake, and along the Vitiaz Trench for the Vitiaz group.

Finally, there are the true "outboard" isolated earthquakes which occur adjacent to, well separated from, and above the deepest regular activity in a Wadati–Benioff zone. At present the type example for this is the Tonga–Fiji region (Chen and Brudzinski, 2001; and Fig. 5.2A). Brudzinski and Chen (2000; 2003) and Chen and Brudzinski (2003) report that the active outboard region is highly anisotropic but that P and S velocities are not higher than in the surrounding mantle. Their interpretation is that the outboard region is a remnant of a previous episode of subduction along the Tonga or Vitiaz trenches, and consists of metastable olivine that is too cold to change to a denser spinel form and thus too buoyant to sink into the lower mantle.

So, do isolated earthquakes only occur within actively subducting lithosphere? More generally, does the existence of an earthquake with a depth of 100 km or more constitute *prima facie* evidence for ongoing subduction? The last two paragraphs suggest that the answer is "No." Rather, temperature, not subduction, is the fundamental parameter that places fundamental constraints on where seismicity may occur. For example, Chen and Molnar (1983) conclude that seismicity is possible at temperatures of 250°–450° and 600°–800° in crustal and mantle material, respectively. They undertook a thorough investigation of the depths of intraplate and intracontinental seismic activity and found that deep crustal seismicity does occur in zones of crustal convergence, in mountain belts and old cratons. There are reports of intermediate-depth (90–150 km) quakes well separated from ongoing subduction beneath northern Africa (Hatzfeld and Frogneux, 1981; Ramdani, 1998) and the Tibetan plateau (Chen and Yang, 2004).[6]

---

[6] However, in both regions there are controversies about whether the reported focal depths are accurate; see Sections 10.21.1 and 10.26.2.

5.1.5 Statistics: another approach to nests and isolated earthquakes

Up till now this chapter has described the clustering visible in cross sections oriented perpendicular to and parallel to the trench axis, natural viewpoints if you think in terms of a subduction model. However, if you have the heart and soul of a statistician you might argue that this is misleading and incomplete; instead you might ask: what properties of deep earthquake clustering are identifiable from a purely statistical description? For example, are deep earthquakes more isolated than shallow earthquakes? Is the absence of discussion in Section 5.1.4 about isolated intermediate-depth earthquakes simply an oversight? Or, are isolated deep-focus earthquakes more common than isolated events with intermediate depths? Is the Bucaramanga nest unique or are there numerous other clusters with similar statistical properties?

Single-link cluster analysis (Frohlich and Davis, 1990) provides an objective statistical procedure for identifying clusters and isolated events in an earthquake catalog. If there are N earthquakes in a catalog, the procedure involves three basic steps.

- First, one constructs a particular set of $N-1$ 'links' joining the N earthquakes into a single large cluster. Constructing a link simply involves identifying two specific earthquakes; however, it is convenient to think of the link as a straight line connecting the two events. For single-link cluster analysis, one chooses the links so that the sum of lengths of the $N-1$ links is a minimum.[7]
- Second, one orders the $N-1$ links from longest to shortest.
- Third, to identify natural clusters one removes links sequentially, always removing the largest remaining link. For example, removing the longest link divides the N earthquakes into two clusters; removing the K longest links divides them into $K+1$ clusters.

Thus, an individual earthquake is highly isolated if it separates into its own "cluster of one" when only a few large links have been removed; a group of earthquakes is highly clustered if they remain linked together as a cluster even after the removal of most of the links (see Fig. 5.11).

To address the questions raised at the beginning of this section I applied single-link cluster analysis to three sets of earthquakes.

- Large earthquakes – 1818 events occurring during 1897–1999 with magnitudes of 7.0 or greater and reported in the Abe, EV Centennial, or Harvard CMT catalogs.
- Harvard CMT catalog selection – 11344 earthquakes occurring in 1977–2003 and with $M_W \geq 5.5$;
- EHB catalog selection – 39062 earthquakes with "better-determined" locations occurring between 1964 and June 2002 and with $m_b \geq 5.0$.

---

[7] To construct the $N-1$ links, one first constructs links between each of the N earthquakes and its nearest neighbor; then, one adds links between each linked group and the nearest-neighboring linked group; this procedure is repeated until $N-1$ links have been identified and the earthquakes form a single linked group.

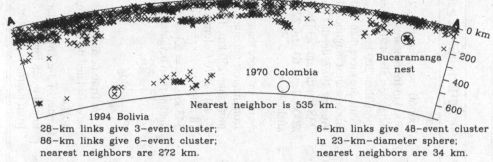

Fig. 5.11 Single-link cluster analysis, and four examples of clustering and isolation among earthquakes in South America. The plot shows north-south cross section 20°S-10°S; the hypocenters evaluated are better-determined events in the EHB catalog with $m_b \geq 5.0$. The 1970 Colombia earthquake is highly isolated as it is 535 km from its nearest neighbor. The 1994 Bolivia earthquake is part of a cluster of six events, each within 86 km of one another, separated by 272 km from the nearest-neighboring event. The Bucaramanga nest is a dense cluster of 48 earthquakes contained in a 23-km diameter sphere, each within 6 km of another nest event, and separated from neighboring earthquakes by 23 km. Finally, at shallow and intermediate depths 1480 earthquakes form an extended "cluster" of events where all lie within 50 km of another event in the cluster.

The first two of these sets should be approximately uniform and complete for the specified time periods and magnitude ranges; however, the locations are not especially accurate and they lack small-magnitude earthquakes that contribute to some of our knowledge about clustering and isolation. The third set is neither uniform nor complete but includes many smaller earthquakes that allegedly have quite accurate locations (Engdahl *et al.*, 1998).

Applying single-link-cluster analysis to these catalogs (Table 5.2) demonstrates unequivocally that deep-focus isolated earthquakes are much less common than isolated shallow-focus events; but highly isolated intermediate-depth earthquakes are rarer still. For example, in the EHB catalog about 60 shallow earthquakes lie 500 km or more from their nearest neighbor, while there is only one comparably isolated deep event, the 31 July 1970 Colombia earthquake. In the catalog of large earthquakes ($M \geq 7$), of those occurring 250 km or more distant from their nearest neighbor, 91 are shallow, 14 are intermediate, and 12 are deep-focus. However, all of the 14 intermediate-depth events occur prior to 1950 and have reported depths of only 60–100 km. Moreover many occur in regions where deep activity seems unlikely, such as the Indian Ocean, near the Galapagos Islands, along the South Atlantic ridge, or in eastern Canada. This suggests that their depths may be inaccurate.

Moreover, large-magnitude deep-focus earthquakes are well-represented among highly isolated events. The 1954 Spanish deep-focus earthquake is the most isolated

Table 5.2. *Properties of isolated earthquakes as evaluated by applying single-link cluster analysis to three catalogs.*

Event	Catalog		
	1818 large earthquakes ($M \geq 7$)	11344 Harvard CMT ($M_w \geq 5.5$)	39062 EHB 'better' ($m_b \geq 5.0$)
Shallow-intermediate-deep comparisons			
Very isolated, shallow ($h < 60$ km)	22 events are more than 750 km from nearest neighbor; 91 events more than 250 km.	138 events are more than 200 km from nearest neighbor.	About 60 events are more than 500 km from nearest neighbor; about 260 events more than 200 km.
Very isolated, intermediate (60 km $\leq h < 300$ km)	14 events are more than 250 km from nearest neighbor; however all will have reported depths of 60–100 km and occur prior to 1950, many in regions where no modern deep quakes are known.	3 events are more than 200 km from nearest neighbor; most isolated is 23 December 2000 with depth of 69 km, 308 km from nearest neighbor.	2 events are more than 200 km from nearest neighbor (10 August 1987, depth 176 km, 238 km distant; 23 December 2000, depth 63 km, 227 km distant).
Very isolated, deep-focus ($h \geq 300$ km)	12 events are more than 250 km from nearest neighbor; including 1921 South America, 1954 Spanish, and 1970 Colombia earthquakes.	8 events are more than 200 km from nearest neighbor; most isolated is 13 May 1979, 259 km from nearest neighbor.	8 events are more than 200 km from nearest neighbor, only 2 more than 250 km. These are 31 July 1970 Colombia and 18 August 1968, with depth 543 km, $m_b = 6.1$, 415 km distant from neighbors.
Celebrated isolated events			
29 March 1954 Spanish	758 km from nearest neighbor; most isolated deep earthquake	Not in catalog	Not in catalog
31 July 1970 Colombia	257 km from nearest neighbor	Not in catalog	535 km from nearest neighbor; most isolated deep-focus quake
9 June 1994 Bolivia	188 km from nearest neighbor	130 km from nearest neighbor.	Member of 3-event cluster linked by 28 km links; 86 km from nearest neighbor.

earthquake in the large-earthquake catalog. In the EHB catalog subset, 10 deep earthquakes are more than 200 km from their nearest neighbors; of these the two largest are also the most isolated, the 31 July 1970 Colombia earthquake at 535 km from its nearest neighbor, and the earthquake of 18 August 1968 ($m_b = 6.1$) at 434 km from its nearest neighbor. The 1994 Bolivia earthquake was one of several events occurring within a region where activity was rare and well-separated from adjacent earthquakes in all directions (see Fig. 5.11).

Among clusters having numerous earthquakes, the single-link analysis confirms that the Bucaramanga nest is highly unusual (Table 5.3); in the EHB catalog no other deep cluster is so localized in space but also well-separated from neighboring events. The most similar group would be a 33-event cluster in Japan at about 65 km depth; both clusters can be contained within 25-km diameter spheres, but the Japanese cluster has fewer earthquakes and is less isolated from neighbors. In contrast, the so-called Vrancea nest doesn't seem particularly unusual; in all three catalogs its earthquakes aren't particularly numerous, it isn't unusually compact, and it isn't remarkably isolated from adjacent seismicity.

Most concentrations of deep seismicity are "distributed," i.e., they occupy a relatively large region that isn't very distant from neighboring seismically active regions. Thus, they don't naturally form distinct clusters since even small increases in link length allow more events to join the group. In all three catalogs the single-link analysis finds intermediate-depth concentrations in the Hindu Kush and the New Hebrides, and deep-focus concentrations in Tonga. Elsewhere it appears that "size matters"; e.g., there are no Bucaramanga earthquakes in the large-event historical catalog and the Bucaramanga nest is not particularly unusual in the Harvard CMT catalog. Similarly, in the Phillippines, the Banda Sea, and the Scotia arc there are intermediate-depth concentrations which appear in the Harvard CMT and EHB catalogs, but which are absent among the large historical earthquakes.

Except in the Tonga–Kermadec region, virtually all intense deep concentrations occur at intermediate depths. However, the single-link cluster analysis indicates that concentrations with numerous events are even more common among shallow-focus earthquakes than among either intermediate- or deep-focus events.

A limitation of single-link clustering is that it concentrates on near-neighbor relationships; instead, one might ask; what is the distribution of distances between all the pairs of earthquakes within a catalog? If each focus is a center and if foci are distributed uniformly and randomly in space, the total number within a distance R of any focus will be proportional to R^δ; the fractal dimension δ is 3 if the quakes occupy a volume in space; δ is 2 if quakes occur on a planar surface such as a Wadati–Benioff zone; and δ is 1 if there are events only along linear features such as the surface traces of faults. Kagan and Knopoff (1980b) and Kagan (1991a) evaluated inter-event distances in the NOAA global catalog and found that between

Table 5.3. *Properties of representative deep earthquake clusters. Entries determined by applying single-link cluster analysis to three catalogs. Clusters are intermediate-depth if mean depth is 60–300 km; deep-focus if mean depth exceeds 300 km.*

Cluster	Catalog		
	1818 large earthquakes ($M \geq 7$)	11344 Harvard CMT ($M_w \geq 5.5$)	39062 EHB "better" ($m_b \geq 5.0$);
	Localized or oft-cited clusters		
Bucaramanga nest; near 6.8° N, 73° W, 160 km depth	Not in catalog: largest Bucaramanga earthquake has $M \sim 5.9$.	A 4-event cluster linked by 22-km links; 147 km from nearest neighbor	A 48-event cluster linked by 6-km links; cluster is 34 km from nearest neighbors and contained within 23-km-diameter sphere.
Vrancea, Romania; near 46° N, 26.5° E	1940, 1977, and 1986 earthquakes form 3-event cluster linked by 42-km links, 499 km from nearest neighbor.	A 4-event cluster linked by 65 km links; 255 km from nearest neighbor.	A 19-event cluster linked by 25-km links; 64 km from nearest neighbor.
Japan; near 41.5° N, 142° E, 65 km depth	Not in catalog	Not in catalog	A 33-event cluster linked by 6-km links; cluster is 10 km from nearest neighbors and contained within 21-km sphere.

'Distributed' concentrations of seismicity

	Clusters of 5-or-more events linked by 50-km links	Clusters of 15-or-more events linked by 30-km links	Clusters of 40-or-more events linked by 20-km links
Banda Sea	None	1 intermediate-depth 16-event cluster.	2 intermediate-depth clusters with 317 events and 109 events.
Hindu Kush	1 intermediate-depth cluster with 19 events.	1 intermediate-depth 43-event cluster.	2 intermediate-depth clusters with 69 events and 102 events.
New Hebrides	2 intermediate-depth clusters with 8 events and 6 events.	2 intermediate-depth clusters with 45 events and 15 events.	4 intermediate-depth clusters with 254 events, 76 events, 52 events, and 43 events.
New Guinea	None	None	2 intermediate-depth clusters with 175 events and 52 events.
Philippines	None	1 intermediate-depth 21-event cluster	6 intermediate-depth clusters with 209 events, 142 events, 83 events, 79 events, 70 events, and 40 events.
South America	1 deep-focus cluster with 5 events.	None	4 intermediate-depth clusters with 104 events, 60 events, 48 events, and 47 events.
Tonga–Kermadec	2 deep-focus clusters, each with 5 events.	5 deep-focus clusters with 98 events, 92 events, 52 events, 26 events, and 15 events.	1 intermediate-depth cluster with 52 events. 5 deep-focus clusters with 443 events, 351 events, 125 events, 52 events, and 46 events.
Other regions	None	Scotia Arc: 1 intermediate-depth 21-event cluster.	Java: 1 intermediate-depth 48-event cluster. Mariana: 1 intermediate-depth 58-event cluster. Scotia Arc: 1 intermediate-depth 144-event cluster.

distances of about 20 km to 1000 km, δ for quakes with focal depths down to 140 km was close to 2. In contrast, δ was about 1.5 for earthquakes with depths exceeding 140 km. They suggested that shallower seismicity was plane-like ($\delta = 2$) because shallower earthquakes occurred on planar faults; the observation that δ is less for earthquakes deeper than 140 km indicates that they are more strongly concentrated in clusters.[8]

5.2 Temporal clustering

5.2.1 What is an aftershock?

What is an aftershock? There is no universally accepted definition; yet all agree that aftershocks are earthquakes that are close in space and time to some designated large earthquake. In practice, there are two approaches to a definition, one grounded in physics and the other in statistics. The first approach is to define (usually tacitly) an aftershock as any quake that follows and is reasonably close geographically to some large quake, no matter how much time has elapsed since it occurred. Thus contemporary earthquakes occurring in southern Alaska occasionally are still called "aftershocks" of the great 1964 Good Friday earthquake, even though 40 years have elapsed. The justification is that for any such earthquake, with a rupture dimension of hundreds of km and a repeat time of a century or more, the physical processes responsible are likely to affect seismic activity over great distances and a long time interval.

The second, statistical approach is to require that a following quake meet some specified test of spatial and temporal closeness to the mainshock. A problem is no matter what test you choose to apply, you will find some spurious aftershocks. Even if a Poisson process generates the quakes' locations, some fraction are "aftershocks" simply because, by chance, they are close in space and in time to the mainshock. Furthermore, many different statistical tests have been proposed. Some include space–time criteria that depend on the mainshock size; others do not and simply isolate various kinds of non-random behavior.[9]

5.2.2 Deep earthquakes don't have aftershocks

So, do deep earthquakes have aftershocks? Historically there was confusion about this. Leith and Sharpe (1936) stated that "aftershocks are rarely observed" for deep

---

[8] Kagan (1991a) also notes that δ depends on the proportion of deformation that is aseismic (a smaller proportion for deep seismicity) as well as the time span of the catalog analyzed.

[9] A third, more sophisticated approach is to fit the times and locations of earthquakes within a stochastic process model (e.g., Kagan and Knopoff, 1976; Zhuang *et al.*, 2004). This obtains the optimum parameters for the model as well as the probabilities that events are independent or are aftershocks.

earthquakes, while Gutenberg and Richter (1938) stated that "their frequency is not significantly different" than for shallow shocks.[10] Subsequently, Shlien and Toksoz (1970) and Utsu (1972) could not distinguish occurrence times of deep earthquakes from a Poisson process. But Isacks *et al.* (1967) reported measurable temporal clustering of deep earthquakes, and both Oike (1971a; 1971b) and Wyss and Shamey (1975) report examples of deep earthquake "multiplets."

The explanation for this confusion is that, unlike shallow earthquakes, intermediate- and deep-focus quakes only rarely have following sequences with numerous, small aftershocks. Page (1968) and Pavlis and Hamburger (1991) require aftershocks to be numerous, to be smaller in magnitude than the main event, and to die off according to Omori's law (see eq. 5.1 in the next section). In fact, intermediate- or deep-focus quakes with more than 20 reported aftershocks are extremely unusual (Wiens and Gilbert, 1996; Wiens *et al.*, 1997); only a handful of deep earthquakes have 15 or more aftershocks locatable using teleseismic data alone.[11]

Furthermore, some extraordinarily large deep earthquakes have no aftershocks at all. These include the 1970 Colombia earthquake (Fig. 2.4, Table 5.2, and Section 10.18.5) that had no following quakes in the ISC catalog in 32 years following its occurrence; and the 1954 Spanish deep earthquake (see Sections 1.3, 10.21, and Table 5.2) with only three small following quakes occurring in 1973, 1990, and 1993. Both the Spanish and Colombia quakes had identifiable subevents during the rupture process, which had a duration of a minute or so; however, neither had any reported "true" aftershocks, occurring more than 60 seconds after the beginning of the rupture (e.g., see Willemann and Frohlich, 1987). I am unaware of any well-recorded magnitude 7.5 shallow earthquakes with no measurable aftershocks. Thus, the statement "deep earthquakes don't have aftershocks" is partially correct, at least in the limited sense of Page (1968); i.e., only rarely are their aftershocks numerous, and when aftershock sequences do exist they are never or almost never like those of shallow earthquakes.

5.2.3 *Deep earthquakes do have aftershocks*

There have been a few deep earthquakes with well-developed aftershock sequences that would satisfy the requirements of Page (1968). In the Hindu Kush, three

---

[10] The 26 May 1932 Tonga earthquake (see Section 10.3.3) clearly influenced Gutenberg and Richter. From inspection of teleseismic data they knew it had at least seven aftershocks, including several that were locatable by the crude global network of that time.

[11] If we apply the ratios method (Fig. 5.13) to the full ISC catalog and define aftershocks as events occurring within the interval T_A such that a Poisson process has one or more events with probability 1%, only four sequences have 15 or more events. These are the intermediate-depth earthquakes of 19 April 1975, 17 February 1999, and 28 March 2000, and the Flores Sea deep-focus earthquake of 17 June 1996.

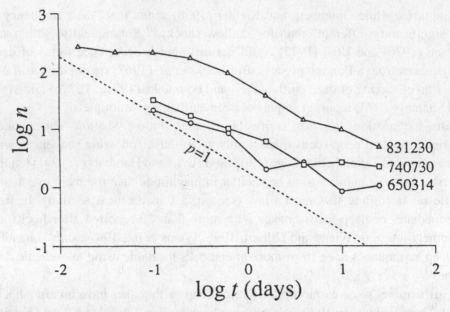

Fig. 5.12 Aftershock decay rates following intermediate-depth Hindu Kush earthquakes occurring in 1965, 1974, and 1983. The vertical axis is number n of aftershocks per unit time, the horizontal axis is elapsed time t since the occurrence of the mainshock. Individual curves are labeled mainshock occurrence dates. The dashed line labeled "$p = 1$" corresponds to a $1/t$ decay process. Figure reproduced from Pavlis and Hamburger (1991).

such earthquakes occurred at nearly the same location in 1965, 1974, and 1983 (Fig. 5.12); all had focal depths of about 210 km and magnitudes m_b of 6.2–6.4 (Lukk, 1968; Pavlis and Hamburger, 1991). Kisslinger and Hasegawa (1991) described well-developed aftershock sequences for the 21 March 1987 earthquake in the central Aleutians ($M_{W(CMT)} = 6.4$; $h_{EHB} = 99$ km) and one 9 January 1987 in northern Honshu ($M_{W(CMT)} = 6.6$; $h_{EHB} = 75$ km). Both exhibited distinct rate changes beginning about 60–120 days following the mainshock, which the investigators thought might be attributable to possible loading due to nearby creep or similar phenomena. Fuchs *et al.* (1979) reported that there were more than 100 aftershocks following the 4 March 1977 earthquake in Vrancea, Romania ($M_{W(CMT)} = 7.5$; $h_{CMT} = 84$km), including more than 50 for which focal depths could be determined. An unusual feature of this sequence is that the aftershock activity spread to the crust beginning about 20 min after the mainshock, with continuing aftershocks at both shallow and intermediate depths, separated by a gap between 40 km and 70 km.

Beneath 400 km depth, earthquakes having numerous aftershocks seem to be even rarer; Wiens and Gilbert (1996) suggested that aftershocks were more common in colder subducting slabs, but reported only five quakes possessing sequences having more than 10 aftershocks. In all but one of these sequences (see Tinker *et al.*, 1998) there was a local network nearby to document the aftershocks; thus one wonders whether many large teleseismically-located earthquakes might possess such sequences which go unrecorded. In Tonga, Wiens *et al.* (1997) did indeed find aftershocks following all nine large ($M_W \geq 6.0$) intermediate- and deep-focus quakes that took place while their local network was in operation. They concluded that the Tongan sequences generally possessed about an order of magnitude fewer aftershocks than did comparably-sized California shallow earthquakes; also, the Tongan quakes exhibited a greater difference between mainshock and aftershock magnitudes.

Several sequences of aftershocks of deep-focus earthquakes are especially notable. After the 9 March 1994 Tongan earthquake ($M_W = 7.6$; $h_{CMT} = 568 \, \text{km}$) a local network recorded 144 aftershocks extending across the entire width of the Wadati–Benioff zone and extending into the surrounding aseismic region (Wiens *et al.*, 1994; Wiens and Gilbert, 1996; McGuire *et al.*, 1997; Wiens and McGuire, 2000). Similarly, Myers *et al.* (1995) observed 89 aftershocks of the 9 June 1994 Bolivian quake ($M_W = 8.2$; $h_{CMT} = 647 \, \text{km}$); here again local network data was available and the locations clearly delineated a larger active region than the supposed Wadati–Benioff zone (Silver *et al.*, 1995). And for the 23 August 1995 Mariana deep earthquake ($h_{CMT} = 599 \, \text{km}$; $M_{W(CMT)} = 7.1$), Wu and Chen (1999) report 15 aftershocks, all located using teleseismic observations.

Because of the rarity of well-developed aftershock sequences for deep earthquakes, the decay of activity has been investigated in only a few cases. The most common empirical model for the activity rate $r(t)$ of aftershock sequences is Omori's (1895) law, a simple power-law decay depending on t, the elapsed time after the initial shock. In its modern form, Omori's law states:

$$r(t) = \frac{K}{(c+t)^p},$$ (5.1)

where c is an offset parameter to account for behavior immediately following the main event and the exponent is called the "p value." Nyffenegger and Frohlich (2000) evaluated p values for nearly all the sequences discussed in this section. For three Hindu Kush sequences[12] they obtained relatively low values of 0.53, 0.51, and 0.83; elsewhere the p values were near 1.0, i.e., similar to values found typically for shallow events. In particular, they found p of 1.04 for the 1994 Tonga sequence, 1.19 for the 1994 Bolivia sequence, 0.90 for the 1987 Aleutian sequence, and 0.92 for the 1987 Honshu sequence.

---

[12] Only the 1983 sequence is the same as in Fig. 5.12; some Hindu Kush sequences are more swarm-like than aftershock-like, and thus different aftershock-identification methods find different sequences.

5.2.4 Statistics: another approach to aftershocks

An alternative to describing particular unusual sequences is to apply statistical tests to identify aftershocks within earthquake catalogs. This section will review several investigations of catalogs, and also explain some statistical methods that are especially useful for evaluating the properties of deep earthquakes. All these investigations compare locations and origin times in the catalog to a Poisson process, and then evaluate properties of sequences of related earthquakes.

To determine how the properties of sequences depend on depth or geographic location, it is desirable to evaluate catalogs that are uniform and complete, or at least nearly so. Thus, Prozorov and Dziewonski (1982) considered events with $m_b \geq 5.0$ in the ISC catalog, and fit sequences to a model where the number of aftershocks was magnitude-dependent. They concluded that aftershocks occurred at depths down to 250 km and from 450–650 km, but noted that "not a single aftershock has been detected . . . in a depth range from 250 to 450 km." Similarly, Kagan and Knopoff (1980a) and Kagan (1991b) evaluated PDE earthquakes with $m_b \geq 5.3$ and Harvard earthquakes with $M_W \geq 6.0$. Kagan found that "the number of dependent events [aftershocks] first decreases steadily as the depth increases; for a depth range 141–280 km the number of aftershocks reaches the minimum. However, for deep events the number of dependent events increases again."

Although catalog completeness and uniformity is desirable, ignoring small-magnitude earthquakes is problematic for deep earthquakes since their aftershocks aren't numerous and easily missed. Thus, among "principal events" large enough to form a nearly complete population, Frohlich (1987) searched for aftershocks within the entire ISC catalog, identifying them with statistical tests designed to be accurate as long as the catalog is uniform locally. In particular, he considered all ISC earthquakes occurring within two years and 100 km off each principal event. Of these, following events comprised an aftershock sequence if they occurred within an interval where, if the activity was a Poisson process, the times of the prior events suggested that no earthquake should occur 99% of the time (see Fig. 5.13). Within this sequence, he identified the largest-magnitude earthquake as the mainshock, prior and following events were fore- and aftershocks, respectively. An updated version of Frohlich's analysis (Fig. 5.14) indicates that aftershocks become steadily less common between the surface and about 150 km. Between 150 km and about 440 km about 10% or less of the principal events have aftershocks, and most had only one. Aftershocks are especially rare between 245–330 km; of 126 principal events in this depth range, only two have aftershocks.[13]

---

[13] With Frohlich's definition one finds aftershocks for 1% of events generated by a Poisson process; thus, finding two aftershocks among 126 principal events does not imply there is any physical process generating aftershock activity.

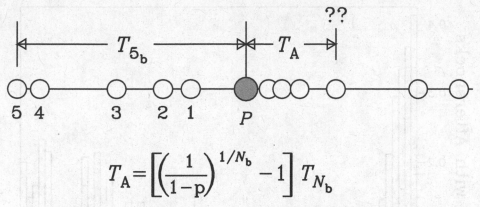

$$T_A = \left[\left(\frac{1}{1-p} \right)^{1/N_b} - 1 \right] T_{N_b}$$

Fig. 5.13 Identifying aftershocks with the ratios method (Frohlich and Davis, 1985). If T_{N_b} is the relative time of the N_bth event prior to a principal event P and p is a probability level, the ratios' method determines a critical time interval T_A that defines which following events are aftershocks with confidence $(1-p)$. In particular, T_A is the interval such that the chance of any event occurring within time T_A is less than p for a Poisson process where N_b events occur in time T_{N_b}. In the figure there are $N_b = 5$ preceding events and thus if $p = 0.01$, the T_A is $0.002T_{5_b}$. The ratios' method is especially sensitive for events which seldom have numerous aftershocks and when catalogs are incomplete but locally uniform, qualities that make it useful for identifying small-magnitude aftershocks of deep earthquakes.

Persh and Houston (2004) have obtained comparable results from an analysis of aftershocks of 290 earthquakes with $M_W \geq 6.3$ occurring in 1977–2002. Because a few unusual large events might dominate aftershock counts, they evaluated after-shock productivity after normalizing for mainshock moment. This analysis also showed a notable paucity of aftershocks between depths of 300–500 km, and then a distinct increase at depths exceeding 550 km. They suggested that this might indicate changes in the rupture mechanism occurring at both 300 and 550 km depth.

However, what is most remarkable is the steady change in character of after-shock activity beneath about 45–50 km depth (Figs. 4.11 and 5.14); beneath this depth aftershocks become significantly less common and those that occur are most often doublets rather than sequences of numerous events. Presumably studies such as Shlien and Toksoz (1970) didn't find temporal clustering for deep seismicity because their statistical methods for determining non-Poisson temporal behavior weren't sufficiently sensitive.

Although deep earthquakes seldom have more than one aftershock, these appear to satisfy an Omori law. In particular, if aftershock sequences decay accord-ing to a power law with exponent p (eq. 5.1), then the sum of activity rates for several sequences also is a power-law with exponent p. Thus Davis and Frohlich (1991) suggested "stacking" times for the occasional aftershocks of deep

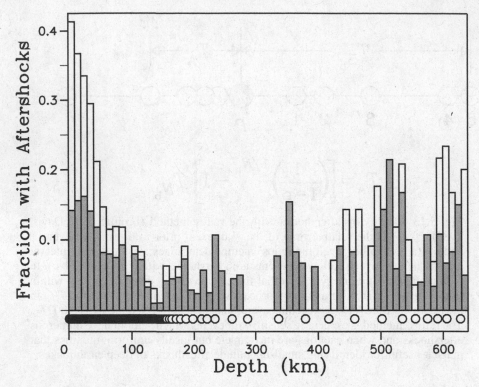

Fig. 5.14 Depth dependence of aftershock occurrence. Vertical axis is the fraction of earthquake sequences having exactly one aftershock (dark bars) or more than one aftershock (white bars). Circles beneath the zero line show the focal depth of every 50th sequence. For these data, mainshocks are earthquakes in the EHB catalog 1964–2002 that have $m_b \geq 5.2$, well-determined focal depths, and which are not aftershocks of previous events. Aftershocks are found by applying a ratios test (Frohlich and Davis, 1985) to relative times of all ISC catalog events occurring within two years and 100 km of each mainshock; these are aftershocks if they occur within a period following the mainshock where the chance of any earthquake occurring is 1 per cent or less. Note that aftershocks are relatively uncommon between depths of ∼100–450 km.

earthquakes (Fig. 5.15); in this way they obtained a p value (0.54 for deep earthquakes in the ISC catalog) even though the individual sequences generally possessed only a single aftershock. This p value is considerably smaller than those reported for most individual sequences in the previous section, and smaller also than the p values of 1.1 and 2.0 that Kagan (1991b) obtained for intermediate- and deep-focus earthquakes in the PDE catalog. At present it isn't clear whether this lower value has physical significance or whether it is an artifact of the stacking process.

When deep aftershocks do occur, what is their spatial relationship with the focal mechanism of the mainshock? Do they tend to occur along a planar rupture surface?

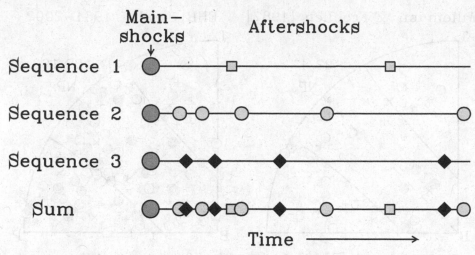

Fig. 5.15 Diagram illustrating how aftershock sequences may be superimposed to study their decay laws (Davis and Frohlich, 1991). If all aftershock sequences in a region follow the same decay law, then the resulting sum of sequences, with the mainshocks shifted to time zero, will have the same decay law. For example, if several sequences follow an exponential decay such that their rates are $c_i\exp(-at)$, the resulting sum of sequences will also follow an exponential decay $A\exp(-at)$ where $A = \Sigma c_i$. This also holds true for power law decays such as Omori's law and the modified Omori's law (eq. 5.1).

Willemann and Frohlich (1987) reviewed the relative locations for reported after-shocks of 59 deep earthquakes, and found no evidence for clustering along nodal planes (Fig. 5.16). They also showed that previous reports of such clustering (e.g., Oike, 1971a) were erroneous, coming about because investigators were apparently unaware that 50% of the focal sphere lies within 16° of one or another of the two nodal planes (see Frohlich and Willemann, 1987). They thus suggested that deep aftershocks did not represent afterslip along a planar fault, but might occur in response to a general redistribution of stress due to the occurrence of the mainshock or ongoing nearby phase transitions.

Nevertheless, a few investigators have reported clustering of aftershocks along nodal planes. Harris (1982) reported this for seven aftershocks of an intermediate-focus earthquake that occurred in New Zealand on 5 January 1973 ($h_{CMT-Hist} = 162\,km$; $M_{W(CMT-Hist)} = 6.6$). For the 9 March 1994 Tongan deep earthquake ($h_{CMT} = 564\,km$; $M_{W(CMT)} = 7.6$), McGuire *et al.* (1997) and Wiens and McGuire (2000) observed that most of the approximately 80 aftershocks locatable by the regional network concentrated along a steeply dipping plane approximately consistent with one of the mainshock's nodal planes. And, for the 23 August 1995 Mariana deep earthquake ($h_{CMT} = 599\,km$; $M_{W(CMT)} = 7.1$), Wu and Chen (1999) reported that the 15 teleseismically located aftershocks appear to lie along a planar

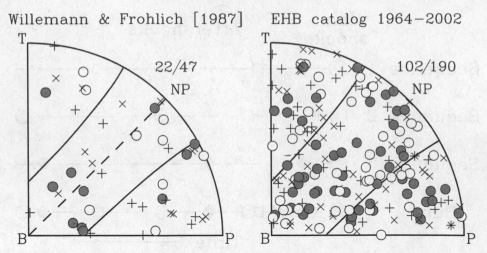

Fig. 5.16 Relation between aftershock directions and mainshock focal mechanisms for deep earthquakes ($h \geq 100$ km). (Left) Data from 47 mainshock-aftershock pairs relocated by Willemann and Frohlich (1987), here replotted after incorporating new focal mechanisms from the CMT and CMT historical catalogs where available. (Right) 190 pairs identified using the ratios method (Fig. 5.13) among well-determined hypocenters in the EHB catalog 1964–2002. +s are pairs separated by 10 km or less; Xs by 10–20 km; open circles by 20–40 km; and filled circles 40 km or more. Each point indicates the orientation of a vector connecting hypocenters of a mainshock-aftershock pair with respect to the mainshock's T, B, and P axis. Half of the focal sphere lies within the curved solid lines surrounding the nodal plane NP (dashed line), i.e., within $15.8°$ of NP. Note that these data do not support the hypothesis that deep aftershocks lie along nodal planes; only about half of the plotted points lie within the curved solid lines (22 of 47 in plot at left; 102 of 190 in plot at right).

surface.[14] In contrast, for the 9 June 1994 Bolivian earthquake ($h_{\text{CMT}} = 647$ km; $M_{\text{W(CMT)}} = 8.2$) only a small subset of the aftershocks, many of which were small ($m_b \sim 2$), coincided with the apparent fault plane. Thus, Myers *et al.* (1995) suggested that no along-plane clustering would have been discernible if local network observations had not been available.

There is thus an apparent contradiction. In those rare cases when deep earthquakes possess many locatable aftershocks, they often occur preferentially along planar surfaces. But no such clustering is verifiable for "typical" deep earthquakes that seldom have more than one or two aftershocks (Fig. 5.16). A reanalysis of the Wiens and McGuire data for the 9 March 1994 Tonga earthquake (Fig. 5.17) indicates that aftershocks do not occur preferentially close to nodal planes of the

---

[14] My own reanalysis of the 1995 Mariana aftershocks did not find them clustered along nodal planes of the focal mechanism reported by Harvard, either for the aftershock locations reported by Wu and Chen (1999) or for the locations of the same events in the EHB catalog.

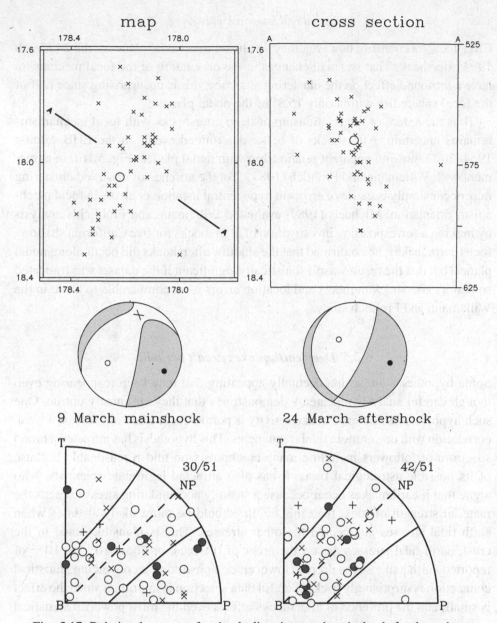

map   cross section

9 March mainshock   24 March aftershock

30/51 NP   42/51 NP

Fig. 5.17 Relation between aftershock directions and mainshock focal mechanism for the Tonga earthquake of 9 March 1994 ($M_w = 7.6$; $h_{CMT} = 568$ km). Map (upper left) and cross section (upper right) show locations of 51 fore- and aftershocks reported by Wiens and McGuire (2000). Their directions relative to the mainshock show only a slight preference to cluster near the nodal planes of the Harvard CMT mechanism (center left); only 30 of the 51 events lie within 15.8° of the nodal planes (lower left). However, a significant fraction (42 of the 51 events) lie within 15.8° of the nodal plane of the 24 March 1994 aftershock (center right and lower right). One interpretation of these data is that the data in Fig. 5.16 might not lie preferentially along nodal planes only because there are small errors or minor variations in the orientations of mainshock mechanisms with respect to a plane along which seismic activity occurs.

CMT focal mechanism; however, they do cluster near nodal planes of the 24 March 1994 aftershock. That is, small changes in the orientation of the focal mechanism have a profound effect on the clustering statistics; this is unsurprising since half of the focal sphere lies within only 15.8° of the nodal planes.

Thus the nature of the relationship of deep aftershocks with focal mechanisms remains uncertain. Aftershocks of better-determined events in the EHB catalog 1964–2002 show no apparent relationship with nodal planes (Fig. 5.16), in agreement with Willemann and Frohlich (1987). Yet the absence of observed clustering may occur simply because of errors in hypocentral location or errors in focal mechanism orientation. Michaels (1989) evaluated Willemann and Frohlich's analysis by making a corresponding investigation of aftershocks for five California shallow-focus earthquakes; he confirmed that the shallow aftershocks did occur along nodal planes, but that the result wasn't statistically significant if the dataset was truncated so that its aftershock numbers and location errors were comparable to those in the Willemann and Frohlich study.

5.2.5 *Deep earthquakes aren't periodic*

Some hypotheses are so intellectually appealing that they keep reappearing even though careful analysis repeatedly demonstrates that they are mostly untrue. One such hypothesis is that earthquake activity is periodic, e.g. showing at least a weak correlation with one or more tidal frequencies. This hypothesis has attracted a broad spectrum of followers including many crackpots who find it reasonable because of its possible astrological basis. It has also attracted legitimate scientists, who argue that if earthquakes occur because a steadily accumulating stress exceeds the material strength of rock, then this failure should be more likely at times when earth tidal stresses augmented the other stresses. This is plausible since in the crust, earth tidal stresses are on the order of 0.1 bars, or about 10^{-3} to 10^{-2} of reported earthquake stress drops. However, demonstrating a convincing statistical connection is surprisingly tricky. Careful data selection is important since the effect is small; and the presence of aftershocks affects even the most powerful statistical tests.

For shallow earthquakes, most careful studies generally conclude either that the correlation doesn't exist (e.g., Heaton, 1982; Vidale *et al.*, 1998) or is very weak (Shirley, 1988; Tanaka *et al.*, 2002). Beeler and Lockner (2003) performed laboratory experiments to evaluate the effect of periodic stresses on rock failure, and concluded that the absence of a correlation between phase of the stress and failure time indicated that the period of the varying stress was shorter than the characteristic nucleation time of the failure process.

But could deep earthquakes be different? Several early papers reported that deep earthquakes were periodic (e.g., Landsberg, 1935; Stetson, 1935; 1937). More recently, a surprise result from the Apollo missions was an unambiguous connection between tidal stresses and the occurrence of deep "moonquakes" situated about 900 km beneath the Moon's surface (Lammlein *et al.*, 1974; Minshull and Goulty, 1988). The relationship was so strong that its presence in the data is obvious without statistical proof.

However, subsequent research has failed to find any substantial connection between terrestrial deep earthquake activity and the tides or any other periodic phenomenon. A landmark paper by Jeffreys (1938) demonstrated methodological problems with most of the previous research concerning earthquake periodicity; when McMurray (1941) reanalyzed the earlier deep earthquake data he found no tidal relationship. Similarly, Heaton (1982) found none, even though considerably more data were available to him. The most thorough analysis was by Curchin and Pennington (1987). Their application of the Rayleigh test (see Fig. 5.18) found semidiurnal periodicity at the 90% confidence level for three geographic regions; however, this was insignificant since they applied the test to 30 geographic regions. Tanaka *et al.* (2002) looked for correlations between the phase of earth and ocean-loading tidal stresses among groups of earthquakes having Harvard CMT focal mechanisms. Although they concluded there was a small correlation for deep earthquakes the significance was marginal (only 2 of 16 groupings statistically significant).

A more reasonable assertion is that, except for the occasional aftershock, deep earthquake occurrence times are remarkably similar to a Poisson process (e.g., see Wyss and Toya, 2000). However, Kagan and Jackson (1991) expressed a slightly different view; after evaluating temporal clustering for both shallow and deep earthquakes in several catalogs they concluded:

We find no evidence of mainshock sequence periodicity for strong shallow, intermediate, or deep earthquakes; no evidence is found even for decreased seismic activity after a strong earthquake. Conversely, analysis of time-distance statistical moments points to clustering as a major long-term feature of the seismic process. After we take into account the effect of [aftershocks], the degree of clustering . . . is the same for earthquakes in different depth ranges.

Yet not all seismologists agree that deep earthquake activity isn't periodic – especially Greek seismologists (e.g., Comninakis and Papazachos, 1980; Papadopoulos and Voidomatis, 1987). Subsequently, Polimenakos (1993) used the Rayleigh test to conclude that there were seasonal variations in the occurrence of intermediate earthquakes in the Hellenic arc; however, the data aren't especially convincing.[15] The investigations that find seasonal periodicity often involve catalogs extending

[15] Polimenakos found a result significant at the 95% confidence level only for intermediate-depth quakes with $M \geq 6.0$, one of 12 groupings studied. Moreover, the data graphed suggest that there may be time-dependent uniformity problems with magnitudes in the 1901–1985 catalog; e.g., although there are 32 quakes, none occur after 1971.

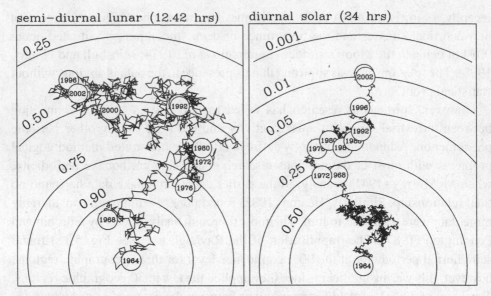

Fig. 5.18 Rayleigh's (1919) "random walk" test applied to evaluate periodicity at two periods for the Bucaramanga, Colombia, earthquake nest. At the origin time t_k of the kth earthquake, one plots a unit-length phase vector, i.e., rotated about an angle $2\pi t_k/T$, where T is the period of interest. The test compares the length R of the sum of these phase vectors to the value expected for a random process. Plotted data are all 1291 Bucaramanga earthquakes reported by the ISC 1964–2002. (Left) Semidiurnal lunar tidal period ($\sim$12.42 hrs): the data indicate no relationship since the sum of phase vectors for 1964–2002 lies between the 25th and 50th percentiles (circles labeled 0.25 and 0.50) for a random process. (Right) Solar diurnal period (24 hrs): the data indicate a diurnal relationship since the sum of phase vectors is at about the first percentile (circle labeled 0.01). Presumably the data reflect weak diurnal periodicity only because detection of seismic signals depends on background noise that varies with time of day.

back to 1900 or earlier; a common flaw is that they make little effort to evaluate the reliability of catalog magnitudes.

Finally, several investigations report even longer-period relationships. Abe and Kanamori (1979) investigated the energy release due to deep seismicity for the period 1904 to 1974 based on Abe's (1981) revised magnitude scale m_B. They observed a significant maximum in energy release occurring around 1910, and also asserted that the energy release curve showed "a good correlation with the Chandler wobble." For nearly the same data Jakubcova and Pick (1987) found a weak correlation ($\sim$0.45) between global seismic energy release from deep ($h \geq 60$ km) earthquakes and variations in the motions of the Sun and outer planets that have a period of 11.9 years. Finally, Enescu and Enescu (1999) argue that solid-earth tides influence strong Vrancea earthquakes; moreover, they allege that about eight centuries of historical data indicate Vrancea earthquakes occur most often during the first eight years of a century. For all these studies the statistical analysis supporting these conclusions isn't convincing, and the reliability of both magnitude and energy data prior to the mid-twentieth century is questionable.

5.3 Correlations with other phenomena

The human eye is very, very good at detecting (or imagining) spatial non-randomness.[16] This section discusses two research areas where there is literature reporting correlations between deep seismic activity and other phenomena; however, the correlations have yet to be confirmed by convincing statistical analysis.

5.3.1 Earthquakes and volcanoes

Island-arc volcanoes generally occur approximately above the 100-km contour of the Wadati–Benioff zone (see Fig. 4.8); but does the occurrence of individual intermediate-depth earthquakes correlate with the location and timing of volcanic eruptions? Carr and Stoiber (1973) asserted that seven of eight recent Central American volcanic eruptions were associated with nests of intermediate-depth (70–110 km) earthquakes, usually located a few tens of km seaward of the volcanoes. They further speculated that the development of "concentrations of intermediate-depth earthquakes may signal that quiescent volcanoes will soon enter a period of renewed activity." Carr (1983) also noted that directly above concentrations of intermediate-depth seismicity there were few volcanoes; he thus suggested that developing melt phases lowers mantle strength and creates the aseismic zone, but that nests develop by stress concentrations at the edges.

Blot (1981a; 1981b) proposed an even more elaborate model. He proposed an equation for the time lag t in days between an intermediate-depth earthquake and a volcanic eruption:

$$t = 740 \log_{10} \left(\frac{h_d}{h_r} \right) + \frac{h_r}{0.3M + 0.4}. \tag{5.2}$$

Here M is earthquake magnitude, and h_d and h_r are respectively the depths of the candidate earthquakes ($\sim$175–300 km) and typical earthquakes at the "root" of the volcano ($\sim$120–160 km). In a retrospective analysis of a volcano in New Zealand, he obtained extraordinary agreement between observed eruption times and those predicted by this equation. Unfortunately his analysis was non-statistical and unconvincing; it is likely that the apparent agreement came about only because the activity rate for both eruptions and earthquakes was high.

---

[16] Possibly this trait evolved because it had some evolutionary advantage. In the search for wild berries or nuts for food, for example, there may be little penalty for a few "false alarms" in your efforts to find food clusters, but you are in serious trouble if you aren't searching for concentrations at all.

Subsequently, Papadopoulos (1986) analyzed activity between 1800 and 1985 in the Hellenic arc; he concluded that three eruption cycles of the volcano Santorini followed "energy highs" in intermediate-depth earthquake activity, with a time lag of about 15 years. However, he provides no statistical arguments in support of this conclusion; there also seems to be uncertainty about the focal depths of earthquakes during the critical periods, all of which took place before 1930. In a followup article Papadopoulos (1987) investigates deep earthquake-volcano correlations at seven different circum-Pacific subduction zones; here he does apply statistics but again the potential for systematic errors is great.

Unfortunately, none of the investigations discussed above are convincing. While there are not necessarily any physical reasons preventing a relationship between volcanic and intermediate-depth seismic activity, in all these studies there are problems both with the earthquake catalogs used and with the statistical methods applied.

5.3.2 Shallow and intermediate earthquakes

Since earthquakes near trenches are an expression of plate motion and regional stress release, it is plausible that the occurrence of a large shallow quake might produce significant stress changes both in the outer rise and downdip portions of a plate, thus causing seismic activity there. Mogi (1973; 1988) analyzed space–time seismicity graphs for several regions, and concluded that seismicity tended to migrate downward in the years following, and sometimes preceding, large subduction zone earthquakes; in some cases this migration extended to 600 km depth. Similarly, Astiz and Kanamori (1986), Astiz *et al.* (1988), Lay *et al.* (1989) and Dmowska and Lovison (1992) reviewed the occurrence of relatively large intermediate quakes in the years prior to and following great shallow earthquakes. For example, Astiz and Kanamori (1986) found that the 1960 Chile quake ($M_W = 9.5$) was preceded by quakes with downdip tensional focal mechanisms in 1934 at 120 km depth and in 1949 at 70 km depth, and then followed by a downdip compressional quake at 150 km in 1971.

The authors of these papers include some of the twentieth century's most venerated seismologists; however, none of them actually apply any statistical tests to assess the relationship – if any – between the shallow and intermediate activity. Data published by Lay *et al.* (1989) (Fig. 5.19) does suggest that there may be a quiescent period of a year or so prior to and following great shallow earthquakes. However, at present there has been more effort generating physics-based models (e.g., Taylor *et al.*, 1996) to explain and quantify the purported correlations than to corroborate that they actually occur. In a study of seismicity in Mexico and South America, Lemoine *et al.* (2002) found no temporal relationship between large thrust earthquakes and intermediate-depth activity.

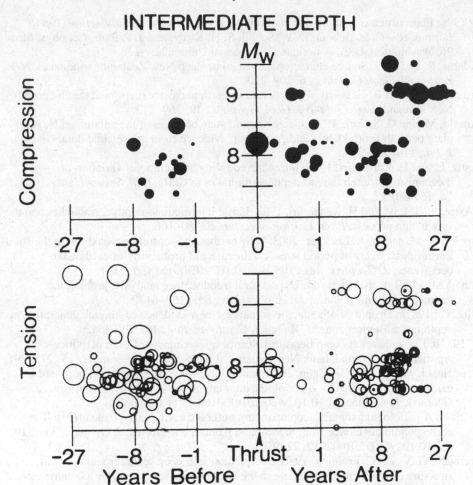

Fig. 5.19 Relative times of intermediate-depth earthquakes occurring downdip of large ($M_w > 7$) shallow subduction-zone thrusts. Vertical axis indicates the moment magnitude of the thrust event; circle sizes are scaled linearly with the magnitude of the intermediate-depth quake. Solid symbols (top) indicate downdip compressional mechanisms while open symbols (bottom) are downdip tensional mechanisms. Note that intermediate-depth activity appears to be lower for approximately one year prior to and one year following large shallow thrusts. Figure reproduced from Lay *et al.* (1989).

5.4 References

Abe, K., 1981. Magnitudes of large shallow earthquakes from 1904 to 1980, *Phys. Earth Planet. Int.*, **27**, 72–92.

Abe, K. and H. Kanamori, 1979. Temporal variation of the activity of intermediate and deep-focus earthquakes, *J. Geophys. Res.*, **84**, 3589–3595.

Abers, G. A., 1992. Relationship between shallow- and intermediate-depth seismicity in the eastern Aleutian subduction zone, *Geophys. Res. Lett.*, **20**, 2019–2022.

1996. Plate structure and the origin of double seismic zones. In *Subduction, Top to Bottom*, eds. G. E. Bebout, D. W. Scholl, S. H. Kirby, and J. P. Platt, Geophys. Mon. 96, Washington, D.C., American Geophysical Union, 223–228.

Adams, R. D., 1963. Source characteristics of some deep New Zealand earthquakes, *New Zealand J. Geol. Geophys.*, **6**, 209–220.

Adams, R. D. and B. G. Ferris, 1976. A further earthquake at exceptional depth beneath New Zealand, *New Zealand J. Geol. Geophys.*, **19**, 269–273.

Araujo, M. and G. Suárez, 1994. Geometry and state of stress of the subducted Nazca plate beneath central Chile and Argentina: evidence from teleseismic data, *Geophys. J. Int.*, **116**, 283–303.

Astiz, L. and H. Kanamori, 1986. Intraplate coupling and temporal variation of mechanisms of intermediate-depth earthquakes in Chile, *Bull. Seismol. Soc. Amer.*, **76**, 1614–1622.

• Astiz, L., T. Lay, and H. Kanamori, 1988. Large intermediate-depth earthquakes and the subduction process, *Phys. Earth Planet. Int.*, **53**, 80–166.

Beeler, N. M. and D. A. Lockner, 2003. Why earthquakes correlate weakly with the solid Earth tides: effects of period stress on the rate and probability of earthquake occurrence, *J. Geophys. Res.*, **108**, doi:10.1029/JB001518, p. 2391.

Bevis, M. and B. L. Isacks, 1984. Hypocentral trend surface analysis: probing the geometry of Benioff zones, *J. Geophys. Res.*, **89**, 6153–6170.

Blot, C., 1981a. Deep root of andecitic volcanoes: new evidence of magma generation at depth in the Benioff zone, *J. Volcanol. Geotherm. Res.*, **10**, 339–364.

1981b. Earthquakes at depth beneath volcanoes, forerunners of their activities. Application to White Island, New Zealand, *J. Volcanol. Geotherm. Res.*, **9**, 277–291.

Brudzinski, M. R. and W.-P. Chen, 2000. Variations in P wave speeds and outboard earthquakes: evidence for a petrologic anomaly in the mantle transition zone, *J. Geophys. Res.*, **105**, doi:10.1029/2000JB900160, 21661–21682.

2003. A petrologic anomaly accompanying outboard earthquakes beneath Fiji-Tonga: corresponding evidence from broad-band P and S waveforms, *J. Geophys. Res.*, **108**, doi:10.1029/2002JB002012, 2299.

Burbach, G. V. and C. Frohlich, 1986. Intermediate and deep seismicity and lateral structure of subducted lithosphere in the circum-Pacific region, *Rev. Geophys.*, **24**, 833–874.

Cahill, T. and B. L. Isacks, 1986. An apparent double-planed Benioff zone beneath northern Chile resulting from misidentification of reflected phases, *Geophys. Res. Lett.*, **13**, 333–336.

Carr, M. J., 1983. Nests of intermediate depth (70–160 km) earthquakes adjacent to active volcanoes during 1963–1982, *J. Volcanol. Geotherm. Res.*, **19**, 349–365.

Carr, M. J. and R. E. Stoiber, 1973. Intermediate depth earthquakes and volcanic eruptions in Central America, 1961–1972, *Bull. Volcanol.*, **37**, 326–337.

Cassidy, J. F. and F. Waldhauser, 2003. Evidence for both crustal and mantle earthquakes in the subducting Juan de Fuca plate, *Geophys. Res. Lett.*, **30**, 1095, doi:10.1029/2002GL015511.

Chatelain, J. L., S. W. Roecker, D. Hatzfeld, and P. Molnar, 1980. Microearthquake seismicity and fault plane solutions in the Hindu Kush region and their tectonic implications, *J. Geophys. Res.*, **85**, 1365–1387.

• Chen, W.-P. and M. R. Brudzinski, 2001. Evidence for a large-scale remnant of subducted lithosphere beneath Fiji, *Science*, **292**, doi:10.1126/science.292.5526.2475, 2475–2479.

2003. Seismic anisotropy in the mantle transition zone beneath Fiji-Tonga, *Geophys. Res. Lett.*, **30**, doi:10.1029/2002GL016330, 1682.

•Chen, W.-P. and P. Molnar, 1983. Focal depths of intracontinental and intraplate earthquakes and their implications for the thermal and mechanical properties of the lithosphere, *J. Geophys. Res.*, **88**, 4183–4214.

Chen, W.-P. and Z. Yang, 2004. Earthquakes beneath the Himalayas and Tibet: evidence for strong lithopheric mantle, *Science*, **304**, doi:10.1126/science.1097324, 1949–1952.

Chiu, J.-M., B. L. Isacks, and R. K. Cardwell, 1991. 3-D configuration of subducted lithosphere in the western Pacific, *Geophys. J. Int.*, **106**, 99–111.

Christensen, D. H. and L. J. Ruff, 1988. Seismic coupling and outer rise earthquakes, *J. Geophys. Res.*, **93**, 13421–13444.

Comninakis, P. E. and B. C. Papazachos, 1980. Space and time distribution of the intermediate focal depth earthquakes in the Hellenic arc, *Tectonophysics*, **70**, doi:10.1016/0040–1951(80)90278–4, T35–T47.

Comte, D., L. Dorbath, M. Pardo, T. Monfret, H. Haessler, L. Rivera, M. Frogneux, B. Glass, and C. Meneses, 1999. A double-layered seismic zone in Arica, Northern Chile, *Geophys. Res. Lett.*, **26**, doi:10.1029/1999GL900447, 1965–1968.

Cummins, P. R., T. Baba, A. Smith, J. Townend, and K. Uhira, 2002. An integrated focal mechanism database for Japanese earthquakes, *Seismol. Res. Lett.*, **73**, 251.

Curchin, J. M. and W. D. Pennington, 1987. Tidal triggering of intermediate and deep focus earthquakes, *J. Geophys. Res.*, **92**, 13957–13967.

Davis, S. D. and C. Frohlich, 1991. Single-link cluster analysis of earthquake aftershocks: decay laws and regional variations, *J. Geophys. Res.*, **96**, 6335–6350.

Dewey, J. W., 1972. Seismicity and tectonics of western Venezuela, *Bull. Seismol. Soc. Amer.*, **62**, 1711–1751.

Dmowska, R. and L. C. Lovison, 1992. Influence of asperities along subduction interfaces on the stressing and seismicity of adjacent areas, *Tectonophysics*, **211**, doi:10.1016/0040–1951(92)90049-C, 23–43.

Enescu, D. and B. D. Enescu, 1999. Possible cause-effect relationships between Vrancea (Romania) earthquakes and some global geophysical phenomena, *Natural Hazards*, **19**, 233–245.

Engdahl, E. R., R. van der Hilst, and R. Buland, 1998. Global teleseismic earthquake relocation with improved travel times and procedures for depth determination, *Bull. Seismol. Soc. Amer.*, **88**, 722–743.

Engdahl, E. R. and C. H. Scholz, 1977. A double Benioff zone beneath the central Aleutians: an unbending of the lithosphere, *Geophys. Res. Lett.*, **4**, 473–476.

Frohlich, C., 1987. Aftershocks and temporal clustering of deep earthquakes, *J. Geophys. Res.*, **92**, 13944–13956.

Frohlich, C. and S. Davis, 1985. Identification of aftershocks of deep earthquakes by a new ratios method, *Geophys. Res. Lett.*, **12**, 713–716.

1990. Single-link cluster analysis as a method to evaluate spatial and temporal properties of earthquake catalogues, *Geophys. J. Int.*, **100**, 19–32.

Frohlich, C. and R. J. Willemann, 1987. Statistical methods for comparing directions to the orientations of focal mechanisms of Wadati–Benioff zones, *Bull. Seismol. Soc. Amer.*, **77**, 2135–2142.

Frohlich, C., S. Billington, E. R. Engdahl, and A. Malahoff, 1982. Detection and location of earthquakes in the central Aleutian subduction zone using island and ocean bottom seismograph stations, *J. Geophys. Res.*, **87**, 6853–6864.

Frohlich, C., K. Kadinsky-Cade, and S. D. Davis, 1995. A reexamination of the Bucaramanga, Colombia, earthquake nest, *Bull. Seismol. Soc. Amer.*, **85**, 1622–1634.

Fuchs, K., K. P. Bonjer, G. Bock, I. Cornea, C. Radu, D. Enescu, D. Jianu, A. Nourescu, G. Merkler, T. Moldoveanu, and G. Tudorache, 1979. The Romanian earthquake of March 4, 1977. II. Aftershocks and migration of seismic activity, *Tectonophysics*, **53**, doi:10.1016/0040-1951(79)90068-4, 225–247.

Fujita, K. and H. Kanamori, 1981. Double seismic zones and stresses of intermediate depth earthquakes, *Geophys. J. Roy. Astron. Soc.*, **66**, 131–156.

Fukao, Y., M. Obayashi, H. Inoue, and M. Nenbai, 1992. Subducting slabs stagnant in the mantle transition zone, *J. Geophys. Res.*, **97**, 4809–4822.

Giardini, D., 1992. Space-time distribution of deep seismic deformation in Tonga, *Phys. Earth Planet. Int.*, **74**, 75–88.

Glennon, M. A. and W.-P. Chen, 1993. Systematics of deep-focus earthquakes along the Kuril–Kamchatka arc and their implications on mantle dynamics, *J. Geophys. Res.*, **98**, 735–769.

Goto, K., H. Hamaguchi, and Z. Suzuki, 1985. Earthquake generating stresses in a descending slab, *Tectonophysics*, **112**, doi:10.1016/0040-1951(85)90175-1, 111–128.

Guest, A., G. Schubert, and C. W. Gable, 2004. Stresses along the metastable wedge of olivine in a subducting slab: possible explanation for Tonga double seismic layer, *Phys. Earth Planet. Int.*, **141**, doi:10.1016/j.pepi.2003.11.1012, 253–267.

Gutenberg, B. and C. F. Richter, 1938. Depth and geographical distribution of deep-focus earthquakes, *Geol. Soc. Amer. Bull.*, **49**, 249–288.

• Hacker, B. R., S. M. Peacock, G. A. Abers, and S. D. Holloway, 2003. Subduction factory – 2. Are intermediate-depth earthquakes in subducting slabs linked to metamorphic dehydration reactions? *J. Geophys. Res.*, **108**, doi:10.0129/JB001129, 2030.

Harris, F., 1982. Focal mechanism of an intermediate-depth earthquake and its aftershocks, *New Zealand J. Geol. Geophys.*, **25**, 109–113.

Hamaguchi, H., K. Goto, and Z. Suzuki, 1983. Double-planed structure of intermediate-depth seismic zone and thermal stress in the descending plate, *J. Phys. Earth*, **31**, 329–347.

Hasegawa, A., N. Umino, and A. Takagi, 1978a. Double-planed structure of the deep seismic zone in the northeastern Japan arc, *Tectonophysics*, **47**, doi:10.1016/0040-1951(78)90150-6, 43–58.

• 1978b. Double-planed deep seismic zone and upper mantle structure in the northeastern Japan arc, *Geophys. J. Roy. Astron. Soc.*, **54**, 281–296.

Hasegawa, A., S. Horiuchi, and N. Umino, 1994. Seismic structure of the northeastern Japan convergent margin: a synthesis, *J. Geophys. Res.*, **99**, 22295–22311.

Hatzfeld, D. and M. Frogneux, 1981. Intermediate depth seismicity in the western Mediterranean unrelated to subduction of oceanic lithosphere, *Nature*, **292**, doi:10.1038/292443a0, 443–445.

Heaton, T. H., 1982. Tidal triggering of earthquakes, *Bull. Seismol. Soc. Amer.*, **82**, 2181–2200.

House, L. A. and K. H. Jacob, 1982. Thermal stresses in subducting lithosphere can explain double seismic zones, *Nature*, **295**, doi:10.1038/295587a0, 587–589.

Hurtado, S. and E. de Jesus, 1999. Estudio de sismicidad historica en la region de Bucaramanga (Colombia), *Revista de la Academia Colombiana de Ciencias Exactas, Fisicas y Naturales*, **23**, 233–248.

Igarashi, I., T. Matsuzawa, N. Umino, and A. Hasegawa, 2001. Spatial distribution of focal mechanisms for interplate and intraplate earthquakes associated with the subducting Pacific plate beneath the northeastern Japan arc: a triple-planed deep seismic zone, *J. Geophys. Res.*, **106**, doi:10.1029/2000JB900386, 2177–2191.

Iidaka, T. and Y. Furukawa, 1994. Double seismic zone for deep earthquakes in the Izu–Bonin subduction zone, *Science*, **263**, 1116–1118.

Iidaka, T. and K. Obara, 1997. Seismological evidence for the existence of anisotropic zone in the metastable wedge inside the subducting Izu–Bonin slab, *Geophys. Res. Lett.*, **24**, doi:10.1029/97GL03277, 3305–3308.

Isacks, B. L. and M. Barazangi, 1977. Geometry of Benioff zones: lateral segmentation and downwards bending of the subducted lithosphere. In *Island Arcs, Deep Sea Trenches, and Back-Arc Basins*, Maurice Ewing Ser., vol. 1, eds. M. Talwani and W. Pitmann III, Washington, DC, American Geophysical Union, 99–114.

Isacks, B. L. and P. Molnar, 1971. Distribution of stresses in the descending lithosphere from a global survey of focal-mechanism solutions of mantle earthquakes, *Rev. Geophys. Space Phys.*, **9**, 103–174.

Isacks, B. L., L. R. Sykes, and J. Oliver, 1967. Spatial and temporal clustering of deep and shallow earthquakes in the Fiji–Tonga–Kermadec region, *Bull. Seismol. Soc. Amer.*, **57**, 935–958.

Jakubcova, I. and M. Pick, 1987. Correlation between solar motion, earthquakes and other geophysical phenomena, *Annales Geophysicae Ser.* B, **5**, 135–141.

• Jeffreys, H., 1938. Aftershocks and periodicity in earthquakes, *Gerlands Beitr. z. Geophysik*, **53**, 111–139.

Kagan, Y. Y., 1991a. Fractal dimension of brittle fracture, *J. Nonlinear Sci.*, **1**, 1–16.

1991b. Likelihood analysis of earthquake catalogs, *Geophys. J. Int.*, **106**, 135–148.

• Kagan, Y. Y. and D. D. Jackson, 1991. Long-term earthquake clustering, *Geophys. J. Int.*, **104**, 117–133.

Kagan, Y. Y. and L. Knopoff, 1976. Statistical search for non-random features of the seismicity of strong earthquakes, *Phys. Earth Planet. Int.*, **12**, 291–318.

1980a. Dependence of seismicity on depth, *Bull. Seismol. Soc. Amer.*, **70**, 1811–1822.

1980b. Spatial distribution of earthquakes: the two-point correlation function, *Geophys. J. Roy. Astron. Soc.*, **62**, 303–320.

Kao, H. and W.-P. Chen, 1994. The double seismic zone in Kuril–Kamchatka: the tale of two overlapping single zones, *J. Geophys. Res.*, **99**, doi:10.1029/93JB03409, 6913–6930.

1995. Transition from interplate slip to double seismic zone along the Kuril–Kamchatka arc, *J. Geophys. Res.*, **100**, doi:10.1029/95JB00239, 9881–9903.

Kao, H. and L.-G. Liu, 1995. A hypothesis for the seismogenesis of a double seismic zone, *Geophys. J. Int.*, **123**, 71–84.

Kao, H. and R.-J. Rau, 1999. Detailed structures of the subducted Philippine Sea plate beneath northeast Taiwan: a new type of double seismic zone, *J. Geophys. Res.*, **104**, doi:10.1029/1998JB900010, 1015–1033.

Kawakatsu, H., 1985. Double seismic zone in Tonga, *Nature*, **316**, doi:10.1038/316053a0, 53–55.

1986a. Double seismic zone kinematics, *J. Geophys. Res.*, **91**, 4811–4825.

1986b. Downdip tensional earthquakes beneath the Tonga arc: a double seismic zone? *J. Geophys. Res.*, **91**, 6432–6440.

Kirby, S. H., W. B. Durham, and L. A. Stern, 1991. Mantle phase transitions and deep-earthquake faulting in subducting lithosphere, *Science*, **252**, 216–225.

Kirby, S., E. R. Engdahl, and R. Denlinger, 1996. Intermediate-depth intraslab earthquakes and arc volcanism as physical expressions of crustal and uppermost mantle metamorphism in subducting slabs. In *Subduction, Top to Bottom*, eds. G. E.

Bebout, D. W. Scholl, S. H. Kirby and J. P. Platt, Geophys. Mon. 96, Washington, DC, American Geophysical Union, 195–214.

Kisslinger, C. and A. Hasegawa, 1991. Seismotectonics of intermediate-depth earthquakes from properties of aftershock sequences, *Tectonophysics*, **197**, doi:10.1016/0040-1951(91)90398-C, 27–40.

Kono, M., Y. Takahashi, and Y. Fukao, 1985. Earthquakes in the subducting slab beneath northern Chile: a double seismic zone, *Tectonophysics*, **112**, doi:10.1016/0040-1951(85)90180-5, 211–225.

Lammlein, D. R., G. V. Latham, J. Dorman, Y. Nakamura, and M. Ewing, 1974. Lunar . seismicity, structure, and tectonics, *Rev. Geophys. Space Phys.*, **12**, 1–21.

Landsberg, H., 1935. Some correlations between the occurrence of deep- and shallow-focus earthquakes, *Trans. Amer. Geophys. Un.*, **16**, 91–93.

Lawson, J. E., 1967. *Location of Intermediate Depth Northern Colombian Earthquakes with a Small Digital Computer*, M. S. thesis, University Tulsa, 347 pp.

Lay, T., 1997. *Structure and Fate of Subducting Slabs*, San Diego, CA., Academic Press, 185 pp.; reprinted from volume 35 of *Advances in Geophysics: Seismological Structure of Slabs*, eds. R. Dmowska and B. Saltzman, 1–185.

Lay, T., L. Astiz, H. Kanamori, and D. H. Christensen, 1989. Temporal variation of large intraplate earthquakes in coupled subduction zones, *Phys. Earth Planet. Int.*, **54**, doi:10.1016/0031-9201(89)90247-1, 258–312.

Leith, A. and J. A. Sharpe, 1936. Deep focus earthquakes and their geological significance, *J. Geol.*, **44**, 877–917.

Lemoine, A., R. Madariaga, and J. Campos, 2002. Slab-pull and slab-push earthquakes in the Mexican, Chilean and Peruvian subduction zones, *Phys. Earth Planet. Int.*, **132**, doi:10.1016/S0031–9201(02)00050-X, 157–175.

Lukk, A. A., 1968. The aftershock sequence of the Dzhurm deep-focus earthquake of 14 March 1965, *Izv., Earth Phys.*, **4**, 83–85.

Lundgren, P. and D. Giardini, 1994. Isolated deep earthquakes and the fate of subduction in the mantle, *J. Geophys. Res.*, **99**, 15833–15842.

•Malavé, G. and G. Suárez, 1995. Intermediate-depth seismicity in northern Colombia and western Venezuela and its relationship to Caribbean plate subduction, *Tectonics*, **14**, doi:10.1029/95TC00334, 617–628.

Matsuzawa, T., N. Umino, A. Hasegawa, and A. Takagi, 1986. Normal fault type events in the upper plane of the double-planed seismic zone beneath the northeastern Japan arc, *J. Phys. Earth*, **34**, 85–94.

McGuire, J. J. and D. A. Wiens, 1995. A double seismic zone in New Britain and the morphology of the Solomon plate at intermediate depths, *Geophys. Res. Lett.*, **22**, doi:10.1029/95GL01806, 1965–1968.

McGuire, J. J., D. A. Wiens, P. J. Shore, and M. G. Bevis, 1997. The March 9, 1994 (M_W 7.6), deep Tonga earthquake: rupture outside the seismically active slab, *J. Geophys. Res.*, **102**, doi:10.1029/96JB03185, 15163–15182.

McLaren, J. P. and C. Frohlich, 1985. Model calculations of regional network locations for earthquakes in subduction zones, *Bull. Seismol. Soc. Amer.*, **75**, 397–413.

McMurray, H., 1941. Periodicity of deep-focus earthquakes, *Bull. Seismol. Soc. Amer.*, **31**, 33–82.

Michaels, A. J., 1989. Spatial patterns of aftershocks of shallow focus earthquakes in California and implications for deep-focus earthquakes, *J. Geophys. Res.*, **94**, 5615–5626.

Minshull, T. A. and N. R. Goulty, 1988. The influence of tidal stresses on deep moonquake activity, *Phys. Earth Planet. Int.*, **52**, 41–55.

Mogi, K., 1973. Relationship between shallow and deep seismicity in the western
Pacific region, *Tectonophysics*, **17**, doi:10.1016/0040-1951(73)90062-0,
1–22.

1988. Downward migration of seismic activity prior to some great shallow earthquakes
in Japanese subduction zones – a possible intermediate-term precursor, *Pure Appl.
Geophys.*, **126**, 447–463.

Myers, S. C., T. C. Wallace, S. L. Beck, P. G. Silver, G. Zandt, J. Vandecar, and E.
Minaya, 1995. Implications of spatial and temporal development of the aftershock
sequence for the M_W 8.3 June 9, 1994 deep Bolivian earthquake, *Geophys. Res. Lett.*,
22, doi:10.1029/95GL01600, 2269–2272.

Nishiyama, S., 1992. Mantle hydrology in a subduction zone: a key to episodic geologic
events, double Wadati–Benioff zones and magma genesis, *Mathematical Seismology
VII*, *Rept. Statl. Math. Inst.*, **34**, 31–67.

Nyffenegger, P. and C. Frohlich, 2000. Aftershock occurrence rate decay properties for
intermediate and deep earthquake sequences, *Geophys. Res. Lett.*, **27**,
doi:10.1029/1998GL010371, 1215–1218.

Oike, K., 1971a. On the nature of the occurrence of intermediate and deep
earthquakes. 2. Spatial and temporal clustering, *Bull. Disaster Prevention Res. Inst.*,
21, 43–73.

1971b. On the nature of the occurrence of intermediate and deep earthquakes. 3. Focal
mechanisms of multiplets, *Bull. Disaster Prevention Res. Inst.*, **21**, 153–178.

Ojeda, A. and J. Havskov, 2001. Crustal structure and local seismicity in Colombia, *J.
Seismology*, **5**, doi:10.1023/A:1012053206408, 575–593.

• Okal, E. A., 2001. 'Detached' deep earthquakes: are they really? *Phys. Earth Planet. Int.*,
127, doi:10.016/S0031-9201(01)00224-2, 109–143.

Okal, E. A. and C. R. Bina, 1994. The deep earthquakes of 1921–1922 in northern Peru,
Phys. Earth Planet. Int., **87**, 33–54.

2001. The deep earthquakes of 1997 in Western Brazil, *Bull. Seismol. Soc. Amer.*, **91**,
161–164.

• Okal, E. A. and S. H. Kirby, 1998. Deep earthquakes beneath the Fiji Basin, SW Pacific:
Earth's most intense deep seismicity in stagnant slabs, *Phys. Earth Planet. Int.*, **109**,
doi:10.1016/S0031-9201(98)001116-2, 25–63.

Okino, K., M. Ando, S. Kaneshima, and K. Hirahara, 1989. The horizontally lying slab,
Geophys. Res. Lett., **16**, 1059–1062.

Omori, F., 1895. On the aftershocks of earthquakes, *J. College Sci. Imperial Univ. Tokyo*,
7, 111–200.

Oncescu, M. C., 1984. Deep structure of the Vrancea region, Roumania, inferred from
simultaneous inversion for hypocenters and 3-D velocity structure, *Annales
Geophysicae*, **2**, 23–27.

Oncescu, M. C. and K. P. Bonjer, 1997. A note on the depth recurrence and strain release
of large Vrancea earthquakes, *Tectonophysics*, **272**,
doi:10.1016/0040-1951(96)00263-6, 291–302.

Page, R., 1968. Focal depths of aftershocks, *J. Geophys. Res.*, **73**, 3897–3903.

Papadopoulos, G. A., 1986. Large intermediate depth shocks and volcanic eruptions in the
Hellenic arc during 1800–1985, *Phys. Earth Planet. Int.*, **43**, 47–55.

1987. Large deep-focus shocks and significant volcanic eruptions in convergent plate
boundaries during 1900–1980, *Tectonophysics*, **138**,
doi:10.1016/0040–1951(87)90041–2, 223–233.

Papadopoulos, G. A. and P. Voidomatis, 1987. Evidence for periodic seismicity in the
inner Aegean seismic zone, *Pure Appl. Geophys.*, **125**, 613–628.

Pavlis, G. L. and M. W. Hamburger, 1991. Aftershock sequences of intermediate-depth earthquakes in the Pamir–Hindu Kush seismic zone, *J. Geophys. Res.*, **96**, 18107–18117.

Peacock, S. M., 2001. Are the lower planes of double seismic zones caused by serpentine dehydration in subducting oceanic mantle? *Geology*, **29**, doi:10.1130/0091-7613(2001)029, 299–302.

Peacock, S. M. and K. Wang, 1999. Seismic consequences of warm versus cool subduction metamorphism: examples from southwest and northeast Japan, *Science*, **286**, doi:10.1126/science.286.5441.937, 937–939.

Pennington, W. D., 1981. Subduction of the eastern Panama Basin and seismotectonics of northwestern South America, *J. Geophys. Res.*, **86**, 10753–10770.

Pennington, W. D., W. D. Mooney, R. van Hissenhoven, H. Meyer, J. E. Ramirez, and R. P. Meyer, 1979. Results of a reconnaissance microearthquake survey of Bucaramanga, Colombia, *Geophys. Res. Lett.*, **6**, 65–68.

Persh, S. E. and H. Houston, 2004. Strongly depth-dependent aftershock production in deep earthquakes, *Bull. Seismol. Soc. Amer.*, **94**, 1808–1816.

Polimenakos, L. C., 1993. Search for a seasonal trend in earthquake occurrence in the Hellenic arc, *Phys. Earth Planet. Int.*, **76**, 253–258.

Protti, M., F. Güendel, and K. McNally, 1995. Correlation between the age of the subducting Cocos plate and the geometry of the Wadati–Benioff zone under Nicaragua and Costa Rica. In *Geologic and Tectonic Development of the Caribbean Plate Boundary in Southern Central America*, ed. P. Mann, (special paper 295), Boulder, CO, Geological Society of America, 309–325.

Prozorov, A. G. and A. M. Dziewonski, 1982. A method of studying variations in the clustering property of earthquakes: application to the analysis of global seismicity, *J. Geophys. Res.*, **87**, 2829–2839.

Pulpan, H. and C. Frohlich, 1985. Geometry of the subducted plate near Kodiak Island and Lower Cook Inlet, Alaska, determined from relocated earthquake hypocenters, *Bull. Seismol. Soc. Amer.*, **75**, 791–810.

Ramdani, F., 1998. Geodynamic implications of intermediate-depth earthquakes and volcanism in the intraplate Atlas mountains (Morocco), *Phys. Earth Planet. Int.*, **108**, doi:10.1016/S0031-9201(98)00106-X, 245–260.

Ranero, C. R., J. P. Morgan, K. McIntosh, and C. Reichert, 2003. Bending-related faulting and mantle serpentinization at the Middle America Trench, *Nature*, **425**, doi:10.1038/nature01961, 367–373.

•Ratchkovski, N. A. and R. A. Hansen, 2002. New evidence for segmentation of the Alaska subduction zone, *Bull. Seismol. Soc. Amer.*, **92**, 1754–1765.

Ratchkovsky, N. A., J. Pujol, and N. N. Biswas, 1997a. Relocation of earthquakes in the Cook Inlet area, south central Alaska, using the joint hypocenter determination method, *Bull. Seismol. Soc. Amer.*, **87**, 620–636.

  1997b. Stress pattern in the double seismic zone beneath Cook Inlet, south-central Alaska, *Tectonophysics*, **281**, doi:10.1016/0040-1951(97)00042-5, 163–171.

Rayleigh, Lord, 1919. On a problem of vibrations, and of random flights in one, two and three dimensions, *Philosophical Mag., Ser. 6*, **37**, 321–347.

Reyners, M. and K. S. Coles, 1982. Fine structure of the dipping seismic zone and subduction mechanics in the Shumagin Islands, Alaska, *J. Geophys. Res.*, **87**, 356–366.

Reyners, M., R. Robinson, and P. McGinty, 1997. Plate coupling in the northern South Island and southernmost North Island, New Zealand, as illuminated by earthquake focal mechanisms, *J. Geophys. Res.*, **102**, doi:10.1029/97JB00973, 15197–15210.

Sacks, I. S., S. Suyehiro, A. Kamitsuki, M. A. Tuve, M. Otsuka *et al.*, 1967. A tentative value of Poisson's coefficient from the seismic "nest of Socampa." In *Annual Report of the Director, Carnegie Inst. Dep. Terr. Magn., 1965–1966*, 43–45.

Samowitz, I. R. and D. W. Forsyth, 1981. Double seismic zone beneath the Mariana island arc, *J. Geophys. Res.*, **86**, 7013–7021.

Santo, T., 1969a. Characteristics of seismicity in South America, *Bull. Earthquake Res. Inst., Tokyo Univ.*, **47**, 635–672.

1969b. Regional study on the characteristic seismicity of the world. Part I. Hindu Kush region, *Bull. Earthquake Res. Inst., Tokyo Univ.*, **47**, 1035–1048.

1970. Regional study on the characteristic seismicity of the world. Part VI. Colombia, Rumania and South Sandwich Islands, *Bull. Earthquake Res. Inst., Tokyo Univ.*, **48**, 1089–1105.

Sato, T., M. Kosuga, and K. Tanaka, 1996. Tomographic inversion for P wave velocity structure beneath the northeastern Japan arc using local and teleseismic data, *J. Geophys. Res.*, **101**, doi:10.1029/96JB01505, 17597–17615.

Savage, J. C., 1969. The mechanics of deep-focus faulting, *Tectonophysics*, **8**, doi:10.1016/0040-1951(69)90085-7, 115–127.

Schneider, J. F., 1984. *The Intermediate-Depth Microearthquakes of the Bucaramanga Nest, Colombia*, Ph.D. dissertation, University of Wisconsin, Madison, 233. pp.

• Schneider, J. F., W. D. Pennington, and R. P. Meyer, 1987. Microseismicity and focal mechanisms of the intermediate-depth Bucaramanga nest, Colombia, *J. Geophys. Res.*, **92**, 13913–13926.

Seno, T. and Y. Yamanaka, 1996. Double seismic zones, compressional deep trench-outer rise events, and superplumes. In *Subduction, Top to Bottom*, eds. G. E. Bebout, D. W. Scholl, S. H. Kirby, and J. P. Platt, Geophys. Mon. 96, Washington, DC, American Geophysical Union, 347–356.

Seno, T., D. Zhao, Y. Kobayashi, and M. Nakamura, 2001. Dehydration of serpentinized slab mantle: seismic evidence from southwest Japan, *Earth Planets Space*, **53**, 861–871.

Shirley, J. H., 1988. Lunar and solar periodicities of large earthquakes: southern California and the Alaska–Aleutian seismic region, *Geophys. J. Int.*, **92**, 403–420.

Shlien, S. and M. N. Toksoz, 1970. A clustering model for earthquake occurrences, *Bull. Seismol. Soc. Amer.*, **60**, 1765–1787.

Silver, P. G., S. L. Beck, T. C. Wallace, C. Meade, S. C. Myers, D. E. James, and R. Kuehnel, 1995. Rupture characteristics of the deep Bolivian earthquake of 9 June 1994 and the mechanism of deep-focus earthquakes, *Science*, **268**, 69–73.

Sleep, N. H., 1979. The double seismic zone in downgoing slabs and the viscosity of the mesosphere, *J. Geophys. Res.*, **84**, 4565–4571.

Smalley, R. F. and B. L. Isacks, 1987. A high-resolution local network study of the Nazca plate Wadati–Benioff zone under western Argentina, *J. Geophys. Res.*, **92**, 13903–13912.

Smith, S. W., J. S. Knapp, and R. C. McPherson, 1993. Seismicity of the Gorda plate, structure of the continental margin, and an eastward jump of the Mendocino triple junction, *J. Geophys. Res.*, **98**, 8153–8171.

Stetson, H. T., 1935. The correlation of deep-focus earthquakes with lunar hour angle and declination, *Science*, **82**, 523–524.

1937. Correlation of frequencies of seismic disturbance with the hour angle of the Moon, *Proc. Amer. Phil. Soc.*, **78**, 411–424.

Sykes, L. R., 1966. The seismicity and deep structure of island arcs, *J. Geophys. Res.*, **71**, 2981–3006.

Taber, J. J., S. Billington, and E. R. Engdahl, 1991. Seismicity of the Aleutian arc. In *Neotectonics of North America*, eds. by D. B. Slemmons, E. R. Engdahl, M. D. Zoback, and D. D. Blackwell, Boulder, CO, Geological Society of America, 29–46.

Tanaka, S., M. Ohtake, and H. Sato, 2002. Evidence for tidal triggering as revealed from statistical analysis of global data, *J. Geophys. Res.*, **107**, doi:10.1029/2001JB001577, 2211.

Taylor, M. A. J., G. Zheng, J. R. Rice, W. D. Stuart, and R. Dmowska, 1996. Cyclic stressing and seismicity at strongly coupled subduction zones, *J. Geophys. Res.*, **101**, doi:10.1029/95JB035618363-8381.

Tinker, M. A., S. L. Beck, W. Jiao, and T. C. Wallace, 1998. Mainshock and aftershock analysis of the June 17, 1996, deep Flores Sea earthquake sequence: implications for the mechanism of deep earthquakes and the tectonics of the Banda Sea, *J. Geophys. Res.*, **103**, doi:10.1029/97JB03533, 9987–10001.

Trifu, C.-I. and M. Radulian, 1991. Frequency–magnitude distribution of earthquakes in Vrancea: relevance for a discrete model, *J. Geophys. Res.*, **96**, 4301–4311.

Tryggvason, E., 1968. Seismicity of the Bucaramanga region, Colombia, *Geol. Soc. Amer. Spec. Pap.*, **101**, 341–342.

Tryggvason, E. and J. E. Lawson, 1970. The intermediate earthquake source near Bucaramanga, Colombia, *Bull. Seismol. Soc. Amer.*, **60**, 269–276.

Tsukahara, H., 1980. Physical conditions for double seismic planes of the deep seismic zone, *J. Phys. Earth*, **28**, 1–15.

Utsu, T., 1972. Aftershocks and earthquake statistics, IV. Analysis of the distribution of earthquakes in magnitude, time, and space with special consideration to clustering characteristics of earthquake occurrence, 2, *J. Faculty Sci. Hokkaido Univ., Ser. 7 Geophys.* **4**, 1–42.

Van der Hilst, R. and P. Mann, 1994. Tectonic implications of tomographic images of subducted lithosphere beneath northwestern South America, *Geology*, **22**, doi:10.1130/0091-7613(1994)022, 451–454.

Van Keken, P. E., B. Kiefer, and S. M. Peacock, 2002. High-resolution models of subduction zones: implications for mineral dehydration reactions and the transport of water into the deep mantle, *Geochem., Geophys., Geosyst.*, **3**, doi:10.1029/2001GC00256, 1056.

Veith, K. F., 1974. *The Relationship of Island Arc Seismicity to Plate Tectonics*, Ph.D. dissertation, 162 pp., Southern Methodist University, Dallas, TX.

Vidale, J. E., D. C. Agnew, M. J. S. Johnston, and D. H. Oppenheimer, 1998. Absence of earthquake correlation with earth tides: an indication of high preseismic fault stress rate, *J. Geophys. Res.*, **103**, 24567–24572.

Wiens, D. A. and H. J. Gilbert, 1996. Effect of slab temperature on deep-earthquake aftershock productivity and magnitude-frequency relations, *Nature*, **384**, doi:10.1028/384153a01, 53–156.

• Wiens, D. A. and J. J. McGuire, 2000. Aftershocks of the March 9, 1994, Tonga earthquake: the strongest known deep aftershock sequence, *J. Geophys. Res.*, **105**, doi:10.1029/2000JB900097, 19067–19083.

Wiens, D. A., J. J. McGuire, and P. J. Shore, 1993. Evidence for transformational faulting from a deep double seismic zone in Tonga, *Nature*, **364**, doi:10.1038/364790a0, 790–793.

Wiens, D. A., J. J. McGuire, P. J. Shore, M. G. Bevis, K. Draunidalo, G. Prasad, and S. P. Helu, 1994. A deep earthquake aftershock sequence and implications for the rupture mechanism of deep earthquakes, *Nature*, **372**, doi:10.1038/372540a0, 540–543.

Wiens, D. A., H. J. Gilbert, B. Hicks, M. E. Wysession, and P. J. Shore, 1997. Aftershock sequences of moderate-sized intermediate and deep earthquakes in the Tonga subduction zone, *Geophys. Res. Lett.*, **24**, doi:10.1029/97GL01957, 2059–2062.

Willemann, R. J. and C. Frohlich, 1987. Spatial patterns of aftershocks of deep focus earthquakes, *J. Geophys. Res.*, **92**, 13927–13943.

Wu, L.-R. and W.-P. Chen, 1999. Anomalous aftershocks of deep earthquakes in Mariana, *Geophys. Res. Lett.*, **26**, doi:10.1029/1999GL900389, 1977–1980.

Wyss, M. and L. J. Shamey, 1975. Source dimensions of two deep earthquakes estimated from aftershocks and spectra, *Bull. Seismol. Soc. Amer.*, **65**, 403–409.

Wyss, M. and Y. Toya, 2000. Is background seismicity produced at a stationary Poissonian rate? *Bull. Seismol. Soc. Amer.*, **90**, 1174–1187.

Yamaoka, K., Y. Fukao, and M. Kumazawa, 1986. Spherical shell tectonics: effects of sphericity and inextensibility on the geometry of the descending lithosphere, *Rev. Geophys.*, **24**, 27–55.

Yamasaki, T. and T. Seno, 2003. Double seismic zone and dehydration embrittlement of the subducting slab, *J. Geophys. Res.*, **108**, doi:10.1029/2002JB001918, 2212.

Zarifi, Z. and J. Havskov, 2003. Characteristics of dense nests of deep and intermediate-depth seismicity, *Adv. Geophys.*, **46**, 237–278.

Zhao, D., A. Hawegawa, and S. Horiuchi, 1992. Tomographic imaging of P and S waves velocity structure beneath northeastern Japan, *J. Geophys. Res*, **97**, 19909–19928.

Zhuang, J., Y. Ogata, and D. Vere-Jones, 2004. Analyzing earthquake clustering features by using stochastic reconstruction, *J. Geophys. Res.*, **109**, doi:10.1029/2003JB002879, B05301.

Part III

The mechanism of deep earthquakes

6

The deep earthquake source

When a deep earthquake happens, what happens? At its focus or "source," how long does it take? And over how big a region in space does it happen? Does it happen smoothly, or in a series of jerks? Does it radiate seismic waves of different strength in different directions? If so, what is the pattern? These are some obvious questions about the nature of the deep earthquake source. This chapter focuses on the observed properties of the deep earthquake source. Then, the next chapter will review the underlying physical processes that may be responsible for deep earthquakes.

Only a few features of an earthquake's source are directly observable from individual seismograms. These include the source duration, the scalar moment, the radiated energy (Fig. 6.1), and sometimes the relative arrival times of various subevents within the rupture process. The uncertainties in these observations can be large; for individual earthquakes different investigators often report values that differ by a factor of two or more. All other source parameters – static stress drop, moment tensor elements, etc. – are derived. Determining these requires the introduction of models that often introduce systematic errors that produce even greater uncertainty. As we evaluate the literature on the deep earthquake source it is important for the reader to keep in mind which source parameters are observed and which are derived.

6.1 Source duration and size

6.1.1 Duration

Measuring the temporal duration of an earthquake isn't as simple as it seems; the problem is that seismic signals possess a "coda," or extended period of motion following the arrival of the principal signal. While the overall reasons for the existence of the coda are well understood – reverberations occurring as the signal bounces

I thank George Choy for reviewing an earlier draft of this chapter.

191

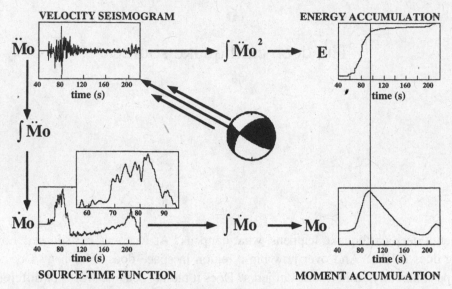

Fig. 6.1   Procedure used by Winslow and Ruff (1999) to determine moment M_o, radiated energy E_{seismic}, and duration t_d. The observed seismogram (top left) is corrected for radiation pattern, geometric spreading, source constants, and instrument response; it has SI units of moment acceleration. The result is integrated once to yield a source-time function (bottom left, with expanded scale). The source-time function is then integrated again to yield moment accumulation (bottom right). The peak in moment accumulation is M_o, and the time from the first break to the peak is t_d. To determine E_{seismic} the velocity seismogram is squared and integrated to yield energy accumulation (top right). E_{seismic} is the value (solid line) reached at time t_d. Figure reproduced from Winslow and Ruff (1999).

around in near-station crustal structure, and multipathing/diffraction effects associated with along-path heterogeneity – unraveling what is source and what is coda isn't easy.

Vidale and Houston (1993) attacked this problem in an especially direct way; they stacked seismograms recorded by several hundred digital stations across the United States (Fig. 6.2). For distant deep earthquakes the source signals were coherent and thus enhanced by the stacking process; the coda signals differed at each station and thus diminished by stacking. For earthquakes with magnitudes M_W between about 6.2 and 7.4, they found that source durations were highly variable; but larger earthquakes generally possessed longer durations. For M_W between 6 and 6.5, the shortest durations seem to be around 3 s; for M_W between 7 and 7.5 the minimum duration was about 7 s. Other investigations of comparable-sized earthquakes (Houston *et al.*, 1998; Bos *et al.*, 1998; Campus and Das, 2000; Houston, 2001; Persh and Houston, 2004) all report similar results, i.e., duration depends strongly on earthquake moment, although for any particular moment observed duration varies by a factor of about four (Fig. 6.3).

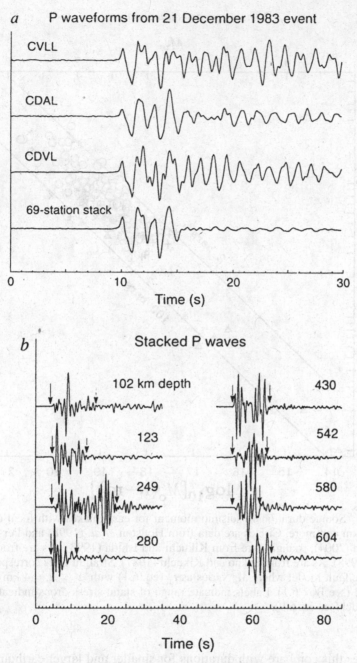

Fig. 6.2 (Top) How stacking seismograms can eliminate coda and show source duration. The top three traces are short-period vertical-component seismograms from stations CVLL, CDAL and CDVL in northern California; the bottom trace is the sum (or stack) of 69 traces, selected from 400 stations in the western United States. Note the well-defined ending of the short-period P signal in the stacked trace. (Bottom) Examples of eight stacks for earthquakes with moments of 1.7–3.2 $\times\ 10^{19}$ newton-m. The arrows indicate the inferred start and end of each rupture; numbers with each trace indicate focal depth. Figure reproduced from Vidale and Houston (1993) with permission from Nature Publishing Group.

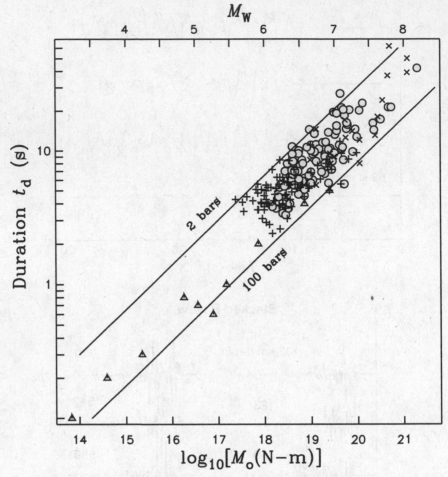

Fig. 6.3 Source duration vs seismic moment for earthquakes with focal depths of 100 km and more. Circles are data from Houston *et al.* (1998) and Persh and Houston (2004); triangles are from Kikuchi and Ishida (1993), +s are from Bos *et al.* (1998), Xs are from Fukao and Kikuchi (1987). Straight lines correspond to a simple fault model where M_o varies as t_d^3 (eq. 6.1) with $V_{rupture} = 4$ km/s and $c = 7/2$ (see Box 6.1). Labels indicate range of static stress drops indicated by this model (see eq. 6.9).

How does this compare with durations for smaller and larger earthquakes? For Tonga deep earthquakes with magnitude m_b between about 3.5 and 5.9, Pennington and Isacks (1979) analyzed analog seismograms at regional stations and determined durations of 1.0–1.5 s, apparently independent of magnitude. However, Kikuchi and Ishida (1993) evaluated digital data and found durations of 0.1–0.2 s for earthquakes in Japan with M_W of 3.2–4.2. Deep earthquakes with magnitudes exceeding 7.4 are rare; however, a few have rupture durations of 30 s or more. For the

Colombia earthquake of 31 July 1970 ($M_{W(CMT-Hist)} = 8.1; h_{CMT-Hist} = 627$ km) Furumoto (1977) found a rupture duration of about 50 s. For the Kuril earthquake of 6 December 1978 ($M_{W(CMT)} = 7.8; h_{CMT} = 181$ km) Brüstle and Müller (1987) determined a rupture duration of about 30 s. For the largest deep earthquake known, the Bolivia event of 9 June 1994 ($M_{W(CMT)} = 8.2, h_{CMT} = 647$) numerous investigators have reported a duration of about 40–50 s (Beck *et al.*, 1995; Ihmlé and Jordan, 1995; Estabrook and Bock, 1995; Goes and Ritsema, 1995; Antolik *et al.*, 1996). Altogether the observations are quite consistent with the assertion that duration t_d varies with $M_o^{1/3}$.

To investigate if there are other sources of the variability evident in Fig. 6.3, we can scale the durations by multiplying by $1/M_o^{1/3}$ to remove the dependence on moment. For M_o between about 10^{18} N-m and 10^{20} N-m where data are plentiful, Vidale and Houston (1993) and Persh and Houston (2004) report that scaled durations were correlatable with focal depth; i.e., they found that earthquakes at depths of 100 km had scaled durations averaging about 11 s, compared to only 6.5 s for those at 600 km. However, Houston *et al.* (1998) found that the average difference was somewhat smaller – about 2 s or 25%. Bos *et al.* (1998) noted that this is comparable to the difference in shear velocity between 100 km and 600 km, and is consistent with models where there is no observable dependence of static stress drop on depth (see Fig. 6.4). For the data in Fig. 6.3, a reasonable fit is given by $M_o[\text{N-m}] = 1.5 \times 10^5 (V_s t_d)^3$, where V_S is the shear velocity in the source region. More precisely, nearly all the observations in Figs. 6.3 and 6.4 satisfy:

$$0.3 \times 10^5 (V_S t_d)^3 < M_o[\text{N-m}] < 15 \times 10^5 (V_S t_d)^3. \qquad (6.1)$$

Furthermore, by stacking moment-scaled source-time functions for 39 earthquakes, Houston *et al.* (1998) found that the essential features of their rupture histories were often quite similar. Although earthquakes often possessed more than one distinct pulse, the averaged moment release rates typically increased steadily to a peak value, then decreased steadily at about half that rate, and finally died out more slowly near the end (Fig. 6.5). Thus although the observed factor of about four variation in source duration is real, the variability in duration for high-moment-release portions of the source is somewhat less.

6.1.2 Spatial extent and rupture velocity

For earthquakes with magnitudes M_W of about 6 or more, it is often possible to pick onset times for distinct subevents within the rupture (Fig. 6.6). If these are visible at several stations one can determine their relative locations. For very large deep earthquakes this demonstrates that ruptures have a spatial extent of at least 60 km (see review by Willemann and Frohlich, 1987). However, determining the

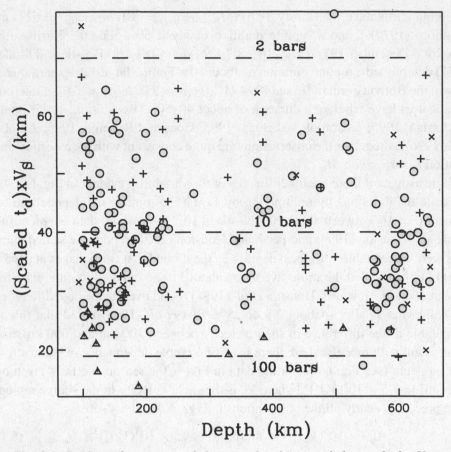

Fig. 6.4 Product of moment-scaled source duration t_d and shear velocity V_s vs focal depth for the data in Fig. 6.3. Data are scaled to correspond to an earthquake with moment M_o of 10^{19} N-m, i.e. scaled duration is observed duration t_d multiplied by $(10^{19}$ N-m$/M_o)^{1/3}$. Dashed lines indicate equivalent static stress drops from eq. 6.9 assuming that $V_{rupture} = 0.8\ V_s$ and $c = 7/2$. Note that there is no obvious dependence of stress drop on depth.

number and timing of subevents is somewhat arbitrary, and thus when different investigators study the same earthquake they do not always obtain identical results. For example, whereas Furumoto (1977) reported seven rupture subevents for the 31 July 1970 Columbia earthquake, Estabrook (1999) could identify only three to five. Moreover, while both agreed that the largest subevent began about 25 seconds after the onset of rupture, Furumoto located it at a distance of 60 km from the hypocenter, while all of Estabrook's models placed it at a distance of 100 km or more. From an analysis of normal modes, Mendiguren and Aki (1978) estimated a rupture length of 123–183 km, and a width of 30 km.

However, the reported subevent locations do provide one means to constrain $V_{rupture}$, the velocity of the rupture front. A comparison of the relative distances and

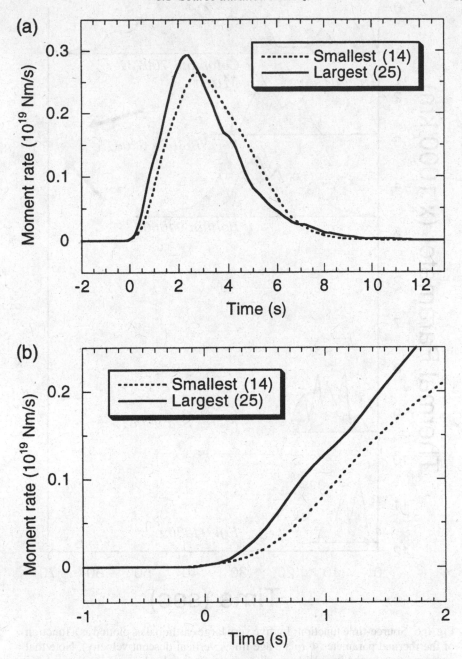

Fig. 6.5 Averages of moment-scaled source-time functions for the 14 smallest and 25 largest earthquakes reported by Houston *et al.* (1998). Amplitudes have been scaled by a factor of $(10^{19}$ N-m$/M_0)^{2/3}$ and time scales have been scaled by factor of $(10^{19}$ N-m$/M_0)^{1/3}$, so that all events have equal weight and average corresponds to an earthquake with moment 10^{19} N-m. The slower start and smoother shape for the smaller earthquakes may be due to the effects of attenuation; otherwise there is little difference between the two groups. Reproduced from Houston *et al.* (1998).

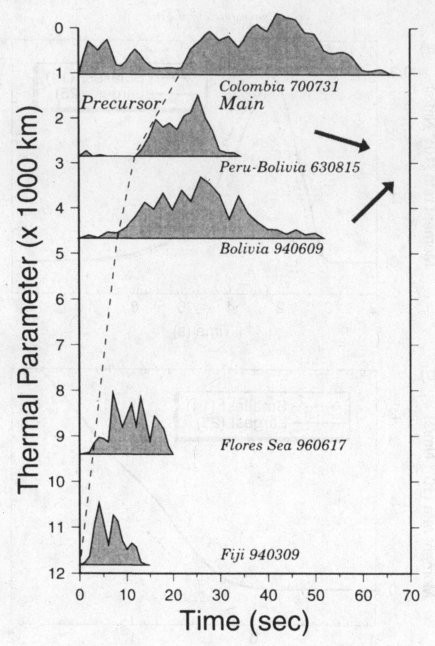

Fig. 6.6 Source-time functions for five very large earthquakes plotted as a function of the thermal parameter Φ (plate age times vertical descent velocity). Note that none are simple and impulsive, as all consist of several subevents, and most have a precursory episode occurring prior to the main rupture. The 1994 Bolivia and 1963 Peru–Bolivia events have about the same thermal parameter (as indicated by the arrows) but are plotted at different positions for clarity. The dashed line separating the precursory and main ruptures suggested to Estabrook (1999) that the presence of higher temperatures and smaller thermal parameters delays the main rupture; probably because the initial rupture velocity is low. Figure reproduced from Estabrook (1999).

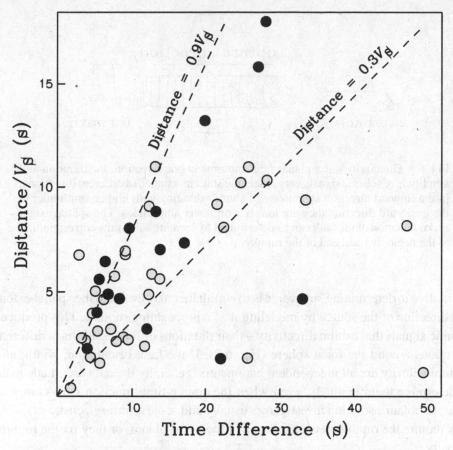

Fig. 6.7 Distance and time difference between subevents within seismic signals from deep earthquakes. Distance is normalized by dividing observed distance by the IASPEI1991 shear wave velocity V_s (Kennett, 1991); dashed lines correspond to a rupture velocities of $V_{rupture}$ of $0.3V_s$ and $0.9V_s$. Lighter circles are data from the review of Willemann and Frohlich (1987); dark circles are additional points from Chen *et al.* (1996), Tibi *et al.* (1999; 2002) and Estabrook (1999).

onset times indicates that nearly all subevents have locations corresponding to a front traveling at a fraction of 0.3 to 0.9 of the shear velocity V_S (Fig. 6.7); thus:

$$0.3V_S < V_{rupture} < 0.9V_S. \tag{6.2}$$

This result is consistent with theoretical analyses of the physics of rupture, which conclude that rupture speeds seldom exceed V_S except in unusual circumstances (e.g., see Dunham *et al.*, 2003; Xia *et al.*, 2004).

While replacing a single point source with a discrete number of subevents makes it possible to fit seismograms with any desired degree of precision, this doesn't necessarily make sense mechanically. It is more plausible that rupture proceeds as a roughly linear front traveling along a more-or-less planar surface. Thus, an

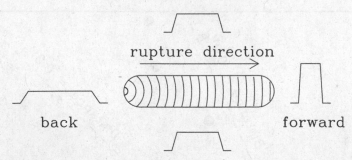

Fig. 6.8 Directivity – if a planar fault ruptures in one direction, the duration and amplitude of seismic signals vary at seismic stations situated in different directions; in the forward direction signals are of shorter duration with higher amplitude; in the backward direction they are longer with lower amplitudes. The spatial extent and orientation of the fault can be determined by locating subevents corresponding to the beginning and end of the rupture.

alternative to determining subevents is to simultaneously evaluate the spatial extent and direction of the source by modeling it as a propagating rupture. This produces seismic signals that exhibit directivity – their durations differ at stations in different directions around the focal sphere (Fig. 6.8). If the fault orientation, width, and rupture velocity are all independent parameters, generally the data aren't adequate to determine them uniquely, even when the source-time function isn't complex. Thus, to obtain a solution investigators usually add some arbitrary constraints; e.g., they require the rupture to occur on one of the nodal planes, or they fix the rupture velocity.

The arbitrariness of these constraints diminishes the significance of comparisons of results reported by different investigators. The literature is fraught with instances where investigators apply less restrictive constraints and determine source models that are significantly different from previous studies. For example, for the 12 May 1990 Sakhalin Island earthquake ($M_W = 7.2$, $h_{CMT} = 613$ km), Kuge (1994) modeled the source as two subevents with nearly identical focal mechanisms, and concluded that the rupture velocity approached the P velocity; subsequently Glennon and Chen (1995) demonstrated that this conclusion wasn't necessary if the two mechanisms were somewhat different.

Nevertheless, a survey of the literature (Table 6.1 and Fig. 6.9) indicates that about 80% of the rupture velocities reported are between 0.3 V_S and 0.9 V_S, in substantial agreement with eq. 6.2 and Fig. 6.7. It also is clear that some of the reported variation is real, at least for the largest events; e.g., while most investigations of the 9 June 1994 Bolivia earthquake find a rupture velocity of about 1–2 km/s, all investigations of the 9 March 1994 Tonga earthquake ($M_{W(CMT)} = 7.6$, $h_{CMT} = 568$ km) obtain velocities of 3–5 km/s.

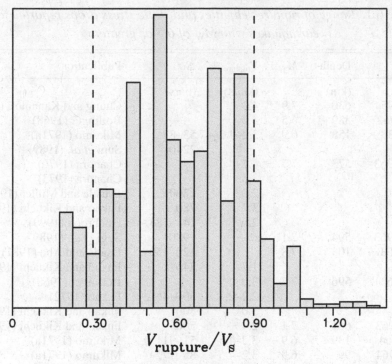

Fig. 6.9 Histogram of rupture velocities reported in investigations that modeled the source process (Table 6.1). Velocities are normalized by the IASPEI1991 shear velocity V_s at the hypocenter. About 80% of velocities reported fall between the dashed lines at $0.3V_s$ and $0.9V_s$.

The variety of models applied to evaluate source properties can be bewildering. These models include:

- A single pulse (1P) – the source model is a single pulse, with the parameters describing it determined from measurements in either the time or frequency domain (e.g., Mikumo, 1971b; Pennington and Isacks, 1979; Oncescu, 1989; Schneider, 1984);
- Subevents in time domain (SE-TD) – investigators pick onset times of individual subevents directly from seismograms, and use them to determine relative locations (e.g., Furumoto, 1977; Brüstle and Müller, 1987; Beck *et al.*, 1995);
- Many pulses on a grid (MP-G) – seismologists construct synthetic seismograms assuming that the source is a discrete number of pulses, usually constrained to lie on a grid, and sometimes exhibiting directivity. They add pulses one at a time until the synthetics have an acceptable fit to observations (e.g., Fukao and Kikuchi, 1987; Glennon and Chen, 1995; Estabrook, 1999);
- Slip on a grid (S-G) – scientists invert waveforms or moment rate functions to determine the pattern of slip in space and time on a grid; to ensure a unique solution they apply various constraints on the direction of slip and the rupture velocity (e.g., Ihmlé, 1998; Antolik *et al.*, 1999).

Table 6.1. *Range of rupture velocities and static stress drops reported for
earthquakes with M_W of 6.3 or greater.*

Date	Depth	M_W	$V_{rupture}$	$\Delta\sigma$	Publication
	(km)		(km/s)	(bars)	
29 Mar 1954	630	7.9	4.3		Chung and Kanamori (1976)
29 Sep 1962	589	6.5	3.0		Bollinger (1968)
4 Jul 1963	158	6.9	3.2–4.5	53–82	Mikumo (1971a)
				2240	Sugi *et al.* (1989)
15 Aug 1963	573	7.7	4.7		Chandra (1970)
			3.3		Chandra (1973)
			2.0	1300	Brüstle and Müller (1987)
			3.3	780	Fukao and Kikuchi (1987)
			2.0	148–183	Estabrook (1999)
25 Aug 1963	594	7.1		1980	Sugi *et al.* (1989)
4 Nov 1963	108	7.8		120–150	Osada and Abe (1981)
			1.7	740	Fukao and Kikuchi (1987)
9 Nov 1963	596	7.7	3.0		Bollinger (1968)
			2.2–2.5	300–460	Fukao (1972)
			3.0	800	Fukao and Kikuchi (1987)
15 Dec 1963	661	7.1	2.4	500	Fukao and Kikuchi (1987)
20 Jan 1964	139	6.9	3.2–4.5	51–91	Mikumo (1971a)
24 Apr 1964	99	6.9	3.2–4.5	85–172	Mikumo (1971a)
26 May 1964	114	7.5	2.0–3.0	120	Abe (1972)
			2.8	1590	Fukao and Kikuchi (1987)
18 Oct 1964	576	7.0		290	Fukao (1970)
18 Mar 1965	219	6.3		75–94	Chung and Kanamori (1980)
				257–642	Chung and Kanamori (1980)
21 Nov 1965	132	6.6	2.5	60	Fukao and Kikuchi (1987)
4 Feb 1966	196	6.4	3.2–4.5	44–78	Mikumo (1971a)
17 Mar 1966	643	6.3		54–964	Chung and Kanamori (1980)
22 Jun 1966	516	7.0	4.6		Oike (1969)
12 Aug 1967	141	6.7		89–100	Chung and Kanamori (1980)
4 Sep 1967	223	6.4		553–4617	Chung and Kanamori (1980)
9 Oct 1967	657	7.3	4.5–5.0	1200	Sasatani (1980)
				4480	Sugi *et al.* (1989)
1 Dec 1967	147	6.4		81–302	Mikumo (1971b)
27 Dec 1967	91	7.0		370	Sugi *et al.* (1989)
28 Feb 1968	349	6.3		118–300	Mikumo (1971b)
11 Mar 1968	87	6.3		20–230	Chung and Kanamori (1980)
14 May 1968	164	6.7		69–283	Mikumo (1971b)
7 Oct 1968	490	7.3		202–620	Mikumo (1971b)
			1.7	920	Fukao and Kikuchi (1987)
19 Jan 1969	254	7.0		81–381	Mikumo (1971b)
10 Feb 1969	670	7.3		940	Sugi *et al.* (1989)
31 Mar 1969	424	6.7	4.7		Oike (1971)
				106–441	Mikumo (1971b)
18 Dec 1969	344	6.4		156–661	Mikumo (1971b)

Table 6.1. (*cont.*)

Date	Depth	M_W	$V_{rupture}$	$\Delta\sigma$	Publication
	(km)		(km/s)	(bars)	
27 May 1970	391	7.2		297–758	Mikumo (1971b)
				1930	Sugi *et al.* (1989)
31 Jul 1970	623	8.1	2.4	260	Furumoto (1977)
			3.8	300	Mendiguren and Aki (1978)
			1.3	680	Fukao and Kikuchi (1987)
			2.0–4.0	394	Estabrook (1999)
30 Aug 1970	650	7.3	4.0	320	Sasatani (1980)
			2.5	830	Fukao and Kikuchi (1987)
29 Jan 1971	525	6.9	4.5	240	Sasatani (1980)
27 Jul 1971	118	7.5		720	Sugi *et al.* (1989)
21 Nov 1971	119	7.1	2.2	540	Fukao and Kikuchi (1987)
24 Dec 1971	99	7.4	1.1	580	Fukao and Kikuchi (1987)
14 Feb 1972	101	7.3		680	Sugi *et al.* (1989)
30 Mar 1972	478	7.3	2.3	460	Sasatani (1980)
				3200	Sugi *et al.* (1989)
4 Apr 1972	399	6.8		160	Fukao and Kikuchi (1987)
11 Jun 1972	333	7.8	2.1	1870	Fukao and Kikuchi (1987)
21 Jul 1973	430	6.3		69–420	Chung and Kanamori (1980)
10 Sep 1973	557	6.7	2.3	440	Koyama (1975)
29 Sep 1973	593	7.8	3.7	539	Koyama (1978)
			3.3	740	Fukao and Kikuchi (1987)
			3.0–4.0	260	Wu and Chen (2001)
28 Dec 1973	540	6.8		960	Sugi *et al.* (1989)
4 Jun 1974	284	6.7		76–353	Chung and Kanamori (1980)
				550	Sugi *et al.* (1989)
8 Nov 1974	122	6.3	3.5–4.0	130	Sasatani (1980)
29 Nov 1974	408	7.1		910	Sugi *et al.* (1989)
22 Feb 1975	374	6.9		96–585	Chung and Kanamori (1980)
			3.0–3.5	610	Sasatani (1980)
				780	Sugi *et al.* (1989)
23 Jan 1976	623	6.8		450	Fukao and Kikuchi (1987)
4 Mar 1977	84	7.5	4.1–5.0		Müller *et al.* (1978)
			3.7	52	Räkers and Müller (1982)
7 Mar 1978	434	7.1	3.6	580	Fukao and Kikuchi (1987)
13 May 1978	440	6.5	2.7	160	Fukao and Kikuchi (1987)
21 Jun 1978	402	6.5	2.7–3.7	21–101	Choy and Boatwright (1981)
6 Dec 1978	181	7.8	4.8	100	Brüstle and Müller (1987)
5 Aug 1979	229	6.5		840	Sugi *et al.* (1989)
13 Apr 1980	166	7.6		890	Sugi *et al.* (1989)
22 Apr 1980	388	6.4		430	Sugi *et al.* (1989)
24 Oct 1980	63	7.2	3.3	400	Brüstle and Müller (1987)
28 Apr 1981	553	6.7		1880	Sugi *et al.* (1989)
28 Sep 1981	309	6.6		580	Sugi *et al.* (1989)
7 Oct 1981	625	6.5		4910	Sugi *et al.* (1989)

(*cont.*)

Table 6.1. (*cont.*)

Date	Depth	M_W	$V_{rupture}$	$\Delta\sigma$	Publication
	(km)		(km/s)	(bars)	
22 Jun 1982	473	7.5		1960	Sugi *et al.* (1989)
6 Sep 1982	156	6.8		1010	Sugi *et al.* (1989)
12 Apr 1983	111	7.0		860	Sugi *et al.* (1989)
1 Jun 1983	186	6.5		380	Sugi *et al.* (1989)
21 Dec 1983	600	6.9		650	Sugi *et al.* (1989)
1 Jan 1984	384	7.2		470	Sugi *et al.* (1989)
6 Mar 1984	446	7.4	4.2–4.8	93	Yoshida (1988)
				570	Sugi *et al.* (1989)
24 Apr 1984	395	6.9	4.0	91	Yoshida (1988)
30 Aug 1986	133	7.2		481–1203	Oncescu (1989)
18 May 1987	73	6.3	1.0–4.0		Glennon and Chen (1995)
7 Sep 1988	491	6.6	3.0–5.0		Glennon and Chen (1995)
12 May 1990	613	7.2	7.0–10.0		Kuge (1994)
23 Jun 1991	570	7.3	3.5	50	Tibi *et al.* (2003)
11 Oct 1992	141	7.4	4.0	230	Tibi *et al.* (2002)
15 Jan 1993	100	7.6		420	Takeo *et al.* (1993)
10 Jan 1994	604	6.9	2.5–4.8	50–150	Antolik *et al.* (1999)
9 Mar 1994	568	7.6	4.0–5.0	260	Goes and Ritsema (1995)
			3.0	140	Lundgren and Giardini (1995)
			3.0–4.5	70	Goes *et al.* (1997)
			4.1	130	McGuire *et al.* (1997)
			5.0	246	Tibi *et al.* (1999)
23 May 1994	70	6.3		15	Quintanar *et al.* (1999)
9 Jun 1994	647	8.2	1.0	1100	Kikuchi and Kanamori (1994)
			1.0–2.0		Beck *et al.* (1995)
			3.0–4.0		Chen (1995)
			1.5–4.0	500	Estabrook and Bock (1995)
			2.0–3.0	2830	Goes and Ritsema (1995)
			1.0	1140	Lundgren and Giardini (1995)
				170–310	Myers *et al.* (1995)
			1.0		Silver *et al.* (1995)
			1.0–3.0	710	Goes *et al.* (1997)
			1.4–2.0	1500	Antolik *et al.* (1996)
			1.1–2.0	700–1900	Ihmlé (1998)
				50–100	Bouchon and Ihmlé (1999)
21 Jul 1994	489	7.3	4.0		Chen *et al.* (1996)
			4.0–5.0	120–300	Antolik *et al.* (1999)
			3.0	280	Tibi *et al.* (2003)
4 Oct 1994	68	8.3		110	Kikuchi and Kanamori (1995)
23 Aug 1995	599	7.1	3.5–5.0	200–480	Antolik *et al.* (1999)
			4.0	90	Tibi *et al.* (2001)
21 Oct 1995	164	7.2		65	Rebollar *et al.* (1999)
			2.5	100	Tibi *et al.* (2002)

Table 6.1. (*cont.*)

Date	Depth	M_W	$V_{rupture}$	$\Delta\sigma$	Publication
	(km)		(km/s)	(bars)	
25 Dec 1995	161	7.1	3.0	70	Tibi *et al.* (2002)
17 Jun 1996	584	7.9	2.0–4.0	160–560	Goes *et al.* (1997)
			4.0–5.0		Tinker *et al.* (1998)
			3.5	70–200	Wiens (1998)
			4.5	100	Antolik *et al.* (1999)
			4.0	209	Tibi *et al.* (1999)
5 Aug 1996	555	7.4	3.2–4.5	100–240	Antolik *et al.* (1999)
			3.0	200	Tibi *et al* (2003)
23 Jan 1997	282	7.1	3.5	80	Tibi *et al.* (2002)
14 Oct 1997	166	7.7	3.0	340	Tibi *et al.* (2002)
28 Oct 1997	119	7.2	2.5	110	Tibi *et al.* (2002)
29 Mar 1998	554	7.2	3.0–4.0	200	Tibi *et al* (2003)
20 Aug 1998	426	7.1	2.0	140	Tibi *et al* (2003)
8 Apr 1999	575	7.1	2.5	230	Tibi *et al* (2003)
			2.0	160	Venkataraman and Kanamori (2004)
6 Aug 2000	411	7.3	2.0	90	Tibi *et al* (2003)
6 Aug 2000	586	7.1	2.0	120	Tibi *et al.* (2003)
28 Jun 2002	582	7.3	2.0	270	Tibi *et al* (2003)

The S-G inversion studies have provided some enlightening insight into what the other methods actually measure. Ihmlé (1998) analyzed the 9 June 1994 Bolivia earthquake using two methods, an S-G method and a MP-G method. He concluded that subevent locations were accurate in the sense that they generally occupied the geographic center of areas that were undergoing significant slip (Fig. 6.10). Paradoxically, however, there was often little or no coterminous slip at the centroid location itself because the main slip front took place on the boundary of a circle that surrounded it. This suggests that parameters such as rupture velocity, fault area, etc., derived from 1P, SE-TD, MP-G methods are not highly accurate – we should avoid comparing values for different earthquakes and interpret them only as approximate representations of fault mechanics.

What can we safely infer about source dimensions from the available research about earthquake sources? For smaller deep earthquakes our best estimates come from the source duration t_d. In particular, if an earthquake nucleates at its hypocenter and a rupture front proceeds linearly away with velocity $V_{rupture}$ for a time t_d, the spatial extent of the source will just be $V_{rupture}t_d$. If we assume that $V_{rupture} < 0.8 V_S$,

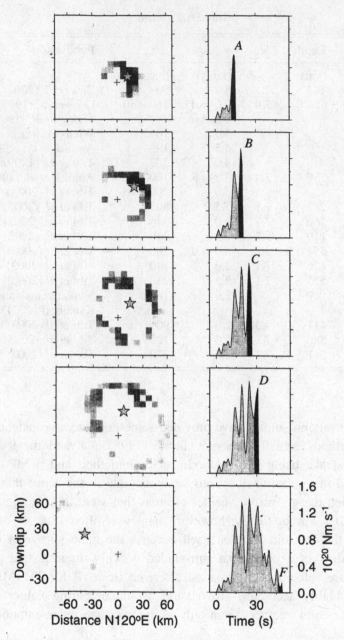

Fig. 6.10 Five snapshots of propagating rupture geometry for the 9 June 1994 Bolivia earthquake. On the right is the developing source-time function, with the 3-s interval corresponding to the snapshot on the left indicated by dark shading. The snapshot shows the slip accumulated during the highlighted time interval; the stars indicate the instantaneous centroid of moment release for the five intervals labeled A, B, C, D, and F. Note that the centroid locations do not always correspond to regions undergoing slip. Figure reproduced from Ihmlé (1998).

then an upper limit on the source extent L_{source} will be:

$$L_{source} = 0.8 V_S t_d. \tag{6.3}$$

Now, between 100 km and 650 km V_S is approximately 5 km/s. Thus, L_{source} is about 4 km for earthquakes with t_d of 1 s and M_W of 5.5, but may be 160–200 km when t_d is 40–50 s and M_W is 7.5–8.3.

For large deep earthquakes the source dimensions determined from directivity or subevent analysis (Fig. 6.6) are often somewhat smaller than that determined by the above $V_{rupture} t_d$ method. For example, for the 9 June 1994 Bolivia earthquake, Houston *et al.* (1998) reported a duration of 42 s while Beck *et al.* (1995) determined that the source was confined to a region with dimensions of approximately 30 km × 50 km. For the Peru–Bolivia earthquake of 15 August 1963 ($M_{W(CMT-Hist)} = 7.7$; $h_{EVC} = 552$ km) the maximum subevent separation was about 60 km (Brüstle and Müller, 1987), but Fukao and Kikuchi (1987) report a duration of 37 s; for the Colombia earthquake of 31 July 1970, Estabrook (1999) reports a duration of 65 s but the spatial extent of the source as determined by Furumoto (1977) was about 70 km. In Willemann and Frohlich's (1987) review of subevent observations, the only event with a larger extent was the 6 December 1978 Kuril earthquake with a spatial extent of 150 km (Brüstle and Müller, 1987). Presumably the spatial extent of large-earthquake sources is smaller than that calculated using eq. 6.3 because the subevents may not all lie along a straight line, because the rupture doesn't necessarily proceed at a constant velocity, and because subevent locations may lie at the centroid of a larger region of slip (see Fig. 6.10).

At present the most plausible estimates of the extent of rupture for very large deep earthquakes come from investigations which use nonlinear inversion methods to determine slip on a grid. For example, for six large earthquakes, Antolik *et al.* (1996; 1999) have determined slip patterns assuming that rupture occurs on a nodal plane and proceeds at a constant velocity. The resulting slip distributions form roughly circular patterns around the nucleation points, although the amount of slip varies considerably in different directions. For four deep-focus earthquakes which had M_W of 7.0 to 7.3 (10 January 1994, 21 July 1994, 23 August 1995, and 5 August 1996) the final rupture dimensions were all between about 15 km and 25 km. In contrast, the rupture lengths for the 9 June 1994 Bolivia (M_W 8.3) and 17 June 1996 Flores Sea ($M_{W(CMT)} = 7.9$, $h_{CMT} = 584$ km) earthquakes (Figs. 6.10 and 6.11) were about 100 km and 120 km, respectively. For all these shocks the pattern of slip consisted of two or more regions with concentrated slip, separated by regions where there was little or no slip; e.g., for the Bolivia earthquake the main rupture occupied a region with dimensions of about 40 km × 60 km.

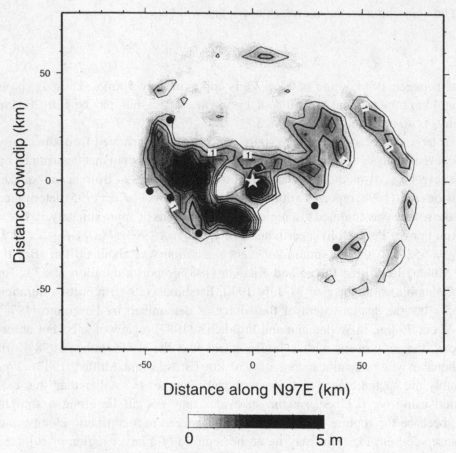

Fig. 6.11 Slip model for the 17 June 1996 Flores Sea earthquake as determined by Antolik *et al.* (1999). They constrained the slip to occur along the south-dipping nodal plane of the focal mechanism, and fixed the rupture velocity at 5.1 km/s. The contour interval is 0.5 m, the star is the hypocenter, and the black filled circles are aftershock locations from Tinker *et al.* (1998). The approximately east–west configuration of slip is parallel to the Wadati–Benioff zone, and thus it appears most of the moment release is contained within the core of the subducting lithosphere. Figure reproduced from Antolik *et al.* (1999).

6.1.3 Radiated energy

Whereas measuring an earthquake's scalar moment M_0 depends on the lower-frequency properties of its seismograms, the higher-frequency signals contain much of its radiated energy $E_{seismic}$. Moreover, mantle attenuation strongly affects the measurement, and so there is even greater station-to-station variability in determinations of $E_{seismic}$ than in M_0. Before broadband digital data were readily available, most investigators derived $E_{seismic}$ from magnitude estimates using various energy–magnitude relationships (e.g., Abe, 1982). However, this wasn't very accurate, especially for deep earthquakes (e.g., see Kikuchi and Fukao, 1988).

An alternative approach was to determine $E_{seismic}$ from measurements of M_o and source duration t_d. For example, Vassiliou and Kanamori (1982) utilized the model result that if the moment release function had a trapezoidal form with rise time t_1 and subsequent duration t_2:

$$E_{seismic} = 2CM_o^2 / \left(t_1^2 t_2\right),$$ (6.4)

where

$$C = 1 / \left(15\pi\rho\, V_P^5\right) + 1 / \left(10\pi\rho V_S^5\right).$$ (6.5)

Although about 95% of the energy is in the S waves,[1] usually the source duration is determined from P waveforms; e.g., Vassiliou and Kanamori (1982) determined t_d from inspection of the initial pulse width of P arrivals. When $E_{seismic}$ is determined from t_d, this implies that graphs of both M_o and $E_{seismic}$ vs t_d would be essentially similar, since both t_1 and t_2 are proportional to t_d and thus both $E_{seismic}$ and M_o are proportional to t_d^3 (see eq. 6.1 and Fig. 6.3). Because duration-derived estimates of $E_{seismic}$ have little information about the higher-frequency properties of the source, they aren't much superior to magnitude-derived estimates; even Vassiliou and Kanamori (1982) stated that they were only "good to an order of magnitude or so."

For this reason, the most credible methods for determining $E_{seismic}$ evaluate the velocity spectra of P and/or S waves (see Fig. 6.1). For analog seismograms this is cumbersome, and thus before digital data became widely available this analysis had been applied to only a few deep earthquakes (e.g., see Wyss, 1970; Choy and Boatwright, 1981; Kikuchi and Fukao, 1988).

However, since 1986 George Choy at the USGS has routinely determined $E_{seismic}$ from broadband velocity spectra for about 100 earthquakes each year[2] and reported the results in monthly bulletins (see Choy and Boatwright, 1995). Subsequently both Newman and Okal (1998) and Winslow and Ruff (1999) have made independent estimates of $E_{seismic}$ for selected events, and both find results that agree generally with Choy's results. For earthquakes with depths greater than 60 km (Fig. 6.12),

---

[1] The two terms in eq. 6.5 correspond to the energy radiated as P and S waves, respectively; substituting $V_S \sim V_P/1.7$ demonstrates that the second term is about 20 times larger than the first.

[2] Considering that the partitioning of energy has fundamental implications about the nature of the mechanical process at the earthquake source, it is remarkable that there have not been more efforts to measure the radiated energy $E_{seismic}$. For more than a decade George Choy has worked virtually alone and with little institutional support to provide measurements of $E_{seismic}$ for larger earthquakes on a routine basis; for these efforts he is to be commended.

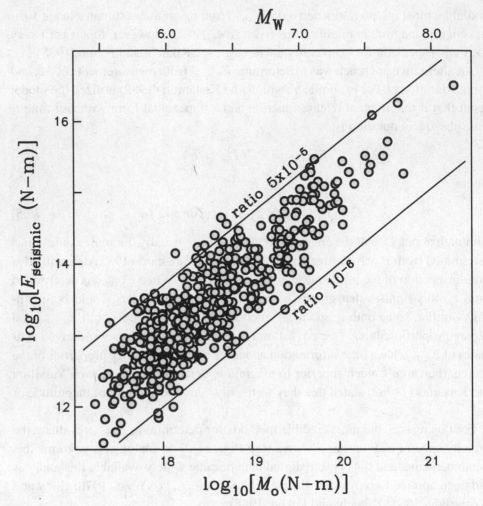

Fig. 6.12 Radiated energy E_{seismic} vs scalar moment M_{o}. Data are 509 earthquakes deeper than 60 km occurring between 1987 and 2004 reported in the USGS Monthly Preliminary Determination of Epicenters (PDE) bulletins. Note that nearly all the data are consistent with ratios of E_{seismic} to M_{o} between 10^{-6} and 5×10^{-5}.

the data are fit by the relationship[3] $E_{\text{seismic}} = 0.8 \times 10^5 \, M_{\text{o}}$, and nearly all the observations satisfy:

$$10^{-6} M_{\text{o}} < E_{\text{seismic}} < 5 \times 10^{-5} M_{\text{o}}. \tag{6.6}$$

---

[3] For earthquakes having focal depths between 160 km and 300 km, for the data in Fig. 6.12 the median of the ratio $E_{\text{seismic}}/M_{\text{o}}$ is or 8×10^{-6}. For earthquakes with focal depths exceeding 300 km the median is slightly less, or 6×10^{-6}.

6.1.4 Derived parameters: static stress drop

Consider a simple model of an earthquake: assume that seismic waves are generated because average slip S occurs over a rectangular fault of length L and width W in a material of rigidity μ. By definition, the change in strain on the fault is just the slip divided by the fault dimension:

$$\Delta\varepsilon = cS/L. \tag{6.7}$$

The constant c is just a fudge factor to account for the fault geometry; it will be on the order of 1.0 but will vary depending on L and W, and whether the fault is rectangular, circular or odd-shaped. The "static stress drop" $\Delta\sigma$ is just the difference between the average shear stress on the fault before and after the earthquake, i.e., $\Delta\sigma = \sigma_{\text{before}} - \sigma_{\text{after}}$. However, by Hook's law this will also equal:

$$\Delta\sigma = \mu\Delta\varepsilon = \mu cS/L = c(\mu AS)/LA = cM_{\text{o}}/LA. \tag{6.8}$$

Here the rigidity μ corresponds to the "spring constant" k in Hook's law, and the strain change $\Delta\varepsilon$ corresponds to the displacement Δx; in the last two equalities we have just multiplied numerator and denominator by fault area A so as to obtain an expression in terms of scalar moment M_{o}, a measurable quantity.

But how can we determine L and A? We might use a directivity analysis. Or, we could determine the area occupied by aftershocks, assuming that this represents rupture area. In the absence of such information, we could make other assumptions; e.g., that L equals the rupture velocity V_{rupture} times the source duration t_{d}; and that A equals L^2; then:

$$\Delta\sigma = cM_{\text{o}}/L^3 = cM_{\text{o}}/(V_{\text{rupture}}t_{\text{d}})^3. \tag{6.9}$$

If we make one additional assumption – that V_{rupture} is a fixed fraction of the shear velocity V_{S} – we can estimate stress drop if we know M_{o} and t_{d}.

Perhaps the most remarkable feature about static stress drop is the vast range in values reported (see Box 6.1). For earthquakes with magnitudes of 6.3 and greater, about 90% of static stress drops reported in the literature are between 50 bars and 1000 bars, although there are outliers (Fig. 6.13 and Table 6.1). Although Table 6.1 seems to imply that different earthquakes have very different static stress drops, the range is little different than that reported for individual events. For example, for the 9 June 1994 Bolivian earthquake Goes and Ritsema (1995) found a static stress drop of 2830 bars, Antolik *et al.* (1996) obtained 1500 bars, Goes *et al.* (1997) reported a stress drop of 710 bars, Myers *et al.* (1995) determined a stress drop of 170–310 bars, and Bouchon and Ihmlé (1999) found it to be 50–100 bars.

What is responsible for these differences? About a factor of four is attributable to different values for the geometry constant c in eq. 6.9; different conclusions about

---

Box 6.1

Why is there such a huge range in reported values for static stress drop?

We derive parameters like the static stress drop $\Delta\sigma$ by measuring features on seismograms and assuming specific models of the rupture process. Often seemingly minor differences in these models can have a huge effect on the values determined for $\Delta\sigma$.

For example, suppose that five seismologists all evaluate seismograms from a large earthquake and conclude that it has a scalar moment M_o of 10^{20} N-m and a source duration t_d of 8 s. At the depth the earthquake occurred the IASPEI1991 shear velocity is 5 km/s.

- Seismologist 1 chooses to apply the most commonly utilized source model where the rupture surface is a circle. For this model the constant c in eq. 6.9 is $7/2$ if L is the circle diameter. Seismologist 1 also assumes that V_{rupture} is $0.8V_S$, and that the fault ruptured across the entire circle diameter so that $V_{\mathrm{rupture}}t_d$ of 32 km corresponds to L; he thus determines $\Delta\sigma$ to be 107 bars.
- Seismologist 2 applies exactly the same source model but makes the quite reasonable assumption that rupture proceeded from the center of the circle to the edge, and thus $V_{\mathrm{rupture}}t_d$ corresponds to the circle radius. She thus finds a $\Delta\sigma$ of 13 bars.
- Seismologist 3 applies the same source model as Seismologist 1 but, after inspecting Fig. 6.9, assumes that V_{rupture} is only $0.6V_S$. He thus believes L is 24 km and finds a $\Delta\sigma$ of 253 bars.
- Seismologist 4 doesn't think a circular source model is appropriate at all. The aftershocks fill a roughly rectangular region with dimensions of 32 km × 20 km. For a rectangular fault model c is $2/\pi$ and from eq. 6.8, $\Delta\sigma$ is 50 bars.
- Seismologist 5 has modeled the rupture in some detail, and has concluded that 50% of the reported moment radiated from a single patch with an area A of 50 km^2. Assuming a circular patch and $L = (4A/\pi)^{1/2}$, this suggests that locally $\Delta\sigma$ was 3445 bars.

---

the fault dimensions and/or area cause even greater differences. And, since most large deep earthquakes can be modeled as spatially distinct, localized subevents separated by low-slip regions, the static stress drop reported depends on whether the analysis strives to find an average, whole-fault value, or the stress drop on the high-slip asperities. For example, for the 9 June 1994 Bolivia earthquake Bouchon and Ihmlé (1999) report that "the maximum stress drop may have reached 1000 bars over a very small area of the fault (about 10 km^2)." Similarly, the static stress drops reported by Sugi *et al.* (1989) are uniformly higher than those reported by others studying the same earthquakes (see Table 6.1), apparently because Sugi's group evaluated subevent- rather than whole-event properties.

For earthquakes with magnitudes smaller than about 6, reported static stress drops are often (but not always) less than those in Fig. 6.13 and Table 6.1. For example (Wyss and Molnar) (1972) used corner frequencies for 19 Tongan deep earthquakes to calculate stress drops of 4–70 bars. For Tongan events with magnitudes between

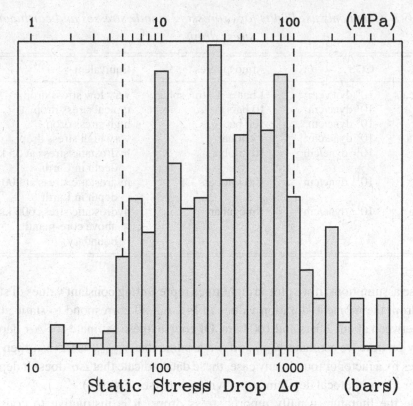

Fig. 6.13 Histogram of static stress drops reported in investigations of earthquakes with magnitudes M_W of 6.3 and greater (Table 6.1). About 80% of reported stress drops fall between the dashed lines at 50 bars and 1000 bars.

3.5 and 5.9, Pennington and Isacks (1979) evaluated pulse durations and concluded that stress drops varied from about 0.1–1.0 bars. Choy and Boatwright (1981) determined stress drops of about 50 bars for subevents of two moderate-sized ($m_b =$ 5.5 and 5.9) earthquakes, one at 520 km depth beneath the Bali Sea, and the other at 371 km depth beneath the Kurils. For 189 microearthquakes in the Bucaramanga nest, Schneider (1984) found static stress drops ranging from 1–248 bars; however, 90% had values less than 20 bars.

Should we thus conclude that small earthquakes tend to have lower static stress drops than large events? For both small and very large earthquakes it is possible to estimate whole-fault static stress drop from measurements of scalar moment M_o and source duration t_d. If we assume that all relevant fault dimensions are equal to the product of t_d and rupture velocity $V_{rupture}$, that $V_{rupture}$ is $0.8\,V_s$, and that the geometric constant c is $7/2$, then:

$$\Delta\sigma = cM_o/L^3 = \left(\frac{7}{2}\right) M_o \bigg/ (0.8V_s t_d)^3 = 6.84\,M_o/(V_s t_d)^3. \qquad (6.10)$$

Table 6.2. *Commonly used units for expressing mantle stresses and earthquake stress drops.*

MKS	CGS	Atmospheres	Equivalent
10^5 Pascals	10^6 dyne/cm^2	1 bar = 1 atmosphere	very low stress drop
1 MPa	10^7 dyne/cm^2	10 bars	typical stress drop
10 MPa	10^8 dyne/cm^2	100 bars	high stress drop
100 MPa	10^9 dyne/cm^2	1 kilobar	very high stress drop
1 GPa	10^{10} dyne/cm^2	10 kilobars	hydrostatic stress at 35 km depth in Earth
10 GPa	10^{11} dyne/cm^2	100 kilobars	hydrostatic stress at 300 km depth in Earth
100 GPa	10^{12} dyne/cm^2	1 megabar	hydrostatic stress 600 km above core-mantle boundary

For these assumptions, if we plot straight lines representing constant values of static stress drop, the moment–duration values in Fig. 6.3 all correspond to static stress drops between about 2 bars and 100 bars. Of course, these estimates of $\Delta\sigma$ depend strongly on how we estimate $V_{rupture}$; if we assume that $V_{rupture}$ is 0.5 V_S, then $\Delta\sigma$ increases by a factor of four. In any case, these data indicate that $\Delta\sigma$ doesn't depend on either M_o or on focal depth in any obvious way (see Fig. 6.4).

While the literature usually reports stress drops, it is instructive to consider earthquake physics in terms of strain ε; i.e., when an earthquake occurs on a fault of length and width L, how much slip S occurs? Suppose quakes only occur when the strain $\varepsilon = S/L$ reaches a certain fraction, say 10^{-5}. Then:

$$M_o = \mu AS = \mu 10^{-5}L^3 = \mu 10^{-5}(0.8V_s t_d)^3 = [10^{-5}\mu(0.8V_s)^3]t_d^3. \quad (6.11)$$

However, this is remarkably like the equation eq. 6.9 describing stress drop $\Delta\sigma$ in terms of t_d; indeed, if we rearrange eq. 6.9:

$$M_o = [\Delta\sigma(0.8V_s)^3/c]t_d^3. \quad (6.12)$$

This suggests that $\Delta\sigma/c$ equals $10^{-5}\mu$. Furthermore, since the constant c is about 7/2 and μ at 400 km depth is about 90 GPa, the stress drop $\Delta\sigma$ is about 32 bars (Table 6.2). Thus, our assumption that earthquakes always occur at a constant strain is equivalent to the assumption that they occur at a constant stress drop. There is, of course, some lack of precision because of uncertainties about the c and about the rupture velocity. For rupture at $0.8V_S$ and c of 7/2, however, the data in Figs. 6.3 and 6.4 imply that earthquakes occur when strain is between about 10^{-6} and 3×10^{-5}.

In addition to static stress drop $\Delta\sigma$, there are other stress drops occasionally reported in the literature. Generally, measuring these requires making very specific assumptions about the source process. For example, if we assume that rupture proceeds outward from the center of a circular fault, Boatwright (1980) determines a quantity he calls the *dynamic stress drop* that is proportional to the initial slope of the moment–rate function. This is of interest because it provides information about fault mechanics in the region where rupture initiates. However, Boatwright's dynamic stress drop has been calculated for only a few earthquakes, and reported values are generally in the same range as the static stress drops (e.g., Choy and Boatwright, 1981; Mori, 1983; Oncescu, 1989; Ruff, 1999).

6.1.5 Derived parameters: apparent stress and effective seismic efficiency

A second stress derivable from seismograms is the apparent stress σ_{apparent}. If E_{total} is the total strain energy released by faulting, we can express E_{total} as the product of the fault area A, the average slip S on the fault surface, and the average shear stress σ_{average} during the rupture, i.e.:

$$E_{\text{total}} = \sigma_{\text{average}} AS. \tag{6.13}$$

For example, if the shear stress changes linearly as the fault slips:

$$\sigma_{\text{average}} = (\sigma_{\text{before}} + \sigma_{\text{after}})/2. \tag{6.14}$$

Since some of the energy may be dissipated by friction on the fault, partial melting of near-source material, or volume changes, only a fraction η of E_{total} is radiated as seismic waves:

$$E_{\text{seismic}} = \eta E_{\text{total}} = \eta \mu \sigma_{\text{average}} AS/\mu = \eta \sigma_{\text{average}} M_{\text{o}}/\mu. \tag{6.15}$$

Thus, while E_{total} and σ_{average} aren't easily measurable, the quantity $\eta \sigma_{\text{average}}$ is – this is the "apparent stress":

$$\sigma_{\text{apparent}} = \eta \sigma_{\text{average}} = \mu E_{\text{seismic}}/M_{\text{o}}, \tag{6.16}$$

and η is the "seismic efficiency."

As with static stress drop, the range of values reported for σ_{apparent} is enormous. For example, Wyss (1970) evaluated E_{seismic} from velocity spectra for 11 deep earthquakes and reported values of σ_{apparent} between 7 bars and 1050 bars,[4] and Oncescu (1989) found σ_{apparent} to be several hundred bars for the Romanian earthquake of 30 August 1986 ($M_{\text{W(CMT)}} = 7.2$, $h_{\text{CMT}} = 133$ km). However, the majority of investigations find σ_{apparent} to be between about 1 bar and 60 bars. The values of $E_{\text{seismic}}/M_{\text{o}}$ reported by Kikuchi and Fukao (1988) and Kikuchi (1992) for

---

[4] However, Wyss determined both E_{seismic} and M_{o} using several alternative methods and was apparently selective when he calcuated σ_{apparent}. Thus, if we use only the radiated energies he determined from velocity spectra and the moments he determined from body waves, the range of values is 1 bar to 294 bars.

Table 6.3. *Average values of* M_o, $E_{seismic}$, *and* t_d *for eight deep-focus earthquakes. Values are as reported in Table 1 of Winslow and Ruff (1999), and as calculated for* $\Delta\sigma$, $\sigma_{apparent}$ *and* η_{eff}. *For each earthquake calculated values in the first line are from eqs. 6.10, 6.16 and 6.19; range of values for* $\Delta\sigma$ *and* η_{eff} *assume* $0.4V_s < V_{rupture} < 0.8V_s$. *Value for* η_{eff} *in subsequent lines is determined from eq. 6.17 using value for* $\Delta\sigma$ *reported in the indicated study.*

Date (study)	M_o (10^{19} N-m)	$E_{seismic}$ (10^{14} J)	t_d (s)	$\Delta\sigma$ (bars)	$\sigma_{apparent}$ (bars)	η_{eff}
1990 May 12	10	21	7.96	82–657	26.4	0.08–0.64
1991 Jun 23	2.8	2.6	16.41	3–22	10.8	1.00–8.0
1992 Jun 11	7.4	8.7	17.35	9–69	9.4	0.27–2.2
1994 Mar 9	16	11	12.23	39–312	8.0	0.05–0.41
Goes *et al.* (1997)				70		0.23
1994 Jun 9	160	320	37.74	12–93	28.8	0.62–4.9
Bouchon and Ihmlé (1999)				50–100		0.58–1.15
Goes *et al.* (1997)				710		0.08
1994 Jul 21	7.7	4.0	10.44	32–259	5.3	0.04–0.33
Antolik *et al.* (1999)				120–300		0.04–0.09
1996 Jun 17	41	55	20.85	19–155	15.8	0.20–1.6
Antolik *et al.* (1999)				100		0.32
Goes *et al.* (1997)				560		0.06
1996 Aug 5	9.8	8.0	10.44	38–303	9.4	0.06–0.50
Antolik *et al.* (1999)				100–240		0.08–0.18

10 deep earthquakes correspond to $\sigma_{apparent}$ between about 1 and 10 bars. For two small deep-focus earthquakes Choy and Boatwright (1981) determined $\sigma_{apparent}$ to be about 60 bars. And the $E_{seismic}$ and M_o data from Winslow and Ruff (1999) (see Table 6.3) correspond to $\sigma_{apparent}$ between about 5 and 30 bars.

By far the most comprehensive data for evaluating $\sigma_{apparent}$ are the values for $E_{seismic}$ and M_o reported by the USGS in the monthly PDE bulletins (Fig. 6.14A). Among more than 500 deep earthquakes (Fig. 6.14B) only two possess a value of $\sigma_{apparent}$ exceeding 60 bars, a value exceeded by numerous shallow earthquakes. For the deep earthquakes the median value is 6 bars and more than 90% are between 1 bar and 30 bars.

There is no obvious way to independently determine η or $\sigma_{average}$ in eq. 6.16. However, if one again assumes that shear stress changes linearly as the fault slips

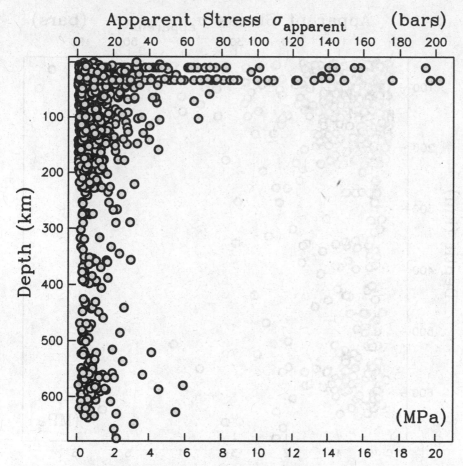

Fig. 6.14A Apparent stress σ_{apparent} vs focal depth for shallow and deep earthquakes. Data are calculated from E_{seismic} and M_{o} using eq. 6.16 for earthquakes occurring between 1987 and 2004 reported in the USGS Monthly Preliminary Determination of Epicenters (PDE) bulletins. At each focal depth we use the PREM value for the rigidity μ.

and that σ_{after} is zero (eq. 6.14), then σ_{average} is $\Delta\sigma/2$ (see Fig. 6.15). The value η_{eff} determined using this assumption is the "effective seismic efficiency":

$$\eta_{\text{eff}} = 2\mu E_{\text{seismic}}/(M_{\text{o}}\Delta\sigma) = 2\sigma_{\text{apparent}}/\Delta\sigma. \qquad (6.17)$$

If we evaluate this using eqs. 6.8 and 6.9 we find:

$$\eta_{\text{eff}} = [2\mu E_{\text{seismic}}LA]/cM_{\text{o}}^2, \qquad (6.18)$$

$$\eta_{\text{eff}} = [2\mu E_{\text{seismic}}(V_{\text{rupture}}t_{\text{d}})^3]/cM_{\text{o}}^2. \qquad (6.19)$$

Most investigators who determine η_{eff} for deep earthquakes usually report values considerably less than 1.0. For example, for ten large earthquakes with focal depths

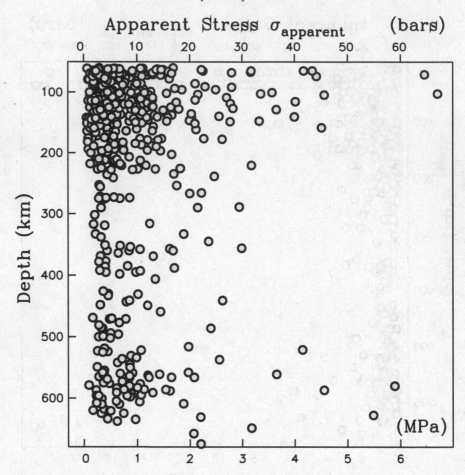

Fig. 6.14B Apparent stress $\sigma_{apparent}$ vs focal depth for deep earthquakes (depth 60 km or greater). Data are as in Figure 6.12, with apparent stress calculated as in Fig. 6.14A.

between 94 and 653 km, Kikuchi (1992) reported values between 0.016 and 0.13. For the earthquakes of 9 March 1994 in Tonga, 9 June 1994 in Bolivia, and 17 June 1996 in the Flores Sea, Estabrook (1999), Winslow and Ruff (1999) and Venkataraman and Kanamori (2004) all found values of η_{eff} between 0.02 and 0.45.

Unfortunately, there are few earthquakes for which $E_{seismic}$, M_o, and either t_d or $\Delta\sigma$ have all been determined. And, many studies reporting η_{eff} don't estimate $E_{seismic}$ from the spectra of broadband data; rather, they estimate it from a moment–magnitude relationship (e.g., Chung and Kanamori, 1980). Finally, any attempt to evaluate η_{eff} using eq. 6.17 will be subject to the same sorts of systematic uncertainties that afflict the determination of $\Delta\sigma$ (see Box 6.1). For example, although Winslow and Ruff (1999) present a table with values for $E_{seismic}$, M_o, and t_d for eight

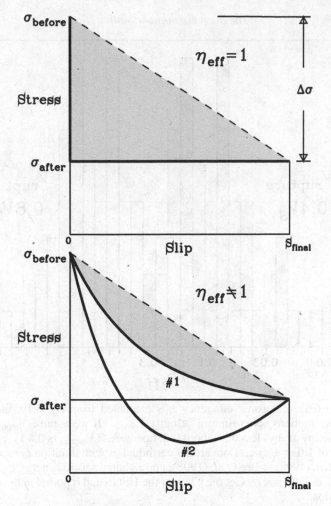

Fig. 6.15 Fault model illustrating relationships between radiated energy E_{seismic}, static stress drop $\Delta\sigma$, seismic efficiency η_{eff} and the stresses σ_{before} and σ_{after} on the fault before and after the earthquake takes place.

(Upper plot) Case where $\eta_{\text{eff}} = 1$. First suppose as rupture begins, the stress on the fault (thick line) drops instantaneously from σ_{before} to σ_{after}. The static stress drop will be $\Delta\sigma = \sigma_{\text{before}} - \sigma_{\text{after}}$, and the energy E_{friction} dissipated by friction on the fault will be $AS_{\text{final}}\sigma_{\text{after}}$ (unshaded rectangular area) where A is the fault area and S_{final} is the average slip. However, since the material surrounding the fault is governed by Hook's law, the stress driving rupture changes linearly with slip (dashed line), the radiated energy E_{seismic} is $AS_{\text{final}} \Delta\sigma/2$ (shaded area). Since the moment M_{o} is μAS_{final}, σ_{apparent} is $\Delta\sigma/2$ and η_{eff} is 1.0 (see eqs. 6.16 and 6.17).

(Lower plot) Case where $\eta_{\text{eff}} \neq 1$. Now suppose that stress on the fault (thick solid line #1) changes smoothly as slip occurs. Once again, E_{seismic} will be proportional to the shaded area between the thick line and the straight dashed line, E_{friction} will be the unshaded area beneath the thick line, and η_{eff} will equal the ratio between the shaded area and the triangular area in the upper graph. Also, in the special case where σ_{after} is zero, η_{eff} is the ratio between E_{seismic} and the total energy dissipated during the earthquake, or $E_{\text{seismic}} + E_{\text{friction}}$. Although η_{eff} in this sketch is less than 1.0, it could exceed 1.0 if during part of the rupture the stress (thick solid line #2) was less than its final value σ_{after}.

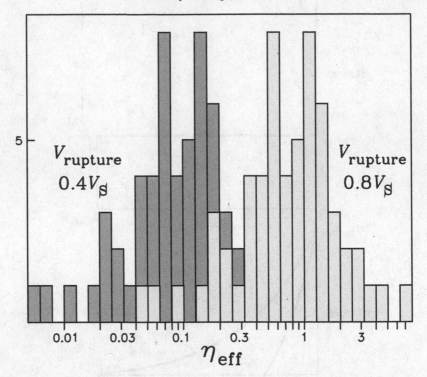

Fig. 6.16 Effective seismic efficiency η_{eff} evaluated from eq. 6.19, using two different assumptions about rupture velocity $V_{rupture}$. If we assume $V_{rupture}$ is 0.4 V_S, η_{eff} is nearly always less than about 0.3; however, if $V_{rupture}$ is 0.8 V_S, η_{eff} often has values of 1.0 or greater. Data are 59 earthquakes with duration t_d reported by Houston *et al.* (1998) or Bos *et al.* (1998), and radiated seismic energy $E_{seismic}$ and moment M_o determined by George Choy at the USGS and reported in the monthly PDE bulletins.

deep-focus earthquakes, they determine η_{eff} only for the three mentioned above. And, for these three, rather than determining $\Delta\sigma$ from eq. 6.10, they use eq. 6.17 and values for $\Delta\sigma$ reported by Goes and Ritsema (1995) and Goes *et al.* (1997). If we instead apply eq. 6.19 to their data, it suggests that in some cases η_{eff} may be 1.0 or even higher (see Table 6.3).

So, is it appropriate to conclude that η_{eff} is typically much smaller than 1.0? The origin of the uncertainty about value η_{eff} is an absence of agreement about how to determine static stress drop $\Delta\sigma$. For example, suppose we apply eqs. 6.9 and 6.17, using $E_{seismic}$ and M_o reported by the USGS and t_d reported by Houston *et al.* (1998) and Bos *et al.* (1998). Then, our conclusions about η_{eff} depend critically on the values chosen for the rupture velocity $V_{rupture}$ (Fig. 6.16): if $V_{rupture}$ is less than half of V_S, then η_{eff} is generally much less than 1.0. However, as $V_{rupture}$ approaches V_S, η_{eff} is often on the order of 1.0 or even greater. Because as yet there is little unanimity in the seismological community about how to determine $\Delta\sigma$, we must conclude that η_{eff} is uncertain as well.

6.1.6 Are deep earthquake sources simple and impulsive?

One observation that seismologists commonly make is that "deep earthquakes produce simple, impulsive P waves." Is this true?[5] Or, more precisely, how is this true?

First, it is not true that deep earthquake ruptures are simple, at least in the sense that their seismic signals consist of a single pulse with no internal structure. Instead, most careful investigations of P-wave signals find that their ruptures usually possess distinct subevents whenever a deep earthquake is large enough (say, $M_W > 5.8$) so that short-period instrumentation and regional attenuation allow one to observe them (e.g., Choy and Boatwright, 1981; Fukao and Kikuchi, 1987; Houston *et al.*, 1998). However, there is limited evidence that deep earthquake ruptures are somewhat simpler than ruptures of comparable-sized shallow earthquakes. Houston (2001) evaluated moment-scaled source-time functions for 255 earthquakes and concluded that sources shallower than 40 km had more zero crossings (i.e., subevents) than those at greater depths.

Second, it is only partly true that deep earthquake sources are impulsive, in the sense that they possess higher amplitude, higher stress drop signals with shorter time durations than similar-sized shallow earthquakes. We have seen above that there is a vast range of static stress drops reported for deep earthquakes; in an average sense the reported values may be slightly higher than for shallow events.[6] However, stress drop is also highly variable for shallow events and may depend on tectonic conditions; e.g., Scholz *et al.* (1986) suggest that shallow intraplate earthquakes typically have static stress drops about six times higher than shallow interplate earthquakes. Thus their stress drops may be comparable to those of deep earthquakes plotted in Figs. 6.3 and 6.4. Houston (2001) did observe depth variation in moment-scaled source-time functions, finding that event durations were briefer at depths exceeding 40 km, and shortest of all beneath 550 km. However, her scaling method did not include the effect of shear velocity variations (see Fig. 6.4) which accounts for much of the difference she observed.

Finally, it is certainly true that deep earthquakes commonly produce very nice seismograms with strong, high-amplitude body wave phase arrivals; and this is true less often for shallow earthquakes. However, this happens because shallow earthquakes occur near the earth's surface in a highly heterogeneous environment where there are generally distinct crustal layers (even oceans); hence, shallow earthquake body wave arrivals are cluttered up with numerous reflected and converted phases from rays that reverberate within these layers. Shallow earthquakes are also more likely to possess aftershock sequences which may not be temporally distinct from

---

[5] This question is important for discussions of the earthquake source process; e.g., both Chung and Kanamori (1980) and Vassiliou and Kanamori (1982) assumed that deep sources were simple and impulsive to derive estimates of radiated energy and static stress drop from measurements of source duration.

[6] Because of the various systematic difficulties in determining static stress drop, I am not yet convinced that typical static stress drops are different for shallow and deep earthquakes.

Table 6.4. *Examples of* 3×3 *symmetric tensors commonly used in geology and physics. Tensors express the relationship between two non-parallel vectors v_1 and v_2, such that $v_2 = Tv_1$.*

v_2	v_1	Tensor name	Usual symbol
Angular momentum	Angular velocity	Moment of Inertia	**I**
Force/area acting on plane	Unit vector perpedicular to plane	Stress tensor	σ
Deformation	Unit direction vector	Strain tensor	ε
Seismic amplitude	Unit direction vector along ray	Moment tensor	**M**

the main rupture. Thus, while it is true that many deep earthquakes often do produce simple, impulsive P waves, this fact alone doesn't tell us much about the mechanical differences between the sources of shallow and deep earthquakes.

6.2 Directional properties of seismic radiation

6.2.1 Possible kinds of earthquake sources – the moment tensor

What are the essential, model-independent features of seismograms that tell us about the nature of the earthquake source? In addition to the scalar moment M_o, the duration t_d, and the radiated seismic energy $E_{seismic}$, there is important information in the radiation pattern – the amplitude variations of the signal in different directions around the focal region.

The moment tensor has proved to be an extraordinarily successful means for describing the different radiation patterns that are possible (Fig. 6.17). Like other oft-used tensors in geology and physics (Table 6.4), the moment tensor is just a 3×3 matrix rule describing the relationship between two vectors – one being a unit vector describing the direction a ray takes as it leaves the focal region, and the other proportional to the amplitude of seismic waves in the three directions in space. Given the moment tensor, one can calculate the amplitudes of seismic signals at different stations surrounding the hypocenter; or, given the amplitudes of P and S waves at enough stations, one can determine the moment tensor (Fig. 6.18).

For all realizable deep earthquake sources[7] the moment tensor **M** is a symmetric matrix and thus has six independent matrix elements m_{ij}. However, for conceptualizing different earthquake sources the m_{ij} are not usually the most useful parameterization of **M**; specifying just its orientation in space requires three

[7] Moment tensors don't represent the seismic radiation pattern which occurs when external forces or torques are applied at the focus. This can't be the case for events such as earthquakes that occur well below the earth's surface. On the surface, however, events such as meteorite impacts or landslides can apply forces at the surface and actually produce such "single-force" radiation patterns (e.g., Ben-Menahem, 1975; Kawakatsu, 1989).

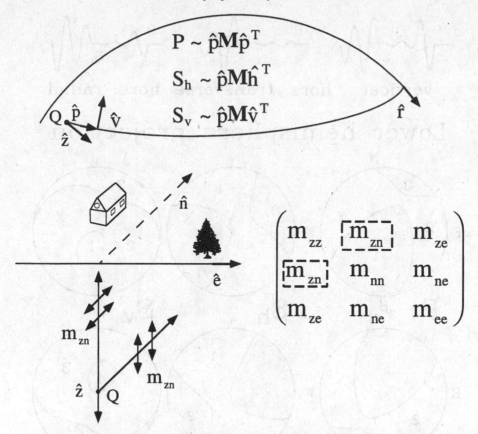

Fig. 6.17 What is a moment tensor **M**? **M** is a 3×3 matrix (lower right) which describes how the amplitude of P and S vary with direction around the hypocenter Q (top). For rays leaving along a direction parallel to a unit vector $\vec{p}$, P amplitudes are proportional to $\vec{p}\mathbf{M}\vec{p}^{\mathrm{T}}$. If $\vec{h}$ is a horizontal unit vector perpendicular to $\vec{p}$ and if $\vec{v}$ is a unit vector perpendicular to $\vec{p}$ and $\vec{h}$, then S_H and S_V amplitudes are proportional to $\vec{p}\mathbf{M}\vec{h}^{\mathrm{T}}$ and $\vec{p}\mathbf{M}\vec{v}^{\mathrm{T}}$, respectively. What do the individual elements of **M** mean? Suppose we represent **M** in a coordinate system with downward vertical-, north-, and east-oriented axes $\vec{z}$, $\vec{n}$, and $\vec{e}$ (lower left). Then, the m_{zn} component of **M** is proportional to the S_V amplitude of the S wave leaving the hypocentral region traveling north; or, the north component of the S wave which leaves the source traveling vertically upwards. Figure reproduced from Frohlich (1996).

parameters.[8] The three remaining parameters, the eigenvalues of **M**, are the principal moments – m_T, m_B, *and* m_P. From these, it is instead often useful to form three other parameters; the first of these is the scalar moment M_o, which defines the overall strength of **M**:

$$M_o = \left[\left(m_T^2 + m_B^2 + m_P^2 \right) / 2 \right]^{1/2} = \left[\Sigma m_{ij}^2 / 2 \right]^{1/2}. \qquad (6.20)$$

---

[8] The CMT catalog kindly specifies the azimuth and plunge angles for the T, B, and P axes, the eigenvectors of **M**. This is, of course, six parameters and not three; this is because they are not all independent.

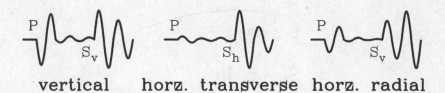

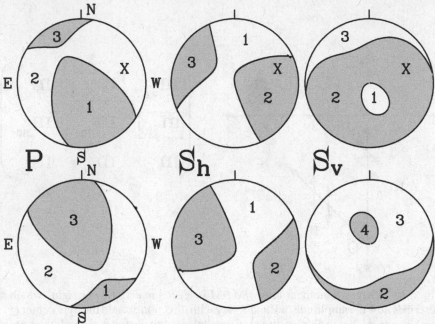

Fig. 6.18 "Beachball" focal mechanisms are useful plots showing how the sign of the first motion of a seismic signal varies with direction as it leaves the hypocentral region. The beachball plots are overhead maps views of a hypothetical sphere surrounding the hypocenter, showing directions where **M** predicts positive (grey) or negative (white) signal amplitudes. Rays leaving vertically plot in the center of the beachball, while rays leaving horizontally plot along the edges. Usually such plots are "lower hemisphere projections," i.e., they depict only the lower inside half of the sphere as viewed from above. "Upper hemisphere projections" depict the upper outside half of the sphere; because of symmetry either projection gives a complete description of the radiation pattern. An individual seismograph station will measure signals which leave from a particular direction on the focal sphere (e.g., see "X" on lower hemisphere projections); at the station the three components of ground motion (vertical; horizontal-transverse; horizontal radial – along $\vec{p}$ in Fig. 6.17) provide data which allow a seismologist to confirm the sign of the first motions of signals leaving the hypocentral region. Except in the irregular cases (see Frohlich, 1996), the nodal lines separating regions of positive and negative signals divide the focal sphere into three regions for P and S_h, and four regions for S_v. The numbers on the beachball plots designate these regions.

The second is the isotropic component M_I:

$$M_I = (m_T + m_B + m_P)/3 = \Sigma m_{ii}/3. \tag{6.21}$$

Finally, the so-called CLVD component f_{clvd} is:

$$f_{clvd} = \max[-(m_B - M_I)/(m_T - M_I); -(m_B - M_I)/(m_P - M_I)]. \tag{6.22}$$

This parameter is zero if **M** is a so-called pure double couple and 0.5 if **M** is a so-called pure compensated linear vector dipole (CLVD); thus, if f_{clvd} is 0.40, we say that the earthquake has an 80% CLVD component.

Do not despair if this seems hopelessly algebraic; there are really only three distinct "pure" source types which are plausible for deep earthquakes – these are an isotropic source, a double couple source, and a CLVD source (Fig. 6.19). A source with an isotropic component is plausible if solid–solid phase transitions cause or accompany deep earthquakes; since conservation of energy could only allow the phase change to occur if it were accompanied by a decrease in volume, this would cause a uniform component of motion towards the focal region – an implosion. The source is a double couple if the radiation is from slip along a planar fault, or, from slip on most types of curved faults (Frohlich, 1990). Finally, a CLVD source is conceivable if the earthquake rupture occurred simultaneously along two or more properly oriented faults; or, for certain other exotic (and, for deep earthquakes, unlikely) occurrences, as when fluids open a tensile crack (Frohlich, 1994).

6.2.2 Do deep earthquakes have isotropic components?

Occasionally in the business of science, a hypothesis is so appealing that people keep testing it again and again, even though each time the test results come up negative. An example is the hypothesis that deep earthquakes have isotropic components caused by sudden, implosive phase transitions within the upper mantle. This hypothesis first appeared about 1930 (e.g., see Stechschulte, 1932; Leith and Sharpe, 1936) when scientists began to comprehend that deep and shallow earthquakes were different mechanically and that there were phase transitions in the mantle at depths where deep earthquakes occur. As data improved it was proposed again in the 1960s (e.g., see Benioff, 1964; Evison, 1963; 1967) and the 1970s (Dziewonski and Gilbert, 1974; Gilbert and Dziewonski, 1975). And, as we shall see below, it is still being considered today.

Some investigations have found evidence for isotropic components, especially studies analyzing P-wave first motion data or body-wave amplitudes. For example, Randall and Knopoff (1970) determined isotropic components between 5% explosive and 10% implosive for five deep earthquakes; Fitch *et al.* (1980) found

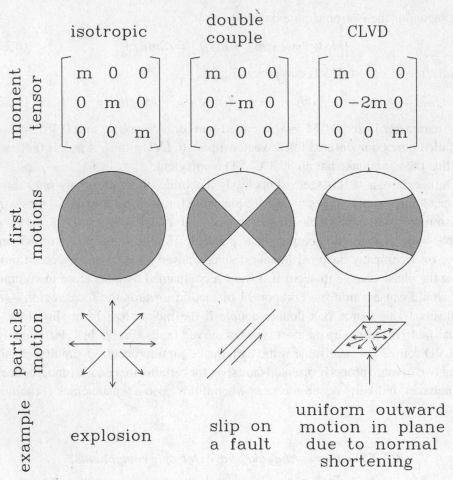

Fig. 6.19   Three "pure" types of mechanisms representable by a moment tensor **M**. For a pure isotropic mechanism such as an implosion or an explosion, the three principal moments of **M** are all equal; the radiation amplitudes are the same in all directions leaving the hypocentral region. For a pure double-couple mechanism one of the principal moments is zero and the other two are equal and opposite; the radiation pattern is quadrantal and is like that produced by slip on a planar fault. For a pure CLVD, two of the principal moments are equal; the third is twice as large but opposite in sign; the radiation pattern is like the rupture of a water balloon, with outward motions along an equatorial region, and inward motions along the poles.

a 9% implosive component for a deep Tonga earthquake. However, more recent studies (e.g., Stimpson and Pearce, 1987) which include uncertainty estimates generally conclude that isotropic components must be less than 10% or so, and, that the uncertainties are large enough so that one cannot confirm the source had an isotropic component.

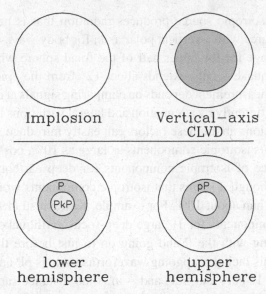

Fig. 6.20 Focal sphere plots illustrating the difficulty of distinguishing isotropic source components. For surface waves, signals from an isotropic implosive source (top left) are similar to those for a vertically-oriented CLVD source (top right); both are generated mainly by the inward-directed horizontal motions (white area) around the edges of the focal sphere. For body waves, stations record the clearest P-wave signals at distances of 30°–95° (darker area, bottom left); however, this covers only a portion of the lower half of the focal sphere. Focal sphere coverage improves somewhat (lighter areas, bottom left and right) if phases besides P are used, such as pP, PcP, and PKP.

Similarly, isotropic components have been found in some investigations of full-waveform data in both the time and frequency domain. Dziewonski and Gilbert (1974) and Gilbert and Dziewonski (1975) reported a large implosive component before two South American deep earthquakes; for the deep-focus Colombia event of 31 July 1970 they concluded it produced more than half the observed scalar moment. But Okal and Geller (1979) argued that an isotropic component was not resolvable, and Russakoff *et al.* (1997) reanalyzed Gilbert and Dziewonski's data and found that there was no evidence for an isotropic component when the analysis incorporated theoretical progress concerning mode splitting and coupling. Similarly, Silver and Jordan (1982) found a large isotropic component for a deep earthquake beneath Honshu (7 March 1978, $M_{W(CMT)} = 7.1$, $h_{CMT} = 434$ km); however, Riedesel (1985) and Kawakatsu (1996) were unable to detect any isotropic component for the event.

Part of the problem with all the above investigations is that it is seldom possible to get coverage of seismic signals over enough of the focal sphere to resolve isotropic components with precision (Fig. 6.20, and see Dufumier and Rivera, 1997). For surface waves, an isotropic source produces radiation that is nearly identical to that for a CLVD source with a vertical polar axis; for body waves, one seldom has much station coverage for the upper half of the focal sphere which – even for a 600 km deep earthquake – only extends about 12° from the epicenter. A further problem is that the measurement depends on comparing signals at different stations; these are affected differently by attenuation and lateral variations in earth structure. Inadequate assumptions about these factors can easily introduce systematic errors that produce spurious isotropic components as large as 10%, possibly even more.

Thus the presence of isotropic components for deep earthquakes still hasn't been confirmed, although it seems that isotropic components would be resolvable if they were larger than about 10%. For example, Kawakatsu (1991; 1996) investigated isotropic components for 19 large deep-focus earthquakes. He used body waveforms beginning with the P and going on to just before the arrival of surface waves; since this includes upgoing waveforms such as pP and sP, downgoing waveforms such as P, PcP, and ScP, and – in some distance ranges – sideways-traveling waves such as PS and PPP. His overall conclusion was that he could distinguish no statistically believable isotropic components. He did find examples of both implosive and explosive sources as large as 8%; but, because his standard deviations were typically between 3% and 12%, no observed isotropic components were significantly different from zero. Hara *et al.* (1996) investigated five deep-focus earthquakes at very low (normal mode) frequencies and reported that "the magnitude of the isotropic component is at most 5 per cent of the deviatoric seismic moment if it exists." None of the investigations of the 9 June 1994 Bolivia earthquake have found a resolvable isotropic component (Hara *et al.*, 1995; 1996; Okal, 1996).

6.2.3 Do deep earthquakes have CLVD components?

Since Knopoff and Randall (1970) and Randall and Knopoff (1970) evaluated body-wave amplitudes of several deep earthquakes and proposed that some have sources with significantly large compensated linear vector dipole (CLVD) components (see Box 6.2), there has been persistent questioning about whether reported CLVD components are real or instead produced by systematic errors in the analysis procedure. Often the discrepancy between various investigators is considerable. For example, for the deviatoric portion of the Colombia earthquake of 31 July 1970, Gilbert and Dziewonski (1975) found a CLVD component that was almost four times larger than the double-couple component ($f_{clvd} = 0.39$). However, Okal and Geller (1979) concluded that the source was indistinguishable from a pure double couple, and Russakoff *et al.* (1997) found values of f_{clvd} of 0.13 or less in all the frequency ranges they analyzed.

Box 6.2
What is a CLVD source?

An implosion or an explosion produces an isotropic source; slip along a planar fault produces a double-couple source; what might produce a CLVD source? Two exotic (but unlikely) possibilities are: solid–solid phase transitions where a sudden rearrangement of crystal structure changes the shape, but not the volume of a material; or the sudden injection of magma or other natural fluids opens a large crack in the earth's interior. A third, much more likely explanation is that two ordinary double-couple earthquakes occur close enough together in space and time so that their seismic signals are superimposed; if these quakes occur on faults with slip and orientation that is just right, they will add together to produce a CLVD source (Fig. 6.21).

A fourth explanation that has been proposed is that CLVD sources are produced by "slip on curved fault surfaces." This is also unlikely, since analysis has shown that slip patterns on many categories of curved faults actually produce double-couple source radiation, especially when there is symmetry either in the shape of the fault surface or in the slip pattern (Fig. 6.22). The fault geometries and slip patterns which do produce sources with significant CLVD components are very unlike what most geologists think of when they visualize a "curved fault surface."

By far the largest objectively-determined catalog of earthquake mechanisms is the Harvard CMT catalog, which at the end of 2004 contained data for 22,365 earthquakes occurring since 1977. Of these, about 5533 had focal depths of 60 km and greater (Fig. 6.23), of which 22% had $f_{clvd} > 0.20$, while 25% had $f_{clvd} < 0.05$. These percentages are only slightly different for the earthquakes shallower than 60 km (19% and 27%). These data clearly indicate that intermediate- and deep-focus earthquakes with reported CLVD components of 40% or more are common, but in this respect they are little different from shallow earthquakes.

Moreover, a few deep earthquakes have been studied in great detail, demonstrating beyond doubt that some undoubtedly possess sources with CLVD components, sometimes as large as 60%–80%. These include two intermediate-depth earthquakes in the Kermadec and Banda Sea areas (26 January 1983, $h_{CMT} = 224$ km, $M_{W(CMT)} = 7.0$; 24 November 1983, $h_{CMT} = 157$ km, $M_{W(CMT)} = 7.4$) investigated by Riedesel (1985), who compared normal mode synthetic seismograms with long-period observations over a range of frequencies. In a similar study, Hara *et al.* (1996) concluded that the 21 July 1994 Japan Sea earthquake ($h_{CMT} = 489$; $M_{W(CMT)} = 7.3$) had a CLVD component of about 30%. Also, Kuge and Kawakatsu (1993) analyzed 21 deep earthquakes which Harvard reported having a CLVD component of 30% or more ($f_{clvd} > 0.15$). They found 17 of these had a CLVD component that was statistically different from zero, and concluded that

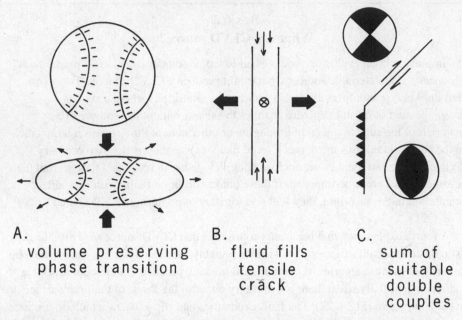

A.
volume preserving
phase transition

B.
fluid fills
tensile
crack

C.
sum of
suitable
double
couples

Fig. 6.21 Possible models for a CLVD source: a pure CLVD source has particle motion that is inward (outward) along one axis, and outward (inward) along
the two remaining axes, with no net volume change. This might occur: (a) if a
phase transition caused a spherical volume to become disk-shaped, with no net
change in volume; (b) when fluid suddenly opens a tensile crack; or (c) if two
suitably-oriented double-couple mechanisms occur simultaneously. Figure reproduced from Frohlich *et al.* (1989).

12 had CLVD components of 40% or more. Kuge and Kawakatsu (1990; 1992)
also evaluated seismograms for three of these earthquakes; for two they were able
to identify specific double-couple subevents within the P arrivals which together
combined to produce the CLVD source (see Fig. 6.24). This is strong evidence
that systematic errors aren't responsible for all reported CLVD sources; some are
clearly real and produced by the superposition of suitably oriented double-couple
subevents.

Yet, as with isotropic sources, the difficulty of obtaining seismic signals from
many directions over the focal sphere makes it hard to demonstrate convincingly that
most reported CLVD source components aren't caused by systematic errors (e.g.
see Henry *et al.*, 2002). Nowadays various organizations independently determine
CMT for most earthquakes with magnitudes of 5.5 and greater; unfortunately, the
correlation between the sizes of the CLVD components they report is depressingly
low (Fig. 6.25). This is perhaps unsurprising, since in order to find the CLVD
component for a moment tensor **M** one must accurately determine the smallest of
the three principal moments of **M**. Indeed, it is difficult to refute the hypothesis that
the vast majority of earthquakes are pure double couples; if so, the observation that

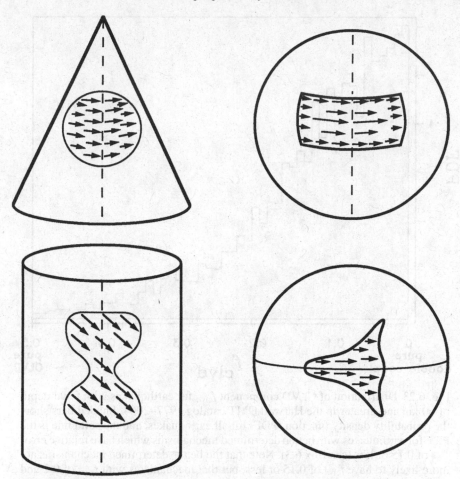

Fig. 6.22 Examples of slip patterns on curved fault surfaces that produce pure double-couple sources, with no CLVD components. The arrowed areas represent the fault surface, and the length and direction of the arrows are proportional to the magnitude and direction of the fault slip. In all cases shown the sources are pure double couples because the fault surfaces possess symmetry about a plane containing the axis of revolution, and because the magnitude and direction of slip is the same for pairs of points having a corresponding relationship to the plane of symmetry. For the sphere at lower right, there is no such symmetry, but the source is a double couple because the slip pattern is symmetric with respect to the equatorial plane. Figure reproduced from Frohlich (1990).

the mean value of f_{clvd} is about 0.1 (Fig. 6.23) simply means that the systematic errors in determining **M**'s principal moments are about 20%.

Moreover, several investigations have constructed synthetic data to estimate the effect of unmodeled, near-source velocity inhomogeneity on determinations of **M** (e.g., see Johnston and Langston, 1984; Woodhouse, 1981; Kuge and Kawakatsu, 1993; Tada and Shimazaki, 1994). For slab-like structures, they generally find that

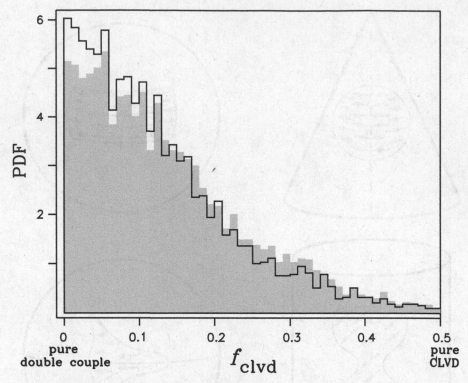

Fig. 6.23 Distribution of CLVD component f_{clvd} for earthquakes with focal depth of 60 km and greater in the Harvard CMT catalog 1977–2004. Shaded bars show the probability density function (PDF) for all earthquakes, and the solid line is the PDF for earthquakes with better-determined mechanisms which have relative error E_{rel} of 0.15 or less (see Box 6.3). Note that the better-determined mechanisms are more likely to have f_{clvd} of 0.15 or less, but that mechanisms with f_{clvd} of 0.2 and greater are numerous in both groups.

when there is reasonable coverage of the focal sphere, spurious CLVD components will have magnitudes similar to that of the unmodeled velocity anomaly. Thus, if the subducting slab is 10% faster than the surrounding mantle or if there is directional anisotropy of 10%, then the spurious CLVD components will be smaller than 10% to 20% of the total moment. Neverthless, while deep earthquakes clearly occur within an inhomogeneous environment – subducting lithospheric slabs – it may be more homogeneous than the crust where ordinary, shallow earthquakes occur.

Because many reported non-double couples may be spurious, some interpretations about their significance may be of doubtful validity. For example, for the USGS moment tensor catalog Kubas and Sipkin (1987) reported that larger earthquakes within the subducted Nazca plate tended to have large non-double-couple

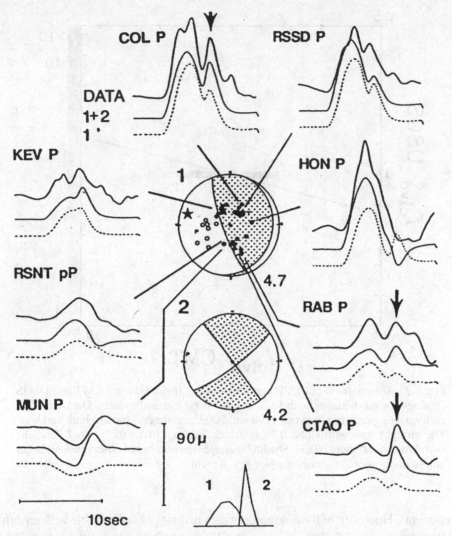

Fig. 6.24 Evidence that subevents with different mechanisms are responsible for CLVD mechanisms. The figure presents observed and synthetic broadband seismograms for the earthquake of 1 January 1984 ($h_{\mathrm{CMT}} = 384\,\mathrm{km}$, $M_{\mathrm{W}} = 7.2$) beneath Japan. At each station, the upper solid waveform is the P or pP displacement seismogram. The middle waveform is a synthetic calculated on the assumption of two subevents with disparate mechanism 1 and 2 shown in the center. The lower waveform is a synthetic calculated on the assumption that the two subevents have the same focal mechanism. The subevent interpretation is essential to an explanation of the large amplitudes of the second subevent at stations to the south and north (see arrows). Initial up and down motions for long-period records are plotted as open and closed circles, respectively, on focal sphere 1, and the star symbol represents the relative location of the second subevent with respect to the location of the initial rupture. The stations are: MUN, Mundaring, Australia; RSNT, Yellowknife, Canada; KEV, Kevo Finland; COL, College Outpost, Alaska; RSSD, Black Hills, South Dakota; HON, Honolulu, Hawaii; RAB, Rabaul, New Guinea; and CTAO, Charters Towers, Australia. Figure reproduced from Kuge and Kawakatsu (1990).

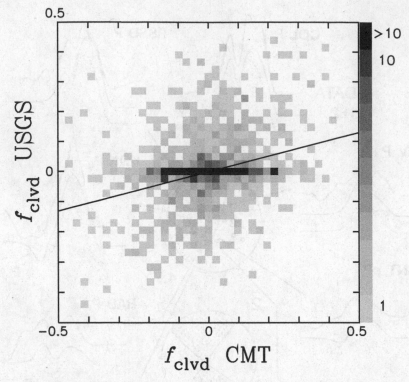

Fig. 6.25 Comparison of CLVD components f_{clvd} in the Harvard CMT and USGS catalogs for earthquakes with focal depths of 60 km and greater. Data are 1017 earthquakes occurring between 1980 and 2003 that are common to both catalogs. The straight line with slope 0.26 is a least-squares fit to the plotted data; the correlation coefficient is 0.31. Shading indicates the number of observations; darker colors imply more observations (see key at right).

components. However, of the nine earthquakes in their published table with reported CLVD components of 40% or more, the Harvard CMT catalog found similarly large values for only two, and one of these was of the opposite polarity. Houston (1993) suggested that lithospheric thickening caused by downdip compression should cause deep-focus earthquakes with moments larger than about 10^{20} N-m to possess large CLVD components. However, her only observations of very deep large earthquakes were Gilbert and Dziewonski's (1975) data from the 31 July 1970 and 15 August 1963 South American earthquakes; as previously mentioned, subsequent analyses have not confirmed that these earthquakes had large CLVD components (Okal and Geller, 1979; Russakoff *et al.*, 1997). And very large, very deep earthquakes occurring subsequent to Houston's study haven't generally possessed large non-double-couple components. Vavryčuk (2004) attributed observations of CLVD mechanisms of deep-focus earthquakes in Tonga to unmodeled anisotropy in the subducting lithosphere, and after performing an inversion concluded that the strength of the anisotropy was about 7% for P and 13% for S. Although this result is plausible, Vavryčuk wasn't selective about choosing better-determined mechanisms for the inversions and thus other interpretations of his observations are equally plausible.

In summary: while it is clear that a few earthquakes do have large CLVD components which aren't attributable to systematic errors, the available modeling studies suggest that for "typical" earthquakes such errors may well cause much of the observed CLVD components. Thus the "true" size of the CLVD component for "typical" earthquakes has yet to be established. Because near-surface reflections and reverberations do not contaminate body-waves from deep- and intermediate-focus events, they are excellent candidates for study of this problem.

6.3 Focal mechanisms and Wadati–Benioff zone geometry

6.3.1 Orientation of double-couple mechanisms

Since deep earthquake sources are not isotropic, are their T, B, and P axes distributed randomly, or is there a pattern in their orientations? If so, what physical models of faulting might explain this pattern? For shallow earthquakes, for example, there is strong clustering of T, B, and P axes (Frohlich, 2001); whether the greatest, intermediate or least principal stress is near-vertical controls whether one finds normal, strike-slip, or thrust earthquakes, i.e., vertical clustering of T, B, or P axes. For deep earthquakes, the stress-free surface that organizes shallow earthquake mechanisms is absent. What, then, controls the orientation of mechanisms for deep earthquakes?

In a classic study, Isacks and Molnar (1971) showed that the local downdip direction of the Wadati–Benioff zone influences the mechanisms for intermediate- and deep-focus earthquakes (see Fig. 5.5). In many geographic regions either T axes or P axes pointed predominantly downdip; presumably this is because of thermal-mechanical forces – as cold, dense lithosphere sinks into the mantle, gravitational driving forces produce zones of tension or compression within the slab interior. Beneath 300 km depth, in all geographic regions the mechanisms were most often of the downdip P type, possibly suggesting that slabs met some kind of high-viscosity barrier as they subducted below about 650 km. Between 100 km and 300 km depth Isacks and Molnar found both predominantly P-type and predominantly T-type regions. The T-type regions often possessed no earthquake activity at depths exceeding 300 km; the interpretation was that here the sinking lithosphere was in tension because the slab was short and subduction had not progressed into the higher-viscosity region of the mantle.

But, what does "predominantly downdip" mean? Does this mean that all deep P axes are within, say, 5° of the downdip direction? Or, since half of a sphere's surface area is within 60° of any axis, does this mean only that slightly more than 50% of all deep P axes are within 60° of downdip? Vassiliou (1984) and Frohlich and Willemann (1987) proposed to answer these questions by evaluating the distributions of directions for T, B, and P axes by applying standard methods, such as Bingham and Anderson–Darling statistics. For example, Apperson and Frohlich (1987) compared T, B, and P axes in the Harvard catalog with the Wadati–Benioff zone orientations determined by Burbach and Frohlich (1986). For deep-focus earthquakes ($h > 300$ km) they confirmed that the P axes clustered in the downdip direction. However, the clustering was somewhat diffuse; the median angle between P and downdip was 29°. Moreover, while B and T did cluster along-strike and normal to the Wadati–Benioff zone, as in the model of Fig. 6.26, the clustering was less intense than for P, with the median angular between B and along-strike being 38°, and between T and the normal being 36°.[9]

The diffuse character of the clustering is partly attributable to systematic errors in determining the relative orientations of T, B and P with respect to the Wadati–Benioff zone. If one removes mechanisms with large CLVD components and large relative errors (see Box 6.3) and ignores earthquakes at the lateral edges of Wadati–Benioff zones or in highly contorted zones, the remaining data are quite consistent with the downdip P, normal T and along-strike B model (see Figs. 6.26 and 6.27). Indeed, within some geographic regions, such as Argentina–Bolivia and central Java, more than 60% of the deep-focus earthquakes have P, B and T axes within 20° of the downdip, along-strike, and slab-normal directions.

For intermediate-depth earthquakes the situation is more complicated (Fig. 6.29). In a few regions, namely Tonga, the northern Izu–Bonin arc, and northern Kamchatka, the focal mechanism orientations are most often like those of deep-focus earthquakes, with a loose clustering of downdip P, normal T, and along-strike B axes. In the inclined-slab regions of South America, most mechanisms have downdip T, normal P, and along-strike B axes. Elsewhere mechanisms with downdip T are most common; however, there is often some complexity in the distributions of B and P. And in a few areas such as Java and the Marianas, P and T are quite scattered, and don't seem to be strongly controlled by the geometry of the Wadati–Benioff zone.

Also, Brudzinski and Chen (2005) found that P and T axes for both intermediate- and deep-focus earthquakes don't cluster much along any direction when the plunge of the Wadati–Benioff zone is less than about 20°. Their analysis focused on the contrast in clustering in South America between the flat-slab and inclined-slab

---

[9] The clustering is not much tighter when less-well-determined mechanisms are removed. For the data in Fig. 6.27, the median distance between T and normal is 33°, between B and along strike it is 38°, and between P and downdip it is 29°. Only 214 of 764 mechanism (28%) have all three axes within 30° of the predicted direction.

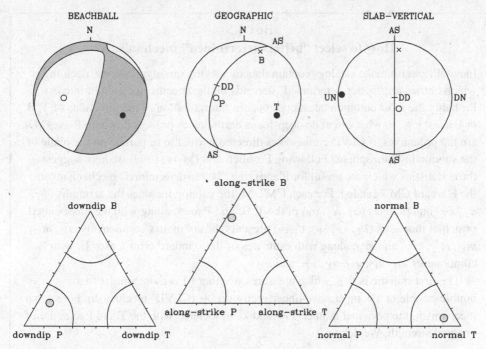

Fig. 6.26 Graphical methods for evaluating the orientation of focal mechanism axes with respect to a Wadati–Benioff zone. For an earthquake focal mechanism (beachball – upper left) we can plot the T, B, and P axes on a lower-hemisphere projection (upper center) along with the curved line representing the plane of the Wadati–Benioff zone. Along this curved line, the points at the edges of the plot correspond to the horizontal along-strike (AS) direction, and the point furthest from the center is the downdip (DD) direction. To compare mechanisms for earthquakes from different Wadati–Benioff zones, it is convenient to rotate the T, B, and P axes to a slab-vertical system (upper right) where the downdip direction plots in the center. For the example plotted, the P axis is directed beneath the slab because it is closer to the downward normal (DN) than the upward normal (UN) direction. An alternative method (lower diagrams) for displaying orientations is to construct three triangle diagrams showing how the mechanism is oriented with respect to the downdip, along-strike, and normal directions. For example, if a_T, a_B, and a_P are the angles between the T, B, and P axes and the along-strike direction; then, $\sin^2 a_T + \sin^2 a_B + \sin^2 a_P = 1$, and Frohlich (2001) presents equations for plotting the mechanism on a triangle diagram. The earthquake plotted as an example (14 October 1997; $h_{CMT} = 166$ km; $M_W = 7.7$) occurred in Tonga, where the Wadati–Benioff zone has a strike direction of 25° and a dip of 48°. The curved lines on the triangle diagrams delineate directions that are within 30° of downdip (lower left), along-strike (lower center), and normal (lower right).

regions, and a corresponding contrast between the deep-focus earthquakes that extend laterally west of Tonga beneath the Fiji Plateau, and those at comparable depths in the inclined seismic zone (see Sections 10.3 and 10.4).

Finally, in several geographic regions there are both downdip P-type and downdip T-type mechanisms (e.g., Tonga, see Fig. 6.29). Sometimes careful relocation

Box 6.3
How to select "better-determined" mechanisms

Inevitably, earthquake catalogs contain data of varying quality; however, deciding which entries are "better-determined" depends on the specific question of interest. Probably the most common questions concern interpretation of the directions of T, B, or P axes; e.g., to what extent do deep-focus earthquakes possess downdip P axes? Or, are the polar axes of CLVD earthquakes directed normal to or parallel to the plane of the subducting lithosphere? Following Frohlich and Davis (1999) we here suggest three statistics which are useful for identifying "better-determined" mechanisms in the Harvard CMT catalog. For each CMT $\mathbf{M}$, the catalog includes the azimuths (ϕ_T, ϕ_B, ϕ_P) and plunges (α_T, α_B, α_P) of the T, B, and P axes, along with their associated principal moments (λ_T, λ_B, λ_P); it also presents the six matrix components m_{11}, m_{12}, m_{13}, m_{22}, m_{23}, and m_{33}, along with estimates of the standard error tensor $\mathbf{U}$, with components u_{11}, u_{12}, u_{13}, u_{22}, u_{23}, and u_{33}.

The first statistic is f_{clvd}, which measures whether $\mathbf{M}$ is more similar to a double-couple or a compensated linear vector dipole (CLVD) mechanism. For such a mechanism, the principal moments λ_T and λ_P associated with the T and P axes are of unequal strength. We define:

$$
\begin{aligned}
f_{clvd} &= |\lambda_B/\lambda_T| \text{ if } |\lambda_T| \geq |\lambda_P| \\
&= |\lambda_B/\lambda_P| \text{ if } |\lambda_T| \leq |\lambda_P|;
\end{aligned}
\tag{6.23}
$$

$f_{clvd} = 0$ for a pure double-couple mechanism, and $f_{clvd} = 0.5$ for a pure CLVD mechanism. As f_{fclvd} increases and a mechanism approaches a CLVD, the orientations of the B axis and either the T or P axis become indeterminate (Fig. 6.28). Thus, unless one is specifically concerned with the orientation of the dipole axis, mechanisms with smaller values of f_{clvd} will be better-determined.

The second statistic is n_{free}, the number of moment tensor elements that are not constrained during the determination of $\mathbf{M}$. When a moment tensor element m_{ij} is constrained, Harvard reports the associated standard errors u_{ij} to be identically zero; thus we determine n_{free} simply by counting the number of non-zero u_{ij} associated with the six m_{ij}. There are theoretical reasons which cause the m_{12} and m_{13} components of $\mathbf{M}$ to be indeterminate as earthquake depth approaches zero; thus, u_{12} and u_{13} are often zero for very shallow earthquakes. The geometric effect of these constraints is to artificially force one of the three principal axes (T, B, or P) to be oriented vertically (Fig. 6.28). Thus, better-determined $\mathbf{M}$ have $n_{free} = 6$; any of the $\mathbf{M}$'s axes can take any orientation in space.

The third statistic is the relative error E_{rel}, which compares the relative sizes of the standard error tensor $\mathbf{U}$ and the moment tensor $\mathbf{M}$:

$$
E_{rel} = \sqrt{\frac{u_{11}^2 + u_{22}^2 + u_{33}^2 + 2u_{12}^2 + 2u_{13}^2 + 2u_{23}^2}{m_{11}^2 + m_{22}^2 + m_{33}^2 + 2m_{12}^2 + 2m_{13}^2 + 2m_{23}^2}}.
\tag{6.24}
$$

Conceptually, E_{rel} is just the norm or "scalar moment" of U divided by the scalar moment of **M**. In most cases E_{rel} is a number between 0 and 1; for CMT in the entire Harvard catalog its median value is 0.13. If one performs tests where one allows the components of **M** to vary within the limits fixed by the standard errors **U** (Frohlich, 1995), one finds that the orientations of the T, B and P axes are better constrained for events with smaller values of E_{rel}.

What statistical requirements must a mechanism satisfy to qualify as "better-determined"? This will depend on how selective one wishes to be. Frohlich and Davis (1999) arbitrarily recommended that: (1) $f_{\text{clvd}} \leq 0.20$; (2) $n_{\text{free}} = 6$; and (3) $E_{\text{rel}} < 0.15$. In the entire Harvard catalog, 46% of the CMT meet all three of these requirements, while 55% satisfy requirements (2) and (3), but not (1).

indicates that the different types are segregated within planar, approximately parallel groups, separated by distances of 10–40 km (see Fig. 5.7 and Section 5.1.2). These "double seismic zones" occur in a number of regions (see Table 5.1), including relatively cold subduction environments such as Japan and Tonga, and warmer regions such as Cascadia and parts of South America. Various explanations have been proposed to explain these double zones (see Section 5.1.2); one suggestion is that they occur because – after the slab is bent at the trench as it subducts – it must become unbent at greater depths, producing paired regions of tension and compression in its two outer surfaces. Other proposed explanations depend on thermal stresses produced by unequal heating of the slab exterior and interior, or on mechanical stresses caused by temperature-dependent phase transitions or dehydration reactions which have gone to completion in only portions of the subducting material.

Thus, while some of the data corroborate the Isacks and Molnar (1971) picture that deep mechanisms have downdip P or T, along-strike B, and slab-normal T or P, there are plenty of exceptions. Indeed, Apperson and Frohlich (1987) found that fewer than 30% of all intermediate- and deep-focus earthquakes had P or T within 30° of downdip, B within 30° of along-strike, and T or P within 30° of the normal. Thus, in some sense the "typical" deep earthquake is anomalous, in that it does not agree strictly with the Isacks and Molnar model of Fig. 5.5.

What is the significance of this variability in focal mechanism geometry in subducting lithosphere? Chen *et al.* (2004) review several alternative models that might explain this, including (1) perturbations of phase boundaries within slabs (Yoshioka *et al.*, 1997); (2) Frank's (1968) "ping-pong" model where slab bending creates lateral membrane stresses; (3) the so-called FK (Fujita and Kanamori, 1981) model where lithospheric age and the rate of convergence controls slab stress; (4) temporal variations in stress caused by large interplate thrust earthquakes (Astiz *et al.*, 1988; Lay *et al.*, 1989), and (5) reactivation of fossil faults originally formed

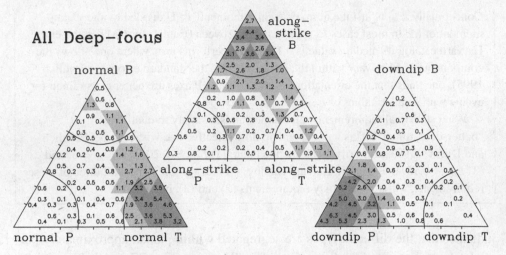

Fig. 6.27 Distributions of T, B, and P axis directions for deep-focus earthquakes. The triangle diagrams (see Fig. 6.26) show directions with respect to Wadati–Benioff zone geometry for 764 deep-focus earthquakes from the Harvard CMT catalog (1977–2004). The plotted mechanisms all satisfy the criteria suggested by Frohlich and Davis (1999) (see Box, 6.3), and are from earthquakes in Wadati–Benioff zones with "A" or "B" quality geometry as assigned by Apperson and Frohlich (1987). Numbers and shading in each triangular subregion compare the ratio of mechanisms observed to the number expected if axial directions are distributed uniformly in space (see Frohlich, 2001). No numbers and white shade indicate no mechanisms observed; darker shades indicate more frequent mechanisms. Note that T, B, and P tend to cluster near the normal, along-strike and downdip directions, respectively.

at the surface (e.g., Jiao *et al.*, 2000). Chen and his coauthors concluded that at intermediate depths the data confirms the influence of Isacks and Molnar's thermal-mechanical force model, perturbations of phase boundaries, and lateral membrane stresses, but that there was little evidence favoring the other proposed models.

In some regions, lateral flow within the mantle may apply lateral stresses to the subducting slab. In western Java, for example, the shallow subduction is highly oblique, and it is easy to imagine that at intermediate depths the subducting slab is being dragged sideways, with the resulting combination of along-strike compression and downdip tension producing a slab "tearing" where the faulting has slab-normal B. Similarly, in Tonga, Giardini and Woodhouse (1986) noted that if the Wadati–Benioff zone underwent a northward shear of about 500 km, it then fit plate reconstructions which match features in the seismicity with surface features such as the Louisville Ridge. They suggested that this might reflect a steady 5 cm/yr southward motion of the deepest part of the slab, relative to the surface. Apperson and Frohlich (1987) did observe that P axes in Tonga were directed about 15° northward

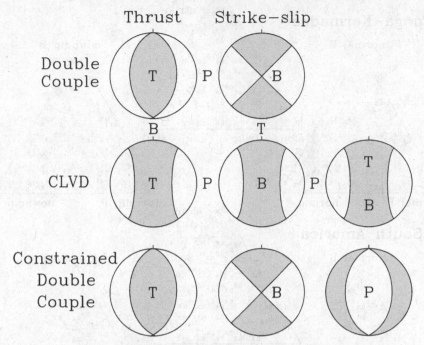

Fig. 6.28 Limitations of focal mechanisms determined from CMT with CLVD components, and constrained CMT. To distinguish thrust (top row, left; vertical T axis) and strike-slip (top row, right; vertical B axis) mechanisms we must know the orientation of the T and B axes; this is not possible for pure ($f_{clvd} = 0.5$) CLVD mechanisms (second row), which have identical radiation patterns regardless of whether the reported T and B axes are vertical, horizontal, or in between. For shallow and/or poorly recorded earthquakes Harvard often constrains the m_{12} and m_{13} components of the CMT to be zero; this means that the only allowable mechanisms have exactly vertical T, B, or P axes (bottom row).

of the downdip direction; this is consistent with Giardini and Woodhouse's suggestion. This northward bias is evident in summary focal mechanism plots for deep-focus Tonga earthquakes in cross sections E, F, and G in Fig. 10.6, but for intermediate events only in cross section G.

6.3.2 Orientation of CLVD mechanisms

What trends are observable for the orientations of T, B, or P axes for earthquakes with large CLVD source components? For mechanisms in the Harvard CMT catalog, Kuge and Kawakatsu (1993) noted that there is a correlation between the state of stress within subducting slabs and the sign of the average CLVD component f_{clvd} (Fig. 6.30). Mechanisms with polar-P components occur more often in Wadati–Benioff zones undergoing in-plate compression, while mechanisms with polar-T

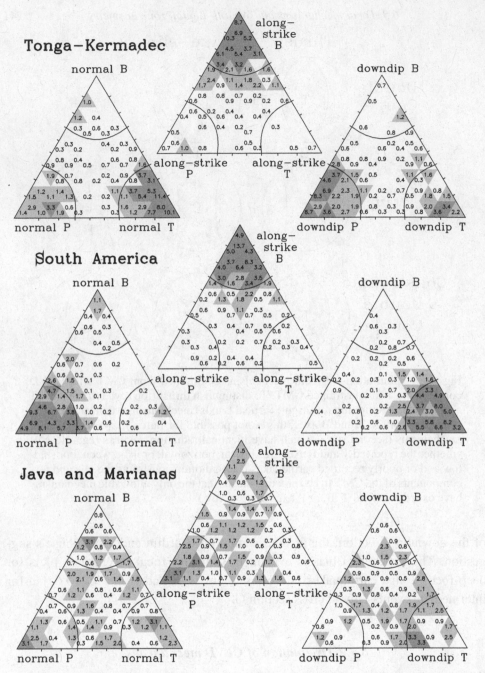

Fig. 6.29 Distributions of T, B, and P axis directions for intermediate-depth earthquakes in three geographic regions (see Fig. 6.26); (top) 281 intermediate earthquakes in Tonga–Kermadec; (middle) 375 intermediate earthquakes in South America; and (bottom) 264 intermediate earthquakes in Java and the Marianas. Mechanisms are Harvard CMT (1977–2004) satisfying the criteria suggested by Frohlich and Davis (1999) (see Box 6.3); mechanisms are from Wadati–Benioff zones assigned "A" and "B" ratings by Apperson and Frohlich (1987).

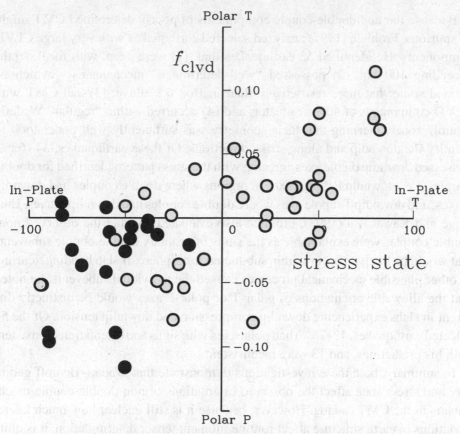

Fig. 6.30 Relationship between mean non-double-couple component f_{clvd} and the stress state of the Wadati–Benioff zone for earthquakes with depths greater than 100 km; dark symbols indicate depths exceeding 300 km. The CLVD component f_{clvd} is $m_B/\max(|m_T|,|m_P|)$. Stress state in each zone is determined from numbers N_T and N_P of T and P axes oriented within 25° of the plane of the Wadati–Benioff zone; the stress state is defined as $100 \times (N_T - N_P)/N_{Total}$. Mechanisms are Harvard CMT (1977–2004) with relative error E of 0.15 and less (see Box 6.3); each plotted point is from a Wadati–Benioff zone subregion assigned an "A" or "B" rating by Apperson and Frohlich (1987).

components occur where there is in-plate tension. Kuge and Kawakatsu performed calculations suggesting that high seismic velocities in the subducting slab may produce spurious CLVD components of up to 40% for double-couples with along-strike B and downdip P or T axes. However, for a few carefully-studied earthquakes analysis shows that appropriately-oriented subevents cause the non-double-couple component (e.g., Fig. 6.24); thus they concluded that the observed correlation might be attributable either to the systematic slab effect or to genuine complexity in the source process.

Because the non-double-couple components of poorly determined CMT might be spurious, Frohlich (1995) analyzed selected earthquakes with very large CLVD components. He identified 55 earthquakes that: (1) were deep, with focal depths exceeding 100 km; (2) possessed "well-determined" mechanisms – which he defined somewhat more restrictively than in Box 6.3 above; (3) had CMT with CLVD components of 40% or greater; and (4) occurred within "regular" Wadati–Benioff zones, meaning that their geometry was sufficiently well understood to identify the downdip and along-strike directions. Of these earthquakes, 34 (62%) possessed downdip dipole axes agreeing with the stress pattern identified for double couples, i.e., downdip P dipole axes in regions where double couples had downdip P axes, and downdip T dipole axes where double couples had downdip T axes. Like Kuge and Kawakatsu (1993), Frohlich also evaluated whether the observed non-double couples were explainable as the sums of ordinary double-couple subevents that would be likely to occur within subducted lithosphere. If slab bending, tearing, or other plausible mechanical processes caused the individual subevents, he noted that the allowable orientations of polar-T or polar-P axes would be distinctly different in slabs experiencing downdip compression and downdip tension. Of the 55 selected earthquakes, 42 (76%) had polar axes with signs and orientations consistent with his predictions, and 13 were inconsistent.

In summary, both these investigations demonstrate that Wadati–Benioff geometry and stress state affect the observed orientations of non-double-couple mechanisms in the CMT catalog. However, because it is still unclear how much lateral variations in earth structure affect routine moment tensor determination, it is quite possible that the observed correlations (e.g., Fig. 6.30) are due to systematic effects and not to intrinsic properties of the deep earthquake source.

6.4 References

Abe, K., 1972. Focal process of the South Sandwich Islands earthquake of May 26, 1964, *Phys. Earth Planet. Int.*, **5**, 110–122.

1982. Magnitude, seismic moment and apparent stress for major deep earthquakes, *J. Phys. Earth*, **30**, 321–330.

Antolik, M., D. Dreger, and B. Romanowicz, 1996. Finite fault source study of the great 1994 deep Bolivian earthquake, *Geophys. Res. Lett.*, **23**, doi:10.1029/96GL00968, 1589–1592.

• 1999. Rupture processes of large deep-focus earthquakes from inversion of moment rate functions, *J. Geophys. Res.*, **104**, doi:10.1029/1998JB00042, 863–894.

Apperson, K. D. and C. Frohlich, 1987. The relationship between Wadati–Benioff zone geometry and P, T and B axes of intermediate and deep focus earthquakes, *J. Geophys. Res.*, **92**, 13821–13831.

Astiz, L., T. Lay, and H. Kanamori, 1988. Large intermediate-depth earthquakes and the subduction process, *Phys. Earth Planet. Int.*, **53**, 80–166.

Beck, S. L., P. Silver, T. C. Wallace, and D. James, 1995. Directivity analysis of the deep Bolivian earthquake of June 9, 1994, *Geophys. Res. Lett.*, **22**, doi:10.1029/95GL01089, 2257–2260.

Benioff, H., 1964. Earthquake source mechanisms, *Science*, **143**, 1399–1406.

Ben-Menahem, A., 1975. Source parameters of the Siberian explosion of June 30, 1908, from analysis and synthesis of seismic signals at four stations, *Phys. Earth Planet. Int.*, **11**, 1–35.

Boatwright, J., 1980. A spectral theory for circular seismic sources: simple estimates of source dimension, stress drop and radiated seismic energy, *Bull. Seismol. Soc. Amer.*, **70**, 1–27.

Bollinger, G. A., 1968. Determination of earthquake fault parameters from long-period P waves, *J. Geophys. Res.*, **73**, 785–807.

Bos, A. G., G. Nolet, A. Rubin, H. Houston, and J. E. Vidale, 1998. Duration of deep earthquakes determined by stacking of Global Seismograph Network seismograms, *J. Geophys. Res.*, **103**, doi:10.1029/98JB01352, 21059–21065.

Bouchon, M. and P. Ihmlé, 1999. Stress drop and frictional heating during the 1994 deep Bolivia earthquake, *Geophys. Res. Lett.*, **26**, doi:10.1029/1999GL005410, 3521–3524.

Brudzinski, M. R. and W.-P. Chen, 2005. Earthquakes and strain in subhorizontal slabs, *J. Geophys. Res.*, **110** doi:10.1029/2004JB003470, B08303.

Brüstle, W. and G. Müller, 1987. Stopping phases in seismograms and the spatio-temporal extent of earthquakes, *Bull. Seismol. Soc. Amer.*, **77**, 47–68.

Burbach, G. V. and C. Frohlich, 1986. Intermediate and deep seismicity and lateral structure of subducted lithosphere in the circum-Pacific region, *Rev. Geophys.*, **24**, 833–874.

Campus, P. and S. Das, 2000. Comparison of the rupture and radiation characteristics of intermediate and deep earthquakes, *J. Geophys. Res.*, **105**, doi:10.1029/1999JB900384, 6177–6189.

Chandra, U., 1970. The Peru–Bolivia border earthquake of August 15, 1973, *Bull. Seismol. Soc. Amer.*, **60**, 636–646.

1973. Source process of a large deep-focus earthquake and its tectonic implications – the western Brazil earthquake of 1963: comments, *Phys. Earth Planet. Int.*, **7**, 115–120.

• Chen, P.-F., C. R. Bina, and E. A. Okal, 2004. A global survey of stress orientations in subducting slabs as revealed by intermediate-depth earthquakes, *Geophys. J. Int.*, **159**, doi-:10.1111/j.1365-246X.2004.02450.x, 721–733.

Chen, W.-P., 1995. En echelon ruptures during the great Bolivian earthquake of 1994, *Geophys. Res. Lett.*, **22**, doi:10.1029/95GL01805, 2261–2264.

Chen, W.-P, L.-R. Wu, and M. A. Glennon, 1996. Characteristics of multiple ruptures during large deep-focus earthquakes. In *Subduction, Top to Bottom*, eds. G. E. Bebout, D. W. Scholl, S. H. Kirby, and J. P. Platt, Geophys. Mon. 96, Washington, DC, American Geophysical Union, 357–368.

Choy, G. L. and J. L. Boatwright, 1995. Global patterns of radiated seismic energy and apparent stress, *J. Geophys. Res.*, **100**, doi:10.1029/95JB01969, 18205–18228.

1981. The rupture characteristics of two deep earthquakes inferred from broadband GDSN data, *Bull. Seismol. Soc. Amer.*, **71**, 691–711.

• Chung, W.-Y. and H. Kanamori, 1976. Source process and tectonic implications of the Spanish deep-focus earthquake of March 29, 1954, *Phys. Earth Planet. Int.*, **13**, 85–96.

1980. Variation of seismic source parameters and stress drops within a descending slab and its implications in plate mechanics, *Phys. Earth Planet. Int.*, **23**, 134–159.

Dufumier, H. and L. Rivera, 1997. On the resolution of the isotropic component in moment tensor inversion, *Geophys. J. Int.*, **131**, 595–606.

Dunham, E. M., P. Favreau, and J. M. Carlson, 2003. A supershear mechanism for cracks, *Science*, **299**, doi:10.1126/science.1080650, 1557–1559.

Dziewonski, A. M. and F. Gilbert, 1974. Temporal variation of the seismic moment tensor and the evidence of precursive compression for two deep earthquakes, *Nature*, **247**, 185–188.

•Estabrook, C. H., 1999. Body wave inversion of the 1970 and 1963 South American large deep-focus earthquakes, *J. Geophys. Res.*, **104**, doi:10.1029/1999JB900244, 28751–28767.

Estabrook, C. H. and G. Bock, 1995. Rupture history of the great Bolivian earthquake: slab interaction with the 660-km discontinuity? *Geophys. Res. Lett.*, **22**, doi:10.1029/95GL02234, 2277–2280.

Evison, F. F., 1963. Earthquakes and faults, *Bull. Seismol. Soc. Amer.*, **53**, 873–891.

1967. On the occurrence of volume change at the earthquake source, *Bull. Seismol. Soc. Amer.*, **57**, 9–25.

Fitch, T. J., D. W. McCowan, and M. W. Shields, 1980. Estimation of the seismic moment tensor from teleseismic body wave data with applications to intraplate and mantle earthquakes, *J. Geophys. Res.*, **85**, 3817–3828.

Frank, F. C., 1968. Curvature of island arcs, *Nature*, **220**, 363.

Frohlich, C., 1990. Note concerning non-double-couple source components from slip along surfaces of revolution, *J. Geophys. Res.*, **95**, 6861–6866.

1994. Earthquakes with non-double-couple mechanisms, *Science*, **264**, 804–809.

1995. Characteristics of well-determined non-double-couple earthquakes in the Harvard CMT catalog, *Phys. Earth Planet. Int.*, **91**, doi:10.1016/0031-9201(95)03031-Q, 213–228.

1996. Cliff's nodes concerning plotting nodal lines for P, Sh and Sv, *Seismol. Res. Lett.*, **67**, 16–24.

2001. Display and quantitative assessment of distributions of earthquake focal mechanisms, *Geophys. J. Int.*, **144**, doi:10.1046/j.1365–246x.2001.00341.x, 300–308.

Frohlich, C. and S. D. Davis, 1999. How well constrained are well-constrained T, B, and P axes in moment tensor catalogs? *J. Geophys. Res.*, **104**, doi:10.1029/1998JB900071, 4901–4910.

Frohlich, C. and R. J. Willemann, 1987. Statistical methods for comparing directions to the orientations of focal mechanisms and Wadati–Benioff zones, *Bull. Seismol. Soc. Amer.*, **77**, 2135–2142.

Frohlich, C., M. A. Riedesel, and K. D. Apperson, 1989. Note concerning possible mechanisms for non-double-couple earthquake sources, *Geophys. Res. Lett.*, **16**, 523–526.

Fujita, K. and H. Kanamori, 1981. Double seismic zones and stresses of intermediate depth earthquakes, *Geophys. J. Roy. Astron. Soc.*, **66**, 131–156.

Fukao, Y., 1970. Focal process of a deep-focus earthquake as deduced from long period P and S waves, *Bull. Earthquake Res. Inst., Tokyo Univ.*, **48**, 707–727.

1972. Source process of a large deep-focus earthquake and its tectonic implications – the western Brazil earthquake of 1963, *Phys. Earth Planet. Int.*, **5**, 61–76.

Fukao, Y. and M. Kikuchi, 1987. Source retrieval for mantle earthquakes by interactive deconvolution of long-period P-waves, *Tectonophysics*, **144**, doi:10.1016/0040-1951(87)90021-7, 249–269.

Furumoto, M., 1977. Spatio-temporal history of the deep Colombia earthquake of 1970, *Phys. Earth Planet. Int.*, **15**, 1–12.

Giardini, D. and J. H. Woodhouse, 1986. Horizontal shear flow in the mantle beneath the Tonga arc, *Nature*, **319**, doi:10.1038/319551a0, 551–555.

Gilbert, F. and A. M. Dziewonski, 1975. An application of normal mode theory to the retrieval of structural parameters and source mechanisms from seismic spectra, *Phil. Trans. Roy. Soc., London*, **A278**, 187–269.

Glennon, M. A. and W.-P. Chen, 1995. Ruptures of deep-focus earthquakes in the northwestern Pacific and their implications on seismogenesis, *Geophys. J. Int.*, **120**, 706–720.

Goes, S. and J. Ritsema, 1995. A broadband P wave analysis of the large deep Fiji Island and Bolivia earthquakes of 1994, *Geophys. Res. Lett.*, **22**, doi:10.1029/95GL02011, 2249–2252.

Goes, S., L. Ruff, and N. Winslow, 1997. The complex rupture process of the 1996 deep Flores, Indonesia earthquake (M_W 7.9) from teleseismic P-waves, *Geophys. Res. Lett.*, **24**, doi:10.1029/97GL01245, 1295–1298.

Hara, T., K. Kuge, and H. Kawakatsu, 1995. Determination of the isotropic component of the 1994 Bolivia deep earthquake, *Geophys. Res. Lett.*, **22**, doi:10.1029/95GL01602, 2265–2268.

1996. Determination of the isotropic component of deep focus earthquakes by inversion of normal-mode data, *Geophys. J. Int.*, **127**, 515–528.

Henry, C., J. H. Woodhouse, and S. Das, 2002. Stability of earthquake moment tensor inversions: effect of the double-couple constraint, *Tectonophysics*, **383**, doi:10.1016/S0040-1951(02)00379-7, 115–124.

Houston, H., 1993. The non-double-couple component of deep earthquakes and the width of the seismogenic zone, *Geophys. Res. Lett.*, **20**, 1687–1690.

2001. Influence of depth, focal mechanism, and tectonic setting on the shape and duration of earthquake source time functions, *J. Geophys. Res.*, **106**, doi:10.1029/2000JB900468, 11137–11150.

Houston, H., H. M. Benz, and J. E. Vidale, 1998. Time functions of deep earthquakes from broadband and short-period stacks, *J. Geophys. Res.*, **103**, doi:10.1029/98JB02135, 29895–29913.

• Ihmlé, P. F., 1998. On the interpretation of subevents in teleseismic waveforms: the 1994 Bolivia deep earthquake revisited, *J. Geophys. Res.*, **103**, doi:10.1029/98JB00603, 17919–17932.

Ihmlé, P. F. and T. H. Jordan, 1995. Source time function of the great 1994 Bolivia deep earthquake by waveform and spectral inversions, *Geophys. Res. Lett.*, **22**, doi:10.1029/95GL01437, 2253–2256.

• Isacks, B. L. and P. Molnar, 1971. Distribution of stresses in the descending lithosphere from a global survey of focal-mechanism solutions of mantle earthquakes, *Rev. Geophys. Space Phys.*, **9**, 103–174.

Jiao, W., P. G. Silver, Y. Fei, and C. T. Prewitt, 2000. Do intermediate- and deep-focus earthquakes occur on pre-existing weak zones? An examination of the Tonga subduction zone, *J. Geophys. Res.*, **105**, doi:10.11029/2000JB900314, 28125–28138.

Johnston, D. E. and C. A. Langston, 1984. The effect of assumed source structure on inversion of earthquake source parameters: the eastern Hispaniola earthquake of 14 September 1981, *Bull. Seismol. Soc. Amer.*, **74**, 2115–2134.

Kawakatsu, H., 1989. Centroid single force inversion of seismic waves generated by landslides, *J. Geophys. Res.*, **94**, 12363–12374.

• 1991. Insignificant isotropic component in the moment tensor of deep earthquakes, *Nature*, **351**, doi:10.1038/351050a0, 50–53.

1996. Observability of the isotropic component of a moment tensor, *Geophys. J. Int.*, **126**, 525–544.

Kennett, B. L. N., 1991. *IASPEI 1991 Seismological Tables*, Canberra, Australian National University, 167 pp.

Kikuchi, M., 1992. Strain drop and apparent strain for large earthquakes, *Tectonophysics*, **211**, doi:10.1016/0040–1951(92)90054-A, 107–113.

Kikuchi, M. and Y. Fukao, 1988. Seismic wave energy inferred from long-period body wave inversion, *Bull. Seismol. Soc. Amer.*, **78**, 1707–1724.

Kikuchi, M. and M. Ishida, 1993. Source retrieval for deep local earthquakes with broadband records, *Bull. Seismol. Soc. Amer.*, **83**, 1855–1870.

Kikuchi, M. and H. Kanamori, 1994. The mechanism of the deep Bolivia earthquake of June 9, 1994, *Geophys. Res. Lett.*, **21**, doi:10.1029/94GL02483, 2341–2344.

1995. The Shikotan earthquake of 4 October 1994: lithospheric earthquake, *Geophys. Res. Lett.*, **22**, doi:10.1029/95GL00883, 1025–1028.

Knopoff, L. and M. J. Randall, 1970. The compensated linear-vector dipole: a possible mechanism for deep earthquakes, *J. Geophys. Res.*, **75**, 4957–4963.

Koyama, J., 1975. Source process of the Vladavostok deep-focus earthquake of September 10, 1973, *Sci. Rept. Tohoku Univ., Ser. 5 – Geophys.*, **23**, 83–101.

1978. Seismic moment of the Vladavostok deep-focus earthquake of September 29, 1973, deduced from P waves and mantle Rayleigh waves, *Phys. Earth Planet. Int.*, **16**, 307–317.

Kubas, A. and S. A. Sipkin, 1987. Non-double-couple earthquake mechanisms in the Nazca plate subduction zone, *Geophys. Res. Lett.*, **14**, 339–342.

Kuge, K., 1994. Rapid-rupture and complex faulting of the May 12, 1990, Sakhalin deep earthquake: analysis of regional and teleseismic broadband data, *J. Geophys. Res.*, **99**, 2671–2686.

Kuge, K. and H. Kawakatsu, 1990. Analysis of a deep "non-double-couple" earthquake using very broadband data, *Geophys. Res. Lett.*, **17**, 227–230.

1992. Deep and intermediate-depth non-double-couple earthquakes: interpretation of moment tensor inversions using various passbands of very broadband seismic data, *Geophys. J. Int.*, **111**, 589–606.

1993. Significance of non-double-couple components of deep and intermediate-depth earthquakes: implications from moment tensor inversions of long-period seismic waves, *Phys. Earth Planet. Int.*, **75**, 243–266.

Lay, T., L. Astiz, H. Kanamori, and D. H. Christensen, 1989. Temporal variation of large intraplate earthquakes in coupled subduction zones, *Phys. Earth Planet. Int.*, **54**, doi:10.1016/0031-9201(89)90247-1, 258–312.

Leith, A. and J. A. Sharpe, 1936. Deep-focus earthquakes and their geological significance, *J. Geol.*, **44**, 877–917.

Lundgren, P. and D. Giardini, 1995. The June 9 Bolivia and March 9 Fiji deep earthquakes of 1994. I. Source processes, *Geophys. Res. Lett.*, **22**, doi:10.1029/95GL02233, 2241–2244.

McGuire, J. J., D. A. Wiens, P. J. Shore, and M. G. Bevis, 1997. The March 9, 1994 (M_W 7.6) deep Tonga earthquake: rupture outside the seismically active slab, *J. Geophys. Res.*, **102**, doi:10.1029/96JB03185, 15163–15182.

Mendiguren, J. A. and K. Aki, 1978. Source mechanism of the deep Colombian earthquake of 1970 July 31 from the free oscillation data, *Geophys. J. Roy. Astron. Soc.*, **55**, 539–556.

Mikumo, T., 1971a. Source process of deep and intermediate earthquakes as inferred from long-period P and S waveforms. 1. Intermediate-depth earthquakes in the southwest Pacific region, *J. Phys. Earth*, **19**, 1–19.

1971b. Source process of deep and intermediate earthquakes as inferred from long-period P and S waveforms. 2. Deep-focus and intermediate-depth earthquakes around Japan, *J. Phys. Earth*, **19**, 303–320.

Mori, J., 1983. Dynamic stress drops of moderate earthquakes of the eastern Aleutians and their relation to a great earthquake, *Bull. Seismol. Soc. Amer.*, **73**, 1077–1097.

Müller, G. K., P. Bonjer, H. Stöckl, and D. Enescu, 1978. The Romanian earthquake of March 4, 1977. I. Rupture process inferred from fault-plane solution and multiple event analysis, *J. Geophys.*, **44**, 203–218.

Myers, S. C., T. C. Wallace, S. L. Beck, P. G. Silver, G. Zandt, J. Vandecar, and E. Minaya, 1995. Implications of spatial and temporal development of the aftershock sequence for the M_W 8.3 June 9, 1994 deep Bolivian earthquake, *Geophys. Res. Lett.*, **22**, doi:10.1029/95GL01600, 2269–2272.

Newman, A. V. and E. A. Okal, 1998. Teleseismic estimates of radiated seismic energy: the E/M_o discriminant for tsunami earthquakes, *J. Geophys. Res.*, **103**, doi:10.1029/98JB02236, 26885–26898.

Oike, K., 1969. The deep earthquake of June 23, 1966 in Banda Sea: a multiple shock, *Bull. Disaster Prevention Res. Inst.*, **19**, 55–66.

1971. On the nature of the occurrence of intermediate and deep earthquakes. 3. Focal mechanisms of multiplets, *Bull. Disaster Prevention Res. Inst.*, **21**, 153–178.

Okal, E. A., 1996. Radial modes from the great 1994 Bolivian earthquake: no evidence for an isotropic component to the source, *Geophys. Res. Lett.*, **23**, doi:10.1029/96GL00375, 431–434.

Okal, E. A. and R. J. Geller, 1979. On the observability of isotropic seismic sources: the July 31, 1970 Colombian earthquake, *Phys. Earth Planet. Int.*, **18**, 176–196.

Oncescu, M. C., 1989. Investigation of a high stress drop earthquake on August 30, 1986 in the Vrancea region, *Tectonophysics*, **163**, doi:10.1016/0040-1951(89)90116-9, 35–43.

Osada, M. and K. Abe, 1981. Mechanism and tectonic implications of the great Banda Sea earthquake of November 4, 1963, *Phys. Earth Planet. Int.*, **25**, 129–139.

Pennington, W. D. and B. L. Isacks, 1979. Analysis of short-period waveforms of P phases from deep-focus earthquakes beneath the Fiji Islands, *Geophys. J. Roy. Astron. Soc.*, **56**, 19–40.

Persh, S. E. and H. Houston, 2004. Deep earthquake rupture histories determined by global stacking of broadband P waveforms, *J. Geophys. Res.*, **109**, 4311, doi:10.1029/2003JB002762 .

Quintanar, L., J. Yamamoto, and Z. Jiménez, 1999. Source mechanism of two 1994 intermediate-depth-focus earthquakes in Guerrero, Mexico, *Bull. Seismol. Soc. Amer.*, **89**, 1004–1018.

Räkers, E. and G. Müller, 1982. The Romanian earthquake of March 4, 1977. III. Improved focal model and moment determination, *J. Geophys.*, **50**, 143–150.

Randall, M. J. and L. Knopoff, 1970. The mechanism at the focus of deep earthquakes, *J. Geophys. Res.*, **75**, 4965–4976.

Rebollar, C. J., L. Quintanar, J. Yamamoto, and A. Uribe, 1999. Source process of the Chiapas, Mexico, intermediate-depth earthquake ($M_W = 7.2$) of 21 October 1995, *Bull. Seismol. Soc. Amer.*, **89**, 348–358.

Riedesel, M. A., 1985. *Seismic Moment Tensor Recovery at Low Frequencies*, Ph.D. dissertation, University of California, San Diego., 245 pp.

Ruff, L. J., 1999. Dynamic stress drop of recent earthquakes: variations within subduction zones, *Pure Appl. Geophys.*, **154**, doi:10.1007/s000240050237, 409–431.

• Russakoff, D., G. Ekström, and J. Tromp, 1997. A new analysis of the great 1970 Colombia earthquake and its isotropic component, *J. Geophys. Res.*, **102**, doi:10.1029/97JB01645, 20423–20434.

Sasatani, T., 1980. Source parameters and rupture mechanism of deep-focus earthquakes, *J. Faculty Sci. Hokkaido Univ., Ser. 7 (Geophys.)*, **6**, 301–384.

Schneider, J. F., 1984. *The Intermediate-Depth Microearthquakes of the Bucaramanga Nest, Colombia*, Ph.D. dissertation, University of Wisconsin, Madison, 233 pp.

Scholz, C. H., C. Aviles, and S. Wesnousky, 1986. Scaling differences between large interplate and intraplate earthquakes, *Bull. Seismol. Soc. Amer.*, **76**, 65–70.

Silver, P. G., S. L. Beck, T. C. Wallace, C. Meade, S. C. Myers, D. E. James, and R. Kuehnel, 1995. Rupture characteristics of the deep Bolivian earthquake of 9 June 1994 and the mechanism of deep-focus earthquakes, *Science*, **268**, 69–73.

Silver, P. G. and T. H. Jordan, 1982. Optimal estimate of scalar seismic moment, *Geophys. J. Roy. Astron. Soc.*, **70**, 755–787.

Stechschulte, V. C., 1932. The Japanese earthquake of March 29, 1928, and the problem of depth of focus, *Bull. Seismol. Soc. Amer.*, **22**, 81–137.

Stimpson, I. G. and R. G. Pearce, 1987. Moment tensors and source processes of three deep Sea of Okhotsk earthquakes, *Phys. Earth Planet. Inter.*, **47**, 107–124.

Sugi, N., M. Kikuchi, and Y. Fukao, 1989. Mode of stress release within a subducting slab of lithosphere: implication of source mechanism of deep and intermediate-depth earthquakes, *Phys. Earth Planet. Int.*, **55**, 106–125.

Tada, T. and K. Shimazaki, 1994. How much does a high-velocity slab contribute to the apparent non-double-couple components in deep-focus earthquakes? *Bull. Seismol. Soc. Amer.*, **84**, 1272–1278.

Takeo, M., S. Ide, and Y. Yoshida, 1993. The 1993 Kushiro-Oki, Japan, earthquake: a high stress-drop event in a subducting slab, *Geophys. Res. Lett.*, **20**, 2607–2610.

Tibi, R., C. H. Estabrook, and G. Bock, 1999. The 1996 June 17 Flores Sea and 1994 March 9 Fiji–Tonga earthquakes: source processes and deep earthquake mechanisms, *Geophys. J. Int.*, **138**, doi:10.1046/j.1365-246x.1999.00879.x, 625–642.

Tibi, R., D. A. Wiens, and J. A. Hildebrand, 2001. Aftershock locations and rupture characteristics of the 1995 Mariana deep earthquake, *Geophys. Res. Lett.*, **28**, doi:10.1029/2001GL013059, 4311–4314.

• Tibi, R., G. Bock, and C. H. Estabrook, 2002. Seismic body wave constraint on mechanism of intermediate-depth earthquakes, *J. Geophys. Res.*, **107**, 2047, doi:10.1029/2001JB000361.

Tibi, R., G. Bock, and D. A. Wiens, 2003. Source characteristics of large deep earthquakes: constraint on the faulting mechanism at great depths, *J. Geophys. Res.*, **108**, doi:10.1029/2002JB001948, 2091.

Tinker, M. A., S. L. Beck, W. Jiao, and T. C. Wallace, 1998. Mainshock and aftershock analysis of the June 17, 1996, deep Flores Sea earthquake sequence: implications for the mechanism of deep earthquakes and the tectonics of the Banda Sea, *J. Geophys. Res.*, **103**, doi:10.1029/97JB03533, 9987–10001.

Vassiliou, M. S., 1984. The state of stress in subducting slabs as revealed by earthquakes analysed by moment tensor inversion, *Earth Planet. Sci. Lett.*, **69**, 195–202.

Vassiliou, M. S. and H. Kanamori, 1982. The energy release in earthquakes, *Bull. Seismol. Soc. Amer.*, **72**, 371–387.

Vavryčuk, V., 2004. Inversion for anisotropy from non-double-couple components of moment tensors, *J. Geophys. Res.*, **109**, doi:10.1029/2003JB002926, B07306.

Venkataraman, A. and H. Kanamori, 2004. Observational constraints on the fracture energy of subduction zone earthquakes, *J. Geophys. Res.*, **109**, 5302, doi:10.1029/2003JB002549.

Vidale, J. E. and H. Houston, 1993. The depth dependence of earthquake duration and implications for rupture mechanisms, *Nature*, **365**, doi:10.1038/365045a0, 45–47.

Wiens, D. A., 1998. Source and aftershock properties of the 1996 Flores Sea earthquake, *Geophys. Res. Lett.*, **25**, doi:10.1029/98GL00417, 781–784.

Willemann, R. J. and C. Frohlich, 1987. Spatial patterns of aftershocks of deep focus earthquakes, *J. Geophys. Res.*, **92**, 13297–13943.

• Winslow, N. W. and L. J. Ruff, 1999. A hybrid method for calculating the radiated energy of deep earthquakes, *Phys. Earth Planet. Int.*, **115**, doi:10.1016.S0031-9201(99)00077-1, 181–190.

Woodhouse, J. H., 1981. The excitation of long period seismic waves by a source spanning a structural discontinuity, *Geophys. Res. Lett.*, **11**, 1129–1131.

Wu, L.-R. and W.-P. Chen, 2001. Rupture of the large (M_W 7.8), deep earthquake of 1973 beneath the Japan Sea with implications for seismogenesis, *Bull. Seismol. Soc. Amer.*, **91**, 102–111.

Wyss, M., 1970. Stress estimates for South American shallow and deep earthquakes, *J. Geophys. Res.*, **75**, 1529–1544.

Wyss, M. and P. Molnar, 1972. Source parameters of intermediate and deep focus earthquakes in the Tonga arc, *Phys. Earth Planet. Int.*, **6**, 279–292.

Xia, K., A. J. Rosakis, and H. Kanamori, 2004. Laboratory earthquakes: the sub-Rayleigh-to-supershear rupture transition, *Science*, **303**, doi:10.1126/science.1094022, 1859–1861.

Yoshida, S., 1988. Waveform inversion for rupture processes for two deep earthquakes in the Izu–Bonin region, *Phys. Earth Planet. Int.*, **52**, 85–101.

Yoshioka, S., R. Daessler, and D. A. Yuen, 1997. Stress fields associated with metastable phase transitions in descending slabs and deep-focus earthquakes, *Phys. Earth Planet. Int.*, **104**, doi:10.1016/S0031-9201(97)00031-9, 345–361.

7

The mechanics of deep earthquakes

•

When a deep earthquake happens, what happens mechanically? Is the rock failure process similar for deep and shallow earthquakes? Or are deep earthquakes an entirely different phenomenon? Since seismic rupture involves the rupture of a crystalline material, what happens at the molecular level?

For the most part scientists who tackle these questions work in rock mechanics laboratories, although seismological observations do place some constraints on the answers. Moreover, at present scientists don't all agree about the answers. This chapter will approach these questions from a historical perspective, focusing on the early proponents for various important ideas, and then summarizing our current state of knowledge.

In general, seismological evidence indicates that temperature and pressure are critical factors affecting where deep earthquakes do and don't occur. Deep earthquakes are almost nonexistent except beneath subduction zones, where cold lithosphere convects downward into the mantle. Moreover, deep-focus earthquakes occur in zones where the subducted lithosphere is coldest, i.e., where the thermal parameter Φ, the product of lithospheric age and vertical descent rate, is highest (see Section 2.4 and Figs. 2.8 and 4.7). Thus, another important question we should ask is: how do temperature and pressure affect the mechanical processes that allow rock to fracture or deform?

7.1 Fracture, friction, and flow

Mechanically, what are shallow-focus earthquakes? A simplistic explanation is that shallow earthquakes represent brittle fracture of rock in response to shear stress. Once the rock fractures and a fault surface develops, additional earthquakes may occur if shear stress overcomes friction and there is new slip on the fault. Thus

I thank David Rubie for his review of an earlier draft of this chapter.

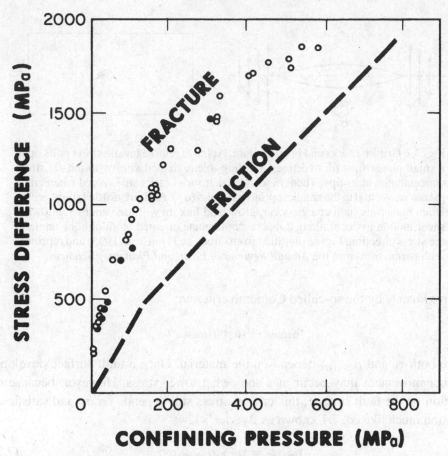

Fig. 7.1 Strength of Westerly granite vs confining pressure. Open and closed circles are results from two different laboratory studies of fracture of intact samples. Note that shear strength (stress difference) increases with confining stress, although the relationship is more complicated than suggested by eq. 7.1. The dashed line summarizes laboratory observations of frictional strength for sliding on an optimally oriented plane. Figure reproduced from Scholz (2002) with permission from Cambridge University Press.

laboratory investigations of earthquake mechanics generally focus on understanding the details of brittle fracture and rock friction. There is an enormous literature on this subject (for a summary see Scholz, 2002).

By the end of the nineteenth century engineers had developed a practical knowledge of rock fracture and friction because the development of modern civilization was accompanied by the construction of ever-more-elaborate buildings, domes, and arches that required a copious amount of quarrying, splitting, shaping, and piling up of rocks. For rocks, experiments indicate that fracture depends both on the shear stress τ and the confining (or hydrostatic) stress σ_h (Fig. 7.1). Moreover, for a particular value of σ_h, brittle fracture occurs only when τ exceeds a value specified

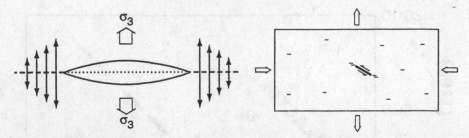

Fig. 7.2 Griffith cracks and brittle failure. At left, a remote tensile stress pulls open a small planar flaw (dotted line), yielding a lens-shaped void with tensile stress concentration at its tips. Theory indicates that such cracks are favored when crack planes are normal to the smallest principal stress (σ_3). At right, brittle failure occurs when numerous such cracks occur, grow, and link up with one another, allowing shear motion to occur along a macroscopic plane oriented at an oblique angle to the stress direction. Figure modified from Green and Houston (1995) and reprinted with permission from the *Annual Reviews of Earth and Planetary Sciences*.

approximately by the so-called Coulomb criterion:

$$\tau_{\text{fracture}} = \tau_0 + \mu_{\text{fracture}}\sigma_h. \tag{7.1}$$

Here both τ_0 and μ_{fracture} depend on the material. Once a fault surface develops, further movement may occur at a somewhat lower stress. However, because of friction on the fault surface, this critical stress still depends on σ_h and satisfies a relation much like eq. 7.1 known as Byerlee's law:

$$\tau_{\text{friction}} = \tau_F + \mu_{\text{friction}}\sigma_h, \tag{7.2}$$

with τ_F and μ_{friction} once again depending on the material. For most materials μ_{friction} is about 0.60–0.85.

In the twentieth century, theoretical studies provided more-or-less acceptable explanations for both eqs. 7.1 and 7.2. Griffith (1924) proposed that brittle fracture occurred when shear stress caused microscopic cracks to form, grow, and ultimately to link up along a rupture surface (Fig. 7.2). He was able to estimate the compressive failure stress in terms of the material tensile strength by evaluating the energy necessary to form new cracks and to extend preexisting cracks. Subsequent investigators modified his theory to allow friction along crack surfaces closed up by compressive stress, and obtained a theoretical failure stress identical in form to eq. 7.1. Similarly, theorists have been fairly successful explaining eq. 7.2 by modeling friction in terms of microscopic surface asperities that are in contact with one another because of normal stress, and respond to shear stress by elastic/plastic yielding or fracture. Friction and fracture are closely related phenomena; at very high confining pressures when a significant amount of the surface is in contact, the frictional strength of faults may approach the fracture strength of intact rock.

Box 7.1

Rock deformation at the molecular level

If enough stress is applied to a rock, it will break or deform. At the molecular level this may involve several different mechanisms. Which mechanism dominates, and how much stress is required to activate it, depends on pressure and temperature, and also on the grain size of minerals in the rock.

Brittle fracture Fracture involves opening microscopic cracks; in individual crystals this requires tensile stresses that separate individual molecular bonds. However, in rocks and other granular materials cracks form preferentially along the grain boundaries since the bonding forces between grains are less than the molecular bonds in intact crystals. The grain boundaries may be especially weak if films of water or other materials have reacted chemically with the grains at any time during a rock's history.

Ductile flow In crystals ductile flow occurs along numerous microscopic dislocations that interrupt the regular lattice of molecules. It requires less stress to shift the location of the relatively weak molecular bonds surrounding these dislocations than to break and move stronger bonds where there are no dislocations in the lattice. Once again, in granular materials the deformation may occur principally along grain boundaries, which are weaker. Finer-grained materials may deform under lower stress because there are more grain boundaries and because relative motion between grains can occur without deforming or breaking large grains.

Because bonds shift more easily as the temperature approaches melting, in crystals ductile flow occurs more easily at higher temperatures. However, in granular materials there are competing effects related to grain size. Typically, grains are smallest when a material is first formed. Over time grains tend to grow larger, especially when temperatures are close to melting and individual molecules are more mobile. Thus, ductile flow is favored when temperatures are close enough to melting so that dislocations shift easily, but far enough from melting so that grain sizes remain small and grow only slowly.

Phase changes Phase changes involve the reorganization of all the bonds between molecules within a crystal. For example, the transformation of α-olivine to γ-spinel occurs because the SiO_4 anions change from hexagonal close packing to cubic close packing, which is accompanied by a 9% reduction in volume.

However, rocks may also deform plastically, undergoing ductile flow rather than brittle fracture (see Box 7.1 and Fig. 1.12). For this process shear stress $\tau_{ductile}$ depends on the strain rate $\dot{\varepsilon}^Y$; for dry olivine experiencing shear stresses less than 100 MPa, Goetze (1978) indicates that

$$\tau_{ductile} = A\dot{\varepsilon}^{1/3}\exp[Q/(3RT)], \qquad (7.3)$$

where A is a constant, Q is an activation energy, and R is the gas constant. Basically, flow begins to dominate when dislocations in the crystalline lattice can propagate as easily as cracks. Since cracking involves both an increase in volume and frictional work, it is inhibited by increasing pressure but is insensitive to temperature. In contrast, because dislocations are lattice defects, their movements depends strongly on temperature and to a lesser extent on pressure. In any case, in the mantle when temperatures are above 600 °C or so, rock generally responds to stress by flow, rather than brittle fracture as described by eqs. 7.1 and 7.2.

7.2 Mechanisms that won't work

7.2.1 Deep earthquakes are just like shallow earthquakes

Why must deep earthquakes and shallow earthquakes be mechanically different? After all, they appear to share many similarities; since about 1965 the most careful studies conclude that most deep and shallow quakes have double-couple radiation patterns and little or no isotropic component (see Section 6.2); both exhibit a considerable range of sizes and include both rare magnitude 8 giants and numerous smaller-magnitude events (see Section 4.4); both have roughly similar source-time functions, rupture velocities, and stress drops (see Section 6.1).

There are several basic reasons why ordinary brittle fracture cannot explain deep earthquakes. First, at depths of 100–650 km the overburden produces hydrostatic stresses σ_h of 3–23 GPa; thus eq. 7.1 indicates that fracture would require shear stresses τ_{fracture} of about the same order of magnitude. Unfortunately, no one has ever proposed any mantle process that could produce such large shear stresses. Second, brittle fracture requires the opening of cracks, which is impossible given the known strength of materials and the magnitude of σ_h. For example, Griggs and Handin (1960) noted:

For deep-focus earthquakes certainly, and most probably even for shallow disturbances (a few tens of kilometers), ordinary Coulomb fracture is impossible. The internal friction of dry rocks under tens or hundreds of thousands of bars pressure would demand impossibly high shearing stresses of many kilobars . . . It is, of course, inconceivable that an open crack could exist at depth, so that the most baffling problem is the nature of a flaw of the Griffith type under these conditions . . . It appears that, in the case of the deepest earthquakes, the fracture theory requires a shear strength of some 10,000 times greater than seems reasonable . . .

Finally, even if some process within the mantle were capable of producing such large shear stresses, there is abundant evidence that they cannot be maintained, as ductile flow will occur. For example, Jeffreys (1924) analyzed the stresses necessary to support the maximum and minimum topographic elevations observed on the Earth's surface. He concluded that the mantle's strength must be low and thus couldn't maintain the stresses necessary for earthquakes. More recently, Wiens and

Stein (1983) found that the transition from brittle fracture to plastic flow corresponds to the maximum depth of earthquakes in oceanic lithosphere. For both young and old lithosphere, the deepest shallow earthquakes appear to occur approximately at the 600 °C isotherm (Fig. 2.5); laboratory experiments suggest at greater depths plastic flow relieves stress, and so no earthquakes occur.

Few scientists today would suggest that the fracture/friction model is simply wrong and doesn't apply to shallow earthquakes; the model has been highly successful at explaining the results of laboratory experiments at temperatures, pressures, and shear stresses that correspond to those in the crust. Thus, since the brittle fracture/friction mechanism can't explain deep earthquakes, deep earthquakes aren't just like shallow earthquakes. But, if not, what is the essential difference? Can we alter the fracture/friction model in some way so that is it applicable at the pressures where deep earthquakes occur? Or must we consider fundamentally different mechanisms to explain deep earthquakes?

7.2.2 Deep earthquakes are implosions accompanying solid–solid phase transitions

In the decade following Wadati's papers confirming the existence of deep earthquakes, analysis of earthquake travel times indicated that there were discontinuous velocity increases within the upper mantle, and the scientific community had concluded that these might represent phase transitions in mantle material (see Box 7.2 and Table 7.1). Indeed, when Jeffreys (1936) presented his paper on travel times and earth structure to the Royal Astronomical Society, a crystallographer named J. D. Bernal[1] immediately suggested that a phase transition in olivine might be the cause. Since high-pressure laboratory apparatus of the time couldn't reach the appropriate mantle pressures and temperatures, he made this inference from his knowledge of germanate analogs to olivine where transitions occurred at lower pressures. The published report of Bernal's suggestion (Anonymous, 1936) is about as clear an explanation as one can find anywhere for the workings of olivine mineralogy and the process of geophysical inference from analog materials:

An explanation of the discontinuity, which Bullen has shown to be also one of density, was desirable; no reasonable new material seemed likely to fit the data, and Dr. Bernal had been asked whether a new state of old material, probably olivine, was likely at high pressure.

---

[1] John Desmond Bernal was an influential but controversial individual who made important scientific contributions in earth science, molecular biology, and information science. Bensman (2001) states that Bernal was a committed Communist who believed that scientific journals would be replaced by a national distribution scheme through which central agencies would distrubute single papers. Bernal's writings concerning scientific publication strongly influenced the career of Eugene Garfield, the founder of the *Science Citation Index*, who dedicated the first issue of *SCI Journal Citation Reports* to Bernal.

Table 7.1. *Upper mantle rock and mineral names commonly used in the rock mechanics literature.*

Name	Composition	Examples; alternative names	Comments
Olivine mineral group			
olivine	$(Mg, Fe)_2SiO_4$	α form; peridot; Ol	Stable at atmospheric pressure, thought to be most abundant mineral in the upper mantle.
forsterite	Mg_2SiO_4	α form; Fo	Magnesium-pure olivine. Named for eighteenth century German naturalist Johann Forster.
fayalite	Fe_2SiO_4	α form; Fa	Iron-pure olivine. Named for its type locality – Fayal Island in the Azores.
wadsleyite	$(Mg, Fe)_2SiO_4$	β-spinel	High-pressure form of olivine with same molecular structure as true spinel (Al_2MgO_4). Named for Australian crystalographer Arthur David Wadsley.
ringwoodite	Mg_2SiO_4	γ-spinel	High-pressure form of forsterite with a structure that somewhat resembles spinel. Named for petrologist Alfred Edward Ringwood.
Olivine-rich rock group			
peridotite			Any olivine-rich, non-feldspathic rock.
dunite			A peridotite that is 90% or more olivine.
harzburgite			A peridotite whose principal second constituent is orthopyroxene. Named for its type locality – Harzburg, Germany.
Other significant mantle minerals			
pyroxene	$ABSi_2O_6$	e.g., clino-pyroxenes diopside, augite, and jadeite	A group of common rock-forming minerals with formula $ABSi_2O_6$. Here, A may be Mg, Fe, Ca, and Na; B may be Mg, Fe, and Al; sometimes Si is partly replaced by Al.
orthopyroxene	$(Mg, Fe)SiO_3$	e.g., enstatite, ferrosilite	Thought to be second most abundant mineral in upper mantle.

Name	Formula	Abbreviation	Description
serpentine	$Mg_3Si_2O_5(OH)_4$		Mineral commonly formed by the hydration of olivine and pyroxene.
garnet	$A_3B_2(SiO_4)_3$		For garnet in the mantle, A is principally Ca and Mg, and B is Al, Fe, and Mn.
majorite	$Mg_3(Fe, Al, Si)_2(SiO_4)_3$		A high-pressure form of garnet. Named for Alan Major.
"perovskite"	$MgSiO_3$	Pv	Beneath 650 km depth, mantle minerals may disassociate to oxides with the same structure as true perovskite ($CaTiO_3$). Named for nineteenth century Russian mineralogist Lev Aleksevich von Perovski.
"ilmenite"	$(Mg, Fe)SiO_3$	akimotoite; Il	Beneath 650 km depth, mantle minerals may disassociate to oxides with the same structure as true ilmenite ($FeTiO_3$). Named for the Ilmenski Mountains in Russia.
magnesiowustite	$(Mg, Fe)O$	Mw	Beneath 650 km depth, mantle minerals may disassociate to this oxide.
stishovite	SiO_2		High-pressure form of quartz, which may occur beneath 650 km depth. Named for Russian mineralogist Sergei Stishov.

Other significant upper mantle rocks

Name	Formula	Abbreviation	Description
basalt			Rock commonly formed by partial melting of mantle material. Transforms to eclogite at pressures and temperatures corresponding to upper mantle.
eclogite			A rock consisting mostly of garnet and omphacitic (Na-, Al-rich) pyroxene.
lherzolite			A rock with substantial amounts of both pyroxenes and olivine.

Synthetic and imaginary rocks and minerals

Name	Formula	Abbreviation	Description
germanate	$(A, B)_nGeO_4$		Any synthetic mineral where the silicon in the silicate (SiO_4) portion has been replaced by germanium (Ge).
pyrolite			A hypothetical lherzolitic rock that A. E. Ringwood proposed to explain the properties of the mantle.
piclogite			A hypothetical eclogitic rock that D. Anderson proposed to explain the properties of the mantle.

Box 7.2

Petrology of the deep earthquake zone – the pyrolite model

What rocks make up the deep earthquake zone? Any proposed petrological model must be consistent with the velocity and density of the mantle as inferred from seismological and gravitational observations. Moreover, a model should incorporate what can be inferred about upper mantle composition from sampling of kimberlite pipes, alkali basalts, and direct tectonic emplacement of upper mantle rocks in the crust.

After considering these constraints, Ringwood (1962; 1970; 1975) proposed that the upper mantle was composed of a hypothetical material he called pyrolite (pyroxene + olivine ± pyrope garnet). According to Ringwood, between 150 km and 350 km the mantle is:

57%	α-$(Mg, Fe)_2SiO_4$	olivine
17%	$(Mg, Fe)SiO_3$	orthopyroxene
12%	$(Ca, Mg, Fe)_2Si_2O_6$-$NaAlSi_2O_6$	omphacitic clinopyroxene
14%	$(Mg, Fe, Ca)_3(Al, Cr)_2Si_3O_12$	pyrope-rich garnet.

Then, between 360 km and 420 km, two phase transformations occur: the pyroxene transforms to garnet, and the α-olivine transforms to a spinel (β) form (Fig. 7.3). Between 420 km and 600 km the composition is:

57%	β-$(Mg, Fe)_2SiO_4$	spinel structure
39%	$(Mg, Fe, Ca)_3(Al, Cr, Fe)_2Si_3O_{12}$-	complex garnet solid
	$(Mg, Fe, Ca)_3(Mg, Fe, Ca)_{Si4}O_{12}$	solution
4%	$NaAlSi_2O_6$	jadeite

At 650 km there are further transformations: the β-spinel transforms to the strontium plumbate structure, the pyrope-rich garnet transforms to perovskite structure, and the jadeite disproportionates to a calcium ferrite structure.

Over the past 35 years Ringwood's pyrolite model has strongly influenced investigators of deep earthquake mechanics; however, competing models have been proposed. For example, Anderson and Bass (1984) proposed that the mantle was made of a $CaO + Al_2O_3$ – rich (eclogitic) assemblage of minerals they called piclogite (picrite + eclogite). The piclogite model has not been particularly popular among either geochemists or petrologists, but is commonly used when investigators wish to consider an alternative to the pyrolite model.

Dr. J. D. Bernal said that at ordinary pressures olivine, which is magnesium orthosilicate with part of the magnesium replaced by ferrous iron, is a hexagonal lattice of oxygen atoms in which the silicon and metals occupy the cavities, somewhat asymmetrically. It appeared possible that when the lattice was much compressed the cavities might become too small to hold the silicon atoms, and a different structure would have to be found. The structure of a crystal was largely a matter of the sizes of the atoms, the cavities having to be neither too large nor too small to hold them. Thus the effect of compressing the lattice as a whole

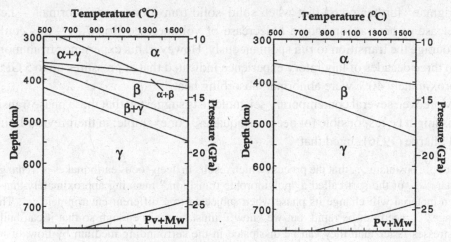

Fig. 7.3 Phase diagram (left) for olivine ($Mg_{0.8}Fe_{0.2}SiO_4$) showing stability fields for olivine (α), modified spinel (β) and spinel (γ), as well as regions where two phases coexist simultaneously. The right plot shows a simplified and idealized phase diagram in which only univariant reactions are considered. Figure reproduced from Kirby *et al.* (1996b).

would be similar to that of inserting a larger atom, such as germanium, in place of silicon. Magnesium germanate had been studied by Goldschmidt and found to exist at ordinary pressures in two forms, one isomorphous with olivine and the other cubic. The latter is about 9 per cent denser, and therefore is the high-pressure form. By analogy it therefore seemed probable that at high pressure olivine would adopt a cubic form. The change of density suggested by the germanate would be the right order of magnitude.

As knowledge of mantle petrology developed and rock mechanics experiments reached higher temperatures and pressures it became plausible that other mantle constituents might undergo phase transitions. Bridgman (1945) observed that:

In the course of my high pressure experimenting I have observed many transitions; in the last few years I have extended the pressure of the observations to [5 GPa], and still more recently, some of the results being as yet unpublished, to [10 GPa]. The pressure range is thus sufficient to justify perhaps the expectation that a statistical study may be suggestive for geology, in spite of the fact that the temperature of the measurements does not extend beyond 200 °C. About one-third of the substances examined to date have transitions . . . Furthermore, many substances have more than two phases. [And] the chance that a substance will exhibit a new phase in a given pressure range is statistically independent of whether it has already exhibited a transition at lower pressures. This, combined with the fact that many transitions thermodynamically possible are doubtless suppressed because of the narrow temperature range of the experiments, suggests the conclusion that the chances are that any material chosen at random will undergo at least one transition in the geological range. The presumption is that the materials in situ in the earth's crust are not in the crystal system familiar in the laboratory.

Bridgman[2] further noted that when solid–solid transitions were "normal" – i.e., a release of heat accompanied a decrease of volume (see Box 7.3) – this could encourage the transition to run spontaneously. However, his experience from more than three decades of laboratory experience indicated that at pressures up to 5 GPa, approximately 40% were abnormal, absorbing heat.[3]

Moreover, several contemporary seismologists suggested that these phase transitions might be responsible for deep earthquakes. For example, in their review Leith and Sharpe (1936) stated that:

Let us . . . postulate . . . that the process which results in deep-focus earthquakes is a change in state . . . of the sort called a "polymorphic transition," meaning approximately that a given material will change its phase when subjected to a different environment . . . The change . . . need not be rapid, but we know it must be rapid enough so that it can build up stresses faster than they can be dissipated in the surrounding medium by flow, or no catastrophic failure would result . . . The surrounding medium will yield elastically . . . until its elastic limit is reached, at which point failure will occur in the normal manner, by fracturing. If we consider the change in environment which causes a polymorphic transition to be one requiring external shearing stress, the net effect is that the resulting earthquake would be indistinguishable in any respect from those due to an external stress system not accompanied by a change in state.

Thus, almost since the discovery of deep earthquakes it has been clear that mantle constituents undergo one or more phase changes at depths of 700 km or less, and that these might explain deep earthquakes (see Section 6.2.2). As Holmes (1931; 1933) had proposed that the mantle underwent convection, it was plausible that downgoing convection cells would provide a steady supply of material that could change phase catastrophically and produce deep earthquakes. This hypothesis has some appeal, and it reappears regularly in the literature, even up to the present day.

For example, on strainmeter records at Nana, Peru, Benioff (1964) observed downward motions prior to a Peruvian deep earthquake that occurred on 19 August 1961. He then concluded that:

In searching for a physically possible mechanism capable of applying a step force at depth . . . I can think only of sudden collapse of a small volume of rock at the focus, such as might be caused by a change of state . . . Owing to the weight of the overburden, the void is filled immediately by a downward motion of the rock mass. For long-period waves at least, such a source is equivalent to a downward step.

---

[2] Percy Williams Bridgman was a remarkable individual. In about 1905 he became interested in high-pressure phenomena. Subsequently he began developing apparatus that permitted him to reach ever-higher pressures in his laboratory at Harvard, allowing him to dominate research concerning the high-pressure properties of materials for more than 50 years. He also published several well-received books on the philosophy of science, and in 1946 was awarded the Nobel Prize in physics. He committed suicide in 1961 when he discovered he had incurable cancer.

[3] This meant that solid–solid transitions contrasted sharply with the melting transitions; except for water, bismuth, antimony, and gallium, for all ordinary materials the melting transition is "normal".

Box 7.3

Solid–solid phase transitions, the Clapeyron equation, and analog materials

As subducting lithosphere travels from the Earth's surface to depths of 700 km or more, minerals may react to increasing pressure P and temperature T by undergoing one or more solid–solid phase transformations (see Table 7.1). Could these solid–solid phase transformations be responsible for deep earthquakes?

If so, a thermodynamic relationship called the Clapeyron equation places severe restrictions on the transformation's properties. Generally transformations such as the olivine–spinel transition involve a reordering of the crystal structure; i.e., the relocation of olivine's oxygen anions from a somewhat distorted hexagonal close packing system to spinel's approximately cubic close packing system. This reordering is accompanied by a change in volume ΔV and requires supplying heat ΔH. The Clapeyron equation[4] states that for the equilibrium phase transformation, the slope (dP/dT) of pressure with respect to temperature satisfies:

$$(dP/dT) = \Delta V/(T\Delta H). \qquad (7.4)$$

Because deep earthquakes occur quite suddenly and involve ruptures that may travel many kilometers in just a few seconds (see Section 6.1.2), the phase transformation that is responsible must involve a decrease in volume ($\Delta V < 0$) and it must release heat ($\Delta H < 0$). Otherwise, the reaction will proceed slowly since to sustain it one must supply both mechanical energy $P\Delta V$ and thermal energy ΔH.

In principal, the Clapeyron equation makes it possible to perform laboratory tests for a solid–solid transition to determine whether either ΔV or ΔH is positive, thus eliminating the transition as a possible cause of deep earthquakes. However, for olivine ($[Mg, Fe]_2 SiO_4$) over much of the twentieth century it wasn't possible to make accurate laboratory measurements of ΔV, ΔH, P, and T at appropriate mantle pressures and temperatures.

For this reason, many of the quantitative experiments have involved synthetic *analog materials* such as $(Mg, Fe)_2 GeO_4$, which has a molecular structure nearly identical to olivine except that the silicon atoms have been replaced by germanium. Often, these so-called germanates have phase diagrams that are similar to diagrams for the silicates except that the solid–solid transitions occur at much lower pressures.

---

[4] Named after Benoit Paul Emile Clapeyron, a French engineer who formulated the equation. In 1849–1850 William Thomson (later, Lord Kelvin) and his father, James, demonstrated that the pressure variation of the melting point of ice agreed with that predicted by Clapeyron's equation. This was one of the first successful applications of thermodynamics to a physical problem. The ice–water transformation, incidentally, would be a poor candidate as an earthquake mechanism since, although it involves a reduction in volume ($\Delta V < 0$), it requires the addition of heat ($\Delta H > 0$).

Similarly, Dziewonski and Gilbert (1974) and Gilbert and Dziewonski (1975) evaluated moment release from the Colombia earthquake of 31 July 1970 and the Peru–Brazil earthquake of 15 August 1963 and concluded that episodes of precursory compression began at least 80s prior to the main rupture. They suggested that this precursor represented an ongoing phase change which induced shear failure in adjacent material.

The observations of Benioff, Dziewonski, and Gilbert lent credibility to the phase transition hypothesis, and provoked a number of scientists to evaluate the physics of sudden phase transitions. Dennis and Walker (1965) pointed out that any phase transition that might cause deep earthquakes would have to be metastable (see Fig. 1.4), as otherwise it could only proceed as fast as P and T conditions changed to allow it to occur.

A phase transition at equilibrium pressure and temperature cannot result in the spontaneous release of energy because the rate of appearance of the new phase is determined by the rate at which energy enters or leaves the system. Thus a sudden release of free energy requires a spontaneous change from the metastable state.

They noted that downward convection of cold material is likely to create a situation that could lead to metastability; since rates of reactions increase with temperature, as the slab warms the reaction could occur.

If the rate constant [of a phase transition] were small, a mineral phase or assemblage could be brought through and past equilibrium without undergoing transformation; it would then be metastable. The rate constant for a reaction depends on temperature. Therefore [as the slab warms] increasing temperature would increase the rate constant and eventually the reaction would proceed spontaneously.

Subsequently, various investigators analyzed the kinetics of the olivine–spinel phase transition, including Kasahara and Tsukahara (1971), Sung (1974), Sung and Burns (1976a; 1976b), Vaisnys and Pilbeam (1976), Poirier (1981), Liu (1983), and Hodder (1984). These investigations were generally inconclusive about whether phase transitions could occur fast enough to generate seismic waves, but they did demonstrate that the reaction rates depended critically on grain size and on the degree of metastability, i.e., how much P and T could rise above the equilibrium values for the phase transition without it actually taking place. The general conclusion was that transitions fast enough to generate earthquake waves might be possible if grain sizes were small and if material remained in a metastable state until $P–T$ conditions were significantly different from the equilibrium values.

In spite of its appeal, the deep-earthquakes-as-implosions hypothesis faced problems from the very beginning. By far the most serious problem is that anyone who inspects seismograms finds that the gross features of the radiation pattern of deep earthquakes simply don't resemble an implosion. In particular, whereas an implosion would generate no S waves and produce dilatant P-waves indicating motion towards the focal region, all well-recorded intermediate- and deep-focus earthquakes exhibit strong shear waves and a more-or-less equal complement of compressional and dilational first motions. This was clear to Honda (1932; 1934), Koning (1942), and other early seismologists who investigated the earthquake source. And subsequent careful investigations haven't found evidence that deep earthquakes have isotropic source components (e.g., Kawakatsu, 1991; 1996; Hara *et al.*, 1996; Okal, 1996). Moreover, Russakoff *et al.* (1997) and Estabrook (1999) reevaluated the 1963 and 1970 earthquakes analyzed by Dziewonski and Gilbert (1974), but concluded that no isotropic component was resolvable when the data were modeled using modern analytical methods. Thus, at present there is simply no observational evidence that deep earthquakes have isotropic components.

The observation that deep earthquakes have strong S waves seems inconsistent with an implosion source; however, this isn't a strong argument since it is well-known that nuclear explosions often generate both S waves and Love waves. For example, Ekström and Richards (1994) evaluated the radiation patterns for 71 Kazakhstan nuclear explosions and found that typically the double-couple moment M_o was about 10–100% as large as the isotropic moment M_I (Fig. 7.4). Presumably, the observed double-couple radiation is attributable to preexisting tectonic stress that is relieved when the explosion occurs.

There have also been other objections to the deep-earthquakes-as-sudden-phase-transitions hypothesis. The basic concern is whether it is possible for phase transitions to occur suddenly over spatial scales as large as 10–100 km, the approximate rupture dimensions of the largest deep earthquakes. Although Vaisnys and Pilbeam (1976) were proponents of the solid–solid phase transition mechanism, they expressed concern that once a phase transition began, the rapid temperature rise might quench the transition. Moreover, the drop in pressure accompanying the volume decrease would tend to return material towards equilibrium and slow the reaction, an effect observed subsequently by Liu *et al.* (1998) and Mosenfelder *et al.* (2000). There are also questions about whether phase transitions are consistent with the observation that earthquakes sometimes happen repeatedly in the same place (Wiens and Snider, 2001) and yet also occur over a broad variety of depths within subducted lithosphere. It would seem that phase transitions would be favored at particular sites where appropriate *P–T* conditions occurred; however, once the transition has taken place and generated an earthquake, it would seem that further earthquakes at that site would be impossible.

Although these objections may not all be valid, it is clear that deep earthquakes are not simply implosions accompanying phase transitions. As noted above and in Section 6.2, their radiation patterns as represented by ordinary seismograms do not exhibit measurable isotropic components; and the occasional reports of very-low

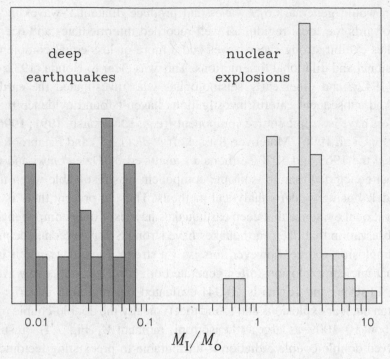

Fig. 7.4 Distributions of the ratio M_I/M_o of isotropic moment M_I and double-couple moment M_o for 19 deep-focus earthquakes and 71 Kazakhstan nuclear explosions. Earthquake data are from Kawakatsu (1996), and explosion data are from Ekström and Richards (1994) and Dreger and Woods (2002). The explosion results demonstrate that explosion/implosion sources can generate double-couple components, and thus the observation that deep earthquakes generate S waves does not rule out an implosion source. However, the existence of an implosion component for deep earthquake sources is unconfirmed since 0.1 is approximately the measurement error for M_I/M_o.

frequency precursory isotropic components haven't been confirmed by additional analysis. Thus, although phase transitions almost certainly do occur in subducting lithosphere, and although these may significantly intensify shear stresses in regions where deep earthquakes occur, some other mechanical process must cause the events that we routinely record on seismograms as deep earthquakes.

A variant of the deep-earthquakes-as-phase-transitions hypothesis is that the phase transitions are so-called martensitic transitions (Lomnitz-Adler, 1990) which have been thoroughly studied because they are of great importance in metallurgy. Martensitic transformations are phase changes between two metastable solid phases in which the parent phase transforms into a new phase having the same chemical composition but a different crystalline structure. Their most outstanding property is that they are diffusionless, which allows the new phase to propagate with a velocity of the same order as the speed of sound in the material. The transformation is accomplished by shearing discrete volumes of material; for a given composition there is a definite orientation between the parent and product crystals. However, since the transformation need not involve large volumes of material it is plausible that it could radiate seismic energy as a double-couple source rather than as an implosion.

Currently there isn't strong support for the hypothesis that deep earthquake ruptures are due to martensitic transitions. Whether the olivine–spinel transformation can proceed by a true martensitic mechanism is doubtful because of the cation shuffling that would be necessary; thus the term "martensitic-like" is often used. For single crystals of olivine, Kerschhofer *et al.* (2000) did observe a transformation that produced the orientation relationships predicted for a martensitic-like process, although this occurred only at pressures exceeding 18 GPa and under highly metastable conditions. In general the crystalography of the host structure controls the shear strains associated with these transformations (e.g., Kirby and Stern, 1993; Kirby *et al.*, 1996b) and these strains thus vary from grain to grain. Through-going shear strains simply don't develop into macroscopic faults. Although Meade and Jeanloz (1989) proposed that the olivine–spinel transformation might be martensitic, Green (1984) reviewed numerous experiments on olivine and olivine analogs and concluded that the transformation was martensitic only under very large shear stresses, such as might occur in diamond anvil experiments. At lower stresses and near the melting point evidence indicated that the transformation occurred via a nucleation growth mechanism. Generally, there seems to be little evidence that the grain-scale crystalographic shear that occurs in martensitic transformations ever produces macroscopic faulting instabilities in metals, ceramics, or minerals.

7.3 Mechanisms that might work

7.3.1 Dehydration embrittlement

The section above describing ordinary brittle fracture never mentioned fluids; and, at depths where deep quakes occur it was deemed impossible for Griffiths cracks to open, grow, and form faults. However, Hubbert and Rubey (1959) noted that pore fluids in the crust had a profound effect on faulting; in particular, along faults fluids with pore pressure P effectively reduce the normal stress σ_n in eqs. 7.1 and 7.2 to $\sigma_n - P$, thus reducing both strength and friction, and allowing faults to slip at much greater depths than would otherwise be possible. Similarly, Griggs and Handin (1960) noted that dehydration reactions might provide a mechanism for releasing stored elastic energy at very high pressures, although they concluded that this was unlikely. *Dehydration embrittlement* is the hypothesis that this occurs;

i.e., that hydrated minerals release fluids when they reach appropriate temperature–pressure conditions in the mantle, thus permitting brittle fracture to occur in the mantle at greater depths than under dry conditions. Nowadays the term dehydration embrittlement is also used when the cracks aren't filled by actual fluid material but instead by fine-grained, low-strength solid byproducts of the dehydration reaction.

Is it plausible that there are fluids in the mantle where deep earthquakes occur? The existence of island arc volcanism above Wadati–Benioff zones is strong evidence that subduction carries hydrated crustal material into the mantle and that dehydration does occur within subducting lithosphere. Petrologists believe that island arc volcanoes represent the dehydration and partial melting of subducted minerals such as amphibolite, serpentine, brucite and forsterite. This process forms magmas that rise and fractionate to produce island arc volcanism (e.g., see Ringwood, 1975).

The earliest support for dehydration embrittlement came from laboratory experiments. Raleigh and Paterson (1965) and Raleigh (1967) investigated the failure of serpentines at pressures up to 0.5 GPa, finding that:

As the temperature is increased, the strength drops in the usual way until at 600°C there is a very marked decrease with further increase in temperature . . . Coinciding with the onset of weakening the mode of failure changes from ductile yielding to shear fracturing, in most cases accompanied by a sudden, small drop in load. Weakened and embrittled specimens when removed from their jackets show signs of dampness, and new phases resulting from the reaction are present . . . [In the Earth this phenomenon] would be expected to give rise to a very marked increase in the incidence of seismic faulting at the depth of dehydration . . . The dehydration products as they are carried down along the limb of the convection cell will be weak and brittle to the depth where all the water is finally lost to the surface; this zone of weakness effectively decouples the immobile continental mantle from the convecting oceanic mantle by faulting within the zone.

Subsequently, Meade and Jeanloz (1991) performed laser-heated diamond-anvil measurements on olivine, pyroxene, and serpentine, and reached pressures corresponding to depths of 650 km. They observed no acoustic emissions in either olivine or pyroxine samples at pressures above ~7 GPa (about 200 km depth). However, in serpentine they observed acoustic emissions at temperature–pressure ranges corresponding roughly to mantle depths of 0–300 km (Fig. 7.5). They presented evidence that dehydration reactions in the serpentine accompanied these emissions. They also observed acoustic emissions in serpentine at pressures corresponding to depths of 300–700 km, but these weren't accompanied by dehydration, and will be discussed later.

However, the early seismological evidence supporting dehydration embrittlement wasn't very compelling. After performing an analysis of *b* values for hypocenters reported by the Japanese Meteorological Agency, Anderson *et al.*

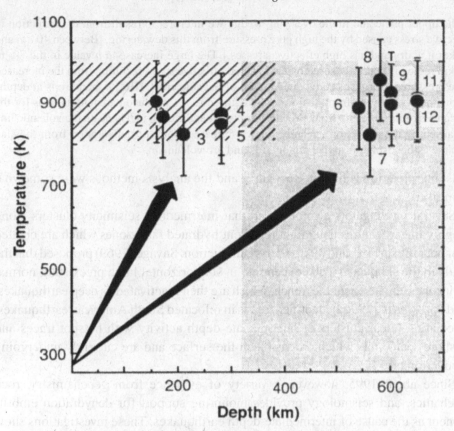

Fig. 7.5 Two failure mechanisms reported in serpentine. The plot shows temperature–pressure regimes corresponding to conditions where Meade and Jeanloz (1991) reported acoustic emissions associated with dehydration (hatched region) and amorphization (shaded region). Points labeled 1–5 indicate conditions in five Wadati–Benioff zones where only intermediate-depth earthquakes occur; points 6–12 are conditions in zones where there are deep-focus earthquakes. Arrows show estimated temperature–pressure paths for the coldest material in the upper 12 km of subducting lithosphere. Figure reproduced from Meade and Jeanloz (1991) with permission from the American Association for the Advancement of Science.

(1980) suggested that variations with depth might be attributable to dehydration reactions in subducting lithosphere. They found that b values were about 1.0 at the surface, but dropped to about 0.6–0.7 at depths of 50–70 km. Then there was a sharp peak of about 1.0 at 80 km, and a drop again at 90 km before going up again beneath 100 km. They concluded that:

Changes caused by dehydration offer a simple explanation for the variations in b with depth in the upper seismic plane. Beginning at the surface, high b in the upper plane and wedge is caused by the expulsion of free water as the slab begins to be underthrust. The high b value in the upper plane and in the overlying wedge we interpret to result from the reduction in effective stress caused by the high pore pressure from this dewatering. [Between 40 km and 70 km it is dry and has high effective stresses.] The large increase in b value in the 70- to 90-km region is perhaps caused by the lowered effective stress resulting from the increased pore pressure in turn caused by dehydration of oceanic crust. [But] b then drops at depths below the dehydration front, [as] water is required to . . . produce parental magmas for the volcanoes of the volcanic front. We thus propose that the drop in b beneath the volcanic line is caused by the advection upward of newly released water of dehydration from the slab into the asthenospheric source region of island arc volcanism.

This interpretation is highly speculative and the analysis methods were somewhat primitive.[5]

Several investigators also observed that intermediate seismicity clusters along roughly linear features that might represent hydrated fault zones which are created at or near the surface, and then undergo subduction. Savage (1969) proposed that the fluids in the subducted slab concentrate in subhorizontal large preexisting normal fault zones, created near the trench, which are then reactivated as deep earthquakes. And Kirby *et al.* (1996a) identified trends in relocated South American earthquakes, associating linear clusters of intermediate-depth activity with hotspot traces and offshore seamounts which are visible at the surface and are currently undergoing subduction (see Fig. 4.5).

Since about 1995, however, a variety of evidence from geochemistry, rock mechanics, and seismology provides mounting support for dehydration embrittlement as the cause of intermediate-depth earthquakes. These investigations show that:

(1) subduction does carry water into the mantle to depths where intermediate-focus earthquakes occur;
(2) a number of hydrated minerals exhibit catastrophic failure at mantle temperatures and pressures; and
(3) the occurrence and depth extent of intermediate Wadati–Benioff zones and double seismic zones coincide with temperature–pressure conditions where mineral dehydration should occur.

First, since the mid-1980s measurements of trace amounts of ^{10}Be in volcanoes at some island arcs (Fig. 7.6) provide direct confirmation that subduction carries significant amounts of surface water to depths of magma genesis (see Morris *et al.*, 2002) in only a few million years. ^{10}Be is produced naturally only by fission of oxygen and nitrogen by cosmic rays in the atmosphere; it attaches to aerosols

---

[5] Recently Wiemer and Wyss (1997) have developed improved methods for evaluating spatial variations in b values, and in data from regional catalogs have observed significant variations over distance scales of only a few km (e.g., Wiemer *et al.*, 1998). It would be instructive to apply these methods to network data from Japan and other subduction zones to reevaluate the results of Anderson and his co-workers.

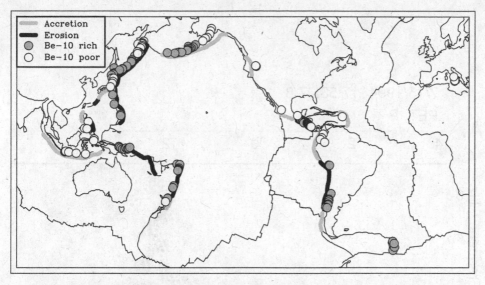

Fig. 7.6 Accretion–erosion of sediments at subduction zones and evidence that hydrated crustal material is carried into the mantle. Subduction zones marked light gray are characterized by subduction accretion; zones marked black by subduction erosion. Filled circles indicate volcanoes where [10]Be measurements indicate surface sediments are subducted to depths of magma genesis; open circles are volcanoes not enriched in [10]Be (data from Morris *et al.*, 2002).

carried to the Earth's surface by precipitation. Its half-life is only 1.5 million years and thus its presence in volcanoes proves that hydrated sediments must reach depths of 100 km or so within 6 million years or less. This suggests that the process that carries water into the mantle is fairly efficient.

In several subduction zones seismological observations show there exists a layered structure coincident with the top surface of the subducting lithosphere; this layer extends to depths of 100–250 km and may contain hydrated subducting material. The key observations are highly dispersed seismic waves (Fig. 7.7) traveling along this layer up the slab to regional stations. Modeling of these phases indicates waveguide thicknesses are ~2–7 km and velocities are comparable to those of blueschists or other hydrated lithospheric material (Abers, 2000; 2005). Abers and others present evidence that these waveguides occur within the central Aleutians, mainland Alaska (Abers and Sarker, 1996), Chile (Martin *et al.*, 2003), Japan (Matsuzawa *et al.*, 1987; Helffrich and Stein, 1993; Ohmi and Hori, 2000; and see Section 10.11.2), Kuril–Kamchatka, the Marianas, and Nicaragua (Abers *et al.*, 2003).

But is it plausible that dehydration embrittlement causes earthquakes well beneath the top surface of the subducting lithosphere? Since Wadati–Benioff zones

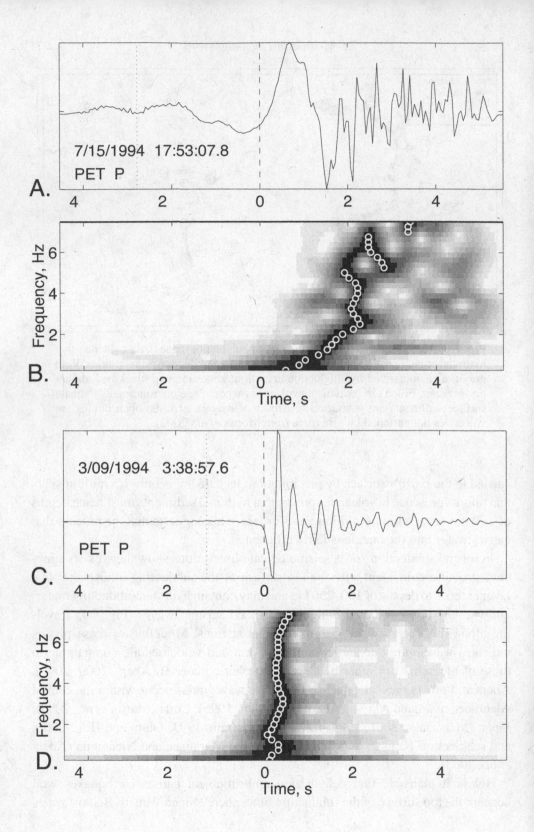

A. 7/15/1994 17:53:07.8 PET P

B. Time, s

C. 3/09/1994 3:38:57.6 PET P

D. Time, s

typically have widths of 10 km or more and double seismic zones are usually separated by distances of 20 km or more, dehydration embrittlement requires that hydrated minerals must occur deep within the subducting slab. Peacock (2001) suggested that the lithosphere may be hydrated well below the slab's top surface because of outer rise earthquakes, which allow "infiltration of seawater several tens of km into the oceanic lithosphere." Using controlled-source multichannel seismic data, Ranero *et al.* (2003) imaged the seafloor on the outer rise of the Middle America Trench and concluded that there were bending-related faults dipping at ~45° which extended at least 20 km into the lithosphere. They noted that this

creates a pervasive tectonic fabric that cuts across the crust, penetrating deep into the mantle. The along-strike length and depth of penetration of these faults are also similar to the dimensions of the rupture area of intermediate-depth earthquakes.

Some of the support for deep hydration of the slab is more speculative. Davies (1999) went so far as to assert that all hydrated crustal material is bound within the subducting slab and that intermediate-depth earthquakes actually drive fluid flow within the slab, thus providing the mechanism by which water enters the mantle wedge and allows magma genesis to occur. Seno and Yamanaka (1996) propose that double seismic zones only occur in zones where shallow compressional outer-rise earthquakes also occur, and attribute both to dehydration embrittlement reactions. Finally, from an analysis of Tongan earthquakes with depths down to about 450 km, Jiao *et al.* (2000) speculate that these also may occur along faults that formed at the outer rise, then becoming reactivated after being subducted into the mantle. The basic observation is that intermediate focal mechanisms have nodal planes that are asymmetric with respect to the downdip direction of the slab. The asymmetry is identical to that observed for outer rise events in the compilation of Christensen and Ruff (1988). At depths exceeding 450 km the P axes are downdip but the nodal planes are more or less randomly scattered, "suggesting a change of mechanism or a loss of slab integrity."

Major support for the dehydration embrittlement hypothesis comes from continuing rock mechanics experiments that confirm and extend the earlier results of Raleigh and Paterson (1965) and Meade and Jeanloz (1991). For example,

---

Fig. 7.7 Seismological evidence for a low-velocity layer at the top of the subducting lithosphere. P phases that travel along the slab surface (A) possess a distinct highly dispersed coda that is not evident for phases that do not sample the slab (C). This is evident both from visual inspection and also from narrow-band filtering (B and D); shading in (B) and (D) shows how spectral content changes over time; white circles indicate peak frequency. Station PET in the Kuril arc recorded both seismograms; earthquake (A) occurred at 118 km depth at a distance of 5.29°; earthquake (C) occurred at 189 km depth at a distance of 1.06°. Figure reprinted from Abers (2000) with permission from Elsevier.

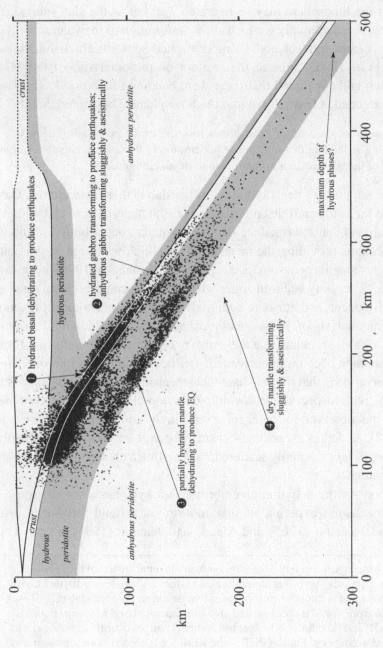

Fig. 7.8 Hacker et al.'s (2003) interpretation of the relationship between petrology and seismicity in northwest Japan where subducting lithosphere is relatively cold. Figure reproduced from Hacker et al. (2003).

Dobson *et al.* (2002) measured acoustic emissions in stressed samples of serpentine at temperature–pressure conditions corresponding to depths of 50–200 km in subduction zones. Jung *et al.* (2004) found that faulting in serpentine was delineated by ultrafine-grained solid reaction products formed during dehydration, and that failure continued even at pressures where the volume change accompanying dehydration became negative. At pressures corresponding to depths of 100–250 km Zhang *et al.* (2004) found that eclogites possessed a faulting instability associated with the precipitation of water at grain boundaries and the associated production of very small amounts of melt.

The most compelling results supporting dehydration embrittlement come from thermopetrological models that appear to explain the presence/absence of and variation among double seismic zones (see Section 5.1.2). The models find that dehydration reactions explain the depth extent and distances separating double seismic zones for a variety of conditions, including regions like Cascadia where the subducting lithosphere is very hot, and like northeast Japan where it is very cold. For example, Hacker *et al.* (2003) propose that the subducting plate has four layers (Fig. 7.8):

(1) hydrated, fine-grained basaltic upper crust dehydrating under equilibrium conditions and producing earthquakes facilitated by dehydration embrittlement;
(2) coarse-grained, locally hydrated gabbroic lower crust that produces some earthquakes during dehydration but transforms chiefly aseismically to eclogite at depths beyond equilibrium;
(3) locally hydrated uppermost mantle dehydrating under equilibrium conditions and producing earthquakes;
(4) anhydrous mantle lithosphere transforming sluggishly and aseismically to denser minerals.

They note that "to our knowledge, no intermediate-depth earthquakes have been reliably located above [the low-velocity layers at the top of the slab]" and propose that "upper seismic zones ... correspond spatially to dehydration of the crust. Lower seismic zones ... correspond spatially to dehydration of the mantle. Seismic gaps between double zones correspond spatially to the thermally cold core of the slab. Aseismic mantle below the zone(s) of active seismicity is anhydrous." They present comparisons between calculations and observations for Cascadia, southwest Japan (Nankai), central Costa Rica, and northeast Japan (Tohoku; see Fig. 7.9).

A number of different groups have made similar calculations and reached similar conclusions (Peacock and Wang, 1999; Kerrick and Connolly, 2001; Peacock, 2001; Seno *et al.*, 2001; Connolly and Kerrick, 2002; Omori *et al.*, 2002), although not all agree with respect to the details of the responsible reactions. For example, Yamasaki and Seno (2003) modeled the thermal and petrological conditions in northeast and

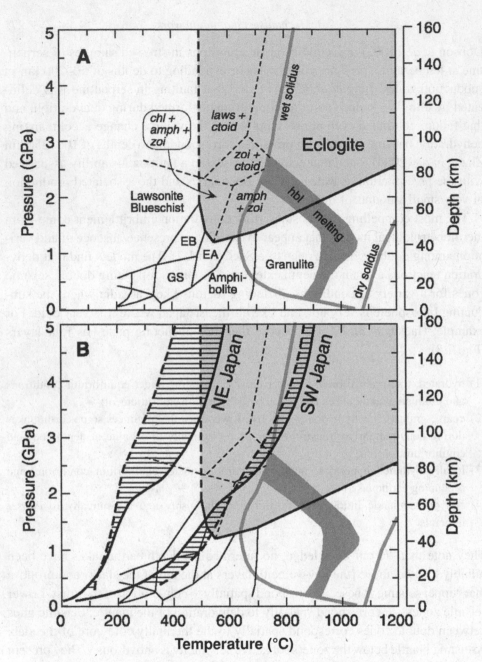

Fig. 7.9 Metamorphic conditions in oceanic crust subducted beneath northeast and southwest Japan.

(Top) Metamorphic facies, hydrous minerals (italics) stable in the eclogite facies (light gray), and partial melting reactions (dark gray) for basaltic compositions.

(Bottom) Calculated *P–T* conditions (horizontal lined areas) for oceanic crust subducting beneath northeast and southwest Japan. The solid line is the top of the subducting oceanic lithosphere; the dashed line is the base of the subducting oceanic crust. Figure reproduced from Peacock and Wang (1999) with permission from the American Association for the Advancement of Science.

southwest Japan, Taiwan, northern Chile, Cape Mendocino (see Section 10.14), and the eastern Aleutians, and state:

We find that the dehydration loci of serpentine have upper and lower branches, and the upper branch is almost in the mantle wedge, except in northeast Japan and eastern Aleutians where it penetrates into the slab. We therefore regard that the upper plane seismicity is representing the dehydration embrittlement of oceanic crust ... We find that for most of the double seismic zones the lower plane seismicity is located along the lower dehydration loci of [mantle] serpentine ...

Although their modeling explained the presence or absence of double zones, in several regions they had some difficulty in explaining the depth extent of the zones.

If dehydration embrittlement so successfully explains many features of intermediate-depth seismicity, could it be responsible for deep-focus earthquakes as well? For several reasons, many (but not all) investigators have concluded that dehydration embrittlement couldn't cause earthquakes at depths exceeding 300–350 km. First, petrological studies of serpentine, forsterite, and other possible mantle constituents generally conclude that most dehydration reactions should go to completion by the time pressures reach about 5–10 GPa (150–300 km) and temperatures reach about 700 °C (e.g., see Ringwood, 1975). Second, the approximately exponential decline in earthquake frequency with depth (see Fig. 4.3) suggests that the underlying cause of the mechanical instability is becoming exhausted. For example, Green and Houston (1995) state that:

The majority of the water that goes down with the slab is in the crust and sediments and is lost to the mantle wedge above the slab where it induces arc and back-arc melting. Most of the remainder is along faults and is slowly cooked out as the slab warms and induces a decreasing population of earthquakes that continues up to about 300 km. Any H_2O remaining in the slab at this depth probably is dissolved in the pyroxenes and olivines and becomes unavailable for faulting. [Dehydration embrittlement cannot be responsible for deep-focus earthquakes since] experimental studies of dehydration reactions suggest it is unlikely that these reactions could be inhibited above about 700 °K, restricting to intermediate depths earthquakes triggered by this mechanism.

Furthermore, while there is evidence that the presence of water decreases the strength of olivine, Chen *et al.* (1998) found it may have little or no effect on the strength of the higher-pressure, spinel forms. They used x-ray diffraction to study rheological properties of both dry and hydrous olivine and its high-pressure polymorphs, reaching $P–T$ conditions of 10 GPa–600 °C for olivine and wadsleyite (β-spinel), and 20 GPa–1000 °C for ringwoodite (γ-spinel). They concluded that

the observations indicate that olivine is much weaker than the other two minerals for this range of conditions, and water weakens olivine dramatically but only slightly weakens wadsleyite and ringwoodite.

On the other hand, Omori *et al.* (2004) argued that dehydration does continue at transition-zone depths for various constituents of peridotite, and suggested that dehydration embrittlement was responsible for both intermediate- and deep-focus seismicity. Jung *et al.* (2004) concluded that "dehydration embrittlement is a viable mechanism for nucleating earthquakes independent of depth, as long as there are hydrous minerals breaking down under a differential stress." Similarly, Zhang *et al.* (2004) performed laboratory experiments on eclogite and reported that:

eclogite lacking hydrous phases but with significant hydroxyl incorporated as defects in pyroxene and garnet develops a faulting instability associated with the precipitation of water at grain boundaries and the production of very small amounts of melt. This new faulting mechanism satisfactorily explains high-temperature earthquakes in subducting oceanic crust and could potentially be involved in much deeper earthquakes in connection with similar precipitation of water in the mantle transition zone (400–700 km depth).

They found that triggering this instability required very small amounts of fluid, and thus suggested it could act where deep-focus quakes occur even if most mineral phases underwent dehydration at shallower depths. The argument that transition zone temperatures are too high for dehydration reactions may also be incorrect; Brearley and Rubie (1990) found that a dehydration reaction involving the nucleation of multiple mineral phases took place only after reaching temperatures considerably higher than the equilibrium temperature.

In any case, the possibility that subduction might carry significant amounts of water down into the mantle has intrigued a number of investigators (e.g., Panero *et al.*, 2003; Komabayashi *et al.*, 2004; Ohtani *et al.*, 2004). Irifune *et al.* (1998) note that:

While serpentine completely dehydrates at temperatures higher than 600°–800°C and at pressures to 12 GPa, phase assemblages including some dense hydrous magnesium silicates were formed at higher pressures. [This] may play an important role in transportation of water into the lower mantle via subduction of cold slabs.

Bina and Navrotsky (2000) calculate that in the cooler families of slabs, the isotherms may well cross the phase boundary between water and ice VII. The release of water from hydrous minerals as a solid would allow its continued transport downward in solid form, rather than its rapid lateral and upward movement as a fluid. It would thus be available for "water-consuming reactions" at greater depths. They note that there is a marked increase in moment release in Tonga at about 400 km depth, and propose that "perhaps significantly greater amounts of water become available near this depth beneath Tonga than elsewhere."

In the past there were objections to the idea that dehydration embrittlement is responsible for intermediate- and deep-focus earthquakes. Griggs and Handin (1960) noted that Griffiths cracks kept open by pore fluids probably couldn't propagate fast enough to allow rupture since water tends to move by diffusion, an inherently slow process. On the other hand, this argument could explain the paucity of aftershocks for intermediate- and deep-focus earthquakes – whereas individual intermediate earthquakes might occur only after pore fluids had migrated to open sufficiently numerous cracks along a region experiencing high shear stress, the rupture wouldn't be sufficient to trigger aftershocks in surrounding regions because, although the rupture intensified shear stresses in surrounding regions, fluid migration couldn't take place fast enough to open cracks and permit aftershocks to occur.

In general, dehydration embrittlement is appealing because it suggests that one mechanism, brittle fracture, is responsible for both shallow and intermediate earthquakes; it explains how deeper-than-normal earthquakes can occur within subduction zones; and it explains many features of double seismic zones. The phase diagrams for various likely constituents in subducting lithosphere are complicated. For example, Omori *et al.* (2002) constructed a phase diagram in the model system $MgO-Al_2-Al_2O_3-SiO_2-H_2O$, appropriately abbreviated MASH, and identified no fewer than 40 reactions that occur at pressures corresponding to depths up to 280 km (also see Fig. 7.10 and Poli and Schmidt, 2002). There is an increasing realization that for dehydration reactions, whether mechanical failure occurs depends on the relative volume change in both the fluid and solid constituents (e.g., Hacker, 1997; Jung *et al.*, 2004). Clearly, rock mechanics laboratories now have numerous candidate phase changes that might provide possible failure mechanisms, especially as Dobson *et al.* (2004) have demonstrated the cability to measure acoustic emissions in samples at pressures of 16 GPa, corresponding to depths of about 500 km.

7.3.2 Shear instabilities and stress-induced melting

Within a decade after Wadati confirmed that deep earthquakes occur, Percy Bridgman performed a series of experiments indicating that ordinary brittle fracture, stick–slip friction, and ductile flow weren't the only possible failure mechanisms. In these experiments, Bridgman place a hardened bar between the faces of the anvils in a high-pressure apparatus, and then inserted thin disks ($\sim$0.01 cm thickness) of sample materials at the top and bottom of the bar. After bringing the apparatus up to pressure he would rotate the bar about the axis of the anvils to create intense shearing stresses in the sample disks. Typically the bar was moved about 60° in five seconds, and the axial stress was about 5 GPa. Bridgman (1936) observed that:

Plastic flow for the majority of substances takes place perfectly quietly and continuously, but in 40 per cent of the cases flow is not perfectly smooth . . . The failure of smoothness may be almost imperceptible, manifested only by a feeling of very fine-grained grinding to the hand on the handle of the rotating wrench; or there may be simply a squeaking during rotation.

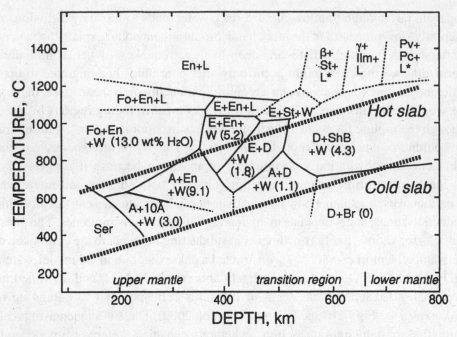

Fig. 7.10 Observed dehydration products from serpentine compared to pressures and temperatures in the upper parts of subducting slabs. Numbers in parentheses indicate the amounts of H_2O (wt%) to be liberated upon transformation from serpentine. In relatively hot slabs serpentine completely dehydrates to form forsterite plus enstatite at depths of about 150 km; in colder slabs dehydration involves transformations to dense hydrous magnesium silicate (DHMS) phases. Abbreviations: A, ShB, D, and E are particular DHMS phases; other minerals are: Br – brucite; En – enstatite; Fo – forsterite; Ilm – ilmenite; L – liquid; Pc – periclase; Pv – perovskite; St – stishovite; W – water (fluid/vapor); β – modified spinel; γ – spinel. Figure reproduced from Irifune *et al.* (1998).

Or there may be a soft chattering, or a harder and more abruptly punctuated chattering. Or there may be a snapping and jumping, sometimes so violent and spasmodic as to make measurements of any quantitative value almost impossible. The phenomena of "snapping" are, I think, of special geological significance. The snapping is merely a manifestation of internal rupture . . . It would seem that this sort of rupture must be a factor in deep-seated earthquakes..

The materials that exhibited this phenomenon generally had low thermal conductivities and high melting temperatures, and included graphite and zinc. Although Bridgman never isolated the exact mechanism or mechanisms of the failure process he observed, apparently the applied shear strains induced dilatency hardening in the test materials, and the increase in temperature induced by frictional heating ultimately allowed failure to occur, either because of catastrophic readjustment of the crystalline structure, or because of local melting.

In any case, Bridgman's results received attention from numerous subsequent investigators interested in the mechanism of deep earthquakes. For example, Orowon (1960) commented on "the remarkable phenomenon" discovered by Bridgman, and stated that:

Earthquakes, except at depths less than about 5–10 km, cannot arise in the manner implied in the classical theory. The only plausible alternative available at present is that they are due to an instability of plastic deformation (creep) . . . If creep produces structural changes that accelerate further creep, the deformation concentrates gradually into thin layers in which high flow rates can develop, and finally even shear melting may occur by the heat development due to plastic deformation.

Orowon noted that accelerating plastic creep explains the occurrence of fore- and aftershocks, and "seems at present the only feasible mechanism of deep and intermediate-depth earthquakes."

There have been only a few laboratory investigations that find shear instabilities in mantle materials. For example, in a diamond anvil experiment at pressures and temperatures corresponding to depths of 300–700 km, Meade and Jeanloz (1991) observed acoustic emissions in samples of serpentine (Fig. 7.5). X-ray diffraction indicated that the failure was accompanied by "amorphization," or a change in the crystal structure. However, Kirby *et al.* (1991) state that Meade and Jeanloz

. . . have not demonstrated that these [amorphous] acoustic events are connected with faulting. [Other laboratory data] and the evidence that transformational faulting in experiments above 300 MPa may be associated with the production of an amorphous phase does suggest that transformational faulting may be responsible for the acoustic emissions that they observe. Deep events occur as much as 50 km into the interiors of subducting plates, not just near the top where hydrous minerals should be most abundant. This observation makes it doubtful that transformation to an amorphous phase is involved in most deep earthquakes.

More recently, Karato *et al.* (1998) investigated failure in samples of $(Mg, Fe)_2 SiO_4$ ringwoodite at 16 GPa and 1600 °K. They found that strength was highly grain-size dependent, with fine-grained regions of the sample being very weak.

Physical arguments have provided the basis for most of the arguments in favor of shear instabilities. Griggs and Handin (1960) and Griggs and Baker (1969) favored shear-induced melting to explain both Bridgman's results and the mechanism of deep earthquakes. They also developed physical models to place constraints on the thickness of the shear zone in terms of the strain rate within the zone and the thermal properties of the material within and surrounding it. Their principal conclusion was that runaway melting was possible only if the thickness exceeded a minimum value (see Fig. 7.11); otherwise the frictional heat generated within the zone was dissipated by thermal conduction into the material surrounding the shear zone, and melting didn't occur. Within the shear zone, Griggs and Baker assumed

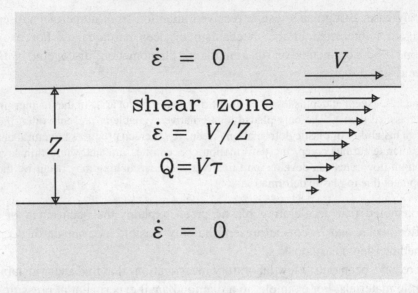

Fig. 7.11 Simple model illustrating that the shear zone must exceed a minimum thickness before thermal instability is possible. Here, across a zone of thickness Z a shear stress τ produces plastic deformation that proceeds at velocity V; this generates heat at rate τV. If the temperature is below melting by an amount ΔT, heat $\rho C_P \Delta T Z$ must be added to achieve melting, where ρ is density and C_P is the specific heat. Thus, in the absence of conduction, melting would occur in time $t_M = \rho C_P \Delta T Z / (\tau V)$. However, the characteristic time for heat loss by conduction is $t_C = Z^2/\chi$, where χ is the thermal diffusivity. This implies melting is likely only if $t_M < t_C$, which implies that $Z > \chi \rho C_P \Delta T / (\tau V)$. That is, Z exceeds 200 m to 32 km if $\chi = 10^{-6} \text{m}^2/\text{s}$, $\rho = 3000 \, \text{kg/m}^3$, $C_P = 800 \, \text{J/kg}^\circ\text{C}$, $\Delta T = 100-400^\circ\text{C}$, $\tau = 100-400 \, \text{MPa}$, and $V = 1-10 \, \text{cm/yr}$.

that the strain rate $\dot{\varepsilon}$ was proportional to $\tau^n \exp(-E/RT)$, where τ was the shear stress, E is an activation energy, T is temperature, and R is the gas constant. Under mantle conditions, the calculations of both Griggs and Handin (1960) and Griggs and Baker (1969) indicated that the minimum thickness of the shear zone was quite large, at least 30 km or more. This did not seem to favor shear-induced melting as the mechanism for deep earthquakes.

However, Ogawa (1987) subsequently developed a more realistic physical model where there was both plastic flow and elastic deformation in the shear zone, which could be either a homogeneous or a heterogeneous material. He found that under appropriate conditions the application of shear stress τ produced a slow increase in temperature in the shear zone; then, as the material weakened the stored elastic energy allowed the temperature to increase explosively. For conditions in the subducting lithosphere he concluded that thermal runaway occurred and produced melt in the shear zone; in particular, he estimated that thermal runaway could occur in zones with thicknesses of 10–100 meters if strain rates $\dot{\varepsilon}$ were on the order of 10^{-12}/s to 10^{-13}/s, and the shear stress τ was 300 MPa. Like Griggs, he noted that conduction at the boundaries prevented thermal runaway when shear zones were too thin. He also observed that thermal runaway wouldn't occur when zones were too thick because the stored elastic energy wasn't sufficiently concentrated to promote the final temperature instability.

Because catastrophic plastic instabilities occur in a variety of metals, polymers, and metallic glasses, Hobbs and Ord (1988) suggested it is plausible they occur in mantle materials, and explored how they might explain various features of Wadati–Benioff zone seismicity. They speculated that:

[Catastrophic] plastic shear is capable of explaining the detailed distribution of intermediate and deep focus earthquakes within subduction zones . . . For a particular strain rate there exists a critical temperature, T_{crit}, which is depth dependent; for temperatures below T_{crit} the material is strain rate softening and, for a soft enough loading system, may undergo catastrophic plastic shear. For temperatures above T_{crit} the material is strain rate hardening and is always stable during plastic shear. The cutoff depth for deep focus earthquakes then corresponds to the transition from strain rate softening to strain rate hardening material, and for commonly accepted geothermal gradients within the slab corresponds to approximately 800 km . . . The differences in foci distribution between subduction zones such as Tonga, New Hebrides, and Peru result from minor differences in the geothermal gradients within the slabs.

Hobbs and Ord further hypothesize that double Wadati–Benioff zones correspond to regions where the temperature of the subducting slab is approximately T_{crit}. The interior of the slab is colder and thus too strong to yield; the exterior is warmer and yields by ductile flow; the double zones merge together at depths of several hundred km as the interior of the slab warms. This paper is intriguing but highly speculative, and not all of its predictions are consistent with observations. For example, they suggest that deep-focus earthquakes should have lower stress drops than intermediate earthquakes; however, the data in Fig. 6.4 don't support this conclusion. And Hobbs and Ord don't consider the effects that phase transitions might have on the distribution of seismic activity.

Karato *et al.* (2001) elaborated on the approach of Hobbs and Ord, and considered a more realistic mineralogy and rheology. They assume that mechanics of the subducting lithosphere are dominated by the properties of dry olivine – its principal constituent – and its high pressure polymorphs, wadsleyite and ringwoodite. They consider available laboratory studies and make theoretical estimates about how the olivine–spinel rheologies depend on temperature, which is, of course, influenced most strongly by lithospheric age and the velocity of subduction. Above 400 km the olivine slab is cold and relatively strong; this explains why most Wadati–Benioff zones are relatively planar at intermediate depths. After olivine transforms to spinel,

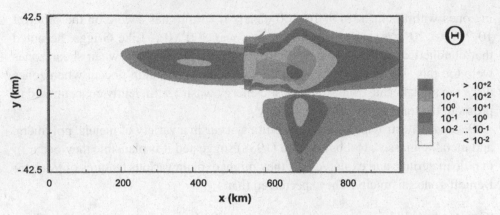

Fig. 7.12 Distribution of the instability parameter Θ within a slab subducting with velocity of 10 cm/yr; x is depth and y is distance from the central cold axis of the subducting slab. Thermal runaway occurs when Θ exceeds 1.0. Note that beneath about 500 km there are two separate regions where thermal runaway is possible, and together these "double seismic zones" span a width of about 40 km. Also, note that thermal runaway is possible at depths of 200–400 km above the olivine–spinel transition zone. Karato *et al.* (2001) calculated these results for a dry olivine–spinel rheology, including the effects of latent heat release and grain-size reduction when the olivine–spinel transition occurs. Figure reproduced from Karato *et al.* (2001).

slab strength changes and grain size is an additional important factor controlling the strength of spinel. Paradoxically, they conclude that lithosphere with the largest-grained, strongest spinel occurs when subduction velocities are about 7 cm/yr; at higher rates the slab is cooler and spinel grains don't grow as quickly, while at lower rates large spinel grains develop but the slab is weaker simply because it is significantly warmer.

While Karato *et al.* (2001) focus attention mostly on the strength and flexural properties of subducted lithosphere, they calculate that shear-induced thermal instabilities provide the most likely explanation for both intermediate- and deep-focus earthquakes. For example, for a subduction velocity of 10 cm/yr (Fig. 7.12) they find that thermal instability should be possible with a zone about 20 km wide extending from about 200 km to 550 km depth; between about 600 km and 650 km the calculations find two separate zones of stability that together have a width of about 40–60 km. They suggest that the double seismic zones observed by Wiens *et al.* (1993) at depths of 350–460 km in Tonga might be expressions of these two zones.

Among seismologists there has been only restrained support for the idea that frictional melting might be an essential feature of earthquake mechanics. Although theoretical studies by McKenzie and Brune (1972) and Richards (1976) indicated that melting was possible if frictional and driving stresses were high enough, it is unclear that this is true for shallow earthquakes, especially as observations indicate few faults contain glassy materials indicative of melting (Sibson, 1973). For the 1994 deep Bolivia earthquake, Kanamori *et al.* (1998), Bouchon and Ihmlé (1999), and Venkataraman and Kanamori (2004) concluded that it was possible that shear melting occurred during the rupture process. However, these investigations used significantly different values for the stress drop $\Delta\sigma$ (55 MPa vs 5–10 MPa), and all suggest that melting occurred only in a very thin layer, with a thickness of a few mm or less. The general assumption in these studies is that shear melting isn't a pervasive feature of deep earthquakes. Rather, it might occur in certain unusual large deep events, or it might accompany rupture initiated by some other process, such as transformational faulting. Thus, it might provide a means for large-earthquake ruptures to occupy larger areas than would otherwise be possible.

Nevertheless, shear instability remains an attractive candidate for deep earthquake mechanism whether it involves actual melting or only an "amorphization" where solid material suddenly loses its granular structure and undergoes a marked loss of strength. Physical modeling arguments provide the strongest evidence favoring shear instability; they demonstrate that in a layer of strain-softening material applied stresses can easily produce thermal runaway. However, the models leave some questions unanswered. Most of the models utilize geometries appropriate for faulting along ruptures parallel to the slab surface, rather than obliquely across the slab as the observations indicate. It isn't entirely clear how instabilities can propagate at such high speeds over regions having dimensions of tens of kilometers. And, if shear instabilities do initiate large deep-focus earthquakes, it is puzzling that ultra-low-frequency analysis does not indicate that there is measurable, accelerating precursory creep prior to their occurrence.

7.3.3 Transformational faulting and anticracks

If a mineral is polymorphic, i.e., if it exists in two or more stable forms, the lower-pressure form is said to exhibit transformational faulting if, in response to an applied shear stress, it faults suddenly by transforming to the higher-pressure form within a fine-grained shear zone. The suggestion that transformational faulting might occur is not new. Indeed, Bridgman (1936) noted that:

There is a theoretical possibility which should be kept in mind. There is no thermodynamic reason why [polymorphic] transitions should not occur under shearing stress which would never occur under any combination of temperature and hydrostatic pressure. In the greatly enhanced range of shearing stress realizable at high pressure one may expect greater opportunity for the existence of this phenomenon than in the comparatively low range of shearing stress limited by plastic flow under ordinary conditions. No examples of phenomena which have been positively identified as being of this character have yet been found . . .

Similarly, Raleigh and Paterson (1965) observed that:

Perhaps [the concept that pore fluids might extend the range of brittle fracture] could even be extended to deep-focus earthquakes if the mantle material there contains a phase, effectively devoid of shear strength (through being very near its melting point or for some other reason), which functions as a fluid phase between the stronger phases present; then the nonhydrostatic stresses would be supported by the stronger phases but cataclastic or brittle fracturing might occur if the weak phase plays a part analogous to that of the water in the serpentinite.

Much later, Steve Kirby and his colleagues performed experiments demonstrating that transformational faulting actually occurred in ice and in tremolite, a variety of amphibole. Kirby (1987) also speculated that transformational faulting might occur in mantle materials and thus might be responsible for deep earthquakes. He noted that an exothermic reaction, such as the olivine–spinel transition, might promote the transformation:

If a transformation produces heat and thus raises the temperature locally near the region being transformed, then this can lead to a transient loss of shear strength or cohesion by promoting the various high-temperature creep mechanisms that can occur which are very temperature sensitive . . . Reduced grain size in regions that have undergone phase changes [may activate] various creep mechanisms that go faster with finer grain size . . . Reconstructive polymorphic phase changes localized along shear zones may be a general phenomenon in crystalline materials stressed non-hydrostatically near transition boundaries. Moreover, [they] may lead to localized shear instabilities manifested by deep earthquakes and produce the observed double-couple first-motion distribution. Finally, if this is the earthquake source mechanism in deeply subducted lithosphere, then earthquakes should shut off when all of the polymorphic phase changes have occurred. This may explain the maximum depth of earthquakes at about 680 km.

Shortly thereafter, Green and Burnley (1989) presented evidence that transformational faulting occurred under specific pressure–temperature conditions in the olivine germanate analog Mg_2GeO_4 (Fig. 7.13). When they applied shear stresses under P–T conditions where, at equilibrium, the olivine would transform to spinel, they observed ductile behavior at both very low and very high temperatures. But, at intermediate temperatures the samples failed catastrophically by faulting; optical and electron microscopy indicated that the faults contained fine grained spinel but were surrounded by regions where the olivine had not yet transformed. Further evaluation of specimens revealed that the fine-grained spinel developed initially within microscopic lens-shaped regions oriented at high angles – nearly 90° – with respect to the largest principal stress (Fig. 7.14). They observed that:

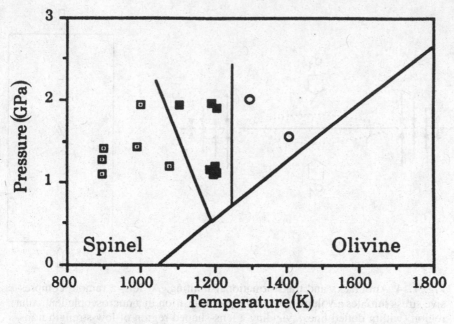

Fig. 7.13 Pressure–temperature conditions and modes of failure for metastable Mg_2GeO_4 undergoing strain rates of $10^{-4}-10^{-5}$ in the laboratory. Solid lines indicate equilibrium phase boundaries. Open squares represent specimens that were strong and ductile; open circles, specimens that were weak and ductile, and closed squares, specimens that failed by transformational faulting. Under conditions where faulting occurs the bulk material remains as metastable olivine, and transforms to spinel only in regions experiencing high shear stress. Reproduced from Green and Burnley (1989) with permission from Nature Publishing Group.

En echelon microfaults occur with stepping in the opposite sense to that commonly observed in brittle faulting. In one experiment, small spinel-lined faults are linked by push-togethers exhibiting enhanced transformation in the form of spinel microlenses, whereas in brittle behavior, faults are linked by pull-aparts.

Green and Burnley (1989) called these microscopic spinel lenses anticracks. Subsequently Green and his collaborators have argued that they function mechanically almost exactly like fluid-filled Griffiths cracks (Burnley *et al.*, 1991; Green and Houston, 1995). That is, anticracks form at oblique angles to the largest principal stress; their lens-shaped geometry induces continuing anticrack growth at their tips; they contain fine-grained material with little or no shear strength, but with the capability of exerting normal stresses on the crack walls. Moreover, as they form they release heat that raises the local temperature, increasing their rate of formation. And, at some critical anticrack density the bulk material loses its ability to support the applied stress, and failure begins by linking up of anticracks and formation of a protofault.

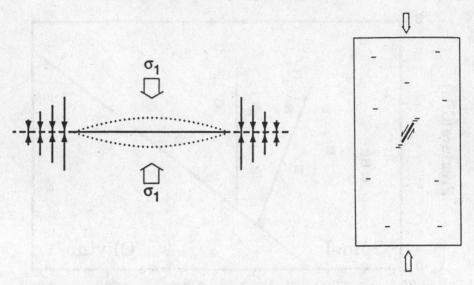

Fig. 7.14 Anticracks and transformational faulting. At left, a remote compressive stress initiates a volume-reducing phase transition in a microscopic lenticular region (within dotted lines), yielding a lens-shaped region of low-strength transformed material having compressive stress concentration at its tips. Because the sign of the stresses is opposite to those in Fig. 7.2, theory indicates that such cracks are favored when crack planes are normal to the largest principal stress (σ_1). At right, transformational faulting occurs when numerous such "anticracks" form, grow, and link up with one another, allowing shear motion to occur along a macroscopic plane oriented at an oblique angle to the stress direction. Figure modified from Green and Houston (1995) and reprinted with permission from the *Annual Reviews of Earth and Planetary Sciences*.

Subsequent laboratory experiments have provided additional support for the anticrack mechanism. Green *et al.* (1990) performed experiments on natural olivine at P–T conditions up to 15 GPa and 1650 °K, and reported anticracks in one specimen that were "morphologically indistinguishable from the lenses found in germanate specimens that faulted." They also recorded acoustic emissions accompanying the deformation (Green *et al.*, 1992). In diamond anvil experiments on fayalite and forsterite, Wu *et al.* (1993) concluded that shear deformation enhanced the kinetics of transformation of silicate olivine (α phase) to modified spinel (β) and spinel (γ) phases, i.e., the transformation took place at lower pressures and temperatures than under equilibrium conditions. By measuring sliding resistance on shear-faulted Mg_2GeO_4 specimens with dimension of about 3×7 mm, Tingle *et al.* (1993) concluded that "despite its remarkable similarity to shear fracture" neither the faulting nor the sliding after faulting was controlled by a brittle process. They observed no detectable acoustic emissions prior to failure; they found that the faulting stress was independent of confining pressure; and that faulting occurred only in a narrow temperature window. They found that the observed pressure and rate dependence of the sliding stress indicated that:

> . . . sliding occurs [not by friction but] by a viscous process induced by the extremely fine grain size of the high density phase that grows in the anticracks and is incorporated into the fault zone during faulting. Thus, the underlying physics of the anticrack mechanism of phase-transformation faulting (and, by extension, the mechanism of deep-earthquake faulting) are distinct from those involved in brittle shear failure.

The anticrack mechanism to explain deep-focus earthquakes still enjoys considerable support. This is partly because of the initial enthusiasm of high-profile proponents such as Harry Green and Steve Kirby (Green and Houston, 1995; Kirby *et al.*, 1996b). And there is intellectual appeal in its analogies to Griffiths cracking and brittle fracture. Indeed, it seems possible that one generic mechanism which relies on the formation and coalescing of microscopic lens-shaped zones may explain earthquakes at all depths: shallow earthquakes – brittle fracture resulting from open cracks created by shear stress; intermediate-focus earthquakes – dehydration embrittlement, with the cracks kept open by water; deep-focus earthquakes – anticracks filled with low-strength, fine-grained material generated by transformational faulting. Although, as noted above by Tingle *et al.* (1993), different physical processes may control the underlying physics of crack generation and growth in different materials and at different depths; all three mechanisms share gross similarities in the process by which catastrophic failure occurs.

Because olivine and spinel must coexist at depths exceeding 400 km, transformational faulting requires the presence of a metastable olivine wedge (Fig. 7.15). Rubie and Ross (1994) evaluated olivine–spinel reaction kinetics and estimated that although the reaction could begin at temperatures of $550 \pm 50\,°C$, substantial amounts of metastable olivine should persist to depths of 500 km or more as long as temperatures at 400 km depth remain below 700 °C. When the olivine–spinel reaction does occur the expected high nucleation and low growth rates were consistent with the formation of very fine-grained spinel, as required for transformational faulting. Pankow *et al.* (2002) and Yoshioka and Murakami (2002) assert that the presence of a metastable wedge should be seismologically detectable in some situations, e.g., by analyzing observations of the P and S coda from quakes in fast-subducting slabs with dip angles of about 30°.

However, not everybody is enamored with the transformational faulting mechanism. Koper *et al.* (1998) conclude that it isn't possible to resolve whether such a wedge exists[6] with travel-time data and presently available observations. And Karato *et al.* (2001) state:

---

[6] Although Iidaka and Suetsugu (1992) and Iidaka and Obara (1997) have interpreted travel-time observations and shear-wave splitting from Japanese earthquakes as supporting the existence of a metastable wedge, their results aren't convincing to this book's author.

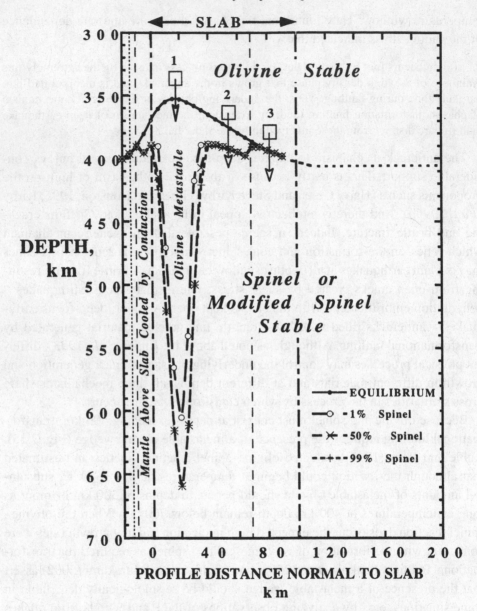

Fig. 7.15 Detailed mineralogical model for the Tonga slab including effects of latent heat release. Three regimes of olivine–spinel reaction occur: in the cold slab interior (square 1) olivine bypasses the equilibrium boundary and survives as metastable olivine until it changes abruptly to spinel; slightly warmer material (square 2) survives briefly and then converts slowly to spinel by a nucleation-controlled mechanism; warmest material (square 3) converts to spinel near the equilibrium boundary at rates comparable to the slab descent rate. Figure reproduced from Kirby *et al.* (1996b).

Some fundamental problems remain with [the transformational faulting] model . . . It is uncertain if a metastable olivine wedge exists or not, particularly in relatively warm slabs such as those beneath South America. In addition, the recent observation of repeating deep earthquakes is difficult to reconcile with the transformational faulting model. [These observations make] this mechanism unlikely to be important.

Mosenfelder *et al.* (2001) argue that current kinetic data indicate that metastable olivine won't survive to the depths of the deepest earthquakes. Similarly, Marton *et al.* (2005) model slab temperatures and conclude that:

In general, the maximum depths of olivine metastability are 160–230 km less than the depths of the deepest seismicity. Even in the extreme case of Tonga, metastable olivine falls short of the deepest earthquakes by 140 km in our model. Even taking the various uncertainties into account, these results indicate that deep-focus earthquakes occur in regions already transformed to wadsleyite or ringwoodite and must therefore be caused by a mechanism other than transformational faulting.

Furthermore, Silver *et al.* (1995) evaluated the rupture process of the 9 June 1994 Bolivia deep earthquake, and concluded:

Seismic data from permanent stations plus portable instruments in South America show that rupture occurred on a horizontal plane and extended at least 30 by 50 km . . . and penetrated through more than a third of the slab thickness. This extent is more than three times that expected for a metastable wedge of olivine at the core of the slab, and thus appears to be incompatible with an origin by transformational faulting.

Similarly, Wiens and Snider (2001) note that:

The large rupture extent of the largest deep earthquakes and the occurrence of aftershocks outside the active slab argue against the transformational faulting model.

Finally, Rubie (1996) discusses various problems with the hypothesis of transformational faulting. He notes that only one study (Green *et al.*, 1990) has documented transformational faulting in samples of silicate olivine at transition zone pressures, and he is unconvinced by the published micrographs purporting to demonstrate that it controls the fault rheology during failure. After performing additional experiments, Dupas-Bruzek *et al.* (1998) were similarly dubious:

We have studied the effect of non-hydrostatic stress on the mechanisms of the olivine–wadsleyite–ringwoodite (α-β-γ) phase transformations and deformation mechanisms of olivine and wadsleyite at high pressure [900 °C and 16 GPa]. Nonhydrostatic stress enhances transformation rates by increasing the density of dislocations which act as nucleation sites for ringwoodite. Although the samples were partially reacted under non-hydrostatic stress, there is no evidence for transformational faulting.

7.4 Are deep quakes caused by more than one mechanism?

Ockham's Razor be damned! Given the vast range of materials, phases, pressures, temperatures, and stresses that occur within Wadati–Benioff zones, why should we expect that a single mechanism is responsible for all deep earthquakes? The idea that there might be multiple mechanisms for deep earthquakes is not new, although the controversy has focused on different issues at different times.

Among the earlier proponents of dual mechanisms were scientists who favored solid–solid phase transitions as the root cause of deep earthquakes. They figured that solid–solid phase transitions might proceed over time scales of minutes to days – fast enough to generate truly enormous shear stresses in neighboring regions, but slow enough so that they didn't radiate seismic energy. If this process was too fast to be relieved by ductile flow, the neighboring regions would then fail catastrophically by some faulting process, producing deep earthquake with double-couple focal mechanisms. For example, Benioff (1964) suggested that:

It may well be that . . . the downward motion of the contracting rock mass may be accompanied by faulting at its periphery. The faulting would be evident in the observed four-lobed initial radiation pattern for the short-period waves, while the gross movement shows up only in the record of long-period strain.

Similarly, while Dziewonski and Gilbert (1974) concluded that phase transitions were responsible for apparent observations of low-frequency compression prior to two South American earthquakes, they favored shear melting to explain the origin of the observed double-couple component of deep earthquakes.

Since about 1970 there has also been a (not unanimous) consensus that different mechanisms were responsible for intermediate- and deep-focus earthquakes. For proponents of the solid–solid phase transition mechanism, this was because for probable mantle constituents, there was no obvious phase transition active at depths of 60–350 km that was as appealing as the olivine–spinel transition, which occurs near 400 km at equilibrium. For proponents of dehydration embrittlement, this was because most minerals hydrated near the Earth's surface should undergo nearly complete dehydration before they reached pressure–temperature conditions corresponding to a depth of 300 km. While the reviews of Green and Houston (1995), Kirby *et al.* (1996a; 1996b) and Green and Marone (2002) all favor dehydration embrittlement as the mechanism responsible for intermediate earthquakes and transformational faulting for deep-focus earthquakes, subsequent publications including some from these authors now provide support for other mechanisms.

However, whether more than one mechanism is necessary to explain deep-focus earthquakes remains as a controversy (see Estabrook, 2004). An essential observation is that a few (but not all) large deep-focus earthquakes appear to have rupture and/or aftershock zones that extend as much as 40–60 km across the width of the subducting lithosphere. This implies that failure occurs within material that experiences temperature variations of 300°–400 °C (Wiens, 2001). This is a problem for proponents of transformational faulting, who expect earthquakes to occur only in a relatively thin metastable wedge of olivine confined to the interior of the slab. Proponents of dehydration embrittlement predict that earthquakes should concentrate near the top surface of the slab where minerals presumably would have experienced more complete hydration when the lithosphere was at the surface. The existence of a second mechanism, such as shear melting, allows both groups to explain rupture and aftershocks that extend through a thicker zone within the subducting slab. Of course, other scientists (e.g., Karato *et al.*, 2001, see quote above) argue that shear melting can explain all deep-focus earthquakes, and thus more than one mechanism isn't required. And Omori *et al.* (2004) contend that dehydration embrittlement explains all deep earthquakes, arguing that other candidate mechanisms can't explain why deep-focus earthquakes can occur in both very cold and very warm environments, such as Tonga and South America.

The observations fueling this controversy are primarily from two earthquakes, the 9 March 1994 Tonga and 9 June 1994 Bolivia events. For the Tonga earthquake, Goes and Ritsema (1995), McGuire *et al.* (1997) and Tibi *et al.* (1999) all evaluated the rupture, and obtain comparable dimensions of 40×30 km, 45×30 km, and 40×20 km, respectively. McGuire *et al.* (1997) and Wiens and McGuire (2000) relocated aftershocks, and found a few that located outside the zone of previous Wadati–Benioff zone activity. For the Bolivia earthquake, Estabrook and Bock (1995) and Silver *et al.* (1995) both evaluated subevents and found that the rupture penetrated at least 35 km of the slab's thickness and had dimensions of 30×50 km or more; this was approximately the aftershock area as determined by Myers *et al.* (1995). Ihmlé (1998) and Antolik *et al.* (1999) found significantly larger areas from a constrained waveform inversion. Thus, for the Tonga and especially for the Bolivia earthquake, the results are probably robust because several independent groups have obtained comparably large rupture areas.

However, the emerging consensus is that many large intermediate- and deep-focus earthquakes have rupture zones with a thickness of 20 km or even less (Wu and Chen, 2001; Tibi *et al.*, 2002; 2003). Given the systematic difficulties that accompany determination of rupture areas, this suggests most intermediate-focus ruptures could easily be accommodated within the thickness of the oceanic crust, and deep-focus ruptures within a metastable wedge. Some of the allegedly exceptional earthquakes such as the 1994 Bolivia and 1970 Colombia earthquakes lie in areas where background seismicity is sparse, and thus it is possible that we don't know the true orientation of the subducting slab, or if it has become thickened or contorted.

7.5 References

Abers, G. A., 2000. Hydrated subducted crust at 100–250 km depth, *Earth Planet. Sci. Lett.*, **176**, doi:10.1016/S0012-821X(00)00007-8, 323–330.

    2005. Seismic low-velocity layer at the top of subducting slabs: observations, predictions, and systematics, *Phys. Earth Planet. Int.*, **149**, doi:10.1016/j.pepi.2004.10.002, 7–29.

Abers, G. A. and G. Sarker, 1996. Dispersion of regional body waves at 100–150 km depth beneath Alaska: in situ constraints on metamorphism of subducted crust, *Geophys. Res. Lett.*, **23**, doi:10.1029/96GL00974, 1171–1174.

Abers, G. A., T. Plank, and B. R. Hacker, 2003. The wet Nicaraguan slab, *Geophys. Res. Lett.*, **30**, doi:10.1029/2002GL015649, 1098.

Anderson, D. L. and J. D. Bass, 1984. Mineralogy and composition of the upper mantle, *Geophys. Res. Lett.*, **11**, 637–640.

Anderson, R. N., A. Hasegawa, N. Umino, and A. Takagi, 1980. Phase changes and the frequency–magnitude distribution in the upper plane of the deep seismic zone beneath Tohoku, Japan, *J. Geophys. Res.*, **85**, 1389–1398.

Anonymous, 1936. Geophysical discussion, *Observatory*, **59**, 265–269.

Antolik, M., D. Dreger, and B. Romanowicz, 1999. Rupture processes of large deep-focus earthquakes from inversion of moment rate functions, *J. Geophys. Res.*, **104**, doi:10.1029/1998JB00042, 863–894.

Benioff, H., 1964. Earthquake source mechanisms, *Science*, **143**, 1399–1406.

Bensman, S. J., 2001. Urquhart's and Garfield's laws: the British controversy over their validity, *J. Amer. Soc. Inf. Sci. Tech.*, **52**, 714–724.

Bina, C. R. and A. Navrotsky, 2000. Possible presence of high-pressure ice in cold subducting slabs, *Nature*, **408**, 844–847, doi:10.1038/35048555.

Bouchon, M. and P. Ihmlé, 1999. Stress drop and frictional heating during the 1994 deep Bolivia earthquake, *Geophys. Res. Lett.*, **26**, doi:10.1029/1999GL005410, 3521–3524.

Brearley, A. J. and D. C. Rubie, 1990. The effects of H_2O on the disequilibrium breakdown of muscovite + quartz, *J. Petrology*, **31**, 925–926.

Bridgman, P. W., 1936. Shearing phenomena at high pressure of possible importance for geology, *J. Geol.*, **44**, 653–669.

• Bridgman, P. W., 1945. Polymorphic phase transitions and geological phenomena, *Amer. J. Sci*, **243A**, 90–97.

Burnley, P. C., H. W. Green, and D. J. Prior, 1991. Faulting associated with the olivine to spinel transformation in Mg_2GeO_4 and its implications for deep-focus earthquakes, *J. Geophys. Res.*, **96**, 425–443.

Chen, J., T. Inoue, D. J. Weidner, Y. Wu, and M. T. Vaughan, 1998. Strength and water weakening of mantle minerals, olivine, wadsleyite and ringwoodite, *Geophys. Res. Lett.*, **25**, doi:10.1029/98GL00043, 575–578.

Christensen, D. H. and L. J. Ruff, 1988. Seismic coupling and outer rise earthquakes, *J. Geophys. Res.*, **93**, 13421–13444.

Connolly, J. A. D. and D. M. Kerrick, 2002. Metamorphic controls on seismic velocity of subducted oceanic crust at 100–250 km depth, *Earth Planet. Sci. Lett.*, **204**, doi:10.1016/S0012–821X(02)00957–3, 61–74.

Davies, J. H., 1999. The role of hydraulic fractures and intermediate-depth earthquakes in generating subduction-zone magmatism, *Nature*, **398**, doi:10.1038/18202, 142–145.

Dennis, J. G. and C. T. Walker, 1965. Earthquakes resulting from metastable phase transitions, *Tectonophysics*, **2**, doi:10.1016/0040-1951(65)90034-X, 401–407.

Dobson, D. P., P. G. Meredith, and S. A. Boon, 2002. Simulation of subduction zone seismicity by dehydration of serpentine, *Science*, **298**, doi:10.1126/science.1075390, 1407–1410.

2004. Detection of microseismicity in multi anvil experiments, *Phys. Earth Planet. Int.*, **143–144**, doi:10.1016/j.pepi.2003.09.023, 337–346.

Dreger, D. and B. Woods, 2002. Regional distance seismic moment tensors of nuclear explosions, *Tectonophysics*, **356**, doi:10.1016/S0040-1951(02)00381-5, 139–156.

Dupas-Bruzek, C., T. G. Sharp, D. C. Rubie, and W. B. Durham, 1998. Mechanisms of transformation and deformation in $Mg_{1.8}Fe_{0.2}SiO_4$ olivine and wadsleyite under non-hydrostatic stress, *Phys. Earth Planet. Int.*, **108**, doi:10.1016/S0031-9201(98) 00086-7, 33–48.

Dziewonski, A. M. and F. Gilbert, 1974. Temporal variation of the seismic moment tensor and the evidence of precursive compression for two deep earthquakes, *Nature*, **247**, 185–188.

Ekström, G. and P. G. Richards, 1994. Empirical measurements of tectonic moment release in nuclear explosions from teleseismic surface waves and body waves, *Geophys. J. Int.*, **117**, 120–140.

Estabrook, C. H., 1999. Body wave inversion of the 1970 and 1963 South American large deep-focus earthquakes, *J. Geophys. Res.*, **104**, doi:10.1029/1999JB900244, 28751–28767.

2004. Seismic constraints on mechanisms of deep earthquake rupture, *J. Geophys. Res.*, **109**, doi:10.1029/2003JB002449, 2306.

Estabrook, C. H. and G. Bock, 1995. Rupture history of the great Bolivian earthquake: slab interaction with the 660 km discontinuity? *Geophys. Res. Lett.*, **22**, doi:10.1029/95GL02234, 2277–2280.

Gilbert, F. and A. M. Dziewonski, 1975. An application of normal mode theory to the retrieval of structural parameters and source mechanisms from seismic spectra, *Phil. Trans. Roy. Soc., London*, **A278**, 187–269.

Goes, S. and J. Ritsema, 1995. A broadband P wave analysis of the large deep Fiji Island and Bolivia earthquakes of 1994, *Geophys. Res. Lett.*, **22**, doi:10.1029/95GL02011, 2249–2252.

Goetze, C., 1978. The mechanics of creep in olivine, *Phil. Trans. Roy. Soc., London*, **A288**, 99–119.

Green, H. H., 1984. How and why does olivine transform to spinel? *Geophys. Res. Lett.*, **11**, 817–820.

Green, H. W. and P. C. Burnley, 1989. A new self-organizing mechanism for deep-focus earthquakes, *Nature*, **341**, doi:10.1038/341733a0, 733–737.

• Green, H. W. and H. Houston, 1995. The mechanics of deep earthquakes, *Ann. Rev. Earth Planet. Sci.*, **23**, doi:10.1146/annurev.ea.23.050195.001125, 169–213.

Green, H. W. and C. Marone, 2002. Instability of deformation, *Rev. Mineral. Geochem.*, **51**, 181–199.

Green, H. W., T. E. Young, D. Walker, and C. H. Scholz, 1990. Anticrack-associated faulting at very high pressure in natural olivine, *Nature*, **348**, doi:10.1038/348720a0, 720–722.

Green, H. W., C. H. Scholz, T. N. Tingle, T. E. Young, and T. A. Koczynski, 1992. Acoustic emissions produced by anticrack faulting during the olivine–spinel transformation, *Geophys. Res. Lett.*, **19**, doi:10.1029/92GL00751, 789–792.

Griffith, A. A., 1924. The theory of rupture. In *Proc. 1st Int. Congr. Appl. Mech.*, eds. C. B. Biezeno and J. M. Burgers. Delft, Tech. Boekhandel en Drukkerij J. Walter Jr., 54–63.

Griggs, D. T. and D. W. Baker, 1969. The origin of deep-focus earthquakes. In *Properties of Matter under Unusual Conditions*, eds. H. Mark and S. Fernbach, New York, NY, Interscience, 389 pp., 23–42.

Griggs, D. and J. Handin, 1960. Observations on fracture and a hypothesis of earthquakes, *Geol. Soc. Amer. Mem.*, **79**, 347–373.

Hacker, B. R., 1997. Diagenesis and the fault-valve seismicity of crustal faults, *J. Geophys. Res.*, **102**, doi.10.1029/97JB02025, 24459–24467.

• Hacker, B. R., S. M. Peacock, G. A. Abers, and S. D. Holloway, 2003. Subduction factory – 2. Are intermediate-depth earthquakes in subducting slabs linked to metamorphic dehydration reactions? *J. Geophys. Res.*, **108**, doi:10.0129/JB001129, 2003.

Hara, T., K. Kuge, and H. Kawakatsu, 1996. Determination of the isotropic component of deep focus earthquakes by inversion of normal-mode data, *Geophys. J. Int.*, **127**, 515–528.

Helffrich, G. and S. Stein, 1993. Study of the structure of the slab–mantle interface using reflected and converted seismic waves, *Geophys. J. Int.*, **115**, 14–40.

Hobbs, B. E. and A. Ord, 1988. Plastic instabilities: implications for the origin of intermediate and deep focus earthquakes, *J. Geophys. Res.*, **93**, 10521–10549.

Hodder, A. P. W., 1984. Thermodynamic constraints on phase changes as earthquake source mechanisms in subduction zones, *Phys. Earth Planet. Int.*, **34**, 221–225.

Holmes, A., 1931. Radioactivity and earth movements, *Transactions of the Geological Society of Glasgow*, **18**, 559–606.

  1933. The thermal history of the Earth, *J. Wash. Acad. Sci.*, **23**, 169–195.

Honda, H., 1932. On the types of the seismograms and the mechanism of deep earthquakes, *Geophys. Mag.*, **5**, 301–326.

  1934. On the mechanism of deep earthquakes and the stress in the deep layer of the earth crust, *Geophys. Mag.*, **8**, 179–185.

Hubbert, M. K. and W. W. Rubey, 1959. Role of fluid pressure in overthrust faulting, *Geol. Soc. Amer. Bull.*, **70**, 115–206.

Ihmlé, P. F., 1998. On the interpretation of subevents in teleseismic waveforms: the 1994 Bolivia deep earthquake revisited, *J. Geophys. Res.*, **103**, doi:10.1029/98JB00603, 17919–17932.

Iidaka, T. and K. Obara, 1997. Seismological evidence for the existence of anisotropic zone in the metastable wedge inside the subducting Izu–Bonin slab, *Geophys. Res. Lett.*, **24**, doi:10.1029/97GL03277, 3305–3308.

Iidaka, T. and D. Suetsugu, 1992. Seismological evidence for metastable olivine inside a subducting slab, *Nature*, **356**, doi:10.1038/356593a0, 593–595.

Irifune, T., N. Kubo, M. Isshiki, and Y. Yamasaki, 1998. Phase transformations in serpentine and transportation of water into the lower mantle, *Geophys. Res. Lett.*, **25**, doi:10.1029/97GL03572, 203–206.

Jeffreys, H., 1924. *The Earth, Its Origin, History and Physical Constitution*, Cambridge, Cambridge University Press, 278 pp.

  1936. The structure of the Earth down to the 20° discontinuity, *Mon. Not. Roy. Astron. Soc., Geophys. Supp.*, **3**, 401–422.

Jiao, W., P. G. Silver, Y. Fei, and C. T. Prewitt, 2000. Do intermediate- and deep-focus earthquakes occur on preexisting weak zones? An examination of the Tonga subduction zone, *J. Geophys. Res.*, **105**, doi:10.11029/2000JB900314, 28125–28138.

Jung, H., H. W. Green, and L. F. Dobrzhinetskaya, 2004. Intermediate-depth earthquake faulting by dehydration embrittlement with negative volume change, *Nature*, **428**, doi:10.1038/nature02412, 545–549.

Kanamori, H., D. L. Anderson, and T. H. Heaton, 1998. Frictional melting during the rupture of the 1994 Bolivian earthquake, *Science*, **279**, doi:10.1126/science.279.5352,839, 839–842.

Karato, S.-I., C. Dupas-Bruzek, and D.C. Rubie, 1998. Plastic deformation of silicate spinel under transition-zone conditions in the Earth's mantle, *Nature*, **395**, doi:10.1038/26206, 266–269.

Karato, S.-I., M. R. Riedel, and D. A. Yuen, 2001. Rheological structure and deformation of subducted slabs in the mantle transition zone: implications for mantle circulation and deep earthquakes, *Phys. Earth Planet. Int.*, **127**, doi:1016/S0031-9201(01)00223-0, 83–108.

Kasahara, J. and H. Tsukahara, 1971. Experimental measurements of reaction rate at the phase change of nickel olivine to nickel spinel, *J. Phys. Earth*, **19**, 79–88.

Kawakatsu, H., 1991. Insignificant isotropic component in the moment tensor of deep earthquakes, *Nature*, **351**, doi:10.1038/351050a0, 50–53.

1996. Observability of the isotropic component of a moment tensor, *Geophys. J. Int.*, **126**, 525–544.

Kerrick, D. M. and J. A. D. Connolly, 2001. Metamorphic devolatilization of subducted oceanic metabasalts: implications for seismicity, arc magmatism and volatile recycling, *Earth Planet. Sci. Lett.*, **189**, doi:10.1016/S0012-821X(01)00347-8, 19–29.

Kerschhofer, L., D. C. Rubie, T. G. Sharp, J. D. C. McConnell, and C. Dupas-Bruzek, 2000. Kinetics of intracrystalline olivine–ringwoodite transformation, *Phys. Earth Planet. Int.*, **121**, doi:10.1016/S0031-9201(00)00160-6, 59–76.

Kirby, S. H., 1987. Localized polymorphic phase transformations in high-pressure faults and applications to the physical mechanism of deep earthquakes, *J. Geophys. Res.*, **92**, 13789–13800.

Kirby, S. H. and L. A. Stern, 1993. Experimental dynamic metamorphism of mineral single crystals, *J. Struc. Geol.*, **15**, 1223–1240.

Kirby, S., W. B. Durham, and L. A. Stern, 1991. Mantle phase changes and deep-earthquake faulting in subducting lithosphere, *Science*, **252**, 216–225.

Kirby, S., E. R. Engdahl, and R. Denlinger, 1996a. Intermediate-depth intraslab earthquakes and arc volcanism as physical expressions of crustal and uppermost mantle metamorphism in subducting slabs. In *Subduction, Top to Bottom*, eds. G. E. Bebout, D. W. Scholl, S. H. Kirby, and J. P. Platt, Geophys. Mon. 96, Washington, DC, American Geophysical Union, 195–214.

• Kirby, S. H., S. Stein, E. A. Okal, and D.C. Rubie, 1996b. Metastable mantle phase transformations and deep earthquakes in subducting oceanic lithosphere, *Rev. Geophys.*, **34**, doi:10.1029/96RG01050, 261–306.

Komabayashi, T., S. Omori, and S. Maruyama, 2004. Petrogenetic grid in the system $MgO-SiO_2-H_2O$ up to 30 GPa, 1600 °C: applications to hydrous peridotite subducting into the Earth's deep interior, *J. Geophys. Res.*, **109**, doi:10.1029/2003 JB002651.

Koning, L. P. G., 1942. On the mechanism of deep-focus earthquakes, *Gerlands Beitr. z. Geophysik*, **58**, 159–197.

Koper, K. D., D. A. Wiens, L. M. Dorman, J. A. Hildebrand, and S. C. Webb, 1998. Modeling the Tonga slab: can travel time data resolve the metastable olivine wedge? *J. Geophys. Res.*, **103**, doi:10.1029/98JB01517, 30079–30100.

Leith, A. and J. A. Sharpe, 1936. Deep-focus earthquakes and their geological significance, *J. Geol.*, **44**, 877–917.

Liu, L., 1983. Phase transformations, earthquakes, and the descending lithosphere, *Phys. Earth Planet. Int.*, **32**, 226–240.

Liu, M., L. Kerschhofer, J. L. Mosenfelder, and D.C. Rubie, 1998. The effect of strain energy on growth rates during the olivine-spinel transformation and implications for olivine metastability in subducting slabs, *J. Geophys. Res.*, **103**, doi:10.1029/98JB00794, 23897–23909.

Lomnitz-Adler, J., 1990. Are deep focus earthquakes caused by a Martensitic transformation? *J. Phys. Earth*, **38**, 83–98.

Martin, S., A. Rietbrock, C. Haberland, and G. Asch, 2003. Guided waves propagating in subducted oceanic crust, *J. Geophys. Res.*, **108**, doi:10.1029/2003JB002450, 2536.

Marton, F. C., T. J. Shankland, D. C. Rubie, and Y. Xu, 2005. Effects of variable thermal conductivity on the mineralogy of subducting slabs and implications for mechanisms of deep earthquakes, *Phys. Earth Planet. Int.*, **149**, doi:10.1016/j.pepi.2004.08.026, 53–64.

Matsuzawa, T., N. Umino, A. Hasegawa, and A. Takagi, 1987. Estimation of thickness of a low-velocity layer at the surface of the descending oceanic plate beneath the northeastern Japan arc by using synthesized PS-wave, *Tohoku Geophys. J.*, **31**, 19–28.

McGuire, J. J., D. A. Wiens, P. J. Shore, and M. G. Bevis, 1997. The March 9, 1994 (M_W 7.6), deep Tonga earthquake: rupture outside the seismically active slab, *J. Geophys. Res.*, **102**, doi:10.1029/96JB03185, 15163–15182.

McKenzie, D. P. and J. N. Brune, 1972. Modeling on fault planes during large earthquakes, *Geophys. J. Roy. Astron. Soc.*, **29**, 65–78.

• Meade, C. and R. Jeanloz, 1989. Acoustic emissions and shear instabilities during phase transformations in Si and Ge at ultrahigh pressures, *Nature*, **339**, doi:10.1038/339616a0, 616–618.

    1991. Deep-focus earthquakes and recycling of water into the Earth's mantle, *Science*, **252**, 68–72.

Morris, J., J. Gosse, S. Brachfeld, and F. Tera, 2002. Cosmogenic Be-10 and the solid Earth; studies in geomagnetism, subduction zone processes, and active tectonics, *Rev. Mineral. Geochem.*, **50**, 207–270.

Mosenfelder, J. L., F. C. Marton, C. R. Ross, L. Kerschhofer, and D. C. Rubie, 2000. Experimental constraints on the depth of olivine metastability in subducting lithosphere, *Phys. Earth Planet. Int.*, **120**, doi:10.1016/S0031-9201(00)00142-4, 63–78.

Mosenfelder, J. L., J. A. D. Connolly, D. C. Rubie, and M. Liu, 2001. Strength of $(Mg,Fe)_2SiO_4$ wadsleyite determined by relaxation of transformation stress, *Phys. Earth Planet. Int.*, **127**, doi:10.1016/S0031-9201(00)00226-6, 165–180.

Myers, S. C., T. C. Wallace, S. L. Beck, P. G. Silver, G. Zandt, J. Vandecar, and E. Minaya, 1995. Implications of spatial and temporal development of the aftershock sequence for the M_W 8.3 June 9, 1994 deep Bolivian earthquake, *Geophys. Res. Lett.*, **22**, doi:10.1029/95GL01600, 2269–2272.

Ogawa, M., 1987 Shear instability in a viscoelastic material as the cause of deep focus earthquakes, *J. Geophys. Res.*, **92**, 13801–13810.

Ohmi, S. and S. Hori, 2000. Seismic wave conversion near the upper boundary of the Pacific plate beneath the Kanto District, Japan, *Geophys. J. Int.*, **141**, doi:10.1046/j.1365–246x.2000.00086.x, 136–148.

Ohtani, E., K. Litasov, T. Hosoya, T. Kubo, and T. Kondo, 2004. Water transport into the deep mantle and formation of a hydrous transition zone, *Phys. Earth Planet. Int.*, **143–144**, doi:10.1016/j.pepi.2003.02.09.015, 255–269.

Okal, E. A., 1996. Radial modes from the great 1994 Bolivian earthquake: no evidence for an isotropic component to the source, *Geophys. Res. Lett.*, **23**, doi:10.1029/96GL00375, 431–434.

• Omori, S., S. Kamiya, S. Maruyama, and D. Zhao, 2002. Morphology of the intraslab seismic zone and devolatilization phase equilibria of the subducting slab peridotite. *Bull. Earthquake Res. Inst., Tokyo Univ.*, **76**, 455–478.

Omori, S., T. Komabayashi, and S. Maruyama, 2004. Dehydration and earthquakes in the subducting slab: empirical link in intermediate and deep seismic zones, *Phys. Earth Planet. Int.*, **146**, doi:10.1016/j.pepi.2003.08.014, 297–311.

Orowon, E., 1960. Mechanism of seismic faulting, *Geol. Soc. Amer. Mem.*, **79**, 323–345.

Panero, W. R., L. R. Benedetti, and R. Jeanloz, 2003. Transport of water into the lower mantle: role of stishovite, *J. Geophys. Res.*, **108**, doi:10.1029/2002JB002053, 2039.

Pankow, K. L., Q. Williams, and T. Lay, 2002. Using shear wave amplitude patterns to detect metastable olivine in subducted slabs, *J. Geophys. Res.*, **107**, doi:10.1029/2001JB000608, 2108.

Peacock, S. M., 2001. Are the lower planes of double seismic zones caused by serpentine dehydration in subducting oceanic mantle? *Geology*, **29**, doi:10.1130/0091–7613(2001)029, 299–302.

Peacock, S. M. and K. Wang, 1999. Seismic consequences of warm versus cool subduction metamorphism: examples from southwest and northeast Japan, *Science*, **286**, doi:10.1126/science.286.5441.937, 937–939.

Poirier, J. P., 1981. On the kinetics of olivine-spinel transition, *Phys. Earth Planet. Int.*, **26**, 179–187.

Poli, S. and M. W. Schmidt, 2002. Petrology of subducted slabs, *Ann. Rev. Earth Planet. Sci.*, **30**, doi:10.1146/annurev.earth.30.091.201.140550, 207–235.

• Raleigh, C. B., 1967. Tectonic implications of serpentinite weakening, *Geophys. J. Roy. Astron. Soc.*, **14**, 113–118.

Raleigh, C. B. and M. S. Paterson, 1965. Experimental deformation of serpentinite and its tectonic implications, *J. Geophys. Res.*, **70**, 3965–3985.

Ranero, C. R., J. P. Morgan, K. McIntosh, and C. Reichert, 2003. Bending-related faulting and mantle serpentinization at the Middle America Trench, *Nature*, **425**, doi:10.1038/nature01961, 367–373.

Richards, P. G., 1976. Dynamic motions near an earthquake fault: a three-dimensional solution, *Bull. Seismol. Soc. Amer.*, **66**, 1–32.

Ringwood, A. E., 1962. A model for the upper mantle, *J. Geophys. Res.*, **67**, 857–866; 4473–4477.

• 1970. Phase transformations and the constitution of the mantle, *Phys. Earth Planet. Int.*, **3**, 109–155.

1975. *Composition and Petrology of the Earth's Mantle*, New York, McGraw Hill, 618 pp.

Rubie, D. C., 1996. Phase transformations in the Earth's mantle. In *High Pressure and High Temperature Research on Lithosphere and Mantle Materials*, eds. M. Mellini, G. Ranalli, C. A. Ricci, and V. Trommsdorff, Proceedings of the International School Earth and Planetary Sciences, Siena, Italy, 41–66.

Rubie, D. C. and C. R. Ross, 1994. Kinetics of the olivine–spinel transformation in subducting lithosphere: experimental constraints and implications for deep slab processes, *Phys. Earth Planet. Int.*, **86**, 223–241.

Russakoff, D., G. Ekström, and J. Tromp, 1997. A new analysis of the great 1970 Colombia earthquake and its isotropic component, *J. Geophys. Res.*, **102**, doi:10.1029/97JB01645, 20423–20434.

Savage, J. C., 1969. The mechanics of deep-focus faulting, *Tectonophysics*, **8**, doi:10.1016/0040-1951(69)90085-7, 115–127.

Scholz, C. H., 2002. *The Mechanics of Earthquakes and Faulting*, (second edn.), Cambridge, U. K., Cambridge University Press, 496 pp.

Seno, T. and Y. Yamanaka, 1996. Double seismic zones, compressional deep trench-outer rise events, and superplumes. In *Subduction, Top to Bottom*, eds. G. E. Bebout, D. W. Scholl, S. H. Kirby, and J. P. Platt, Geophys. Mon. 96, Washington, DC, American Geophysical Union, 347–356.

Seno, T., D. Zhao, Y. Kobayashi, and M. Nakamura, 2001. Dehydration of serpentinized slab mantle: seismic evidence from southwest Japan, *Earth Planets Space*, **53**, 861–871.

Sibson, R. H., 1973. Interactions between temperature and pore fluid pressure during earthquake faulting and a mechanism for partial or total stress relief, *Nature Phys. Sci.*, **243**, 66–68.

• Silver, P. G., S. L. Beck, T. C. Wallace, C. Meade, S. C. Myers, D. E. James, and R. Kuehnel, 1995. Rupture characteristics of the deep Bolivian earthquake of 9 June 1994 and the mechanism of deep-focus earthquakes, *Science*, **268**, 69–73.

Sung, C., 1974. The kinetics of high pressure phase transformations in the mantle: possible significance for earthquake generation, *Proc. Geol. Soc. China*, **17**, 67–84.

Sung, C. and R. G. Burns, 1976a. Kinetics of high-pressure phase transformations: implications to the evolution of the olivine–spinel transition in the downgoing lithosphere and its consequences on the dynamics of the mantle, *Tectonophysics*, **31**, doi:10.1016/0040-1951(76)90165-7, 1–32.

1976b. Kinetics of the olivine–spinel transition: implications to deep-focus earthquake genesis, *Earth Planet. Sci. Lett.*, **32**, 165–170.

Tibi, R., C. H. Estabrook, and G. Bock, 1999. The 1996 June 17 Flores Sea and 1994 March 9 Fiji–Tonga earthquakes: source processes and deep earthquake mechanisms, *Geophys. J. Int.*, **138**, doi:10.1046/j.1365-246x.1999.00879.x, 625–642.

Tibi, R., G. Bock, and C. H. Estabrook, 2002. Seismic body wave constraint on mechanisms of intermediate-depth earthquakes, *J. Geophys. Res.*, **107**, 2047, 10.1029/2001JB000361.

Tibi, R., G. Bock, and D. A. Wiens, 2003. Source characteristics of large deep earthquakes: constraint on the faulting mechanism at great depths, *J. Geophys. Res.*, **108**, doi:10.1029/2002JB001948, 2091.

Tingle, T. N., H. W. Green, C. H. Scholz, and T. A. Koczynski, 1993. The rheology of faults triggered by the olivine–spinel transformation in Mg_2GeO_4 and its implications for the mechanism of deep-focus earthquakes, *J. Struc. Geol.*, **15**, 1249–1256.

Vaisnys, J. R. and C. C. Pilbeam, 1976. Deep-earthquake initiation by phase transformations, *J. Geophys. Res.*, **81**, 985–988.

Venkataraman, A. and H. Kanamori, 2004. Observational constraints on the fracture energy of subduction zone earthquakes, *J. Geophys. Res.*, **109**, 5302, doi:10.1029/2003JB002549.

Wiemer, S. and M. Wyss, 1997. Mapping the frequency–magnitude distribution in asperities: an improved technique to calculate recurrence times? *J. Geophys. Res.*, **102**, doi:10.1029/97JB00726, 15115–15128.

Wiemer, S., S. R. McNutt, and M. Wyss, 1998. Temporal and three-dimensional spatial analysis of the frequency–magnitude distribution near Long Valley caldera, California, *Geophys. J. Int.*, **134**, doi:10.1046/j.1365-246x.1998.00561.x, 409–421.

• Wiens, D. A., 2001. Seismological constraints on the mechanism of deep earthquakes: temperature dependence of deep earthquake source properties, *Phys. Earth Planet. Int.*, **127**, doi:10.1016/S0031-9201(01)00225-4, 145–163.

Wiens, D. A. and J. J. McGuire, 2000. Aftershocks of the March 9, 1994, Tonga earthquake: the strongest known deep aftershock sequence, *J. Geophys. Res.*, **105**, doi:10.1029/2000JB900097, 19067–19083.

Wiens, D. A. and N. O. Snider, 2001. Repeating deep earthquakes: evidence for fault reactivation at great depth, *Science*, **293**, doi:10.1126/science.1063042, 1463–1466.

Wiens, D. A. and S. Stein, 1983. Age dependence of oceanic intraplate seismicity and implications for lithospheric evolution, *J. Geophys. Res.*, **88**, 6455–6468.

Wiens, D. A., J. J. McGuire, and P. J. Shore, 1993. Evidence for transformational faulting from a deep double seismic zone in Tonga, *Nature*, **364**, doi:10.1038/364790a0, 790–793.

Wu, L. R. and W.-P. Chen, 2001. Rupture of the large ($M_w > 7.8$) deep earthquake of 1973 beneath the Japan sea, with implications for seismogenesis, *Bull. Seismol. Soc. Amer.*, **91**, 102–111.

Wu, T.-C., W. A. Bassett, P. C. Burnley, and M. S. Weathers, 1993. Shear-promoted phase transitions in Fe_2SiO_4 and Mg_2SiO_4 and the mechanism of deep earthquakes, *J. Geophys. Res.*, **98**, 19767–19776.

• Yamasaki, T. and T. Seno, 2003. Double seismic zone and dehydration embrittlement of the subducting slab, *J. Geophys. Res.*, **108**, doi:10.1029/2002JB001918, 2212.

Yoshioka, S. and T. Murakami, 2002. The effects of metastable olivine(α) wedge in subducted slabs on theoretical seismic waveforms of deep earthquakes, *J. Geophys. Res.*, **107**, doi:101029/2001JB001223, 2365.

Zhang, J., H. W. Green, K. Bozhilov, and Z. Jin, 2004. Faulting induced by precipitation of water at grain boundaries in hot subducting oceanic crust, *Nature*, **428**, doi:10.1038/nature02475, 633–636.

Part IV

Why bother about deep earthquakes?

8

Are deep earthquakes useful?

Are deep earthquakes useful? That is, do they provide us with information that helps solve problems of practical importance to humankind? Or, are deep earthquakes just another abstruse phenomenon, justifying attention from scientists only "because they are there"?

This chapter's main thesis is that a disproportionate fraction of what we know about the structure and dynamics of the Earth's interior comes specifically from observations of intermediate- and deep-focus earthquakes. Thus, this chapter briefly reviews several topics where deep earthquakes provide the essential information that constrains our present knowledge.

8.1 Structure of the Earth

8.1.1 Gross earth models and the structure of the core

What do we know about the structure of the Earth, and from whom did we learn it? Between 1930 and 1945 three individuals, Harold Jeffreys, Beno Gutenberg, and Kiyoo Wadati were responsible for many of the seismological observations used to construct travel time tables and develop models of gross earth structure (see Frohlich, 1987). One thing is very clear if you reread their papers published in journals such as *Gerlands Beiträge zur Geophysik, Geophysical Magazine*, and *Monthly Notices of the Royal Astronomical Society, Geophysical Supplement* – records of deep earthquakes were fundamentally important in the construction of these models.

For example, Jeffreys (1935) notes that deep earthquakes are superior to shallow earthquakes for determining travel times for three reasons: the absence of surface waves simplifies the recognition of PcP and ScS, the S phase is easier to identify,

I thank Steve Grand for his review of an earlier draft of this chapter.

and the "P movement is strong" (impulsive), unlike that of many shallow shocks. He thus states:

These features make it possible to obtain information from deep-focus shocks that has not yet been available in normal ones.

Deep- and intermediate-focus earthquakes have been especially important for constraining models of structure in the core. Thus, in an early paper presenting a core model and reporting travel times for core phases, Gutenberg and Richter (1938) state:

Additional data on travel times and apparent velocities have been taken from a number of recent shocks, particularly from large deep-focus shocks south of Borneo.

In a paper interpreting PcP and ScS times, Jeffreys (1939a) notes with characteristic modesty that his determination of 3473 ± 2.5 km is "the best attainable" estimate for the core's radius. In describing his data he states that his tables for PcP and ScS are:

... based on the readings of ScS by Scrase and Stechschulte for the [deep-focus] earthquakes of 1931 February 20 and 1928 March 29.

and:

Gutenberg and Richter give empirical times for PcP and ScS, derived from deep focus earthquakes and adopted to a surface focus.

Bolt (1959) states that:

The Jeffrey–Bullen tables depend largely on PKP readings from the Solomon Islands earthquake of 1932 January 9. They were subsequently checked from observations of the Celebes Sea earthquake of 1934 June 29.

Both of these earthquakes were deep-focus events (Jeffreys, 1939b; 1942). Subsequent researchers relied heavily on data from intermediate- and deep-focus earthquakes as they refined the core models proposed by Jeffreys and Gutenberg. For example, Nuttli (1954) used ten earthquakes – seven with depths of 70 km or greater – to remeasure the core radius and evaluate P-structure at the core–mantle boundary. Bolt (1964), Adams and Randall (1964), and Ergin (1967) each proposed core models with two or more discontinuous velocity changes within the outer core (Fig. 8.1); half of the 26 earthquakes used by Bolt were deep, as were 11 of 12 for Adams and Randle and 5 of 6 for Ergin. Buchbinder (1971) and Müller (1973; 1975) were unable to confirm the presence of these discontinuities; and 7 of the 8 earthquakes used as data by Buchbinder were deep, as were 4 of 6 for Müller.

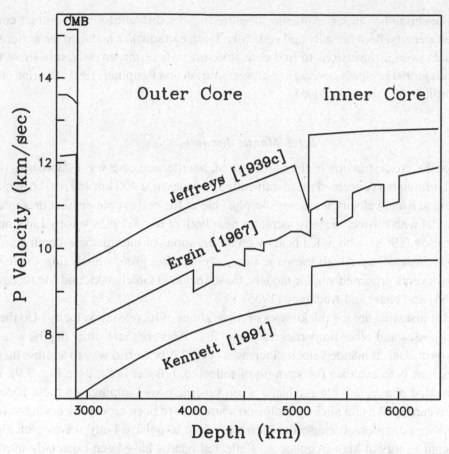

Fig. 8.1 Three models of core P velocity. Signals from deep earthquakes have provided much of the data that constrain models of the core. Note that the more recent IASPEI 1991 model of Kennett (1991) is less complex than earlier models of Jeffreys (1939c) and Ergin (1967). In this figure, the scale on the vertical axis is for Ergin's model; for clarity, graphs for Jeffreys' and Kennett's models are offset by 1.5 km/sec. CMB indicates core–mantle boundary.

Since about 1975, many of the most convincing studies of core structure have relied on the comparison of observed and synthetic seismograms from deep earthquakes. For example, Choy (1977) used SmKS phases from two deep earthquakes to constrain the velocity gradient in the outermost core. Cormier (1981) used Buchbinder's (1971) amplitude data to suggest that there might be anelastic, partially molton material in the inner core. Choy and Cormier (1982) modeled data from a single deep-focus earthquake to investigate the velocity structure of the outermost inner core; Cormier and Choy (1986) used 18 earthquakes – 16 with depths of 150 km or greater – to conclude that lateral heterogeneity within the inner core is "small." Garnero and Helmberger (1995) used three deep events to conclude that uncertainties in lower mantle structure make it difficult to resolve inner core heterogeneity, both laterally and vertically. Deep earthquakes make up the majority of data even in inversions to find core structure, where the investigators strive to obtain global ray-path coverage (e.g., see Souriau and Poupinet, 1991; Souriau and Roudil, 1995; Yu *et al.*, 2005).

8.1.2 Mantle discontinuities

After the gross structure of the Earth's crust, mantle, and core was established, the next refinement concerned the identification of zones near 400 km and 650 km depth where seismic velocity increases sharply. One of the earliest papers that presented a model with distinct velocity increases near both of these depths was by Lukk and Nersesov (1965), who relied heavily on observations of intermediate-depth Hindu Kush earthquakes. About the same time several other groups analyzing shallow-focus events proposed similar models; these included Golenetskii and Medvedeva (1965) and Niazi and Anderson (1965).

But just what are the thicknesses of these zones of increasing velocity? Do their thicknesses and other properties vary laterally? And are there other depths where less-pronounced mantle velocity increases occur? A powerful way to address these questions is to evaluate the strength of reflected, refracted (e.g., see Fig. 3.9), or converted phases that are produced when seismic waves impinge on these zones. Deep earthquake and nuclear explosion sources have been especially desirable for these investigations because they often produce impulsive body waves with significant energy at high frequencies. Reflected phases have been especially useful for estimating the thickness of these transition zones and for determining whether their occurrence is global or regional. Discontinuous or very thin velocity increases produce very strong reflections even at high frequencies, whereas reflections will be weak if the velocity increase extends over a large depth range. For example, some seismograms exhibit distinct phases arriving in the $2\frac{1}{2}$ minutes prior to PKPPKP purportedly caused by reflections from the underside of "sharply defined" velocity increases at depths of about 650 km and 400 km (see Fig. 8.2). Deep earthquakes have provided many of the best-recorded examples of these phases (Engdahl and Flinn, 1969; Teng and Tung, 1973; Nakanishi, 1988; Benz and Vidale, 1993).

More recently, Revenaugh and Jordan (1991b) have performed a highly systematic investigation of reflected phases. They collected records from 130 earthquakes with focal depths exceeding 75 km and searched for mantle velocity increases by evaluating ScS reverberations (Fig. 8.3), i.e., phases such as ScS, ScSScS, ScSScSScS, etc. They then employed modern filtering and stacking methods to

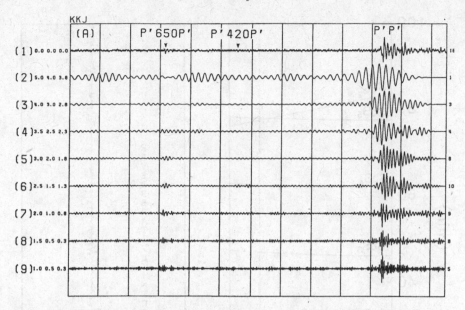

Fig. 8.2 P′P′ precursors. Top trace shows 200 seconds prior to PKPPKP arrival at station KKJ in Japan from the Fiji earthquake of 26 May 1986 (h_{CMT} = 568 km; $M_{W(CMT)}$ = 7.1). The remaining traces are band-pass filtered in the frequency ranges indicated by the first two numbers at left to enhance reflections from mantle boundaries at depths of 650 km and 420 km. Vertical lines indicate 20 s intervals. Figure reproduced from Nakanishi (1988) with permission from Blackwell Publishing, Ltd.

enhance S_H reflections from mantle velocity increases and found strong reflections originating at depths of 414 ± 2 km and 660 ± 2 km in 18 different "seismic corridors" in the western and central Pacific. In many of the seismic corridors they also found evidence of less pronounced reflections originating at depths of 520 km, 710 km, and 900 km. Although other investigations have found velocity increases at similar depths (e.g., Whitcomb and Anderson, 1970; Datt and Muirhead, 1976; Muirhead and Hales, 1980), they are not evident everywhere (e.g., see Revenaugh and Jordan, 1991a; Cummins *et al.*, 1992).

Observations of S-to-P conversions ostensibly produced by structure near 400 km and 650–700 km have been used to investigate the local thickness and lateral variation of these zones (Faber and Müller, 1980; Barley *et al.*,1982; Vidale and Benz, 1992; Niu and Kawakatsu, 1995; 1997; Flanagan and Shearer, 1998; Castle and Creager, 2000). For example, Bock and Ha (1984) and Yamazaki and Hirahara (1994) each analyzed S-to-P converted phases from about 15 Fiji–Tonga deep earthquakes at arrays in Australia and Japan, respectively, and reached similar conclusions. Bock and Ha concluded that the "transition zone near 700 km is only

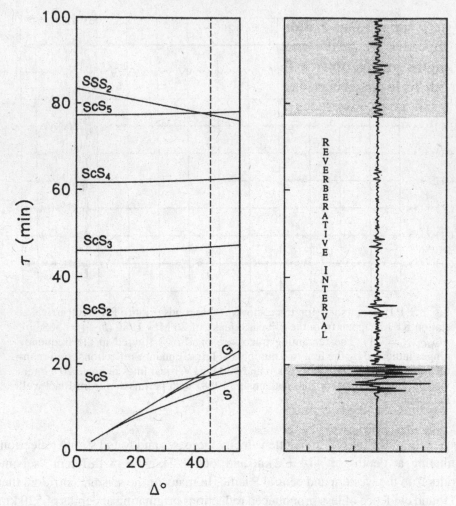

Fig. 8.3 ScS reverberations. For S_H waves at epicentral distances less than about 45°, there is a relatively quiet interval between the arrival of the minor arc Love waves (G) and the first major-arc arrival (core-diffracted SSS_2). The most prominent phases are ScSScS (ScS_2), ScSScSScS (ScS_3), etc., which can be analyzed to search for abrupt velocity increases within the mantle. The seismogram at right is from a Tongan intermediate-depth earthquake recorded in Hawaii. Figure reproduced from Revenaugh and Jordan (1991a).

a few km thick," while Yamazaki and Hirahara found that the "thickness of both discontinuities is at most 5 km." Similarly, Estabrook *et al*. (1994) analyzed S-to-P converted phases from a 360 km-deep Japanese earthquake recorded at the Grafenberg array in Germany. They concluded that structure near the Japanese subduction zone at a depth of about 660 km caused the converted phases; because the phases showed little variation across the array they concluded that the nearby subduction had little effect on the depth of the reflector. Finally, Tono *et al.* (2005) analyzed core-reflected shear phases produced by the 28 June 2002 Vladivostok earthquake ($h_{CMT} = 582$ km; $M_{W(CMT)} = 7.3$) to evaluate how subduction affects topography on the 410 km and 660 km discontinuities. Observed time differences in sScS, sScSS410S and sScSS660S phases across regional stations in Japan indicated that the 660-km discontinuity is systematically depressed by about 15 km along the bottom of the horizontally extending aseismic slab, with a transition from normal to depressed levels over a horizontal distance of less than about 200 km. They found that the 410-km reflection was elevated, in qualitative agreement with expectations for the equilibrium olivine–spinel transition.

Intermediate- and deep-focus earthquakes also provide the most useful data for probing the structure of the lowermost mantle, just above the core-mantle boundary (see the review by Garnero, 2000). For example, Julian and Sengupta (1973) used P-wave residuals from 47 deep earthquakes to show that there was significant global lateral heterogeneity in the lowermost few hundred km of the mantle. Moreover, they found distinctly positive residuals (lower velocities) associated with source-station ray paths that bottomed beneath Hawaii, an area of proposed convective plume activity. Subsequently, Lay and Helmberger (1983) observed distinct triplications in the S_H arrivals from 17 deep-focus and two intermediate-focus earthquakes at distances between about 70° and 80°. They proposed that these were caused by a 2.5–3.0% shear-velocity increase 280 km above the core–mantle boundary, which might "be a sharp first-order discontinuity, or may extend over a transition zone no more than 50 km thick."[1] However, it seems unlikely that this discontinuity exists everywhere – for example, Schlittenhardt *et al.* (1985) did not observe the triplications in record sections from nine deep-focus events. Nevertheless, these papers have generated considerable interest in investigating the structure and lateral variation of the lowermost mantle. Revenaugh and Jordan's (1991c) reflectivity studies of deep sources provide evidence that the lowermost mantle is heterogeneous, as they find evidence for the discontinuity in some, but not all, of their reflection stacks. Analysis of deep earthquakes also provides indications that portions of the lowermost mantle are highly anisotropic (e.g., Kendall and Silver, 1996; Matzel *et al.*, 1996; Pulliam and Sen, 1998; see Fig. 8.4) and may contain localized zones where there is partial melting (Vidale and Hedlin, 1998).

8.1.3 The Earth's crust and receiver functions

Even for a simple, impulsive source, near-surface structure may produce considerable complexity in teleseismic body wave arrivals. Soon after the discovery of

---

[1] Because of Thorne Lay's authorship of this paper, many seismologists now jocularly refer to this as the "Thorne layer."

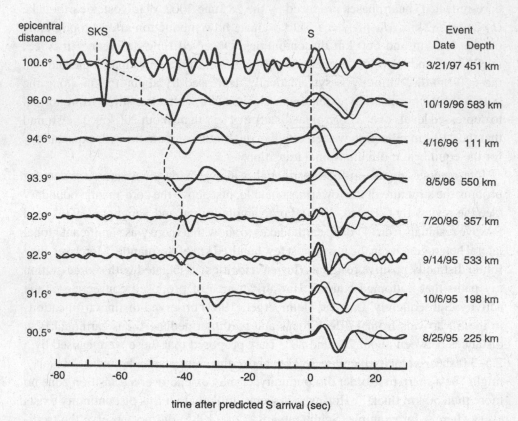

epicentral distance
SKS
S
Event Date Depth

100.6°   3/21/97 451 km

96.0°   10/19/96 583 km

94.6°   4/16/96 111 km

93.9°   8/5/96 550 km

92.9°   7/20/96 357 km

92.9°   9/14/95 533 km

91.6°   10/6/95 198 km

90.5°   8/25/95 225 km

-80   -60   -40   -20   0   20

time after predicted S arrival (sec)

Fig. 8.4 Anisotropy at the core–mantle boundary. S_V-polarized (thick lines) and S_H-polarized (thin lines) seismograms windowed around the S arrival for eight Tonga–Fiji earthquakes recorded at HKT (Hockley, Texas). At these distances, S travels within a few hundred km of the core–mantle boundary; thus the observation that S_V arrives several seconds earlier than S_H suggests that the mantle may be anisotropic there. Between 94° and 100°, Pulliam and Sen (1998) found that about 2.0 sec of splitting remained even after correcting for anisotropy beneath HKT. Figure reproduced from Pulliam and Sen (1998) with permission from Blackwell Publishing, Ltd.

deep earthquakes, seismologists realized that properly interpreting this complexity could provide information about crustal structure. For example, Sharp (1935) and Robertson (1937) evaluated seismograms from a deep-focus earthquake in Japan and an intermediate quake in Guatemala, respectively, and interpreted phases arriving up to 20 seconds after the initial P as reverberations in a crust having two layers and P-to-S conversions at the crust–mantle boundary. From arrivals at ten stations

Sharp concluded that the thickness of continental crust was 32 km, quite similar to the value of 29 km determined by Robertson from her observations at Florissant, Missouri.

Phinney (1964) proposed a more powerful and broadly applicable method for determining crustal structure; since Owens *et al.* (1984) this has been called the receiver-function method. When a teleseismic P wave is incident on the crust from below, crustal layering produces P-to-S and S-to-P conversions; these reverberations produce a P coda with radial and vertical components that are significantly different. This makes it possible to separate effects produced by crustal layering from those produced by complexity in the source-time function, and thus by applying inverse methods to P-coda records one can determine the best-fitting layered crustal structure.

Burdick and Langston (1977) observed that there are often practical difficulties in applying receiver function methods, and argued that "to avoid these difficulties it is advantageous to use observations of deep earthquakes ($h \sim 600 \, \text{km}$) of intermediate size at epicentral ranges of 60° to 80°." Subsequently, most receiver-function investigations preferentially select signals from intermediate- or deep-focus earthquakes (e.g., Langston, 1979). Moreover, deep sources make it possible to apply receiver-function methods even to regional earthquake signals (e.g., see Zhang and Langston, 1995). As digital data has become widely available, receiver-function methods have become quite popular; since 1995 more than 100 receiver-function papers have appeared.

8.1.4 Lateral variations in mantle structure: tomography

Deep earthquakes are also useful for evaluating lateral variations of velocity and attenuation within the mantle. For example, Sipkin and Jordan (1980a) evaluated shear velocity variations using travel times from ScS, ScSScS, etc. In a companion study they also used ScS observations to evaluate variability in attenuation (Sipkin and Jordan, 1980b). For a review concerning lateral variations of seismic velocity within the upper mantle, see Nolet *et al.* (1994).

During the 1970s seismologists began to have access to computers with enough speed and memory to invert matrices with dimensions of 100×100 or greater; this inspired Aki and Lee (1976) to apply tomographic methods to local-network travel-time data to determine more detailed, three-dimensional velocity models. Their approach is to divide the mantle beneath a network into several layers of rectangular blocks, and then for some reasonable starting model trace rays from earthquakes to stations, keeping track of how much time rays spend in each block. Of course, the inversion to determine structure is a highly unstable process unless

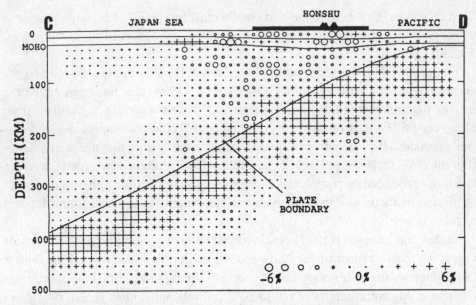

Fig. 8.5 Travel-time tomography near subduction zones. Zhao and Hasegawa (1993) used 50,000 arrival times from about 1200 shallow and deep Japanese earthquakes to obtain this tomographic image of the P velocity beneath central Japan. The sizes of the + and o symbols designate velocities ranging from 6%faster to 6%lower than normal mantle. For this tomographic inversion the starting model included a discontinuous velocity increase across the inclined plate boundary. Figure reproduced from Zhao and Hasegawa (1993).

numerous rays having different source locations traverse each block. Thus, detailed information about velocity variations at depths of 300–700 km is only available from networks near subduction zones, where deep earthquakes provide ray families having the desired geometric variations.

There are only a few subduction zones where there are enough regional seismo-graph stations to apply these so-called tomographic methods. The most detailed studies concern the upper mantle beneath Japan, where there are subcrustal earth-quakes and an extensive network of local stations that has been collecting travel-time data for a longer time period than anywhere else (Fig. 8.5). For example, Hirahara (1977) undertook one of the earliest studies, using data from 20 deep earthquakes; and there have been numerous subsequent studies (e.g., see Zhao *et al.*, 1992; 1994; Zhao and Hasegawa, 1993). Elsewhere, Koch (1985) and Fan *et al.* (1998) used sources in the Vrancea nest to determine mantle structure beneath Romania; Kuge and Satake (1987) investigated structure beneath North Island, New Zealand; Myers *et al.* (1998) and Graeber and Asch (1999) investigated the mantle in South America; and Deal *et al.* (1999) and Roth *et al.* (1999) have imaged the mantle beneath Tonga.

As methods to stabilize tomographic inversions have improved, several investigators have applied them to data from teleseismic catalogs. Usually, however, obtaining reliable detail about mantle structure requires deep sources. For example, van der Hilst and Engdahl (1991) found that in Wadati–Benioff zones where sufficient numbers of pP data are available, their inclusion in tomographic modeling significantly improved the quality and reliability of tomographic images.

8.2 Dynamics of the crust and mantle

8.2.1 Deep subduction and the plate tectonic revolution

The scientific community's acceptance of plate tectonics between about 1965 and 1975 undoubtedly represents the greatest paradigm shift in the earth sciences since the eighteenth century when James Hutton and later Charles Lyell introduced uniformitarianism to geology. From a seismological perspective, this at first would seem to be entirely a shallow-earthquake matter, since the relative motions between lithospheric plates at mid-ocean ridges, transform margins, and deep-ocean trenches are all expressed by shallow seismic activity. But, if lithospheric plates can move on a round planet, somewhere they must converge. And then, where does the lithosphere go?

Because of this question, the scientists involved in developing plate tectonics focused considerable attention on deep earthquakes and Wadati–Benioff zones (see Section 3.4). For example, Oliver and Isacks (1967) investigated the propagation of seismic waves from deep earthquakes in Tonga to regional stations and found that the upper surface of the Wadati–Benioff zone was coincident with the upper surface of a zone of low attenuation having a thickness of about 100 km. Moreover, they concluded that their observations were

consistent with theories of mantle convection with down-going currents in the vicinity of island arcs. If low attenuation of seismic waves correlates with strength, this structure suggests that the lithosphere has been thrust or dragged down beneath the Tongan arc.

In their subsequent landmark paper, Isacks et al. (1968) noted that Utsu (1967) reports similar results concerning attenuation beneath Wadati–Benioff zones. They also refer to focal mechanism investigations by Honda et al. (1957) and Ritsema (1965) which show that generally either the P or T axes of intermediate- and deep-focus earthquakes are aligned along the downdip direction of the Wadati–Benioff zone. Isacks and Molnar (1971) subsequently confirmed the downdip orientation for P or T axes of deep earthquakes in many of the world's island arc systems (see Fig. 5.5).

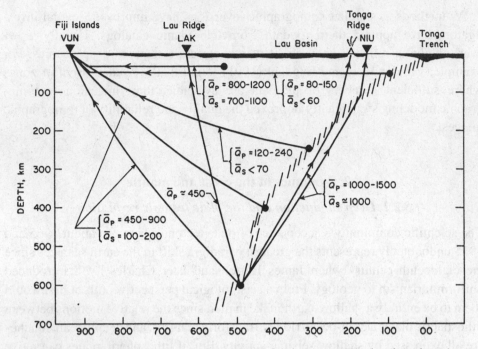

Fig. 8.6 Attenuation in the mantle near subduction zones. At local stations in Tonga and Fiji, Barazangi and Isacks (1971) evaluated the frequencies of short-period P and S arrivals from earthquakes within the Wadati–Benioff zone. This schematic cross section summarizes the results of their study. Attenuation was highest ($Q \sim 60-150$) in the mantle beneath the volcanic arc, and lowest ($Q \sim$ 1000 or more) along paths within subducted lithosphere. Figure reproduced from Barazangi and Isacks (1971).

Deep earthquakes have continued to be the best source of information about the mantle above the Wadati–Benioff zone, as well as about the properties of subducted lithospheric material. Barazangi and Isacks (1971) showed that there was very low attenuation within the mantle above the Tonga Wadati–Benioff zone, particularly near the volcanic arc (Fig. 8.6). Utsu (1971) reported that the structure of the Japanese arc was similar in most essential respects to that in Tonga. Pascal *et al.* (1973) found that there apparently wasn't a high-velocity, low-attenuation zone connecting the 600-km deep cluster of earthquakes in the New Hebrides with the intermediate-depth activity there. In contrast, Isacks and Barazangi (1973) reported that there was a high-velocity slab connecting intermediate- and deep-focus activity in South America. In Tonga, Huppert and Frohlich (1981) obtained precise measurements of the seismic velocity within subducted material using differential travel times of P phases that traveled laterally along the slab (Fig. 8.7).

But the question still remained, where does the lithosphere go? Does it meet a barrier near 700 km where the deepest Wadati–Benioff zones terminate? Or

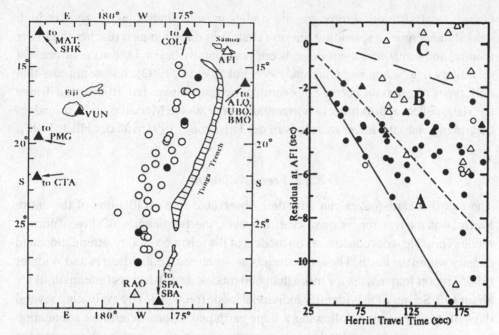

Fig. 8.7 P-velocity within subducted lithosphere. Huppert and Frohlich (1981) evaluated P arrivals from 39 relocated intermediate-depth Tonga–Kermadec earthquakes (circles on map at left) at stations RAO and AFI (open triangles on map). At both RAO and AFI (graph at right), P arrivals were consistently earlier than predicted by the Herrin velocity model, suggesting that the P velocity within the subducted lithosphere was about 8% faster (solid line) than within normal mantle. For these earthquakes, arrivals at AFI and RAO sample a roughly horizontal path close to the Wadati–Benioff zone between the two stations. Huppert and Frohlich relocated the earthquakes using data from local station VUN and ten teleseismic stations (solid triangles, with arrows indicating azimuth to station). On the graph at right, circles and triangles correspond to initial and secondary arrivals; solid symbols represent unambiguous arrivals. Figure reproduced from Huppert and Frohlich (1981).

does it continue on subducting into the lower mantle. Jordan (1977) evaluated travel times for S and ScS phases from the 29 January 1971 earthquake at a depth of about 540 km beneath the Sea of Okhotsk, and concluded that high-velocity lithospheric material extended to a depth of at least 1000 km. Creager and Jordan (1984; 1986) obtained similar results from deep-focus earthquakes beneath several other subduction zones. However, it remained controversial whether lithospheric subduction continued beneath 700 km into the lower mantle, especially since several investigators were critical of Jordan and Creager's methodology (e.g., see Zhou et al., 1990).

However, recent investigations seem to confirm that high-velocity material penetrates into the lower mantle at least beneath some subduction zones (e.g., Ding and

Grand, 1994). But other tomographic studies indicate that often the zone of high velocity undergoes a spreading out or a change in direction as it reaches the lower mantle, and sometimes even extends out horizontally beyond the zone of deepest seismicity (e.g., Zhou and Clayton, 1990; Fukao *et al.*, 1992). It now appears that both types of behavior occur. For example, beneath Japan, Izu–Bonin, and Tonga the slabs deflect approximately horizontally; beneath the Marianas and Kermadecs they plunge into the lower mantle (van der Hilst *et al.*, 1991; van der Hilst, 1995).

8.2.2 Free oscillations

The fact that earthquakes can stimulate observable free oscillations of the entire Earth is of interest for its own sake; however, the frequencies of these "normal modes" provide strict constraints on models of the seismic velocity, attenuation, and density within the Earth. These constraints are numerous; e.g., Masters and Widmer (1995) report frequencies for more than 600 modes, and Q-factors (attenuation) for about 140. Seismologists identify individual mode frequencies by evaluating several days of seismic records following a large earthquake; they determine attenuation by comparing mode amplitudes in successive time periods and by measuring the widths of the peaks.

There are two reasons why one might conclude incorrectly that deep earthquakes aren't very important for free oscillation research. First, very large earthquakes stimulate free oscillations best, and the majority of very large earthquakes have shallow foci. Second, many free oscillation research publications hardly mention focal depth, and instead focus either on mathematical considerations (e.g., Lapwood and Usami, 1981; Dahlen and Tromp, 1998) or on the mode frequencies themselves and their implications for earth structure (e.g., Romanowicz and Berger, 2000). Often focal depth is "hidden" even in papers analyzing data from single earthquakes; for example, Masters and Gilbert (1981) evaluated overtones from the 22 June 1977 Tonga earthquake but failed to note that its focal depth was 60–70 km. Ten of the 33 earthquakes used by Resovsky and Ritzwoller (1998) in their normal-mode analysis of earth structure are intermediate- or deep-focus; however, this fact isn't mentioned in the paper, as the authors list only the dates of the events they use.

However, deep earthquakes are actually superior to shallow-focus events for evaluating certain features of earth structure. As explained by Masters and Widmer (1995):

Deep earthquakes are incapable of exciting fundamental modes which normally dominate the seismogram and obscure the lower amplitude overtones. Theoretical work by Gilbert in the early 70's led to compact expressions for mode excitation and array processing algorithms which were eventually applied to hand-digitized WWSSN recordings of the [31 July 1970 650 km deep] Colombian earthquake. [This led] to a dataset which is essentially that used in the construction of current earth models [i.e., PREM].

In the language of free oscillations, "fundamental modes" are those with no nodes between the Earth's surface and the core, and which sum to produce surface waves. The "overtones" are modes with one or more radial nodes; these often have non-negligible amplitudes in the Earth's interior, making them sensitive to deep structure (Fig. 8.8). Thus, since the 31 July 1970 Columbian and the 9 June 1994 Bolivian earthquake are two of the 15 largest earthquakes to occur since 1965, it is no accident that deep-focus earthquakes provide the essential data for the free oscillation studies of the Earth's inner core (e.g., He and Tromp, 1996; Durek and Romanowicz, 1999). Moreover, deep earthquakes seldom possess large-magnitude aftershocks that complicate the analysis of mode die-off to evaluate the Earth's attenuation structure.

8.2.3 Anisotropy and flow in the mantle

Refraction studies using explosions as well as analysis of surface waves provided much of the earliest evidence indicating that parts of the upper mantle were anisotropic (see review by Silver, 1996). However, there was ambiguity because at least three different phenomena can produce bulk anisotropic behavior. These are:

(1) thin layering of otherwise isotropic materials;
(2) oriented open or fluid-filled cracks in otherwise isotropic materials; and
(3) "true" anisotropy in oriented crystals of anisotropic materials such as olivine, which, in its crystalline form, has a velocity anisotropy of about 20%

For investigations of the mantle, true anisotropy is the most interesting because it is potentially an indicator of flow direction – preferential orientation of the individual crystal may occur due to strain and flow of the mantle if individual crystals have an elongated shape.

Thus, a number of investigators began studying short-period body waves to search for mantle anisotropy, especially after digital data became widely available. For example, Ando and Ishikawa (1982) analyzed S-wave polarization in Japan from seven deep-focus and four intermediate local earthquakes, and concluded that there were some highly anisotropic zones in the mantle beneath Honshu. However, in this and several subsequent studies there was ambiguity in the interpretation because the split arrivals might be caused by anisotropy anywhere along the source–station path, and in addition by near-source P-to-S conversions or by near-station S-to-P-to-S reverberations.

There are fewer of these ambiguities if the experimental design employs deep-focus sources that produce arrivals with near-vertical incidence. Thus Ando (1984) analyzed ScS data at stations within 30° of 20 deep-focus and eight intermediate earthquakes occurring around the rim of the Pacific. He observed shear-wave

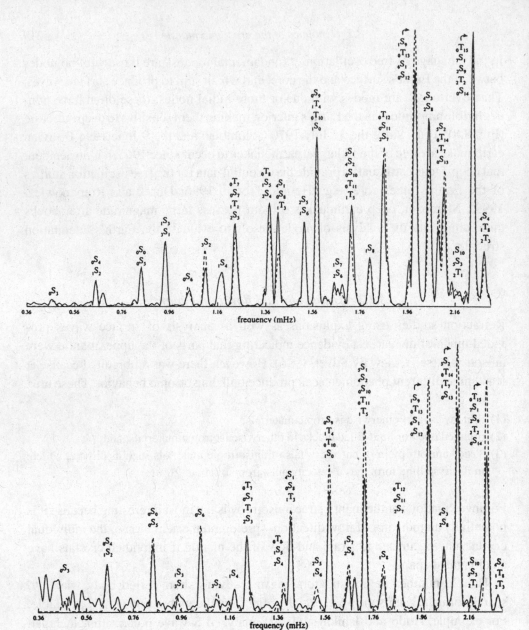

Fig. 8.8 Free oscillation overtones. Radial-component amplitude spectra for a 35 hour time series following: (top) the 9 June 1994 Bolivia deep-focus earthquake ($h_{CMT} = 647\,km$; $M_{W(CMT)} = 8.2$) at station PAS; and (bottom) the 17 February 1996 Irian Jaya earthquake ($h_{CMT} = 72\,km$; $M_{W(CMT)} = 8.2$) at station PTGA. Observed and theoretical spectra are solid and dashed lines, respectively; labels $_nS_1$ and $_nT_1$ indicate spheroidal and toroidal modes, with n being 0 for fundamental modes, and 1 or greater for overtones. For the deeper Bolivian earthquake, note that some overtone peaks are more than six times larger than the peaks for the fundamental modes $_0S_0$, $_0S_5$ and $_0S_6$. This is at least twice as great as the corresponding ratio for the shallower Irian Jaya earthquake. Figure reproduced from Dahlen and Tromp (1998) with permission from Princeton University Press.

splitting, indicating that there was some form of anisotropy in the mantle; the polarization of the faster waves was generally parallel to plate motion near the receiver points. In a further refinement, Silver and Chan (1991) restricted their observations to SKS and SKKS phases from about 50 deep- and intermediate-depth earthquakes. Because these phases traveled as compressional waves within the core, they emerged as "pure" S_V signals at the core-mantle boundary, insuring that any subsequent S_H polarization was attributable to the core-to-surface part of the ray path. Silver and Chan concluded that in subduction zones such as Alaska the anisotropy was indicative of present deep flow in the asthenosphere, caused by the subduction process. However, in stable continental regions they found that there was "fossil" anisotropy, presumably remnants of previous tectonic episodes.

Subsequently numerous investigators have utilized S and SKS phases from deep earthquakes to evaluate anisotropy and infer the flow within the mantle. Near subduction zones, these include Sandoval *et al.* (1994) beneath the Himalayas, Fouch and Fischer (1996) in the northwest Pacific, Fischer *et al.* (2000) in Tonga, Yang *et al.* (1995) in Alaska, Russo and Silver (1994), Shih *et al.* (1991) and Kaneshima and Silver (1995) in South America, Russo *et al.* (1996) in the Caribbean, and Gledhill and Stuart (1996) in New Zealand. In other tectonic environments these include Helffrich (1995) in the United Kingdom, and Özalaybey and Savage (1995) in the western United States.

8.3 The earthquake source

8.3.1 Earthquakes are double couples

Today we believe that the radiation pattern from earthquakes is, for the most part, that of a so-called "double-couple" source. What are the principal observations that led to this conclusion?

The best early observations were from deep earthquakes in Japan.[2] Honda (1933; 1934a; 1934b) and Honda and Miura (1938) made careful measurements of P and S motions for several earthquakes with focal depths near 320 km. For P these indicated that motions towards and away from the source region formed the familiar quadrantal pattern (see Fig. 3.12). Moreover, the Japanese found that polarization directions of S waves were also consistent with the double-couple, or so-called type II model. For example, for the deep earthquake of 20 February 1931, Kawasumi and Yosiyama (1934) collected seismograms from 91 stations throughout the world, and concluded that the data were consistent with a source having:

---

[2] Perry Byerly, in his analysis of the Montana earthquake of 1925 (Byerly, 1926), seems to be the first person to plot the distribution of first motions from an earthquake.

. . . four points on the Earth's surface where S-waves do not appear (where only P-waves appear with maximum amplitude). Moreover, the initial motions of the S-wave seem to diverge [from these points]. We may therefore conclude that this earthquake occurred according to the mechanism of model B, possibly due to two pairs of doublet without moment.

Honda's conception of earthquake focal mechanisms is quite similar to what we believe today, and some of the figures he published seem almost modern (see Fig. 3.13).

Elsewhere, however, seismologists preferred the single-couple or so-called type I model. Strangely, this dispute lasted for about thirty years and wasn't resolved in favor of the double-couple model until the mid-1960s. For example, Ritsema (1958) investigated 59 earthquakes, about half of which were intermediate- or deep-focus events, in the Sunda, Celebes, Philippines, New Guinea, and Solomons arcs. He concluded that the data indicated a distinct preference for the single-couple model. Similarly, Stauder and Adams (1961) concluded that:

. . . in earthquakes of faulting origin

(1) a single dipole with moment serves as an adequate model for the first motion in P and S waves,
(2) the results of analysis of S-wave data are in substantial agreement with the results of fault-plane analysis from the first motion of P, [and]
(3) the S-wave data permit determination of the line of motion at the focus of an earthquake; this selects the fault plane from the two planes obtained in P-wave analysis, making such work more meaningful.

There were several reasons why this controversy wasn't resolved in favor of the double-couple source until the 1960s.

(1) Single- and double-couple sources had identical distributions for the first-motions of P waves; their radiation pattern differed only for shear waves.
(2) The quality of many of the shear-wave observations available at this time wasn't very good. Moreover, because the initial S motion is often buried in the coda of P or obscured by S-to-P converted arrivals, the polarization of S is often difficult to ascertain.
(3) In contrast to earlier periods (see Section 3.2), seismologists generally agreed that slip on faults caused shallow earthquakes; the single-couple models were intuitively attractive as they seemed to arise naturally from the fault-slip concept.
(4) Many influential seismologists outside of Japan (notably Perry Byerly, John Hodgeson, and A. R. Ritsema) plotted source motions using somewhat cumbersome graphical methods (Fig. 8.9); although these seemed natural to observers at the surface they tended to complicate interpretation of the nature of the source. Interestingly enough, the earlier Japanese and some Russian investigators generally used first-motion plots that projected the source motions around the focal sphere and thus look much like the beachball plots in favor today.

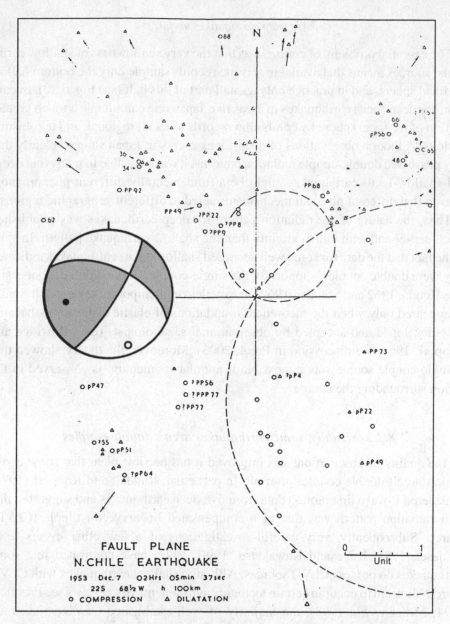

Fig. 8.9 Ingram's (1957) focal mechanism for the intermediate-depth earthquake that occurred in Chile on 7 December 1953 ($h_{Abe} = 110$ km; $m_{B(Abe)} = 7.2$), plotted using Byerly's extended distance method; inset at left is the same mechanism in the more familiar beachball format. In Byerly's method (e.g., Byerly, 1955), rays leaving the focus with an inclination $i°$ from the downward vertical plot a distance $\cot(i)$ from the center of the graph, with an azimuth corresponding to the event–station azimuth. It can be shown that great circles on the focal sphere (and hence nodal planes) form circles on such a plot. Here, individual compressions and dilatations are indicated with circles and triangles, labeled with the phase (P, pP, or PP) and a station number; the large dashed circles are the nodal planes chosen by Ingram. Figure reproduced from Ingram (1957), with beachball inset added.

The essential problem, of course, was that the very shallowness of shallow earthquake sources means that available seismic records sample only the bottom half of the focal sphere, and in practice only a small part of this half (see Fig. 6.20). In contrast, for deep-focus earthquakes in areas like Japan one can sample a much greater portion of the focal sphere by combining records of local, regional, and teleseismic stations. Although observations of deep-focus events in Japan showed clearly that they possessed double-couple radiation patterns, it was plausible to many that deep- and shallow-focus earthquakes might be a fundamentally different phenomenon, or even that different physical mechanisms acted in different geographic regions.

Thus, the nature of the radiation pattern for deep earthquakes was established much earlier and with more certainty than the shallow earthquake pattern. In spite of the fact that the data for some well-recorded shallow-focus earthquakes indicated they were double couples, support for the single-couple source didn't die out easily (see Honda, 1962 and Stauder, 1962 for excellent contemporary reviews). It finally disappeared only when the theoretical foundations of elastic dislocations became fully developed and accepted by observational seismologists (e.g., Burridge and Knopoff, 1964; see discussion in Pujol, 2003). Moreover, the theory showed that a single-couple source was impossible if angular momentum is conserved in the region surrounding the source.

8.3.2 Actually, some earthquakes aren't double couples

As the quality of observations has improved it has become clear that some earthquakes aren't double couples after all. In particular, Randall and Knopoff (1970) considered P-wave first motion data from five deep earthquakes and suggested that their radiation pattern was that of a compensated linear vector dipole (CLVD) source. Subsequently, very careful investigations of a few other events (e.g., Riedesel, 1985; Kuge and Kawakatsu, 1990; 1992) have confirmed that some earthquakes do possess CLVD sources. Although shallow earthquakes with CLVD sources do seem to occur in certain tectonic environments (for reviews see Frohlich, 1994; Miller *et al.*, 1998), the majority of well-established CLVD sources are intermediate- or deep-focus events, since, as explained above, these permit observations with the most complete focal sphere coverage.

A more difficult question concerns whether some earthquakes have an implosive component. Once again, because of their larger focal-sphere coverage the majority of papers on this question use data from deep earthquakes (e.g., Fig. 7.4). Although some highly respected geophysicists have reported isotropic components for a few deep earthquakes (e.g., Gilbert and Dziewonski, 1975; Silver and Jordan, 1982), the most careful recent studies do not confirm their presence (e.g., Kawakatsu, 1991; Hara *et al.*, 1995; Okal, 1996; Russakoff *et al.*, 1997).

8.3.3 Richter's magnitude

Undoubtedly, Charles Richter's most famous contribution to seismology is the magnitude scale. Most seismologists probably wouldn't consider magnitude as his most important scientific contribution – that would either be his series of papers coauthored with Beno Gutenberg on global seismicity (see Gutenberg and Richter references in Chapter 3) or his classic textbook on seismology (Richter, 1958). However, magnitude is the only earthquake statistic familiar to most non-scientists; and, because news accounts of earthquakes commonly mention the Richter scale, Charles Richter is the only seismologist that most people can name.

Most seismologists don't know that Kiyoo Wadati and his then-ongoing work on deep earthquakes strongly influenced Richter's development of the magnitude scale. In the second paragraph of Richter's (1935) paper defining magnitude, he states he "is not aware of any previous approach to this problem along the course taken in this paper, except for the work of Wadati." Then, on the next page:

The procedure used was suggested by a device of Wadati (1931), who plotted the calculated earth amplitudes in microns for various Japanese stations against their epicentral distances. He employed the resulting curves to distinguish between shallow and deep earthquakes . . .

Wadati's (1931) paper was the third in his landmark series of papers entitled "Shallow and deep earthquakes" which confirmed the existence of deep earthquakes (see Section 3.3 and Frohlich, 1987). In this paper he presents semi-log plots of amplitude vs distance curves out to 1000 km for five deep-focus, five intermediate, and 31 shallow-focus earthquakes. For the shallow-focus events the graphed curves for individual events (see Fig. 8.10) are essentially parallel to one another, with the 1 September 1923 Kwanto earthquake ($M_S = 8.2$) having the largest amplitudes, and the other events having amplitudes up to two orders of magnitude smaller. Richter's (1935) paper presents essentially similar data for earthquakes with a range of sizes recorded at stations in California, although for reasons that are unclear, he presents no graphs in his paper, only tables.

8.4 Perhaps deep earthquakes aren't the whole answer

8.4.1 Earth structure and dynamics

In this chapter I have argued that deep earthquakes provide much of the essential data that constrain our knowledge of the structure and dynamics of the Earth's crust, mantle, and core and also the earthquake source itself. Since the possibly unconvinced reader may well ask, "What else is there?" it is only fair to explicitly describe features of the Earth that we did not learn about primarily by analyzing

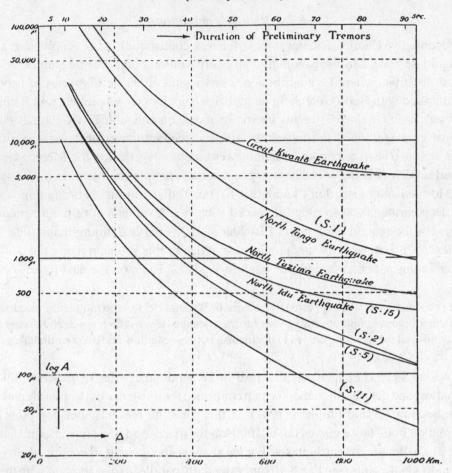

Fig. 8.10 Wadati's (1931) graph of maximum ground motion (microns) vs distance (bottom scale) and (S-P) duration (top scale) for seven shallow earthquakes. Note that all the lines are approximately parallel, with amplitudes for the Great Kwanto earthquake (1 September 1923 $M_S = 8.2$) being approximately two orders of magnitude greater than for event S11 (31 May 1930), which caused minimal damage. This paper, in which Wadati contrasted differences in amplitude–distance curves for shallow and deep earthquakes, strongly influenced Richter's (1935) paper in which he invented the magnitude scale.

deep earthquakes. That is, suppose this chapter's title wasn't "Are deep earthquakes useful?" but instead "When are deep earthquakes *not* useful?"

The most influential early determinations of mantle velocity depended primarily on observations of shallow earthquakes. To determine seismic velocity, the Herglotz–Wiechert method requires an accurate knowledge of the slope of the travel-time curve. Thus, to obtain meaningful slopes while avoiding uncertainties caused by unknown origin times and by lateral variations in structure, it was important to have P observations recorded over a broad range of distances. This required records from stations near the epicenters and also extending out to as great a distance as possible. In the 1920s and 1930s few earthquakes were satisfactory because stations were rare, the event had to be large and occur within a large continent. Otherwise there would certainly be distance ranges lacking observations.

Similarly, shallow-focus earthquakes provided important data that led to the conclusion that there were velocity increases within the upper mantle. Byerly (1926, 1935) evaluated travel times from two earthquakes in Montana and Nevada and noted that there were noticeable changes in the slopes near distances of 17° and 28°. Thus, Jeffreys (1936; 1937) evaluated these data as well as travel times from deep-focus events and concluded that the "20 degree discontinuity" must be due to an increase in seismic velocity near a depth of 470 km; he also made note of Bernal's suggestion that this increase might be associated with the olivine–spinel phase transition (Anonymous, 1936). Between 1965 and 1975 several array studies of slowness, relying primarily on data from shallow earthquakes and nuclear explosions, confirmed the existence of two zones of increase at depths near 400 km and 650 km (e.g., Niazi and Anderson, 1965; Johnson, 1967; Archambeau *et al.*, 1969; Simpson *et al.*, 1974).

Although deep earthquakes have been important in refining the structure of the Earth's core, they weren't involved in its initial discovery or in initially establishing the outer core's liquidity (see Brush, 1980). Oldham (1906) first provided convincing evidence for the existence of a central core by interpreting travel time curves of P and S waves from the devastating shallow earthquake of 12 June 1897 in India. It was two decades later before Jeffreys (1926) established the liquidity of the core, and at the time the seismological data supporting this were considered less convincing than certain tidal and geodetic observations. Moreover, whether the core was liquid remained controversial – it was another three decades before Gutenberg (1957) was willing to state this in print.[3] Lehmann (1936) proposed the existence of the solid inner core after analyzing data from a shallow New Zealand earthquake that occurred on 16 June 1929.

Although it is true that deep earthquakes are especially useful as sources for receiver-function studies of crustal structure, most of what we know about the Earth's crustal structure comes from shallow-earthquake or controlled-source (explosion) data (e.g., see Mooney *et al.*, 1998). The two most widely used methods for determining crustal structure involve the analysis of surface waves and the interpretation of travel times from refracted body waves. For obvious reasons, both of these methods generally utilize data from shallow sources.

[3] It is puzzling why it took so long to convince Gutenberg that the outer core was liquid. For example, Honda (1934b) evaluated S and ScS waves for the Japan Sea deep earthquake of 13 November 1932 recorded at stations in Japan, and demonstrated that the amplitudes of ScS could only be explained by a liquid core.

8.4.2 "Hidden" deep earthquakes?

Yet although deep earthquakes aren't everything, there are certain subjects in seismology where authors often fail to mention that deep earthquakes provide a significant proportion of their data. We have already noted that free oscillation research often depends heavily on such "hidden" deep earthquakes. Receiver-function studies of crustal structure also often fail to explicitly state the focal depths of data sources, even though deep earthquakes tend to be chosen preferentially as sources because their body waves are relatively impulsive and uncontaminated by near-surface crustal reverberations. For example, in a survey of receiver-function papers that appeared between 1997 and 2000 in the *Journal of Geophysical Research*, I identified nine papers that made no mention of the depths of their source data.

Experiments where the design requires precise absolute travel-time picks also favor deep sources. For example, inversions that utilize catalog arrival-time data to simultaneously determine locations and earth structure require earthquakes with well-constrained focal depths that provide impulsive body-waves phases; these are often deep events. For example, Bijwaard *et al.* (1998) used ISC hypocenters relocated by Engdahl *et al.* (1998) as the basis for their global tomography study, and Deal *et al.* (1999) use the same dataset for a regional tomographic study of the mantle beneath Tonga. Neither paper explicitly discusses the depths of their source data; however, of the earthquakes relocated by the Engdahl group, 38%of those with magnitudes exceeding 5.0 and rms residuals of 1.0 or less had focal depths of 60 km or more, compared to 31%for the entire ISC catalog.   [4]

Similarly, various other topics, including investigations of anisotropy and of core–mantle boundary structure, require accurately determined small relative travel-time differences and thus favor the selection of deep sources. In Fouch and Fischer's (1998) study of anisotropy beneath the Marianas, nearly all the earthquakes used are deep- and intermediate-focus events; yet this "hidden" depth information is only evident if one bothers to obtain the non-published data tables available from the American Geophysical Union as ftp files or on diskettes. Levin *et al.* (1999) use SKS and SKKS phases to investigate anisotropy beneath the Appalachians in the United States and the Urals in Russia, but fail to discuss the depth of the sources that provide their data.

Another topic of recent interest is the anisotropic structure of the Earth's inner core and the possibility that this provides an indicator of differential rotation between the mantle and inner core. Many of the papers describing this phenomenon

---

[4] This book's author determined these results. For the 1964–1997 period, the ISC reports 43,856 earthquakes with mb of 5.0 or greater; of these, 13,733 had focal depths of 60 km or more. For the same period the Engdahl *et al.* (1998) catalog has 24,661 earthquakes with mb of 5.0 or greater and rms residuals of 1.0 s or less; of these, 9458 had focal depths of 60 km or more.

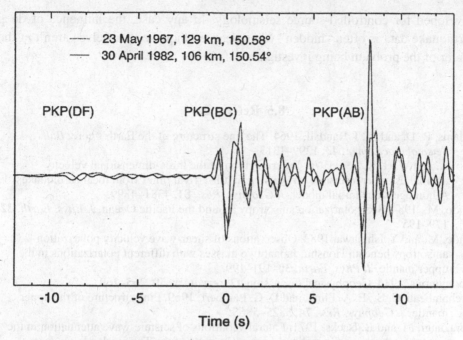

Fig. 8.11 Evidence for inner core rotation? Overlay of two short-period PKP seismograms recorded at station COL in Alaska from intermediate-depth earthquakes occurring 15 years apart in the South Sandwich Islands. When the records are aligned on the PKP(AB) and PKP(BC) phases, which travel completely within the outer core, it is evident that the arrival times of the inner core phase PKP(DF) differ by about 0.4 s. A hypothesis to explain this observation is that the inner core is anisotropic, and during the 15 year interval it has rotated ∼15° so that PKP(DF) travels along a faster path. Figure reproduced from Song and Richards (1996) with permission from Nature Publishing Group.

do indicate that deep sources provide most of the differential travel-time residuals they analyze (e.g., Creager, 1992; Song and Helmberger, 1993; 1995; Song and Richards, 1996; see Fig. 8.11). However, others don't explicitly state the depths of their data sources and instead simply describe their phase picks as "clear," "unambiguous," or "high-quality," (e.g., Creager, 1997; 1999; Song, 2000).

Surveying the very recent literature suggests that the number of papers analyzing "hidden" deep earthquakes has been increasing. Why is this? One explanation is that computers now allow scientists to undertake modeling studies that were previously impossible. Thus, in many papers where depth is "hidden," authors have utilized a metadata approach, assembling data from several previous investigations and focusing on modeling rather than the details of data collection. Also, since about 1990 it has become easier to acquire and manipulate large suites of digital waveforms, and thus seismologists increasingly utilize imaging methods originally developed for controlled-source seismology. In any case, the influence of deep earthquake data is often "hidden" when deep earthquake themselves aren't at the center of the problem being investigated.

8.5 References

Adams, R. D. and M. J. Randall, 1964. The fine structure of the Earth's core, *Bull. Seismol. Soc. Amer.*, **43**, 1299–1313.

Aki, K. and W. H. K. Lee, 1976. Determination of the three-dimensional velocity anomalies under a seismic array using first P-arrival times from local earthquakes; 1. A homogeneous initial model, *J. Geophys. Res.*, **81**, 4381–4399.

Ando, M., 1984. ScS polarization anisotropy around the Pacific Ocean, *J. Phys. Earth*, **32**, 179–195.

Ando, M. and Y. Ishikawa, 1982. Observations of shear-wave velocity polarization anisotropy beneath Honshu, Japan: two masses with different polarizations in the upper mantle, *J. Phys. Earth*, **30**, 191–199.

Anonymous, 1936. Geophysical discussion, *Observatory*, **59**, 265–269.

Archambeau, C. B., E. A. Flinn, and D. G. Lambert, 1969. Fine structure of the upper mantle, *J. Geophys. Res.*, **74**, 5825–5865.

Barazangi, M. and B. Isacks, 1971. Lateral variations of seismic-wave attenuation in the upper mantle above the inclined earthquake zone of the Tonga island arc; deep anomaly in the upper mantle, *J. Geophys. Res.*, **76**, 8493–8516.

Barley, B. J., J. A. Hudson, and A. Douglas, 1982. S to P scattering at the 650 km discontinuity, *Geophys. J. Roy. Astron. Soc.*, **69**, 159–172.

Benz, H. M. and J. E. Vidale, 1993. Sharpness of upper-mantle discontinuities determined from high-frequency reflections, *Nature*, **365**, doi:10.1038/365147a0, 147–150.

Bijwaard, H., W. Spakman, and E. R. Engdahl, 1998. Closing the gap between regional and global travel time tomography, *J. Geophys. Res.*, **103**, doi:10.1029/98JB02467, 30055–30078.

Bock, G. and J. Ha, 1984. Short-period S-P conversion in the mantle at a depth near 700 km, *Geophys. J. Roy. Astron. Soc.*, **77**, 593–615.

Bolt, B., 1959. Travel-times of PKP up to 145°, *Geophys. J. Roy. Astron. Soc.*, **2**, 190–198.
    1964. The velocity of seismic waves near the Earth's center, *Bull. Seismol. Soc. Amer.*, **54**, 191–208.

Brush, S. G., 1980. Discovery of the Earth's core, *Amer. J. Phys.*, **48**, 705–724.

Buchbinder, G. G. R., 1971. A velocity structure of the Earth's core, *Bull. Seismol. Soc. Amer.*, **61**, 429–456.

Burdick, L. J. and C. A. Langston, 1977. Modeling crustal structure through the use of converted phases in teleseismic body-waveforms, *Bull. Seismol. Soc. Amer.*, **67**, 677–691.

Burridge, R. and L. Knopoff, 1964. Body force equivalents for seismic dislocations, *Bull. Seismol. Soc. Amer.*, **54**, 1875–1888.

Byerly, P., 1926. The Montana earthquake of June 28, 1925, G. M. C. T., *Bull. Seismol. Soc. Amer.*, **16**, 209–265.
    1935. The first preliminary waves of the Nevada earthquake of December 20, 1932, *Bull. Seismol. Soc. Amer.*, **25**, 62–80.
    1955. Nature of faulting as deduced from seismograms, *Geol. Soc. Spec. Paper* **62**, 74–86.

Castle, J. C. and K. C. Creager, 2000. Local sharpness and shear wave speed jump across the 660-km discontinuity, *J. Geophys. Res.*, **105**, doi:10.1029/1999JB900424, 6191–6200.

Choy, G. L., 1977. Theoretical seismograms of core phases calculated by frequency-dependent full wave theory, and their interpretation, *Geophys. J. Roy. Astron. Soc.*, **51**, 275–312.

Choy, G. L. and V. F. Cormier, 1982. The structure of the inner core inferred from short-period and broadband GDSN data, *Geophys. J. Roy. Astron. Soc.*, **72**, 1–21.

Cormier, V. F., 1981. Short-period PKP phases and the analastic mechanism of the inner core, *Phys. Earth Planet. Int.*, **24**, 291–301.

Cormier, V. F. and G. C. Choy, 1986. A search for lateral heterogeneity in the inner core from differential travel times near PKP-D and PKP-C, *Geophys. Res. Lett.*, **13**, 1553–1556.

Creager, K. C., 1992. Anisotropy of the inner core from differential travel times of the phases PKP and PKIKP, *Nature*, **356**, doi:10.1038/356309a0, 309–314.

  1997. Inner core rotation rate from small-scale heterogeneity and time-varying travel times, *Science*, **278**, doi:10.1126/science.278.5341.1284, 1284–1288.

  1999. Large-scale variations in inner core anisotropy, *J. Geophys. Res.*, **104**, doi:10.1029/1999JB900162, 23127–23139.

Creager, K. C. and T. H. Jordan, 1984. Slab penetration into the lower mantle, *J. Geophys. Res.*, **89**, 3031–3049.

  1986. Slab penetration into the lower mantle beneath the Mariana and other island arcs of the northwest Pacific, *J. Geophys. Res.*, **91**, 3573–3589.

Cummins, P. R., B. L. N. Kennett, J. R. Bowman, and M. G. Bostock, 1992. The 520 discontinuity? *Bull. Seismol. Soc. Amer.*, **82**, 323–336.

Dahlen, F. A. and J. Tromp, 1998. *Theoretical Global Seismology*, Princeton, NG, Princeton University Press, 1025 pp.

Datt, R. and K. J. Muirhead, 1976. Evidence for a sharp increase in P-wave velocity at about 770 km depth, *Phys. Earth Planet. Int.*, **13**, 37–46.

Deal, M. M., G. Nolet, and R. D. van der Hilst, 1999. Slab temperature and thickness from seismic tomography. 1. Method and application to Tonga, *J. Geophys. Res.*, **104**, doi:10.1029/1999JB900255, 28789–28802.

Ding, X.-Y. and S. P. Grand, 1994. Seismic structure of the deep Kurile subduction zone, *J. Geophys. Res.*, **99**, 23767–23786.

Durek, J. J. and B. Romanowicz, 1999. Inner core anisotropy inferred by direct inversion of normal mode spectra, *Geophys. J. Int.*, **139**, doi:10.1046/j.1365-246x.1999.00961.x, 599–622.

Engdahl, E. R. and E. A. Flinn, 1969. Seismic waves reflected from discontinuities within the upper mantle, *Science*, **163**, 177–179.

Engdahl, E. R., R. van der Hilst, and R. Buland, 1998. Global teleseismic earthquake relocation with improved travel times and procedures for depth determination, *Bull. Seismol. Soc. Amer.*, **88**, 722–743.

Ergin, K., 1967. Seismic evidence for a new layered structure of the Earth's core, *J. Geophys. Res.*, **72**, 3669–3687.

Estabrook, C. H., G. Bock, and R. Kind, 1994. Investigation of mantle discontinuities from a single deep earthquake, *Geophys. Res. Lett.*, **21**, doi:10.1029/94GL01080, 1495–1498.

Faber, S. and G. Müller, 1980. Sp phases from the transition zone between the upper and lower mantle, *Bull. Seismol. Soc. Amer.*, **70**, 487–508.

Fan, G., T. C. Wallace, and D. Zhao, 1998. Tomographic imaging of deep velocity structure beneath the Eastern and Southern Carpathians, Romania: implications for continental collision, *J. Geophys. Res.*, **103**, doi:10.1029/97JB01511, 2705–2724.

Fischer, K. M., E. M. Parmentier, A. R. Stine, R. Alexander, and E. R. Wolf, 2000. Modeling anisotropy and plate-driven flow in the Tonga subduction zone back arc, *J. Geophys. Res.*, **105**, doi:10.1029/1999JB900441, 16181–16191.

Flanagan, M. P. and P. M. Shearer, 1998. Topography on the 410-km seismic velocity discontinuity near subduction zones from stacking of sS, sP, and pP precursors, *J. Geophys. Res.*, **103**, doi:10.1029/98JB00595, 21165–21182.

Fouch, M. J. and K. M. Fischer, 1996. Mantle anisotropy beneath northwest Pacific subduction zones, *J. Geophys. Res.*, **101**, doi:10.1029/96JB00881, 15987–16002.

1998. Shear wave anisotropy in the Mariana subduction zone, *Geophys. Res. Lett*, **25**, doi:10.1029/98GL00650, 1221–1224.

Frohlich, C., 1987. Kiyoo Wadati and early research on deep focus earthquakes: introduction to special section on deep and intermediate focus earthquakes, *J. Geophys. Res.*, **92**, 13777–13788. Reprinted in *History of Geophysics*, **4**, 166–177, 1990.

1994. Earthquakes with non-double-couple mechanisms, *Science*, **264**, 804–809.

Fukao, Y., M. Obayashi, H. Inoue, and M. Nenbai, 1992. Subducting slabs stagnant in the mantle transition zone, *J. Geophys. Res.*, **97**, 4809–4822.

Garnero, E. J., 2000. Heterogeneity of the lowermost mantle, *Ann. Rev. Earth Planet. Sci.*, **28**, doi:10.1146/annurev.earth.28.1.509, 509–537.

Garnero, E. J. and D. V. Helmberger, 1995. On seismic resolution of lateral heterogeneity in the Earth's outermost core, *Phys. Earth Planet. Int.*, **88**, 117–130.

Gilbert, F. and A. M. Dziewonski, 1975. An application of normal mode theory to the retrieval of structural parameters and source mechanisms from seismic spectra, *Phil. Trans. Roy. Soc., London*, **A278**, 187–269.

Gledhill, K. and G. Stuart, 1996. Seismic anisotropy in the fore-arc region of the Hikurangi subduction zone, New Zealand, *Phys. Earth Planet. Int.*, **95**, doi:10.1016/0031-9201(95)03117-0, 211–225.

Golenetskii, S. T. and G. Y. Medvedeva, 1965. On discontinuities of the first kind in the Earth's upper mantle, *Izv. Akad. Nauk SSSR, Phys. Solid Earth*, **5**, 318–322.

Graeber, F. M. and G. Asch, 1999. Three-dimensional models of P wave velocity and P-to-S velocity ratio in the southern central Andes by simultaneous inversion of local earthquake data, *J. Geophys. Res.*, **104**, doi:10.1029/1999JB900037, 20237–20256.

Gutenberg, B., 1957. The boundary of the Earth's inner core, *Trans. Amer. Geophys. Un.*, **38**, 750–753.

Gutenberg, B. and C. F. Richter, 1938. P′ and the Earth's core, *Mon. Not. Roy. Astron. Soc., Geophys. Supp.*, **4**, 363–372.

Hara, T., K. Kuge, and H. Kawakatsu, 1995. Determination of the isotropic component of the 1994 Bolivia deep earthquake, *Geophys. Res. Lett.*, **22**, doi:10.1029/95GL01602, 2265–2268.

He, X. and J. Tromp, 1996. Normal-mode constraints on the structure of the Earth, *J. Geophys. Res.*, **101**, doi:10.1029/96JB01783, 20053–20082.

Helffrich, G., 1995 Lithospheric deformation inferred from teleseismic shear wave splitting observations in the United Kingdom, *J. Geophys. Res.*, **100**, doi:10.1029/95JB01572, 18195–18204.

Hirahara, K., 1977. A large-scale three-dimensional seismic structure under the Japan Islands and the Sea of Japan, *J. Phys. Earth*, **25**, 393–417.

Honda, H., 1933. Notes on the mechanism of deep earthquakes, *Geophys. Mag.*, **7**, 257–267.

1934a. On the amplitude of the P and the S waves of deep earthquakes, *Geophys. Mag.*, **8**, 153–164.

1934b. On the ScS waves and the rigidity of the Earth's core, *Geophys. Mag.*, **8**, 165–177.

1962. Earthquake mechanism and seismic waves, *Geophys. Notes, Tokyo Univ.*, **15**, Supp., 1–97.

Honda, H. and T. Miura, 1938. On the amplitude of the P and the S waves of deep earthquakes (third paper), *Geophys. Mag.*, **11**, 299–305.

Honda, H., A. Masatsuka, and K. Emura, 1957. On the mechanism of the earthquakes and the stresses producing them in Japan and its vicinity, *Sci. Rept. Tohoku Univ., ser. 5 – Geophys.*, **8**, 186–205.

Huppert, L. N. and C. Frohlich, 1981. The P velocity within the Tonga Benioff zone determined from traced rays and observations, *J. Geophys. Res.*, **86**, 3771–3782.

Ingram, R. E., 1957. Fault plane of the Chile earthquake, December 7, 1953, *Bull. Seismol. Soc. Amer.*, **47**, 281–285.

Isacks, B. L. and M. Barazangi, 1973. High frequency shear waves guided by a continuous lithosphere descending beneath western South America, *Geophys. J. Roy. Astron. Soc.*, **33**, 129–139.

Isacks, B. and P. Molnar, 1971. Distribution of stresses in the descending lithosphere from a global survey of focal-mechanism solutions to mantle earthquakes, *Rev. Geophys. Space Phys.*, **9**, 103–174.

Isacks, B. L., J. Oliver, and L. R. Sykes, 1968. Seismology and the new global tectonics, *J. Geophys. Res.*, **73**, 5855–5899.

Jeffreys, H., 1926. The rigidity of the Earth's central core, *Mon. Not. Roy. Astron. Soc., Geophys. Supp.*, **1**, 371–383.

1935. Some deep-focus earthquakes, *Mon. Not. Roy. Astron. Soc., Geophys. Supp.*, **3**, 310–342.

1936. The structure of the Earth down to the 20° discontinuity, *Mon. Not. Roy. Astron. Soc., Geophys. Supp.*, **3**, 401–443.

1937. The structure of the Earth down to the 20° discontinuity (second paper), *Mon. Not. Roy. Astron. Soc., Geophys. Supp.*, **4**, 13–39.

1939a. The times of PcP and ScS, *Mon. Not. Roy. Astron. Soc., Geophys. Supp.*, **4**, 537–547.

1939b. The times of the core waves, *Mon. Not. Roy. Astron. Soc., Geophys. Supp.*, **4**, 548–561.

1939c. The times of the core waves (second paper), *Mon. Not. Roy. Astron. Soc., Geophys. Supp.*, **4**, 594–615.

1942. The deep earthquake of 1934 June 29, *Mon. Not. Roy. Astron. Soc., Geophys. Supp.*, **5**, 33–36.

Johnson, L. R., 1967. Array measurements of P velocities in the upper mantle, *J. Geophys. Res.*, **72**, 309–325.

Jordan, T. H., 1977. Lithospheric slab penetration into the lower mantle beneath the Sea of Okhotsk, *J. Geophys.*, **43**, 473–496.

Julian, B. R. and M. K. Sengupta, 1973. Seismic travel time evidence for lateral inhomogeneity in the deep mantle, *Nature*, **242**, 443–447.

Kaneshima, S. and P. G. Silver, 1995. Anisotropic loci in the mantle beneath central Peru, *Phys. Earth Planet. Int.*, **88**, doi:10.1016/0031-9201(94)02981-G, 257–272.

Kawakatsu, H., 1991. Insignificant isotropic component in the moment tensor of deep earthquakes, *Nature*, **351**, doi:10.1038/351050a0, 50–53.

Kawasumi, H. and R. Yosiyama, 1934. On the mechanism of a deep-seated earthquake as revealed by the distribution of the initial motions at stations throughout the world, *Proc. Imperial Acad. Japan*, **10**, 345–348.

Kendall, J.-M. and P. Silver, 1996. Constraints from seismic anisotropy on the nature of the lowermost mantle, *Nature*, **381**, doi:10.1038/381409a0, 409–412.

Kennett, B. L. N., 1991. IASPEI 1991 *Seismological Tables*, Canberra, Australia, Australian National University, 167 pp.

Koch, M., 1985. Nonlinear inversion of local seismic travel times for the simultaneous determination of the 3D velocity structure and hypocenters – application to the seismic zone in Vrancea, *J. Geophys.*, **56**, 160–173.

Kuge, K. and H. Kawakatsu, 1990. Analysis of a deep "non-double-couple" earthquake using very broadband data, *Geophys. Res. Lett.*, **17**, 227–230.

1992. Deep and intermediate-depth non-double-couple earthquakes: interpretation of moment tensor inversions using various passbands of very broadband seismic data, *Geophys. J. Int.*, **111**, 589–606.

Kuge, K. and K. Satake, 1987. Lateral segmentation within the subducting lithosphere: three dimensional structure beneath the North Island, New Zealand, *Tectonophysics*, **139**, doi:10.1016/0040-1951(87)90098-9, 223–237.

Langston, C. A., 1979. Structure under Mount Rainier, Washington, inferred from teleseismic body waves, *J. Geophys. Res.*, **84**, 4749–4762.

Lapwood, E. R. and T. Usami, 1981. *Free Oscillations of the Earth*, Cambridge, U.K., Cambridge University Press, 243 pp.

Lay, T. and D. V. Helmberger, 1983. A lower mantle S-wave triplication and the shear velocity structure of D″, *Geophys. J. Roy. Astron. Soc.*, **75**, 799–837.

Lehmann, I., 1936. P′, *Int. Geod. Geophys. Union, Assoc. Seismol. Pubs., Ser. A*, **14**, 87–115.

Levin, V., W. Menke, and J. Park, 1999. Shear wave splitting in the Appalachians and the Urals: a case for multilayer anisotropy, *J. Geophys. Res.*, **104**, doi:10.1029/1999JB900168, 17975–17993.

Lukk, A. A. and I. L. Nersesov, 1965. Structure of the upper mantle as shown by observations of earthquakes of intermediate focal depth, *Doklady Akad. Nauk. SSR*, **162**, 14–16.

Masters, G. and F. Gilbert, 1981. Structure of the inner core inferred from observations of its spheroidal shear modes, *Geophys. Res. Lett.*, **8**, 569–571.

Masters, T. G. and R. Widmer, 1995. Free oscillations: frequencies and attenuations. In *Global Earth Physics: a Handbook of Physical Constants*, ed. T. J. Ahrens, Washington, DC, American Geophysical Union, 104–125.

Matzel, E., M. K. Sen, and S. P. Grand, 1996. Evidence for anisotropy in the deep mantle beneath Alaska, *Geophys. Res. Lett.*, **23**, doi:10.1029/96GL02186, 2417–2420.

Miller, A. D., G. R. Foulger, and B. R. Julian, 1998. Non-double-couple earthquakes. 2. Observations, *Rev. Geophys.*, **36**, 551–568.

Mooney, W. D., G. Laske, and T. G. Masters, 1998. CRUST 5.1: a global crustal model at 5° × 5°, *J. Geophys. Res.*, **103**, doi:10.1029/1999JB900168, 727–747.

Muirhead, K. J. and A. L. Hales, 1980. Evidence for P wave velocity discontinuities at depths greater than 650 km in the mantle, *Phys. Earth Planet. Int.*, **23**, 304–313.

Müller, G., 1973. Amplitude studies of core phases, *J. Geophys. Res.*, **78**, 3469–3490.

1975. Further evidence against discontinuities in the outer core, *Phys. Earth Planet. Int.*, **10**, 70–73.

Myers, S. C., S. Beck, G. Zandt, and T. Wallace, 1998. Lithospheric-scale structure across the Bolivian Andes from tomographic images of velocity and attenuation for P and S waves, *J. Geophys. Res.*, **103**, doi:10.1029/98JB00956, 21233–21252.

Nakanishi, I., 1988. Reflections of P'P' from upper mantle discontinuities beneath the Mid-Atlantic Ridge, *Geophys. J. Int.*, **93**, 335–346.

Niazi, M. and D. L. Anderson, 1965. Upper mantle structure of western North America from apparent velocities of P waves, *J. Geophys. Res.*, **70**, 4633–4640.

Niu, F. and H. Kawakatsu, 1995. Direct evidence for the undulation of the 660-km discontinuity beneath Tonga: comparison of Japan and California array data, *Geophys. Res. Lett.*, **22**, doi::10.1029/94GL03332, 531–534.

1997. Depth variation of the mid-mantle seismic discontinuity, *Geophys. Res. Lett.*, **24**, doi:10.1029/97GL00216, 429–432.

Nolet, G., S. P. Grand, and B. L. N. Kennett, 1994. Seismic heterogeneity in the upper mantle, *J. Geophys. Res.*, **99**, 23753–23766.

Nuttli, O. W., 1954. The P phase and the Earth's core, *Trans. Amer. Geophys. Un.*, **35**, 962–968.

Okal, E. A., 1996. Radial modes from the great 1994 Bolivian earthquake: no evidence for an isotropic component to the source, *Geophys. Res. Lett.*, **23**, doi:10.1029/96GL00375, 431–434.

Oldham, R. D., 1906. The constitution of the interior of the Earth, as revealed by earthquakes, *Quart. J. Geol. Soc. London*, **62**, 456–475.

Oliver, J. and B. Isacks, 1967. Deep earthquake zones, anomalous structures in the upper mantle, and the lithosphere, *J. Geophys. Res.*, **72**, 4259–4275.

Owens, T. J., G. Zandt, and S. R. Taylor, 1984. Seismic evidence for an ancient rift beneath the Cumberland Plateau, Tennessee: a detailed analysis of broadband teleseismic P waveforms, *J. Geophys. Res.*, **89**, 7783–7795.

Özalaybey, S. and M. K. Savage, 1995. Shear-wave splitting beneath western United States in relation to plate tectonics, *J. Geophys. Res.*, **100**, doi:10.1029/95JB00715, 18135–18149.

Pascal, G., J. Dubois, M. Barazangi, B. L. Isacks, and J. Oliver, 1973. Seismic velocity anomalies beneath the New Hebrides island arc: evidence for a detached slab in the upper mantle, *J. Geophys. Res.*, **78**, 6998–7004.

Phinney, R. A., 1964. Structure of the Earth's crust from spectral behaviour of long-period body waves, *J. Geophys. Res.*, **69**, 2997–3017.

Pujol, J., 2003. The body force equivalent to an earthquake: a tutorial, *Seismol. Res. Lett.*, **74**, 163–168.

Pulliam, J. and M. K. Sen, 1998. Seismic anisotropy in the core–mantle boundary region, *Geophys. J. Int.*, **135**, doi:10.1046/j.1365–246x.199800612.x, 113–128.

Randall, M. J. and L. Knopoff, 1970. The mechanism at the focus of deep earthquakes, *J. Geophys. Res.*, **75**, 4965–4976.

Resovsky, J. S. and M. H. Ritzwoller, 1998. New and refined constraints on three-dimensional earth structure from normal modes below 3 mHz, *J. Geophys. Res.*, **103**, doi:10.1029/97JB02482, 783–810.

Revenaugh, J. and T. H. Jordan, 1991a. Mantle layering from ScS reverberations: 1. Waveform inversion of zeroth-order reverberations, *J. Geophys. Res.*, **96**, 19749–19762.

1991b. Mantle layering from ScS reverberations: 2. The transition zone, *J. Geophys. Res.*, **96**, 19763–19780.

1991c. Mantle layering from ScS reverberations: 4. The lower mantle and core-mantle boundary, *J. Geophys. Res.*, **96**, 19811–19824.

Richter, C. F., 1935. An instrumental earthquake magnitude scale, *Bull. Seismol. Soc. Amer.*, **25**, 1–32.

1958. *Elementary Seismology*, San Francisco, W. H. Freeman, 768 pp.

Riedesel, M. A., 1985. *Seismic Moment Tensor Recovery at Low Frequencies*, Ph. D. dissertation, University of California San Diego, 245 pp.

Ritsema, A. R., 1958. On the focal mechanism of southeast Asian earthquakes, *Publ. Dominion Obs., Ottawa*, **20**, 341–368.

1965. The mechanism of some deep and intermediate earthquakes in the region of Japan, *Bull. Earthquake Res. Inst., Tokyo Univ.*, **43**, 39–52.

Robertson, F., 1937. Evidences from deep-focus earthquakes for the crustal structure of Missouri, *Bull. Seismol. Soc. Amer.*, **27**, 241–244.

Romanowicz, B. and L. Berger, 2000. Anomalous splitting of free oscillations: a reevaluation of possible interpretations, *J. Geophys. Res.*, **105**, doi:10.1029/2000JB900144, 21559–21578.

Roth, E. G., D. A. Wiens, L. M. Dorman, J. Hildebrand, and S. C. Webb, 1999. Seismic attenuation tomography of the Tonga–Fiji region using phase pair methods, *J. Geophys. Res.*, **104**, doi:10.1029/1998JB900052, 4795–4809.

Russakoff, D., G. Ekström, and J. Tromp, 1997. A new analysis of the great 1970 Colombia earthquake and its isotropic component, *J. Geophys. Res.*, **102**, doi:10.1029/97JB01645, 20423–20434.

Russo, R. M. and P. G. Silver, 1994. Trench-parallel flow beneath the Nazca plate from seismic anisotropy, *Science*, **263**, 1105–1111.

Russo, R. M., P. G. Silver, M. Franke, W. B. Ambeh, and D. E. James, 1996. Shear-wave splitting in northeast Venezuela, Trinidad, and the eastern Caribbean, *Phys. Earth Planet. Int.*, **95**, doi:10.1016/0031-9201(95)03128-6, 251–275.

Sandoval, E. A., J. F. Ni, and T. M. Hearn, 1994. Seismic azimuthal anisotropy beneath the Pakistan Himalayas, *Geophys. Res. Lett.*, **21**, doi:10.1029/94GL01386, 1635–1638.

Schlittenhardt, J., J. Schweitzer, and G. Müller, 1985. Evidence against a discontinuity at the top of D″, *Geophys. J. Roy. Astron. Soc.*, **81**, 295–306.

Sharpe, J. A., 1935. Motion on the surface of the earth in the compressional phase of a deep-focus earthquake, and the effects of a layered crust, *Bull. Seismol. Soc. Amer.*, **25**, 199–222.

Shih, X. R., J. F. Schneider, and R. P. Meyer, 1991. Polarities of P and S waves, and shear wave splitting observed from the Bucaramanga nest, Columbia, *J. Geophys. Res.*, **96**, 12069–12082.

Silver, P. G., 1996. Seismic anisotropy beneath the continents: probing the depths of geology, *Ann. Rev. Earth Planet. Sci*, **24**, doi:10.1146/annurev.earth.24.1.385, 385–432.

Silver, P. G. and W. W. Chan, 1991. Shear wave splitting and subcontinental mantle deformation, *J. Geophys. Res.*, **96**, 16429–16454.

Silver, P. G. and T. H. Jordan, 1982. Optimal estimate of scalar seismic moment, *Geophys. J. Roy. Astron. Soc.*, **70**, 755–787.

Simpson, D. W., R. F. Mereu, and D. W. King, 1974. An array study of P-wave velocities in the upper mantle transition zone beneath north-eastern Australia, *Bull. Seismol. Soc. Amer.*, **64**, 1757–1788.

Sipkin, S. A. and T. H. Jordan, 1980a. Multiple ScS travel times in the western Pacific: implications for mantle heterogeneity, *J. Geophys. Res.*, **85**, 853–861.

1980b. Regional variation of Q_{ScS}, *Bull. Seismol. Soc. Amer.*, **69**, 1055–1079.

Song, X., 2000. Joint inversion for inner core rotation, inner core anisotropy, and mantle heterogeneity, *J. Geophys. Res.*, **105**, doi:10.1029/1999JB900436, 7931–7943.

Song, X. D. and D. V. Helmberger, 1993. Anisotropy of Earth's inner core, *Geophys. Res. Lett.*, **20**, 2591–2594.

1995. Depth dependence of anisotropy of Earth's inner core, *J. Geophys. Res.*, **100**, doi:10.1029/10.1029/95JB00244, 9805–9816.

Song, X. and P. G. Richards, 1996. Seismological evidence for differential rotation of the Earth's inner core, *Nature*, **382**, doi:10.1038/382221a0, 221–224.

Souriau, A. and G. Poupinet, 1991. A study of the outermost liquid core using differential travel times of the SKS, SKKS and S3KS phases, *Phys. Earth Planet. Int.*, **68**, 183–199.

Souriau, A. and P. Roudil, 1995. Attenuation in the uppermost inner core from broadband GEOSCOPE PKP data, *Geophys. J. Int.*, **123**, 572–587.

Stauder, W., 1962. The focal mechanism of earthquakes, *Adv. Geophys.*, **9**, 1–76.

Stauder, W. and W. M. Adams, 1961. A comparison of some S-wave studies of earthquake mechanisms, *Bull. Seismol. Soc. Amer.*, **51**, 277–292.

Teng, T. and J. P. Tung, 1973. Upper-mantle discontinuity from amplitude data of P'P' and its precursors, *Bull. Seismol. Soc. Amer.*, **63**, 587–597.

Tono, Y., T. Kunugi, Y. Fukao, S. Tsuboi, K. Kanjo, and K. Kasahara, 2005. Mapping of the 410- and 660-km discontinuities beneath the Japanese islands, *J. Geophys, Res.*, **110**, doi:10.1029/2004JB003266, B03307.

Utsu, T., 1967. Anomalies in seismic wave velocity and attenuation associated with a deep earthquake zone (I), *J. Faculty Sci. Hokkaido Univ. ser.* **7** (*Geophys*), **3**, 1–25.

1971. Seismological evidence for anomalous structure of island arcs with special reference to the Japanese region, *Rev. Geophys. Space Phys.*, **9**, 839–890.

Van der Hilst, R., 1995. Complex morphology of subducted lithosphere in the mantle beneath the Tonga Trench, *Nature*, **374**, doi:10.1038/374154a0, 154–157.

Van der Hilst, R. and E. R. Engdahl, 1991. On ISC PP and pP data and their use in delay-time tomography of the Caribbean region, *Geophys. J. Int.*, **106**, 169–188.

Van der Hilst, R., E. R. Engdahl, W. Spakman, and G. Nolet, 1991. Tomographic imaging of subducted lithosphere below northwest Pacific island arcs, *Nature*, **353**, doi:10.1038/353037a0, 37–43.

Vidale, J. E. and H. M. Benz, 1992. Upper-mantle seismic discontinuities and the thermal structure of subduction zones, *Nature*, **356**, doi:10.1038/356678a0, 678–683.

Vidale, J. E. and M. A. H. Hedlin, 1998. Evidence for partial melt at the core–mantle boundary north of Tonga from the strong scattering of seismic waves, *Nature*, **391**, doi:10.1038/35601, 682–685.

Wadati, K., 1931. Shallow and deep earthquakes (3rd paper), *Geophys. Mag.*, **4**, 231–283.

Whitcomb, J. H. and D. L. Anderson, 1970. Reflection of P'P' seismic waves from discontinuities in the mantle, *J. Geophys. Res.*, **75**, 5713–5728.

Yamazaki, A. and K. Hirahara, 1994. The thickness of upper mantle discontinuities, as inferred from short-period J-array data, *Geophys. Res. Lett.*, **21**, doi:10.1029/94GL01418, 1811–1814.

Yang, X., K. M. Fischer, and G. A. Abers, 1995. Seismic anisotropy beneath the Shumagin Islands segment of the Aleutian–Alaska subduction zone, *J. Geophys. Res.*, **100**, doi:10.1029/95JB01425, 18165–18177.

Yu, W.-C., L. Wen, and F. Niu, 2005. Seismic velocity structure in the Earth's outer core, *J. Geophys. Res.*, **110**, doi:10.1029/2003JB002928, B02302.

Zhang, J. and C. A. Langston, 1995. Constraints on oceanic lithosphere structure from deep-focus regional receiver function inversions, *J. Geophys. Res.*, **100**, doi:10.1029/95JB02512, 22187–22196.

Zhao, D. and A. Hasegawa, 1993. P wave tomographic imaging of the crust and upper mantle beneath the Japan Islands, *J. Geophys. Res.*, **98**, 4333–4353.

Zhao, D., A. Hasegawa, and S. Horiuchi, 1992. Tomographic imaging of P and S wave velocity structure beneath northeastern Japan, *J. Geophys. Res.*, **97**, 19909–19928.

Zhao, D., A. Hasegawa, and H. Kanamori, 1994. Deep structure of Japan subduction zone as derived from local, regional, and teleseismic events, *J. Geophys. Res.*, **99**, 22313–22329.

Zhou, H.-W. and R. W. Clayton, 1990. P and S wave travel time inversions for subducting slab under the island arcs of the northwest Pacific, *J. Geophys. Res.*, **95**, 6829–6851.

Zhou, H.-W., D. L. Anderson, and R. W. Clayton, 1990. Modeling of residual spheres for subduction zone earthquakes. 1. Apparent slab penetration signatures in the NW Pacific caused by deep diffuse mantle anomalies, *J. Geophys. Res.*, **95**, 6799–6827.

9

Answered and unanswered questions

Realistically, what should you learn from this book? Almost no one reads a scientific monograph from cover to cover; and, even when they do, there can be so much focus on details that it's easy to lose track of what's important. Thus, in this chapter I review what a century of research has taught scientists about deep earthquakes, and then discuss some cultural issues that surround the business that is deep earthquake research.

9.1 What we know and what we don't know

9.1.1 What we have learned in the last 100 years

What do we know today about deep earthquakes that we didn't know when Fusakichi Omori was pondering the peculiarities of the 21 January 1906 Japanese earthquake (see the Preface)? Omori was then well into his career and, since he died in 1923,[1] didn't survive to see much of the progress that this book chronicles. It is interesting to imagine what Omori and Harold Jeffreys (1891–1989) might have thought about this book.[2] Jeffreys, of course, had a long and distinguished career in geophysics, as he published his first scientific paper in 1915, his last paper in 1987 (Jeffreys and Shimshoni, 1987), and grappled with several of the research questions that this book chronicles. Each chapter in this book presents information that would have intrigued Omori and about which Jeffreys would have had some firm opinions. What have we learned in 100 years?

I thank Xaq Frohlich for his review of an earlier draft of this chapter.

[1] Omori's death occurred two months after the great Kwanto earthquake of 1923, which killed approximately 140,000 people. This earthquake must have personally devastated Omori, who had publicly opposed Akisune Imamura's forecast of such an event. Omori was in Australia when the earthquake occurred and reportedly watched a seismograph as it recorded the event; surely he realized immediately that it was in or near Japan and knew that Imamura's forecast had come true.

[2] In spite of their scientific contributions there apparently are no book-length biographies of either Omori or Jeffreys.

Chapter 1. With respect to focal depth, a century of research has established that not all earthquakes are alike – while the majority occur in the Earth's crust, others regularly occur within the mantle. Nowhere do they occur at depths exceeding about 700 km; yet in several geographic regions very large earthquakes occur down to depths of about 650 km. The occurrence of one such earthquake in 1954 beneath Spain suggests that similar, future events may occur in regions where none have been historically observed.

We can give Harold Jeffreys a significant amount of credit for the confidence we have in stating these assertions. He spent much of the 1930s developing earth models and mathematical methods to accurately locate earthquakes; the depths he found for several large deep-focus earthquakes were remarkably accurate and both the models and location methods he developed were used throughout most of the twentieth century. Jeffreys himself considered his earth model and the associated travel time tables he developed with Keith Bullen as one of his three most important scientific contributions (see interview with Spall, 1980).[3]

Chapter 2. Mechanically, earthquakes with depths exceeding about 60 km must be fundamentally different from the familiar shallow earthquakes that occur in the Earth's crust. However, there is abundant evidence at all depths that temperature exercises control over the mechanical failure of rock under stress and thus limits where deep earthquakes occur.

Jeffreys was perhaps the first scientist to realize that deep earthquakes were a mechanical puzzle (see Section 3.3.2). In the 1920s he initially supported the hypothesis that some earthquakes were deep; then, he changed his mind and stated that this was impossible, noting that isostasy proved that the mantle was plastic and couldn't maintain the stresses necessary for fracture. As more data became available in the 1930s he changed his mind again and accepted the observations indicating deep earthquakes did occur. However, he was acutely aware that contemporary laboratory rock mechanics experiments showed that increasing depth also increased the shear stress required for failure. In a contemporary review, Jeffreys (1939a) concluded that thermal stresses were the most likely cause of deep earthquakes, noting that cooling of Earth's outer layers has proceeded to a depth of about 700 km.

Chapter 3. Omori undoubtedly would have been proud to learn that his countrymen, Kiyoo Wadati and Hirokichi Honda, would be among the twentieth century's most accomplished seismologists, and that their work would explain many features of the 1906 earthquake that confused him. Omori would have been disappointed to learn that many western scientists call the regions where deep earthquakes occur Benioff zones, rather than Wadati zones.

---

[3] The other two were the "definite statement in 1926 that the Earth's core is liquid," and studies of P waves recorded in Europe and Japan at distances of 10° and less that confirmed that mantle structure varied laterally.

It isn't clear that Jeffreys fully appreciated Wadati's scientific contribution. In his 1939 review Jeffreys did reference Wadati's (1928) paper. And, in an interview with Spall (1980) he states that the discovery of deep earthquakes is the most important advance in seismology in his lifetime. But, in this interview Jeffreys doesn't mention Wadati; he credits Turner for the initial discovery of deep earthquakes, and Stonely and Scrase for providing "altogether satisfactory" evidence of their existence (see Chapter 3.2.2).

Chapter 4. Omori would have been intrigued by the apparent power-law depth dependence for intermediate earthquakes (Fig. 4.3) and by the power-law distribution of earthquake magnitudes (Fig. 4.14). Today we credit Omori (1895) for proposing the power-law decay of aftershocks with time. Once again, Omori might be chagrined that we now call the power law of magnitudes the Gutenberg–Richter law, as this ignores the earlier contributions of Japanese seismologists (Ishimoto and Iida, 1939).

Chapter 5. Some deep earthquakes are geographically isolated and others occur in clusters; however, this statement applies to shallow-focus earthquakes as well and so neither the traits of spatial clustering nor isolation clearly distinguish the two populations. However, while some deep earthquakes do possess aftershocks, earthquakes with any or many aftershocks are far more common among shallow-focus events.

I suspect that Harold Jeffreys would agree with these assertions about clustering. However, he would be unimpressed with the level of statistical analysis seismologists typically apply to data nowadays. Jeffreys is as highly regarded among statisticians as geophysicists, as he is one of the founding fathers of Bayesean statistics (e.g., Jeffreys, 1939b).

Chapter 6. The similarities between the source properties of shallow and deep earthquakes are more remarkable than their differences. Nothing in their radiation patterns clearly separates the populations of deep and shallow earthquakes. And, when corrected for intrinsic size and for shear velocity in the source region, their populations exhibit significant overlap with respect to their durations, stress drops, complexity of source-time functions, and ratios of scalar moments to radiated energies (see Sections 6.1.3 to 6.1.5).

Omori would have been impressed by Chapter 6, as he would have appreciated the entire framework seismologists have developed to describe earthquake sources. He would perhaps be relieved to learn that modern seismologists would share his belief that the sense of motion for P and S waves from the 1906 earthquake contained important information.

Chapter 7. As late as the 1950s, no one had proposed a viable mechanism to explain deep earthquakes, and laboratory rock-mechanics experiments were not possible at the temperatures and pressures where deep-focus occur. The question then was: if deep-focus earthquakes aren't implosions, what possibly could they be? However, subsequently there has been progress; scientists have proposed three distinct families of explanations: dehydration embrittlement, thermal shear instabilities, and transformational faulting. There is support for each from laboratory experiments at appropriate temperatures and pressures, and the question now is: which mechanism applies, at what depth; and, is there more than one mechanism at work?

Whether Harold Jeffreys would be impressed with Chapter 7 is unclear. Jeffreys was a famously independent thinker; he is well known for not accepting plate tectonics, even as he remained active enough in geophysics to continue publishing into his nineties. The scientific community today still can't agree what causes deep earthquakes; there is little reason to expect Jeffreys would agree we were on the right track.

Chapter 8. One of seismology's best-kept secrets is that understanding earthquakes isn't the principal objective of most earthquake research; rather, the goal is to understand the structure and dynamics of the Earth itself. In this effort, Chapter 8 asserts that observations about deep earthquakes have been disproportionately important.

What in Chapter 8 would most impress Jeffeys and Omori? I suggest that Jeffreys would be most impressed by the application of tomography to map the amount and extent of lateral variation in seismic velocity in the Earth's mantle. Omori, who late in his career devoted considerable effort to studying Japanese volcanoes, would likely have been captivated by plate tectonics and the opportunity it offered for understanding the spatial relationships between earthquakes and volcanoes.

9.1.2 *What I have learned in the last 20 years*

What do I personally know about deep earthquakes that I didn't know 20 years ago, when I first imagined this book? Much that might impress Omori or Jeffreys is now familiar to contemporary scientists. But each chapter of this book contains at least one observation that intrigues me and is new to me since this project began. What about you?

Chapter 1. Before writing this book I had never heard of the 1939 Chillan, Chile, earthquake and did not know how deadly it was (see Section 1.2). I still am puzzled why so little has been published about it. I was unaware how hazardous intermediate-depth earthquakes are in general, especially in South America and (possibly) in the northwestern United States.

Chapter 2. I did not fully understand the relationship between temperature and the thermal parameter Φ (the product of the vertical component of subduction velocity and the age of t_{Age} of the plate when it begins subducting). I did not understand that if conduction of heat through the surface of a lithospheric plate is the only control on its temperature, at depth h the quantity Φ/h is simply the ratio of two times: the time t_{Age} the plate spends at the surface prior to subduction, and the time $t_{sub \to h}$ it spends beneath the surface after subduction (see Fig. 2.6 and eq. 2.6).[4]

Chapter 3. Before reading numerous scientific papers written before about 1925, I didn't know that then even the depth of ordinary "shallow" earthquakes was uncertain. I didn't understand that many geologists thought coseismic fault scarps might simply be the incidental effects of disturbances originating at much greater depths (see Section 3.2).

Upon reading papers written between 1925 and 1960, I learned that Kiyoo Wadati discovered Benioff zones. I also learned that Hugo Benioff's explanation for Wadati–Benioff zones (see Fig. 3.7) now seems far less insightful that Wadati's (see Fig. 3.6 and Section 3.4). I learned also that Hirokichi Honda and his Japanese colleagues understood earthquake focal mechanisms better and earlier than seismologists elsewhere, especially in the United States (see Section 3.5.3 and Figs. 3.12 and 3.13).

Chapter 4. For earthquakes with focal depths deeper than about 50 km, I was surprised to learn that there is little or no evidence that maximum earthquake size depends on depth. In particular, most of the observed variation may come about simply because events with depths between about 200–500 km are less common (see Fig. 4.16). Hence, over the one-century period of available observations, in many depth ranges the largest possible earthquakes may not have occurred.

Before I constructed regional plots of the rate of moment release rate vs depth (Fig. 4.13), I was unaware how much smaller effective slip rates ($\sim$0.01–0.5 cm/yr) were than typical plate convergence rates (>3 cm/yr). This supports the hypothesis that deep earthquakes are in some sense merely an incidental consequence of subduction, occurring only when conditions of stress, temperature and mantle composition are favorable.

Chapter 5. Prior to my personal study of the intermediate-depth earthquake nest near Bucaramanga, Colombia, I did not know how remarkably compact it was, nor how spatially isolated it is with respect to other seismic activity (see Section 5.1.3). In both respects it stands apart from other notable deep earthquake clusters.

Only while writing Chapter 5 did I realize that aftershocks are nonexistent (or nearly so) for earthquakes with focal depths between about 245 and 330 km (Section 5.2.4 and Fig. 5.14). Earthquakes with depths ranging from 70 to 200 km and 500 to 650 km occasionally have aftershocks, but whatever allows these to happen, happens rarely at the depths between.

Chapter 6. Only while writing this book did I clearly comprehend the distinction between "observed" and "derived" parameters describing the source. In particular, there are only a relatively small number of observed quantities such as phase

---

[4] Although numerous scientists have used the thermal parameter Φ, I may be the first to express Φ/h as the ratio of t_{Age} and $t_{sub \to h}$.

arrival times, scalar moment, and radiated energy that we measure more-or-less directly from seismograms. In contrast, derived parameters such as static stress drop, apparent stress, and effective seismic efficiency are mathematical combinations of observed quantities that depend on specific physical models. The derived quantities tend to have huge uncertainties because the models themselves are uncertain and because combining several observations further compounds uncertainty. This may be responsible for much (but not all) of the reported variations in static stress drop (see Box 6.1).

Within the last 20 years I became intrigued with observations indicating that many deep earthquakes in the Harvard CMT catalog had large non-double-couple components, i.e., they were compensated linear vector dipoles (CLVD). I now believe that most (but not all) of these reported CLVD are not real but instead attributable to systematic and modeling errors in the determination of CMT. When CLVD are not spurious I suspect that most are accompanied by multiplicity of the source, i.e., a source consisting of two or more subevents, each with largely double-couple (but unlike) mechanisms (see Section 6.2.3).

Chapter 7. What physical process is responsible for deep-focus earthquakes? The so-called anticrack model is the most recent proposal (see Section 7.3.3). This model suggests that, under appropriate conditions of stress and temperature, microscopic lenses of fine-grained spinel transform within olivine; a deep earthquake occurs when these link up along a macroscopic fault and failure occurs. This is an appealing model as it conceptually combines the Griffiths crack model that explains shallow earthquakes with the persistent hypothesis that polymorphic phase transitions cause deep earthquakes. Even if the anticrack model is wrong it is brilliant. But is it true? Laboratory experiments do provide some support for the anticrack model, but what is absent is a comprehensive body of seismicity observations supporting the model.

In contrast, there is rapidly emerging observational support for the forty-year-old proposal that intermediate-depth earthquakes are attributable to the dehydration of minerals. Dehydration embrittlement also provides a way to extend the depths where cracking is possible; it too receives support from laboratory experiments. What is especially convincing is that the seismicity in mantle environments as thermally different as northern Japan and the northwestern United States is consistent with dehydration reactions expected for specific minerals.

Chapter 8. Prior to writing this book, I was unaware that deep earthquakes provided especially favorable sources for receiver-function studies of the crust. I also was only vaguely aware that shallow earthquakes don't stimulate certain normal modes of the Earth (those having a node at the surface). Thus I was unaware that observations of normal modes of deep earthquakes were especially important as they provide independent information about earth structure. I was also unaware of Kiyoo Wadati's influence on Charles Richter as he formulated the Richter scale of magnitude in the 1930s.

Chapter 10. Prior to writing this book I didn't fully comprehend how much more information we have for intermediate-depth seismicity beneath Japan than for any other region; I was also unaware of the enormous literature concerning Romanian earthquakes. I have come to the realization that the literature for each geographic region has its own unique intellectual focus and character.

9.1.3 Questions still unanswered

What are the questions that will occupy the coming era of deep earthquake research? It's somewhat dangerous to address this because we scientists are generally short-sighted about what will be discovered, over-optimistic about what will be accomplished, and over-invested in the particular topics on which we as individuals are most familiar. With that disclaimer I suggest we will see significant new progress on the following questions.

Under what conditions are deep earthquakes possible?

- What environmental conditions, i.e., stress, temperature and mineralogy, govern where deep earthquakes do and do not occur?[5]
- What geological conditions, e.g., subducted bathymetric features or the subduction of hydrated sediments, enhance or inhibit the occurrence of deep earthquakes?[6]
- What conditions permit large, isolated, very deep earthquakes to occur? That is, where else might we expect events like the 1954 Spanish and 1970 Columbia earthquake?
- Are some deep or intermediate-depth earthquakes unrelated to subduction?[7]

What is the physical mechanism of deep earthquakes?

- What is the physical mechanism (or mechanisms) of intermediate-depth earthquakes?[8]
- What is the mechanism (or mechanisms) of deep-focus earthquakes?
- Why are there no earthquakes beneath a depth of about 700 km?

Why are deep and shallow earthquakes so different? Why are they so similar?

- At various depths and in different environments, what limits the maximum size of deep earthquakes?

---

[5] At the time of writing enormous progress is being made on this question, both in the laboratory and from analysis of seismicity (see Section 7.3.1). One might have posed this question 20 years ago, but there is new reason to be optimistic that we will answer it.

[6] There are tantalizing results from South America suggesting that subduction of bathymetric features is important (see Fig. 4.5) but this is unconfirmed elsewhere, even in Japan where the seismicity data is of superior quality.

[7] This question may be ill-posed. Some deep earthquakes clearly do occur in regions where there is no active subduction ongoing at the surface. Where such intermediate earthquakes occur, often the crust and upper mantle is cool and the lithosphere is thickening or foundering. Is 'foundering' just another kind of subduction?

[8] There has been increasing evidence in support of dehydration embrittlement to explain earthquakes at depths down to about 200 km beneath several subduction zones. However, this doesn't necessarily mean that the same mechanism is at work elsewhere, particularly at depths exceeding 200 km.

- Why do deep earthquakes tend to have few aftershocks? Why are there virtually none between depths of 245 and 330 km?
- What is the relationship between the main- and aftershocks of deep earthquakes? Why are aftershocks often not clustered along mainshock nodal planes?
- Can we demonstrate other ways besides aftershocks for which shallow and deep earthquakes are categorically different?[9]

One reason I am optimistic that we will find answers for some of these questions is that deep earthquake researchers are only beginning to take full advantage of broad-band digital observations. Prior to about 1990 there simply wasn't very much digital data available; moreover, while seismologists were enthusiastic because digital data made it simpler to plot readable seismograms, we most often analyzed these seismograms using methods developed for analog data. Subsequently, however, seismologists have begun to evaluate suites of seismograms using digital filtering, stacking, and imaging methods similar to those used routinely for petroleum exploration. The result is that we now can resolve earth structure and relative hypocentral locations on a much finer scale than was possible previously. At the time of writing we still appear to be in an era when the amount of available digital data is growing exponentially.[10] I am hopeful that programs such as EarthScope will serve to maintain this growth for some time.

I am also optimistic about future contributions from laboratory rock mechanics experiments. Scientists at about several different institutions now have the capability to evaluate rock failure in macroscopic samples at pressures and temperatures corresponding to a depth of 700 km in the mantle. By analyzing quenched samples that have undergone or are about to undergo failure, laboratory researchers can take "snapshots" of the microscopic mineralogical changes that accompany failure. This places important constraints on speculations about the physical mechanism that permits deep earthquakes to occur. Fifty years ago, no reasonable physical explanation for deep earthquakes had been proposed. By fifteen years ago, there were at least three somewhat reasonable explanations, each with confident and articulate proponents. Now we are poised to begin elucidating and eliminating some of these proposed mechanisms.

9.1.4 Pitfalls

However, there are pitfalls that afflict the progress of science; as we attack outstanding questions about deep earthquakes it's worth remembering some of our

---

[9] Several papers suggest that shallow and deep earthquakes have different static stress drops, however, there is significant overlap in values and the systematic problems are significant. My sense is that we may ultimately confirm there are statistically meaningful differences at least locally, even if not globally.

[10] The volume of digital seismic data stored at the IRIS Data Management Center has grown from about 5 terabytes in 1997 to 55 terabytes in June 2004, consistent with a doubling time of only 2.2 years.

past mistakes. One pitfall is that some highly appealing hypotheses are simply wrong, or at least mostly wrong. Yet they are so appealing that scientists keep proposing and testing them, even though each time the data provides little or no support. An example is the hypothesis that deep earthquakes are implosions caused by polymorphic phase transitions. It was variously proposed (and/or rejected) in the 1930s by Gutenberg, by Leith and Sharp, and by Bridgman, in the 1940s by Bridgman (again), and in the 1970s by Evison, Dziewonski and Gilbert. It is possible this hypothesis will be partially vindicated if consensus emerges in support of the anti-crack mechanism; it is even possible that someone may demonstrate that a volume change of a few percent accompanies some deep earthquakes (although demonstrating this won't be easy). Nevertheless, considering the absence of observational support, this hypothesis' persistence is amazing.[11]

A second common pitfall occurs when we adhere to a hypothesis far longer than we should, even after contrary observational or theoretical evidence is put forth. Sometimes this pitfall is unavoidable, especially since scientists know that evidence can be wrong or afflicted by systematic errors. An example is the controversy between about 1935 and 1965 concerning whether earthquake sources were single or double couples (or both). In retrospect, it is clear that Japanese scientists such as Hirokichi Honda had the best data and the clearest understanding of the source; moreover, the level of physics to demonstrate that radiations from dislocations weren't single couples was well within the capabilities of physicists at this time. So, how could so many papers report single-couple sources (e.g., see Section 10.7.2); why did it take until about 1965 to resolve this question? No doubt we will fall prey to such pitfalls again.

A third pitfall is failure to use statistics to frame and test hypotheses. This is unfortunate, especially since the fundamental purpose of statistics is to help scientists test hypotheses.[12] Deep earthquake research has included several examples of this pitfall. One is the assertion that deep earthquake aftershocks cluster along nodal planes (see Section 5.2.4); this may ultimately be proven, but ordinary plots are misleading and so far statistical tests don't confirm it. Another example is the assertion that the occurrence times of large shallow and intermediate-depth earthquakes aren't independent (see Section 5.3.2). Again, this is possible and even plausible but so far hasn't been tested.

---

[11] A second example of an appealing but largely wrong idea is that solid earth tides control the occurrence times of earthquakes, and thus damaging earthquakes ought to be more likely when various celestial bodies are appropriately aligned. Like the phase-transition/implosion hypothesis, we ultimately may learn there is some basis for this (e.g., see Beeler and Lockner, 2003), but it is clear that in most situations the influence of tides on earthquakes is subtle, at most.

[12] The Scientific Method sounds straightforward enough: Observe; form a hypothesis; then, collect data to prove or disprove the hypothesis. The problem is: do the data support the hypothesis or not? That's why statistics got invented, and why it is unfortunate that we scientists don't use them more often.

The final pitfall is, of course, the misuse of statistics to prove hypotheses. Publications that misuse statistics are so prevalent that for the most part we simply ignore them; but, occasionally even experts blunder. For example, Gardner and Knopoff (1974) published a widely-referenced paper with a question for a title – "Is the sequence of earthquakes . . . with aftershocks removed, Poissonian?" This paper is celebrated partly because its entire abstract – a single word – is "Yes." Yet the authors and many who cited it apparently failed to realize that every sequence will come to resemble a random sequence or any other model of sequence, for that matter, if one selectively removes the most contradictory data points – in this case, aftershocks.

9.2 How we came to know what we think we know

In the twentieth and twenty-first centuries most scientists are paid professionals. Then, why do some choose to study deep earthquakes? And, what institutions encouraged this activity? More generally, what factors have controlled the development of our knowledge about deep earthquakes?

9.2.1 Publications and society

One plausible hypothesis is that innovative ideas and significant events control the growth of scientific knowledge. This implies that when an important paper appears or an unusually significant earthquake occurs, it will be followed, with a lag of perhaps two or three years, by a visible increase in the number of publications. Presumably these following papers represent the scientific community's response as scientists explore the implications of the influential paper or event. Thus, in a graph of publications vs time (Fig. 9.1) we should expect peaks or changes in slope that reflect these seminal papers or events.

What are the significant events in deep earthquake publication history? A few years following the appearance of Wadati's (1928) paper, there is a marked increase in the number of publications; this continues from about 1932 to 1939. There is a near-absence of publications between 1943 and 1950, presumably because the scientific world's attention focused on World War II. Some research activity resumes, beginning about 1952, but doesn't reach pre-war levels until 1964. During the postwar period the Cold War caused concern about the detection and monitoring of nuclear explosions; the U.S. response was to provide more funding for research on earthquakes and earth structure, and to participate in the installation of an extensive global seismograph network, the World Wide Standard Seismograph Network (WWSSN). This funding and the seismograms from the WWSSN provided encouragement for a growing number of seismologists, contributing to the steady increase in publication after 1964.

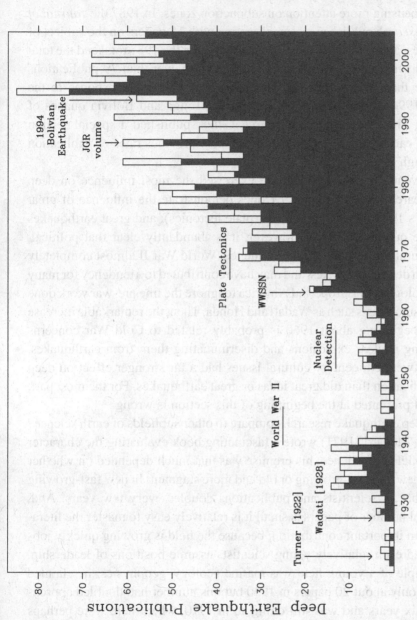

Fig. 9.1 Papers published since 1920 about deep earthquakes or closely related phenomena, or papers where one third or more of the events studied have reported focal depths exceeding 60 km. Solid bars indicate papers identified in a thorough search of 19 selected journals; the contemporary names of these journals are: *Bulletin of the Disaster Prevention Research Institute, Bulletin of the Earthquake Research Institute, Bulletin of the Seismological Society of America, Geophysical Journal International, Geophysical Magazine, Geophysical Research Letters, Gerlands Beiträge zur Geophysik, Journal of the Faculty of Science of Hokkaido University, Series 7 (Geophysics), Journal of Geophysics, Journal of Geophysical Research, Journal of Physics of the Earth, Journal of Seismology, New Zealand Journal of Geology and Geophysics, Physics of the Earth and Planetary Interiors, Pure and Applied Geophysics, Reviews of Geophysics, Tectonics, Tectonophysics, and Tohoku Geophysical Journal.* Open bars are papers found in other journals without a systematic search.

After 1964, Fig. 9.1 exhibits three other peaks for which we might assign a cause. There is a peak lasting about three years that begins in 1969; this coincides with the period when the majority of geoscientists began accepting plate tectonics, and were thus focusing more attention on subduction zones. In 1987 the *Journal of Geophysical Research* published a special section with 15 papers on the subject of deep earthquakes, almost exactly the difference between the 1987 total and the total in previous and subsequent years. Finally, in 1995 no fewer than 79 publications appeared, more than in any previous year. This was a specific response to the occurrence in 1994 of two very large quakes, the Tonga and Bolivia quakes of 9 March and 9 June. *Geophysical Research Letters* published a special section on the Bolivian earthquake in 1995, and higher-than-previous rates of publication continued through 1998.

So, which events or publications have exercised the most influence on deep earthquake research history? Figure 9.1 does demonstrate the influence of great papers (Wadati's 1928 paper), great ideas (plate tectonics), and great earthquakes (the remarkable quakes of 1994). However, it is abundantly clear that political, social, and cultural forces have greater influence. World War II almost completely stifled interest in deep earthquakes, and may have contributed to a tendency for many post-war seismologists in Europe and America to ignore the fine pre-war work done by Japanese seismologists such as Wadati and Honda. Then, the remarkable increase in publication beginning about 1960 is probably related to Cold War concerns about monitoring nuclear explosions and discriminating them from earthquakes. Clearly, in the twentieth century cultural issues had a far stronger effect on deep earthquake publication than did great ideas or great earthquakes. For the most part, the hypothesis I presented at the beginning of this section is wrong.

How does deep earthquake research compare to other subfields of earth science? Some time ago, Menard (1971) wrote a fascinating book evaluating the character of different subfields of science; his premise was that much depended on whether subfields were new and fast-growing or old and more stagnant. In new fast-growing fields the number of scientists and publications doubles every few years. And, because the total number of papers is small it is relatively easy to master the literature and make an important contribution; because the field is growing quickly jobs are available and even relatively young scientists assume positions of leadership. Menard's example of a young field was marine geology–geophysics; he claimed that there were only about 20 papers in 1930 but this number had doubled approximately every six years and was still doing so in 1970, when there were perhaps 3000 papers.

In contrast, in older subfields the number of scientists remains approximately constant or grows very slowly, the literature is huge and impossible to master, and leadership positions are difficult to attain, becoming available largely through

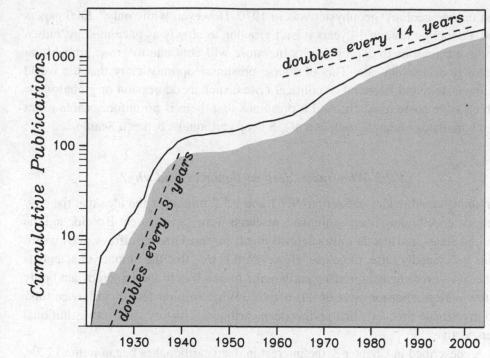

Fig. 9.2 Cumulative number of papers published since 1922. The solid line and grey fill show the two categories described in Fig. 9.1. Dashed lines show growth rates corresponding to doubling rates of three and 14 years, respectively.

retirements and attrition. Menard's example of an older field was vertebrate paleontology. Here, the first papers were published in the middle of the sixteenth century, and there were about 1000 papers by 1800. Between 1800 and 1850, when evolution was a hot topic, publications doubled every fifteen years; about 1870 doubling slowed to 35 years; by 1927 there were more than 50,000 papers and today there are many more. Clearly, nobody can read and master this much information. And, becoming an acknowledged leader or even just getting a job is especially difficult for paleontologists.

To assess the growth of deep earthquake research using Menard's approach I replot the data in Fig. 9.1 as a cumulative semilog plot, beginning with Turner's (1922) paper (Fig. 9.2). The growth rates are clearly different in different eras. Between 1922 and 1935 the total publications increased from two to about 50, doubling approximately every three years; this leveled off during the WWII period. Another period of growth began about 1965; however, the rate has reduced somewhat and since 1980 the doubling time is about 14 years.

Thus, the data suggest that deep earthquake research has matured, but isn't yet a truly old subfield. But it no longer is the truly young field that it was before 1935 and marine geology–geophysics was in 1970. However, with "only" 2000 papers and a doubling time of 14 years it isn't growing as slowly as paleontology, either. There are reasons to expect that its literature will continue to grow, probably at a slowly-decreasing rate. This of course presumes optimistically that the world manages to avoid financial or political crises, such as depression or global wars, which stifle basic research. And it presumes that there is no unforeseeable event that stimulates research, such as a M_W 8 deep earthquake beneath Seattle.

9.2.2 *Who studies deep earthquakes, and why?*

An uninformed reader of Sections 9.1.1 and 9.1.2 might get the idea that the only nations encouraging deep earthquake research were Japan, Great Britain, and the United States, and that the only scientists involved were Omori, Jeffreys, and myself. This is decidedly false, of course. However, it is true that the number of scientists who have ever engaged in deep earthquake research is in the hundreds, and only a few nations sponsor most of this activity. What cultural factors influence this? To investigate this, let's first review deep earthquake history from an institutional perspective.

As described in Chapter 3, the interest in deep earthquakes began in the 1920s, mostly in Japan and in Great Britain (Figs. 9.3 and 9.4).[13] In the 1930s there was broad interest in several countries and at various institutions. In Japan, Wadati and his Tokyo colleagues continued their research, and the Dutch began sending seismologists to their colony, Indonesia. In the United States, Gutenberg and the CalTech group were the most active, but there were also papers published by seismologists at Berkeley and St. Louis University, and by Bridgman at Harvard concerning rock mechanics.

Deep earthquake research virtually stopped during World War II; afterwards it began anew first in countries relatively less affected by the war (United States, Holland, etc.), and then after 1960 in Japan, Great Britain, Germany, and elsewhere. In Japan, deep earthquake research has been steady since, and now is well established at several institutions besides Tokyo University. Overall, since about 1970 the research has become much more broadly international; there are papers about deep earthquakes from institutions on every continent except Africa and Antarctica. These include institutions in Australia, Chile, Colombia, France,

---

[13] The scientists mentioned in Figs. 9.3, 9.4 and 9.5 generally have published several papers about deep earthquakes, Wadati–Benioff zones, or phenomena investigated using deep earthquake data. Some scientists are missing because I was unable to find the necessary information; e.g., Tetsuo Santo's bibliography in the *International Handbook of Earthquake and Engineering Seismology* fails to mention when he received his Ph. D., although it does note that he held Japan's national high jump record at 2.02 m from 1940 to 1958. Undoubtedly, I have inadvertently left out other fine scientists who should have been included. If this is you, please accept my apology, and send me some reprints.

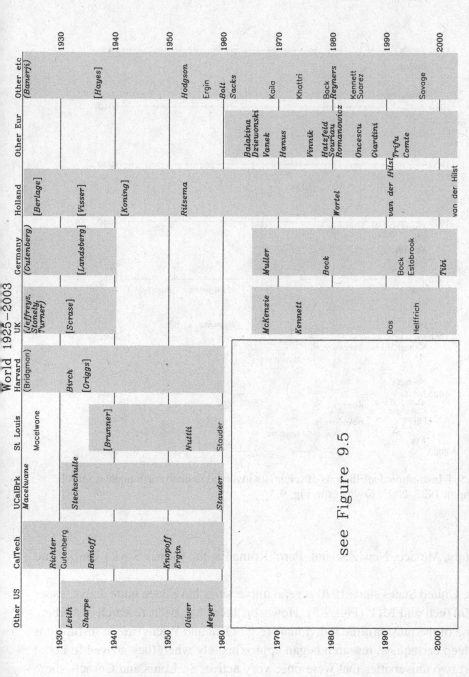

Fig. 9.3 Institutional affiliations of scientists involved in deep earthquake research in 1925–2003 (Japan excluded). Names in italics indicate institution and year of award for Ph.D. degree. Names without italics indicate year of arrival at other institutions employing individuals for significant periods (~10 years or more); names in brackets are individuals with no Ph.D. or for which information is uncertain. For each institution or country, grey shading indicates periods when there was some ongoing deep earthquake research. Abbreviations: Eur – Europe; UCalBrk – University of California at Berkeley; UK – United Kingdom.

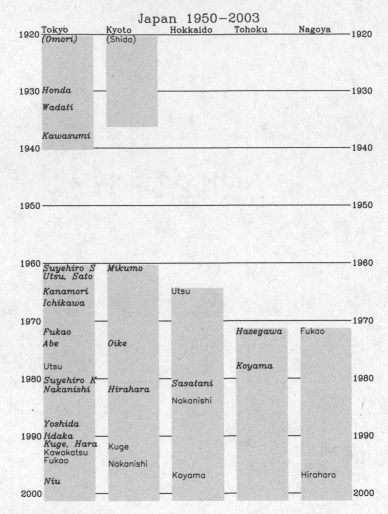

Fig. 9.4 Institutional affiliations of scientists involved in deep earthquake research in Japan 1923–2003. Key is as in Fig. 9.3.

India, Italy, Mexico, New Zealand, Peru, Romania, the former Soviet Union, and Spain.

In the United States since 1970 several universities have been quite active, especially CalTech and MIT (Fig. 9.5). However, there has been research elsewhere and some of the most prolific individuals (e.g., Okal and Wiens) are at institutions where deep earthquake research began approximately when they arrived in about 1980. At two universities that were once very active, St. Louis and Cornell, there is no longer much ongoing deep earthquake research. And, at Berkeley, Columbia and Harvard there is steady interest but deep earthquakes aren't a primary focus.

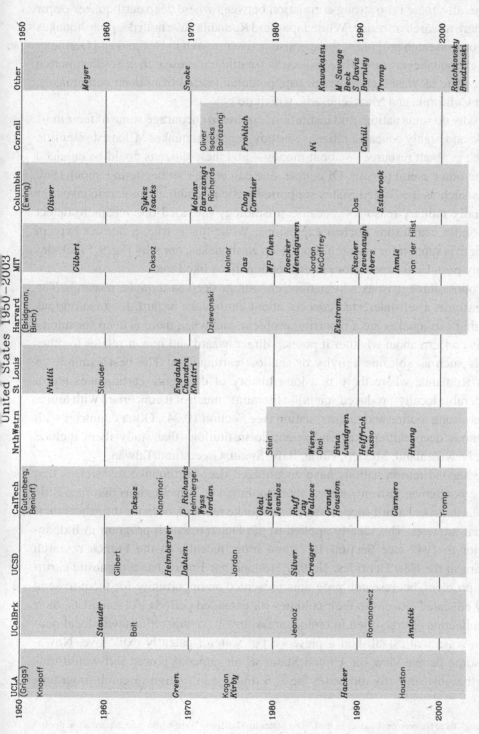

Fig. 9.5 Institutional affiliations of scientists involved in deep earthquake research in the United States 1950–2003. Key is as in Fig. 9.3. Additional abbreviations: MIT – Massachusetts Institute of Technology, NrthWstrn – Northwestern University, UCLA – University of California at Los Angeles, UCSD – University of California at San Diego.

Generally, there is no strong correlation between where deep earthquakes occur and where researchers reside. While Japan and Romania have both deep earthquakes and strong research programs, there is also active research in Holland and Germany where earthquakes are rare. There is less deep earthquake research in South America and the state of Washington, which face potential hazard from deep earthquakes, than in California and Massachusetts, which do not.

So, why do some nations and institutions choose to encourage some of their most capable and highly educated citizens to study deep earthquakes? Clearly, scientific research is about resources – about money – and these citizens could be engaged in some other useful pursuit. Of course, one can argue that beginning about 1960 nations such as the United States supported basic research about earthquakes for essentially military reasons – because seismology provided one means to detect and monitor clandestine nuclear explosions. While this is true, it doesn't explain why there is support for research about deep earthquakes, nor why Fig. 9.1 includes so many post-1960 papers from scientists in nations like Japan.

I propose three reasons why nations encourage deep earthquake research. The first reason is a self-interested concern about earthquake hazard. In some regions damaging earthquakes are a recurring problem, and when there is deep seismicity there is concern about whether it poses a direct hazard and how it relates to other hazards such as volcanic activity or shallow earthquakes. The best example of this is Romania, where there is a long history of damaging earthquakes and a considerable locally-produced scientific literature, much of it concerned with topics such as strong motion and microzonation (see Section 10.24). Other countries with indigenous deep earthquakes and scientific institutions that study them include Japan, New Zealand, Mexico, Chile, Italy, Spain, Greece and Taiwan.[14]

The second reason nations have encouraged deep earthquake research is that science is often an instrument of colonial policy; i.e., some nations that face little or no earthquake hazard support deep earthquake research because it furthers their imperial interests. This clearly applied to the Dutch research program in Indonesia prior to 1949 (see Section 10.7) and more recently to the French research program in the New Hebrides. Neither Holland nor France has a domestic earthquake problem, but research programs presented an opportunity for them to send highly educated citizens to their colonies for extended periods. As scientists, their data collection efforts, often in remote areas, involved interactions with local peoples in ways that established a presence but weren't patently exploitive. Nowadays some people view the United States as an imperial power and would thus argue that this similarly motivates the U.S. interest in foreign research programs.

[14] Although there are deep earthquakes in the United States in Alaska and Washington state, hazard alone doesn't explain why the U.S. engages in deep earthquake research.

National colonial interests may partially explain why scientists like Jeffreys in Great Britain and Gutenberg (originally) in Germany initially engaged in seismological research.

However, a third reason there is deep earthquake research is because basic research enhances national status. Thus nations like the United States, Great Britain, France, and Germany support basic research partly because they view themselves, and aspire to be viewed by others, as world leaders whose interests aren't or aren't simply imperialistic. This operates at the institutional level as well; i.e., universities that aspire to be recognized internationally as elite institutions will support research on topics like deep earthquake research. Examples of elite universities in geographic regions with minimal earthquake hazard but which nevertheless foster deep earthquake research include Columbia in the United States and the University of Cambridge in Great Britain.[15]

Of course, many nations and institutions support deep earthquake research for more than one of these reasons. For example, all three reasons apply to Japan. Japan has damaging earthquakes as well as indigenous deep earthquakes. In the first half of the twentieth century Japan had colonial aspirations; indeed, it was the 'Imperial' Earthquake Investigation Committee that published Omori's (1907) paper. And, there is copious evidence that Japan utilizes earthquake research to enhance its international status. Indeed, how else can one explain why Japanese seismologists have published much of their research in English since before 1900?[16] Finally, the reasons aren't distinct; e.g., activity that one person views as imperialism is considered benevolent leadership by another.

9.2.3 Deep earthquakes and the history of science

How does the century-long research effort that this book summarizes fit into the history of science? I have identified seven books that focus on the development of solid-earth geophysics in the twentieth century (see Table 9.1). In this section I briefly comment on how these books treat deep earthquake research.

I thank Jacqueline Henkel for stimulating discussions that much improved this section.

[15] When my own institution, The University of Texas at Austin, was created by constitutional edict in 1876, the edict stated that "the Legislature shall . . . organize [and] support . . . a university of the first class." Whether the University has attained elite status internationally is open to question; however, I couldn't have written *Deep Earthquakes* here if it didn't aspire to be elite.

[16] Even considering Japan's international aspirations, their long habit of foreign-language publication in seismology is truly remarkable and I don't fully understand it. My friend Junji Koyama suggests it is partly a historical accident that came about because nineteenth century Japanese seismologists worked closely with and were taught by John Milne and other British scientists. Subsequently, foreign language publication became a matter of personal pride for many university scientists, setting them apart from individuals who published only in Japanese. It appears that this has broad-based institutional and government support, as the scientific journals published by the Japanese Meteorological Agency and by Tokyo, Hokkaido, and Tohoku University all are largely in English.

Of course, each of the books has a somewhat different perspective. Hallam (1973) and Marvin (1973) appeared approximately when the majority of earth scientists first accepted plate tectonics. These books and Glen (1982), which came about a decade later, concentrate on summarizing the evidence supporting plate tectonics. Although all mention the philosopher Thomas Kuhn, the inventer/promoter of the concept of scientific paradigms (Kuhn, 1962), none of these books concentrate on philosophical issues. This contrasts with Le Grand (1988), Oreskes (1999) and especially Stewart (1990), who analyze the debate over continental drift to evaluate the cultural process by which scientific change occurs. Finally, Oldroyd (1996) is concerned with philosophical issues but considers a variety of topics other than continental drift, which occupies a single chapter.

From my perspective as a seismologist who has spent a career immersed in deep earthquake research, three features of these books are noteworthy. First, it is remarkable how little attention deep earthquakes receive. None of these books cite or mention Wadati's discovery of deep earthquakes (see Table 9.1) nor the prior controversies about the depth of "normal" earthquakes (see Sections 3.2 and 3.3); none of them mention the still-unresolved puzzle about the physical mechanism of deep quakes (see Chapter 7). Indeed, earthquake seismology doesn't receive much attention generally[17] except for a few topics; e.g., nearly all the books describe Sykes' (1967) research showing that earthquake mechanisms at plate boundaries are consistent with plate motions.

The second noteworthy feature of these books is that they tend to focus on continental drift and plate tectonics while largely ignoring progress in related scientific topics such as deep earthquakes. This may be partly attributable to the influence of Kuhn; i.e., even though some of the books' authors are critical of Kuhn's theories they still choose to concentrate on the pervasive intellectual changes that he calls paradigm shifts rather than describing the processes of accumulation, accretion, etc., which he describes as "normal science."

In this context I observe that the timing of the discovery of deep earthquakes disqualifies it as a paradigm shift or even a "microshift." That is, before Wadati's 1928 paper the depth of "normal" earthquakes was an open question (see Section 3.2). Thus, establishing that some were deep occurred at the same time as the realization that the majority of earthquakes were shallow. Similarly, the realization that brittle fracture couldn't explain the deeper ones occurred at nearly the same time that scientists like A. A. Griffiths and Percy Bridgman developed a basic understanding of rock failure. Thus, this again is not a shift; what is instead remarkable about the

---

[17] Many articles about earth science history also fail to mention earthquake seismology. For example, Doel (2003) states that in the post WWII era "a blizzard of military money descended on geophysics" and then goes on to describe how it affected research in physical oceanography, upper atmospheric science, geodesy, geomagnetism, and the polar regions. However, he doesn't mention earthquake seismology, even though he does note that detection of atomic testing led to "enhancing research programs in atmospheric acoustics."

Table 9.1. *Citations to key publications (first entry) and mention in the text of author's contribution on this topic (second entry) as recorded in books about contemporary Earth science history*

Idea Book	Discovery of deep earthquakes Cites Wadati (1928)	Discovery of inclined deep seismic zones (Wadati–Benioff zones) Cites Wadati (1935)	Mechanisms of shallow earthquake at different kinds of plate boundaries		Philosophy of science Cites Kuhn (1962)
			Cites Benioff (1954)	Cites Sykes (1967)	
Hallam (1973)[a]	No/No	No/No	No/Yes	No/Yes	Yes/Yes
Marvin (1973)[b]	No/No	No/No	Yes/Yes	Yes/Yes	Yes/Yes
Glen (1982)[c]	No/No	No/Yes	Yes/Yes	Yes/Yes	Yes/Yes
LeGrand (1988)[d]	No/No	No/No	No/No	No/No	Yes/Yes
Stewart (1990)[e]	No/No	No/No	Yes/Yes	Yes/Yes	Yes/Yes
Oldroyd (1996)[f]	No/No	Yes/Yes	Yes/Yes	No/No	Yes/Yes
Oreskes (1999)[g]	No/No	No/No	No/No	Yes/Yes	Yes/Yes

[a] The only article cited by Hallam concerning the properties of shallow earthquakes is Isacks *et al.* (1968). He calls inclined earthquake zones "Benioff zones".

[b] Marvin mentions H. H. Turner as the discoverer of deep earthquakes, and subsequently discusses only Gutenberg and Richter concerning the existence of and properties of deep earthquakes. He calls inclined earthquake zones "Benioff zones".

[c] Glen mentions inclined earthquake zones exactly twice, calling them "Benioff–Wadati zones". He mentions K. Wadati and Bruno (sic) Benioff exactly once. He mentions Thomas Kuhn only in footnotes.

[d] LeGrand does cite a 1968 article by Sykes on the same topic as Sykes' 1967 paper.

[e] Stewart calls inclined earthquake zones "Benioff zones".

[f] Oldroyd calls inclined earthquake zones "Benioff–Wadati zones".

[g] Oreskes doesn't mention Wadati and describes Gutenberg as "famous for his pioneering work on microseisms and deep earthquakes."

mechanism question is that it is still unresolved today even though it was posed so soon after the discovery of deep earthquakes and so soon after the development of the Griffith fracture model.

The third notable feature of these books is that they mostly ignore scientific activity outside of the United States, Great Britain, and Germany. Especially, why do these books devote so little attention to the Japanese, who have been at the forefront of deep earthquake research on almost every major question one can think of *except* for the acceptance of plate tectonics (see Section 10.11.2)? For example, all but Oldroyd's book fail to cite Wadati's (1935) paper describing inclined seismic zones. This is remarkable considering that Wadati's paper explicitly suggests inclined zones support Wegener's ideas on continental drift (see Section 3.4.2), and considering that Wadati's paper appeared well before Benioff (1954). None of the books describes these inclined zones as "Wadati zones" or even "Wadati–Benioff zones"; instead they are "Benioff zones" or "Benioff–Wadati zones".[18]

One reason that science historians may ignore Japanese scientists is that many influential earth scientists also have failed to credit Japanese research properly, and this oversight also afflicts standard reference materials. For example, Benioff (1954) does not cite Wadati, and Isacks *et al.* (1968) do not cite either his 1928 or 1935 paper. As noted previously, Harold Jeffreys credits his British colleague, H. H. Turner, for the discovery of deep earthquakes (Spall, 1980). The most recent edition of Arthur Holmes' classic geology text (Duff, 1993) discusses Benioff zones and mentions Benioff (1954) but never Wadati or his research. In a widely-referenced encyclopedia of the earth sciences (Good, 1998), Hirokichi Honda's name does not appear, Fusakichi Omori is mentioned only as a builder/designer of seismographs, and Kiyoo Wadati is mentioned only as the coauthor of a 1962 report on earthquake prediction. None are mentioned by the contributers to Oldroyd's (2002) volume on major contributions in twentieth century geology, even though the volume contains papers devoted to the work of Arthur Holmes and Marie Tharp, and a paper on the contributions of Russian geologists to the development of plate tectonics. Japanese scientists or scientific institutions aren't mentioned by either Brush and Gillmor (1995) or Doel (1997) in their reviews describing the development of twentieth century geophysics.

I thus suggest that there are gaps in the historians' treatment of twentieth century earth science that represent an opportunity; undoubtedly science historians will agree with me that there is still history to be written. Although the story of continental drift and plate tectonics is captivating, there are other earth science topics

---

[18] It is intriguing to compare history's treatment of Kiyoo Wadati with that of Gregor Mendel. Unlike Wadati, Mendel was relatively unknown to his contemporaries, he was not employed by a well-established scientific institution, nor did he publish in English or in an internationally distributed journal. Yet, nowadays we generally speak of "Mendelian" genetics rather than applying the names of the scientists who rediscovered his results. And Mendel's story, unlike Wadati's, has been discussed extensively by science historians.

which deserve a historian's attention, and many which would benefit from a more international perspective. One such topic is the story of how our understanding of deep earthquakes has developed, and how in different nations concerns about hazard, basic research, and security influenced this.

9.3 References

Beeler, N. M. and D. A. Lockner, 2003. Why earthquakes correlate weakly with the solid earth tides: effects of period stress on the rate and probability of earthquake occurrence, *J. Geophys. Res.*, **108**, doi:10.1029/JB001518, p. 2391.

Benioff, H., 1954. Orogenesis and deep crustal structure – additional evidence from seismology, *Geol. Soc. Amer. Bull.*, **65**, 385–400.

Brush, S. G. and C. S. Gillmor, 1995. Geophysics, in *Twentieth Century Physics, Vol. III*, eds. L. M. Brown, A. Pais and B. Pippard, New York, American Institute of Physics Press, 1943–2016.

Doel, R. E., 1997. The earth sciences and geophysics, in *Science in the Twentieth Century*, eds. J. Krige and D. Pestre, Amsterdam, The Netherlands, Harwood Academic Publishing, 391–416.

2003. Constituting the postwar earth sciences: the military's influence on the environmental sciences after 1945, *Social Studies of Science*, **33**, 635–666.

Duff, D., 1993. *Holmes' Principles of Physical Geology*, (4th edn), London, U. K., Chapman and Hall, 791 pp.

Gardner, J. K. and L. Knopoff, 1974. Is the sequence of earthquakes in southern California with aftershocks removed, Poissonian? *Bull. Seismol. Soc. Amer.*, **64**, 1363–1367.

Glen, W., 1982. *The Road to Jaramillo: Critical Years of the Revolution in Earth Science*, Stanford, CA, Stanford University Press, 459 pp.

Good, G. A. (ed.), 1998. *Sciences of the Earth; an Encyclopedia of Events, People, and Phenomena*, New York, Garland Publishers, 901 pp.

Hallam, A., 1973. *A Revolution in the Earth Sciences*, Oxford, U. K., Clarendon Press, 127 pp.

Isacks, B., J. Oliver, and L. R. Sykes, 1968. Seismology and the new global tectonics, *J. Geophys. Res.*, **73**, 5855–5899.

Ishimoto, M. and K. Iida, 1939. Observations of earthquakes registered with microseismographs constructed recently (in Japanese), *Bull. Earthquake Res. Inst., Tokyo Univ.*, **17**, 443–478.

Jeffreys, H., 1939a. Deep-focus earthquakes, *Ergebnisse Kosmischen Physik*, **4**, 75–105.
1939b. *Theory of Probability*, Oxford, U. K. Clarendon Press, 380 pp.

Jeffreys, H. and M. Shimshoni, 1987. On regional differences in seismology, *Geophys. J. Roy. Astron. Soc.*, **88**, 305–309.

Kuhn, T., 1962. *The Structure of Scientific Revolutions*, Chicago, University Chicago Press, 172 pp.

Le Grand, H. E., 1988. *Drifting Continents and Shifting Theories*, Cambridge, U. K., Cambridge University Press, 313 pp.

Marvin, U. B., 1973. *Continental Drift: the Evolution of a Concept*, Washington, DC, Smithsonian Institute Press, 239 pp.

• Menard, H. W., 1971. *Science: Growth and Change*, Cambridge, MA, Harvard University Press, 215 pp.

Oldroyd, D. R., 1996. *Thinking About the Earth: a History of Ideas in Geology*, Cambridge, MA, Harvard University Press, 410 pp.

(ed.), 2002. *The Earth Inside and Out: Some Major Contributions to Geology in the Twentieth Century*, London, U. K., Geological Society, Spec. Publ. 192, 369 pp.

Omori, F., 1907. Seismograms showing no preliminary tremor, *Bull. Imperial Earthquake Investigation Committee*, **1**, No. 3, 145–154.

1895. On the aftershocks of earthquakes, *J. College Sci. Imperial Univ., Tokyo*, **7**, 111–200.

Oreskes, N., 1999. *The Rejection of Continental Drift: Theory and Method in American Earth Science*, New York, NY, Oxford University Press, 420 pp.

Spall, H., 1980. Sir Harold Jeffreys, *Earthquake Information Bull.*, **12**, 48–53.

Stewart, J. A., 1990. *Drifting Continents and Colliding Paradigms: Perspectives on the Geoscience Revolution*, Bloomington, IN, Indiana University Press, 285 pp.

Sykes, L. R., 1967. Mechanism of earthquakes and the nature of faulting on the mid-ocean ridges, *J. Geophys. Res.*, **72**, 2131–2153.

Turner, H. H., 1922. On the arrival of earthquake waves at the antipodes, and on the measurement of the focal depth of an earthquake, *Mon. Not. R. Astron. Soc., Geophys.* Supp. **1**, 1–13.

Wadati, K., 1928. Shallow and deep earthquakes, *Geophys. Mag.*, **1**, 161–202.

1935. On the activity of deep-focus earthquakes in the Japan Islands and neighbourhoods, *Geophys. Mag.*, **8**, 305–325.

Part V

Geographic summary

10

A geographic summary of deep earthquakes

Introduction

This chapter summarizes the geographic distribution of the world's deep earthquakes by presenting information about known or alleged deep seismicity for each of 27 geographic regions (Fig. 10.1) as well as a catch-all "other" region (region 28) and the Earth's Moon (region 29). For each region it discusses the geometry of the Wadati–Benioff zone, describes the pattern of focal mechanisms, briefly reviews the literature, and provides information about "significant" earthquakes, i.e., unusually large or well-studied events. In most of the 29 regions the occurrence of deep earthquakes is undeniable. However, in a few regions (e.g., region 1 near the Hjort Trench) deep earthquakes have been reported and the regional tectonics suggest their occurrence is possible, but their existence is unconfirmed.

Catalogs

In this chapter I regularly compare locations reported in several different catalogs; to avoid repetitive citations I abbreviate these as follows.

– "GR" designates the Gutenberg and Richter (1954) catalog for earthquakes occurring between 1904 and 1952. Because GR applied uniform standards to determine epicenters, focal depths, and magnitudes, this has been the most widely used catalog for information about large earthquakes occurring prior to 1952. For deep earthquake magnitudes Gutenberg (1945) devised a scale depending on body waves that produced values that corresponded with M_S for shallow shocks. The 1954 version of this catalog was the final revision of several compilations that GR had published earlier (e.g., Gutenberg and Richter, 1938; 1939).

I thank Bob Engdahl for his comments on an earlier draft of Chapter 10.
I thank Ray Willemann for reviewing an earlier draft of this section.

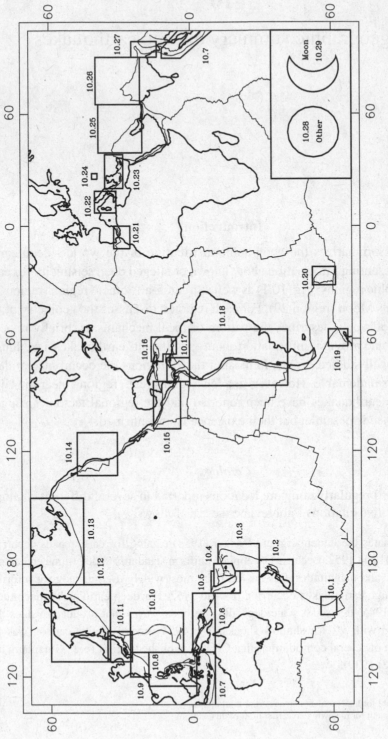

Fig. 10.1 World map summarizing the geographic regions discussed in Sections 10.1 to 10.29 of this chapter, including the 27 regions within the indicated polygons as well as "Other" and the Earth's Moon.

- "Rothé" designates the Rothé (1969) catalog for earthquakes occurring between 1953 and 1965. Rothé's specific intent was to update the GR catalog. However, the methods applied for assigning locations and magnitudes appear to be somewhat different; e.g., Rothé used magnitudes determined by Duda (1965) based on seismograms from Uppsala, Sweden. In many cases Duda's magnitudes are significantly different from those reported elsewhere.
- "ISC" designates the catalog of the International Seismological Centre for earthquakes occurring from 1964 to the present. During this period the ISC catalog reported locations for between 10,000 and 65,000 earthquakes annually. It attempts to be thorough rather than prompt, as it provides relocations using all phases that are available approximately one to two years after earthquakes occur. The ISC catalog reports alternative locations as well as the phase arrivals used to relocate events, and is thus often the best source to evaluate the reliability of questionable locations.
- "Harvard" or "CMT" designates the Harvard Centroid Moment Tensor (CMT) catalog for earthquakes occurring from 1977 to the present (Dziewonski and Woodhouse, 1983). Harvard routinely matches observed digital waveforms at periods of 45 s and greater to determine scalar moments and moment tensors for 500–1000 earthquakes each year. It thus provides focal mechanisms for most earthquakes having moment magnitudes M_W of about 5.5 or greater. The methods for determining CMT have evolved between 1977 and the present and differ somewhat for earthquakes with large and moderate magnitudes (see Dziewonski *et al.*, 1999 for a list of references).
- "CMT Historical" designates CMT for earthquakes occurring prior to 1977. The majority of these are from Huang *et al.* (1994; 1997), Huang and Okal (1998) and Chen *et al.* (2001), who compared available analog and synthetic waveforms and determined CMT for 215 large deep earthquakes occurring between 1907 and 1976.
- "Abe" designates the list of large earthquakes occurring between 1896 and 1980 for which Abe (1981; 1984) redetermined magnitudes m_B using long-period body waves. For deep earthquakes occurring prior to 1977 with magnitude of 7 and greater, Abe's magnitudes are one of the most consistent compilations available.
- "EHB" designates the catalog of selected ISC events occurring 1964–2004 and relocated using procedures described by Engdahl *et al.* (1998). The EHB procedures are designed to be sensitive to focal depth, and generally its hypocenters are more reliable than those in the ISC catalog. The EHB catalog includes some hypocenters that the location procedure has flagged for having questionable depths or locations; I have removed these and they do not appear in the maps or cross sections. The EHB catalog has undergone two major revisions since it first appeared, and thus there is more than one version; I obtained the version used in this book from E. R. Engdahl in March 2005.
- "EVC" or "EV Centennial" designates the composite catalog for the period 1900–1999 compiled by Engdahl and Villaseñor (2002). This incorporates data from all the above catalogs and presents new relocations for all events in the 1918–1942 and 1956–1963 periods and many of the largest events in the 1943–1952 period.

Key for maps and cross sections

For most of the regions I include maps and cross sections of earthquakes from the Harvard, EHB, and EV Centennial catalogs. Except where the captions state otherwise, the following key applies to all figures in this chapter:

– Filled circles are earthquakes in the Harvard catalog;[1]
– Open circles are earthquakes with magnitudes of 7.0 or greater in either the EV Centennial or CMT Historical catalogs;
– Crosses are earthquakes from the EHB catalog;
– Squares are other earthquakes mentioned in the text;
– Earthquakes labeled with dates have magnitudes of 7.5 or greater and/or are mentioned in the text.

In each map dashed lines indicate the plate convergence direction as determined from the poles of DeMets *et al.* (1994). On these maps thin lines delineate shorelines and the locations of plate boundaries as reported by Coffin *et al.* (1998); on some maps there are 100-km interval depth contours of the Wadati–Benioff zone as reported by Burbach and Frohlich (1986). Thicker lines labeled with letters indicate the location and orientation of cross sections; polygonal boundaries indicate the events included in each cross section and in the focal mechanism summary plots.

In geographic regions where deep earthquakes are very numerous, the plots include only selected events. When the captions for maps and cross sections specify "selected" earthquakes, for events in the Harvard catalog I use the criteria suggested by Frohlich and Davis (1999); i.e., "selected" earthquakes have mechanisms with a relative error (defined as the scalar moment of the error tensor divided by the scalar moment of the earthquake) of 0.15 or less, and a CLVD component f_{CLVD} of 0.20 or less (40% of the theoretical maximum). In the EHB catalog, "selected" earthquakes have locations determined using 20 or more P arrivals and an rms error of 1.0 s or less. All of the EHB locations plotted, selected and otherwise, have good depth control, i.e., they had an azimuthal gap of 180° or less and the catalog notes specify them as HEQ, DEQ, or FEQ locations.

Focal mechanisms In each geographic region equal-area plots summarize T, B, and P axis orientations for focal mechanisms in the Harvard CMT catalog. In each plot concentric small circles mark dips of 30° and 60°; in regions where the geometry of the Wadati–Benioff zone is approximately planar I plot a great circle (curved line) showing this plane. These equal-area plots summarize only Harvard CMT

---

[1] When a epicenter is in the EHB as well as the Harvard or EV Centennial catalogs, I generally plot it at the EHB location.

data; i.e., they do not include mechanisms determined from P-wave first motions or composites of local network observations.

Significant earthquakes

For "significant" earthquakes I include a paragraph summarizing why each is important and present a focal mechanism if a believable one is available. Most significant earthquakes are the principal subject of several journal articles; however, some are simply very large or otherwise interesting historically. My selection discriminates against large earthquakes that aren't discussed much in the literature; thus, readers who can't afford to miss such events would do well to review Okal (1992) and Huang and Okal (1998) who have reevaluated the magnitudes and moments of many historical deep earthquakes. All focal mechanisms presented as insets to the text are from the Harvard CMT and CMT Historical catalogs unless indicated otherwise.

Thermal parameter

For each geographic region I present estimates of the thermal parameter Φ, the product of t_{Age}, the lithosphere's age when subduction occurs, and $V_\perp$, the vertical component of the subduction velocity (see discussion in Section 2.4). Determining $V_\perp$ requires knowledge of the dip angle θ of the Wadati–Benioff zone, the obliquity ϕ of the convergence, and the overall convergence rate V (i.e., $V_\perp = V \cos\phi \sin\theta$; see Fig. 10.2). Moreover, Φ may be different for lithosphere subducted to different depths, since, except in unusual situations, the age of oceanic lithosphere that disappears at the trench doesn't remain constant. For simplicity, in this chapter I use the present-day value for t_{Age} to calculate Φ. This means that my estimates for Φ are highly approximate, especially since in a few arcs regional GPS measurements of V and ϕ are available and their values differ from those predicted by the DeMets *et al.* (1994) NUVEL-1 model. For inferring t_{Age} I use the isochrons of Müller *et al.* (2002).

References for the Introduction

• Abe, K., 1981. Magnitudes of large shallow earthquakes from 1904 to 1980, *Phys. Earth Planet. Int.*, **27**, 72–92.

  1984. Complements to "Magnitudes of large shallow earthquakes from 1904 to 1980", *Phys. Earth Planet. Int.*, **34**, 17–23.

Burbach, G. V. and C. Frohlich, 1986. Intermediate and deep seismicity and lateral structure of subducted lithosphere in the circum-Pacific region, *Rev. Geophys.*, **24**, 833–874.

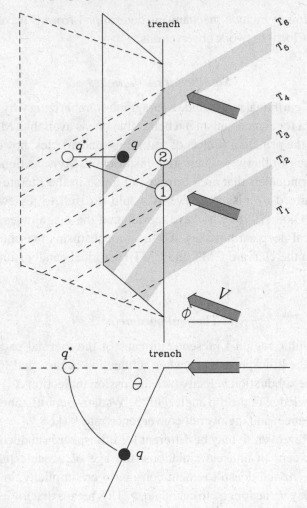

Fig. 10.2 How oblique subduction affects the determination of the thermal parameter Φ. If two plates converge at rate V and the obliquity ϕ is nonzero, the earthquake occurring at q (solid circle) passed through point 1 as it entered the trench; thus, $V_\perp$ is $V \cos\phi \sin\theta$. Note that the lithospheric age near the hypocenter corresponds to that of seafloor now between isochrons T_5 and T_6, rather than seafloor now at points 1 or 2.

• Chen, P.-F., M. Nettles, E. A. Okal, and G. Ekström, 2001. Centroid moment tensor solutions for intermediate-depth earthquakes of the WWSSN-HGLP era (1962–1975), *Phys. Earth Planet. Int.*, **124**, doi:10.1016/S0031-9201(00)00220-X, 1–7.

Coffin, M. F., L. M. Gahagan, and L. Lawver, 1998. *Present-Day Plate Boundary Digital Data Compilation*, University of Texas Institute for Geophysics Tech. Rept. No. 174, 5 pp.

DeMets, C., R. G. Gordon, D. F. Argus, and S. Stein, 1994. Effect of recent revisions to the geomagnetic reversal timescale on estimates of current plate motions, *Geophys. Res. Lett.*, **21**, doi:10.1029/94GL02118, 2191–2194.

Duda, S. J., 1965. Secular seismic energy release in the circum-Pacific belt, *Tectonophysics*, **2**, doi:10.1016/0040-1951(65)90035-1, 409–452.

• Dziewonski, A. M. and J. H. Woodhouse, 1983. Studies of the seismic source using normal-mode theory. In *Earthquakes: Observation, Theory and Interpretation*, eds. H. Kanamori and E. Boschi, pp. 45–137, New York, North-Holland.

Dziewonski, A. M., G. Ekström, and N. N. Maternovskaya, 1999. Centroid-moment tensor solutions for October–December, 1998, *Phys. Earth Planet. Int.*, **115**, doi:10.1016/S0031-9201(99)00062-X, 1–16.

• Engdahl, E. R. and A. Villaseñor, 2002. Global seismicity: 1900–1999. In *International Handbook of Earthquake and Engineering Seismology*, San Diego, CA, Academic Press for International Association of Seismology and Physics of the Earth's Interior, 665–690.

• Engdahl, E. R., R. van der Hilst, and R. Buland, 1998. Global teleseismic earthquake relocation with improved travel times and procedures for depth determination, *Bull. Seismol. Soc. Amer.*, **88**, 722–743.

Frohlich, C. and S. D. Davis, 1999. How well constrained are well-constrained T, B, and P axes in moment tensor catalogs? *J. Geophys. Res.*, **104**, doi:10.1029/1998JB900071, 4901–4910.

Gutenberg, B., 1945. Magnitude determination for deep-focus earthquakes, *Bull. Seismol. Soc. Amer.*, **35**, 117–130.

Gutenberg, B. and C. F. Richter, 1938. Depth and geographical distribution of deep-focus earthquakes, *Geol. Soc. Amer. Bull.*, **49**, 249–288.

   1939. Depth and geographical distribution of deep-focus earthquakes (second paper), *Geol. Soc. Amer. Bull.*, **50**, 1511–1528.

   • 1954. *Seismicity of the Earth and Associated Phenomena*, Princeton, NJ, Princeton University Press, 310 pp.

• Huang, W.-C. and E. A. Okal, 1998. Centroid moment tensor solutions for deep earthquakes predating the digital era: discussion and inferences, *Phys. Earth Planet. Int.*, **106**, doi:10.1016/S0031-9201(97)00111-8, 191–218.

Huang, W.-C., G. Ekström, E. A. Okal, and M. P. Salganik, 1994. Application of the CMT algorithm to analog recordings of deep earthquakes, *Phys. Earth Planet. Int.*, **83**, 283–297.

Huang, W.-C., E. A. Okal, G. Ekström, and M. P. Salganik, 1997. Centroid moment tensor solutions for deep earthquakes predating the digital era: the World-Wide Standardized Seismograph Network dataset (1962–1976), *Phys. Earth Planet. Int.*, **99**, doi:10.1016/S0031-9201(96)03177-9, 121–129.

   • 1998. Centroid moment tensor solutions for deep earthquakes predating the digital era: the historical dataset (1907–1961), *Phys. Earth Planet. Int.*, **106**, doi:10.1016/S0031-9201(98)00081-8, 181–190.

Müller, R. D., C. Gaina, W. Roest, S. Clark, and M. Sdrolias, 2002. The evolution of global oceanic crust from Jurassic to present day: a global data integration, *EOS, Trans. Amer. Geophys. Un.*, **83**, Supp., F700.

• Okal, E. A., 1992. Use of the mantle magnitude M_m for the reassessment of the moment of historical earthquakes. II. Intermediate and deep events, *Pure Appl. Geophys.*, **139**, 59–85.

Rothé, J. P., 1969. The *Seismicity of the Earth, 1953–1965*, Paris, France, United Nations Educational, Scientific and Cultural Organization (UNESCO), 336 pp.

10.1 Hjort Trench

10.1.1 Regional tectonics and deep seismicity

The Hjort Trench is an arc-shaped bathymetric low extending from about 56°S to 60°S along the Australia–Pacific plate boundary (see Fig. 10.3). It is a few hundred km north of the Antarctic–Australia–Pacific triple junction; the entire plate boundary from the triple junction to the Puysegur Trough along southwestern New Zealand is called the Macquarie Ridge Complex. Evidence of several kinds indicates that convergence occurs along the Hjort Trench (Massell *et al.*, 2000; Meckel *et al.*, 2003): there is the deeper-than-regular bathymetry of the trench itself, there is flexure evident in the Australian plate, multi-channel seismic lines reveal reverse faulted sediments in the trench, and Australia–Pacific plate motion vectors take on an increasing trench-perpendicular component.

Searches of the Abe, ISC, and other catalogs for this region turn up a handful of earthquakes with reported focal depths between 60 km and 160 km. Only two of these are well-recorded enough to assess the reliability of the depth. For the M_S 7.2 earthquake of 11 June 1970, both the ISC and Abe report a depth of 64 km. However, the EHB catalog reports a depth of 34 km, and Ruff *et al.* (1989) stated:

One curious feature of the ISC parameters . . . is the focal depth of 64 km, determined by reported times of pP-P. Looking at the seismograms, it seems rather unlikely that a pP phase could be reliably identified. We have used the [Rarotonga] P wave to perform the focal depth test as this station is relatively insensitive to small changes in the focal mechanism. It is immediately obvious that the focal depth is above 40 km, thus the ISC depth is incorrect. The simplicity criterion indicates that either 3 or 5 km is the best point source depth. Thus this is a shallow event, typical of strike-slip events in the Macquarie Ridge and elsewhere.

In the Hjort region, Ruff *et al.* (1989) and Frohlich *et al.* (1997) both find that reliably-determined double-couple focal mechanisms are nearly pure strike-slip, with little or no thrust component.

The ISC reported a focal depth of 160 km for an earthquake occurring on 13 March 1979 with m_b 4.8 and an epicenter on the Australian–Antarctic plate boundary, well away from the Hjort Trench. Thirty-one stations reported phases; however, 19 of these were PKP phases including six with residuals of 18 s or greater, and all but two of the remaining arrivals were from New Zealand and Australia within an azimuthal swath of 70°. No stations reported pP phases. This is not a compelling focal depth.

Since none of the reportedly deep Hjort earthquakes have well-constrained depths, the existence of deep activity here isn't established. However,

I thank Mike Coffin for reviewing an earlier draft of this section.

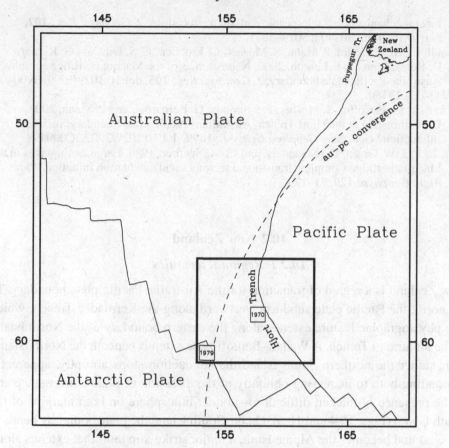

Fig. 10.3 Plate geometry near the Hjort Trench south of New Zealand. Near the Hjort Trench, Australia–Pacific plate motion (dashed line) has a significant plate-boundary-perpendicular component. Squares labeled with dates are earthquakes mentioned in the text.

seismotectonically this region is similar to Fiordland, New Zealand – in both regions large shallow earthquakes occur, there is oblique convergence, and a distinct trench. Since – unlike Fiordland – there haven't yet been local stations in the Hjort region, it is unlikely that deep activity will be confirmed there unless an earthquake occurs which is suitable for analysis with waveform modeling. If subduction does occur the thermal parameter Φ is quite small – less than 1000 km.

10.1.2 Hjort references

Frohlich, C., M. F. Coffin, C. Massell, P. Mann, C. L. Schuur, S. D. Davis, T. Jones, and G. Karner, 1997. Constraints on Macquarie Ridge tectonics provided by Harvard focal mechanisms and teleseismic earthquake locations, *J. Geophys. Res.*, **102**, doi:10.1029/96JB03408, 5029–5041.

Massell, C., M. F. Coffin, P. Mann, S. Mosher, C. Frohlich, C. S. Duncan, G. Karner, D. Ramsay, and J.-F. Lebrun, 2000. Neotectonics of the Macquarie Ridge complex, Australia–Pacific plate boundary, *J. Geophys. Res.*, **105**, doi:10.1029/1999JB900408, 13457–13480.

Meckel, T., M. F. Coffin, S. Mosher, P. Symonds, G. Bernardel, and P. Mann, 2003. Underthrusting at the Hjort Trench, Australian–Pacific plate boundary: incipient subduction? *Geochem. Geophys. Geosyst.*, **1099**, doi:10.1029/2002GC000498.

Ruff, L. J., J. W. Given, C. O. Sanders, and C. M. Sperber, 1989. Large earthquakes in the Macquarie Ridge Complex: transitional tectonics and subduction initiation, *Pure Appl. Geophys.*, **129**, 71–129.

10.2 New Zealand

10.2.1 Regional tectonics

New Zealand is a region of transition for the Australia–Pacific plate boundary. To the north, the Pacific plate subducts westward along the Kermadec Trench, which as a physiographic feature extends along the eastern boundary of the North Island as the Hikurangi Trench. A Wadati–Benioff zone extends beneath the North Island; then, under the northern South Island the subduction stops abruptly, apparently responding both to increasing obliquity of the direction of plate convergence and to the presence of thicker, difficult-to-subduct lithosphere on both margins of the South Island (Figs. 10.4 and 10.5). On the South Island the plate boundary bends to the west and becomes the Alpine Fault, a major strike-slip fault that extends along the western edge of the island. However at the southwestern end of the South Island in Fiordland there is another trench, the Puysegur Trough, and beneath Fiordland there are again intermediate-depth earthquakes; both of these features occur in response to ongoing highly oblique eastward subduction. South of New Zealand the plate boundary trends southward and becomes the Macquarie Ridge, which today is a predominantly strike-slip feature.

10.2.2 Intermediate-focus seismicity beneath the North Island

Beneath the North Island the intermediate-depth earthquake activity is an extension of that in Tonga–Kermadec, i.e., there is no distinct gap in the seismicity between the two regions. However, the occurrence rates for New Zealand earthquakes are much lower than in Tonga. And, over the past 25 years the maximum size of

I thank Martin Reyners and Martha Savage for reviewing an earlier draft of this section.

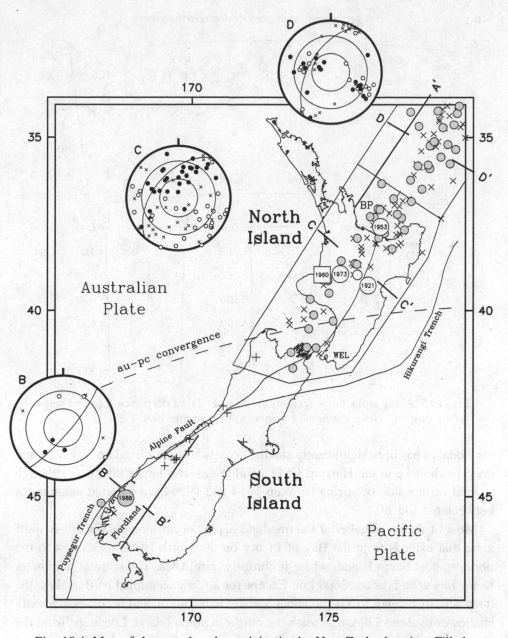

Fig. 10.4 Map of deep earthquake activity in the New Zealand region. Filled circles, open circles, and crosses indicate deep earthquakes in the Harvard CMT, EV Centennial, and EHB catalogs (see explanation in the introduction to this Chapter). The open square labeled '1960' is a sequence mentioned in the text, and the "+" symbols are small-magnitude deep events reported by Downes (1994) and Kohler and Eberhardt-Phillips (2003). Solid lines labeled A–A' to D–D' indicate the orientation of cross sections A through D; other lines indicate plate boundaries and surround events included in sections C and D. Equal-area plots summarize orientations of T (solid circles), B (X), and P (open circles) axes. The dashed line shows the direction of Australia–Pacific (au–pc) plate convergence. Abbreviations: 'BP' – Bay of Plenty; 'WEL' – Wellington.

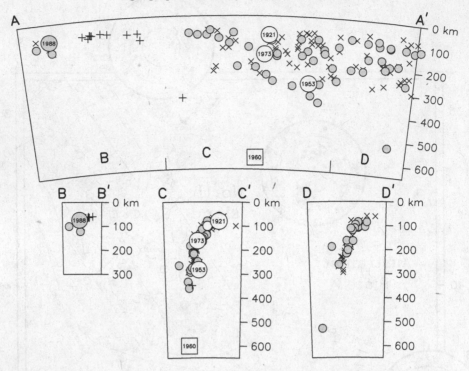

Fig. 10.5 Along-strike cross sections A–A' to D–D' of deep New Zealand seismicity (see Fig. 10.4). Earthquake symbols are as in Fig. 10.4.

earthquakes has been significantly smaller – outside of Fiordland no New Zealand deep earthquake in the Harvard CMT catalog has M_W larger than 6.1, although several earthquakes occurring between 1914 and 1976 have reported magnitudes between 6.4 and 7.0.

Most of the earthquakes at intermediate depths occur along a Wadati–Benioff zone that extends from the Bay of Plenty on the North Island southwest to the northernmost South Island, where it abruptly terminates. The thermal parameter Φ for this zone is about 3000 km. Except for a few exceptional earthquakes, the deep activity extends to a depth about 350 km in the north, and becomes gradually shallower to about 230 km beneath the northern South Island. Locations from the New Zealand National Seismograph Network indicate that the Wadati–Benioff zone extends at least 100 km southwest of region C in Fig. 10.4, and there are reports of occasional small-magnitude earthquakes with depths of 60–100 km beneath the central South Island (Reyners, 1987; Kohler and Eberhart-Phillips, 2003). There is debate about how and if subduction terminates beneath the South Island (see Reyners, 2005).

The dip of the Wadati–Benioff zone varies both laterally and vertically. At a depth of 100 km the apparent dip is about 55° beneath the Bay of Plenty, but becomes less steeply dipping beneath the south-central North Island. Locations determined using both the regional and global network data indicate the Wadati–Benioff zone steepens noticeably beneath about 125–150 km depth, and at greater depths it is nearly vertical. It is possible this steepening is a location artifact caused by the influence of readings from the New Zealand network (see McLaren and Frohlich, 1985 and Fig. 5.6) i.e., the steepening beneath 125 km is not evident in the relocations of either Adams and Ware (1977) or Spencer and Gubbins (1980), who both calculated travel-time residuals assuming the presence of a faster-than-normal-mantle inclined slab.

10.2.3 Deep-focus seismicity

Occasionally true deep-focus earthquakes do occur beneath the North Island. Most occur at depths of 400–500 km, as do similar earthquakes that occur sporadically further north in the Kermadec region.

However, in 1960 two earthquakes occurred with depths of about 600 km and magnitudes of about 6 (Figs. 10.4 and 10.5), and subsequently the New Zealand National Seismograph Network has recorded several small similar earthquakes; relocations indicate all lie within 30 km of one another at nearly the same depth. There has been debate about whether the subducting slab is continuous to this depth or whether instead the earthquakes occur in a detached slab fragment. In any case, the isolation and depth of these earthquakes makes them analogous to other notable isolated deep-focus earthquakes, such as the 1954 Spanish and 1970 Colombia events.

Finally, New Zealand seismologists report several small earthquakes located at 300–350 km beneath the northernmost South Island, well below the other intermediate-depth activity there. These earthquakes had magnitudes M_L of only 3.4–3.5, so it is possible the locations are in error; however, both Downes (1994) and Anderson and Webb (1994) believe the locations are accurate.

10.2.4 Fiordland intermediate-focus seismicity

Beneath Fiordland intermediate-depth earthquakes occur to depths of 150–180 km, with the highest activity between about 50 km and 140 km. The dip of the Wadati–Benioff zone varies from near-vertical in the north to about 60°–70° near the southwestern tip of the South Island. The thermal parameter Φ for Fiordland activity is about 1250 km.

Although three relatively large earthquakes with depths of 60–103 km occurred here in 1938, 1965, and 1988, Fiordland deep earthquakes with magnitudes of 5.0 or greater aren't common. Thus there are only a handful of Fiordland events in most global compilations; it is likely that the geometry and even the existence of this Wadati–Benioff zone would be in doubt if it were not for the presence of the New Zealand regional network.

10.2.5 Literature review

As New Zealand has had both seismic stations and seismologists throughout most of the twentieth century, the intermediate-depth earthquakes on the North Island drew scientific attention fairly early (e.g., see Hayes, 1936; 1943; Eiby, 1964). This attention increased when plate tectonics focused interest on island arc systems (e.g., Hamilton and Gale, 1968; 1969). Mooney (1970) noted that P-wave frequencies from these earthquakes were highly variable, and indicated that there was a highly attenuating region above the Wadati–Benioff zone. Both Adams and Ware (1977) and Spencer and Gubbins (1980) determined velocity and locations jointly for North Island earthquakes recorded by the New Zealand National Seismograph Network. They obtained the best results with an intermediate-depth subducted slab that had a velocity 8–11% faster than the overlying mantle, and found that this slab model permitted earthquakes deeper than about 100 km to remain in a planar, non-vertical Wadati–Benioff zone.

Subsequently, Reyners (1983) and Kuge and Satake (1987) argued that the subducting slab on the North Island is not intact, but instead consists of lateral segments separated by active faults; however, this isn't obvious in the locations of Anderson and Webb (1994). And Robinson (1986) presented locations and focal mechanisms for earthquakes beneath Wellington that indicate that there is a double seismic zone there at depths of 40–80 km.[2] There are various studies summarizing information about deep focal mechanisms beneath the North Island (Dziak and Wesnousky, 1990; Harris, 1982a; 1982b; Webb and Anderson, 1998; Reyners and Robertson, 2004). Unlike Tonga–Kermadec earthquakes, these mostly possess T axes in the plane of the Wadati–Benioff zone.

Because of New Zealand's regional network and the well-established seismological tradition, its Wadati–Benioff zone is as well studied as any other than Japan's. Reyners (1989) presents a map and cross section of 24 years of data for earthquakes with magnitude exceeding 4.0. Anderson and Webb (1994) present an even more thorough description of New Zealand seismicity, including maps and cross sections of selected events recorded over a three-year period by the New Zealand Network after it was upgraded in the late 1980s. These investigations clearly show the geometry, extent, and variations in activity rate for deep earthquakes beneath

---

[2] Martha Savage and Martin Reyners indicate that unpublished studies show there is a double zone above about 150 km beneath much of the North Island, although, as in Japan, activity in the lower zone varies markedly along strike.

the North Island, and the cessation of this activity beneath the northern South Island (see Kohler and Eberhart-Phillips, 2003; Reyners and Robertson, 2004). Audoine *et al.* (2000; 2004), Eberhart-Phillips and Reyners (1997) and Eberhart-Phillips and Henderson (2004) use regional network data to investigate shear-wave anisotropy and velocity structure using tomographic methods. Ansell and Bannister (1996) present depth contours for the geometry of the Wadati–Benioff zone on the North Island at shallow depths – above 60 km; Du *et al.* (2004) perform relocations for seismicity beneath Wellington, where there is a well-developed double seismic zone.

Adams (1963) was the first to call attention to the isolated deep-focus hypocenters with reported depths of 570–612 km beneath the North Island. Subsequently, Adams and Ferris (1976) and Boddington *et al.* (2004) have studied seven similar earthquakes having magnitudes between 3.8 and 4.9. At regional stations these exhibit a distinct phase arriving between P and S, interpreted as a P-S or S-P conversion that occurs near the bottom of the intermediate-depth seismicity at about 200 km depth. These authors and Okal (2001), who evaluated mantle Q for these deep-focus earthquakes, conclude that they occur within a slab fragment detached from the slab at intermediate depths. However, Barazangi *et al.* (1973) found high-frequency S waves arriving at the Wellington (WEL) station from the 1960 events and concluded that the slab is continuous. And Martin Reyners (personal communication) states that recent unpublished tomographic images demonstrate conclusively that the subducted slab is vertical at 300 km depth and connects directly to the activity at 600 km depth.

Because it is relatively remote, the Fiordland region has enjoyed poorer station coverage than the remainder of New Zealand. Although Hayes (1936) located the shock of 8 December 1934 at 45.4° S 166.4° E and stated it was "felt throughout southern New Zealand" and "deep," Fiordland intermediate seismicity wasn't investigated thoroughly prior to the studies of Hamilton and Evison (1967) and Smith (1971). Davey and Smith (1983) concluded that Fiordland subduction has been ongoing for about 7 Ma, and suggested that the intermediate-depth seismicity arises from a fragment of the Australian plate subducted at the Puysegur Trough and subsequently moved northwestward by the transcurrent motion between the two plates. Smith and Davey (1984) relocated 13 years of New Zealand network data, and found that the deepest Fiordland earthquakes are about 150 km deep and occur within a zone that extends laterally about 200 km along New Zealand's southwest coast. In 1993 there was a 24-station temporary network deployed for three months in Fiordland. Eberhart-Phillips and Reyners (2001) applied tomographic methods to 311 well-recorded earthquakes in this dataset, and concluded that the subducted lithosphere is "twisted" as it undergoes northwestward subduction along New Zealand's southwest coast, producing the Wadati–Benioff zone's change in dip from about 60° to near vertical. Reyners *et al.* (2002) used these data to determine focal mechanisms and reported that the deeper events were mostly thrusts, i.e., with approximately downdip T axes.

Finally, there is persistent concern about seismic hazard in New Zealand and in this context there is a tradition of statistical analyses of the regional network catalog. For example, several investigators have noted differences between shallow and deep earthquakes with respect to the spatial and temporal clustering of fore- and aftershocks (Vere Jones *et al.*, 1964; Savage and Rupp, 2000; Stock and Smith, 2002). Also, Harte and Vere-Jones (1999) have evaluated systematic differences between locations determined by the New Zealand network and those in the PDE catalog.

10.2.6 Significant earthquakes

There are no New Zealand deep earthquakes known with magnitudes larger than about 7.2.

29 September 1953: $h_{EVC} = 312$ km; $M_{W(CMT-Hist)} = 7.2$. This earthquake, with m_B of 7.0, is the largest deep New Zealand event in the Abe catalog. Huang and Okal (1998) report that it is the largest historical earthquake worldwide in the depth range 275–325 km; the inset shows the focal mechanism they determined.

23 March 1960: $h_{Adams} = 607$ and 612 km; $M_L = 6.2$. Adams (1963) discusses the occurrence of two deep-focus earthquakes occurring $4\frac{1}{2}$ minutes apart on this date, both with similar magnitudes and focal depths. Both earthquakes had nearly the same epicenter and were comparable in size to two intermediate-depth earthquakes that occurred one minute apart four days later. At New Zealand stations the first motions for the intermediate- and deep-focus earthquakes had opposite polarities, indicating that they had different focal mechanisms; however, apparently no-one has prepared focal mechanisms for the deep-focus events. The reported magnitudes for all these earthquakes may be overestimated – Rothé assigned both deep-focus events magnitudes of 5.5, and both Gibowicz (1972) and Anderson and Webb (1994) suggest that the New Zealand network systematically overestimates magnitudes for local deep earthquakes, with ($M_L - m_b$) typically being about 0.9–1.0.

5 January 1973: $h_{CMT-Hist} = 165$ km; $M_{W(CMT-Hist)} = 6.6$. Harris (1982b) determined a focal mechanism for this earthquake, which occurred beneath the south-central part of the North Island, causing minor damage. He considered it unusual because in the 25 days after it occurred it generated seven aftershocks locatable by the New Zealand network. Its location is just downdip of the earthquake of 28 June 1921 evaluated by Doser and Webb (2003), who assigned an M_W of 6.75 and a depth of 70–80 km, and also the shallow-focus Hawke's Bay earthquake of 2 February 1931 ($M_{S(Abe)} = 7.8$), which reportedly killed about 250 people.

3 June 1988: $h_{CMT} = 62$ km; $M_{W(CMT)} = 6.7$. Reyners *et al.* (1991) investigated this Fiordland event, which is the largest New Zealand deep earthquake in the Harvard catalog. It produced intensities of MMI V and greater throughout much of the South Island; in Fiordland it caused rockfalls and minor landslides, one which blocked a major tourist highway between Te Anau and Milford Sound. It possessed what appear to be more than 100 locatable aftershocks that formed a cigar-shaped zone extending between about 40–80 km depth. The Harvard focal mechanism (see inset) had a near-vertical nodal plane that was perpendicular to the plane of the Wadati–Benioff zone. For these reasons Reyners *et al.* (1991) concluded that the earthquake represented a tear in the subducting slab. In support of this conclusion they observed that the *b* value in the New Zealand catalog was 0.91 for earthquakes near this event, compared to 1.29 for deep earthquakes in the adjacent region to the southwest.

This 1988 earthquake may be typical of other past earthquakes that had nearly the same epicenter, including some that may have had even larger magnitudes. Abe's catalog reports an earthquake on 16 December 1938 with a depth of 60 km and m_B of 6.9; Doser *et al.* (1999) concluded that its epicenter and focal mechanisms were similar to the 1988 earthquake, but gave the 1938 earthquake a magnitude M_S of 7.1 and a shallower focal depth of 47 km. Reyners *et al.* (1991) also mention earthquakes with magnitudes M_L between 6 and 7 and depths of 50–80 km occurring on 10 February 1939, 17 February 1943 and 2 August 1943. Hamilton and Evison (1967) discuss another with a magnitude M_L of 6.2 and a reported depth of 103 km occurring on 20 May 1965.

10.2.7 New Zealand references

Adams, R. D., 1963. Source characteristics of some deep New Zealand earthquakes, *New Zealand J. Geol. Geophys.*, **6**, 209–220.

Adams, R. D. and B. G. Ferris, 1976. A further earthquake at exceptional depth beneath New Zealand, *New Zealand J. Geol. Geophys.*, **19**, 269–273.

Adams, R. D. and D. E. Ware, 1977. Subcrustal earthquakes beneath New Zealand; locations determined with a laterally inhomogeneous velocity model, *New Zealand J. Geol. Geophys.*, **20**, 59–83.

• Anderson, H. and T. Webb, 1994. New Zealand seismicity: patterns revealed by the upgraded National Seismograph Network, *New Zealand J. Geol. Geophys.*, **37**, 477–493.

Ansell, J. H. and S. C. Bannister, 1996. Shallow morphology of the subducted Pacific plate along the Hikurangi margin, New Zealand, *Phys. Earth Planet. Int.*, **93**, doi:10.1016/0031-9201(95)03085-9, 3–20.

Audoine, E., M. K. Savage, and K. Gledhill, 2000. Seismic anisotropy from local earthquakes in the transition region from a subduction to a strike-slip plate boundary, New Zealand, *J. Geophys. Res.*, **105**, doi:10.1029/1999JB900444, 8013–8033.

    2004. Anisotropic structure under a back arc spreading region, the Taupo volcanic zone, New Zealand, *J. Geophys. Res.*, **109**, doi:10.1029/2003JB002932, B11305.

Barazangi, M., B. L. Isacks, J. Oliver, J. Dubois, and G. Pascal, 1973. Descent of lithosphere beneath New Hebrides, Tonga–Fiji and New Zealand: evidence for detached slabs, *Nature*, **242**, 98–101.

Boddington, T., C. J. Parkin, and D. Gubbins, 2004. Isolated deep earthquakes beneath the North Island of New Zealand, *Geophys. J. Int.*, **158**, doi:10.1111/j.1365-246X.2004.02340.x, 972.

Davey, F. J. and E. G. C. Smith, 1983. The tectonic setting of the Fiordland region, south-west New Zealand, *Geophys. J. Roy. Astron. Soc.*, **72**, 23–38.

Doser, D. I. and T. H. Webb, 2003. Source parameters of large historical (1917–1961) earthquakes, North Island, New Zealand, *Geophys. J. Int.*, **152**, doi:10.1046/j.1365-246X.2003.01895.x, 795–832.

Doser, D. I., T. H. Webb, and D. E. Maunder, 1999. Source parameters of large historical (1918–1962) earthquakes, South Island, New Zealand, *Geophys. J. Int.*, **139**, doi:10.1046/j.1365-246x.1999.00986.x, 769–794.

Downes, G., 1994. Exceptionally deep earthquakes in the northern South Island, *New Zealand J. Geol. Geophys.*, **37**, 127–130.

Du, W.-X., C. H. Thurber, M. Reyners, D. Eberhart-Phillips, and H. Zhang, 2004. New constraints on seismicity in the Wellington region of New Zealand from relocated earthquake hypocenters, *Geophys. J. Int.*, **158**, doi:10.1111/j.1365-246X.2004.02366.x, 1088.

Dziak, R. P. and S. G. Wesnousky, 1990. Body-waveforms and source parameters of some moderate-sized earthquakes near North Island, New Zealand, *Tectonophysics*, **180**, doi:10.1016/0040-1951(90)90313-W, 273–286.

Eberhart-Phillips, D. and C. M. Henderson, 2004. Including anisotropy in 3-D velocity inversion and application to Marborough, New Zealand, *Geophys. J. Int.*, **156**, doi:10.1011/j.1365-246x.2003.02044.x, 237–254.

Eberhart-Phillips, D. and M. Reyners, 1997. Continental subduction and three-dimensional crustal structure: the northern South Island, New Zealand, *J. Geophys. Res.*, **102**, doi:10.1029/96JB03555, 11843–11861.

• 2001. A complex, young subduction zone imaged by three-dimensional seismic velocity: Fiordland, New Zealand, *Geophys. J. Int.*, **146**, doi:10.1046/j.0956-540x.2001.01485.x, 731–746.

Eiby, G. A., 1964. The New Zealand subcrustal rift, *New Zealand J. Geol. Geophys.*, **7**, 109–133.

Gibowicz, S. J., 1972. The relationship between teleseismic body-wave magnitude m and local magnitude M_L from New Zealand earthquakes, *Bull. Seismol. Soc. Amer.*, **62**, 1–11.

Hamilton, R. M. and F. F. Evison, 1967. Earthquakes at intermediate depths in south-west New Zealand, *New Zealand J. Geol. Geophys.*, **10**, 1319–1329.

Hamilton, R. M. and A. W. Gale, 1968. Seismicity and structure of North Island, New Zealand, *J. Geophys. Res.*, **73**, 3859–3876.

1969. Thickness of the mantle seismic zone beneath the North Island of New Zealand, *J. Geophys. Res.*, **74**, 1608–1613.

Harris, F., 1982a. Focal mechanisms of subcrustal earthquakes in the North Island, New Zealand, *New Zealand J. Geol. Geophys.*, **25**, 325–334.

1982b. Focal mechanism of an intermediate-depth earthquake and its aftershocks, *New Zealand J. Geol. Geophys.*, **25**, 109–113.

Harte, D. and D. Vere-Jones, 1999. Differences in coverage between the PDE and New Zealand local earthquake catalogues, *New Zealand J. Geol. Geophys.*, **42**, 237–253.

Hayes, R. C., 1936. Normal and deep earthquakes in the south-west Pacific, *New Zealand J. Sci. Tech.*, **17**, 691–701.

1943.The subcrustal structure in the New Zealand region from seismic data, *Bull. Seismol. Soc. Amer.*, **33**, 75–79.

Huang, W.-C. and E. A. Okal, 1998. Centroid moment tensor solutions for deep earthquakes predating the digital era: discussion and inferences, *Phys. Earth Planet. Int.*, **106**, doi:10.1016/S0031-9201(97)00111-8, 191–218.

Kohler, M. D. and D. Eberhart-Phillips, 2003. Intermediate-depth earthquakes in a region of continental convergence: South Island, New Zealand, *Bull. Seismol. Soc. Amer.*, **93**, 85–93.

Kuge, K., and K. Satake, 1987. Lateral segmentation within the subducting lithosphere: three dimensional structure beneath the North Island, New Zealand, *Tectonophysics*, **139**, doi:10.1016/0040-1951(87)90098-9, 223–237.

McLaren, J. P. and C. Frohlich, 1985. Model calculations of regional network locations for earthquakes in subduction zones, *Bull. Seismol. Soc. Amer.*, **75**, 397–413.

Mooney, H. M., 1970. Upper mantle inhomogeneity beneath New Zealand: seismic evidence, *J. Geophys. Res.*, **75**, 285–309.

Okal, E. A., 2001. 'Detached' deep earthquakes: are they really? *Phys. Earth Planet. Int.*, **127**, doi:10.016/S0031-9201(01)00224-2, 109–143.

Reyners, M., 1983. Lateral segmentation of the subducted plate at the Hikurangi margin, New Zealand: seismological evidence, *Tectonophysics*, **96**, doi:10.1016/0040-1951(83)90218-4, 203–223.

1987. Subcrustal earthquakes in the central South Island, New Zealand, and the root of the southern Alps, *Geology*, **15**, doi:10.1130/0091-7613(1987)15, 1168–1171.

• 1989. New Zealand seismicity 1964–1987: an interpretation, *New Zealand J. Geol. Geophys.*, **32**, 307–315.

2005. The 1943 Lake Hawea earthquake – a large subcrustal event beneath the Southern Alps of New Zealand, *New Zealand J. Geol. Geophys.*, **48**, 147–152.

Reyners, M. and E. J. Robertson, 2004. Intermediate depth earthquakes beneath Nelson, New Zealand, and the southwestern termination of the subducted Pacific plate, *Geophys. Res. Lett.*, **31**, doi:10.1029/2003GL019201.

Reyners, M., K. Gledhill, and D. Waters, 1991. Tearing of the subducted Australian plate during the Te Anau, New Zealand earthquake of 1988 June 3, *Geophys. J. Int.*, **104**, 105–115.

Reyners, M., R. Robinson, A. Pancha, and P. McGinty, 2002. Stresses and strains in a twisted subduction zone – Fiordland, New Zealand, *Geophys. J. Int.*, **148**, doi:10.1046/j.1365-246x.2002.01611.x, 637–648.

Robinson, R., 1986. Seismicity, structure and tectonics of the Wellington region, New Zealand, *Geophys. J. Roy. Astron. Soc.*, **87**, 379–409.

Savage, M. K. and S. H. Rupp, 2000. Foreshock probabilities in New Zealand, *New Zealand J. Geol. Geophys.*, **43**, 461–469.

Smith, W. D., 1971. Earthquakes at shallow and intermediate depths in Fiordland, New Zealand, *J. Geophys. Res.*, **76**, 4901–4907.

Smith, E. G. C. and F. J. Davey, 1984. Joint hypocentre determination of intermediate depth earthquakes in Fiordland, New Zealand, *Tectonophysics*, **104**, doi:10.1016/0040-1951(84)90106-9, 127–144.

Spencer, C. and D. Gubbins, 1980. Travel-time inversion for simultaneous earthquake location and velocity structure determination in laterally varying media, *Geophys. J. Roy. Astron. Soc.*, **63**, 95–116.

Stock, C. and E. C. G. Smith, 2002. Adaptive kernel estimation and continuous probability representation of historical earthquake catalogs, *Bull. Seismol. Soc. Amer.*, **92**, 904–912.

Vere-Jones, D., S. Turnovsky, and G. A. Eiby, 1964. A statistical survey of earthquakes in the main seismic region of New Zealand; Part 1 – Time trends in the pattern of recorded activity, *New Zealand J. Geol. Geophys.*, **7**, 722–744.

Webb, T. H. and H. Anderson, 1998. Focal mechanisms of large earthquakes in the North Island of New Zealand: slip partitioning at an oblique active margin, *Geophys. J. Int.*, **134**, doi:10.1046/j.1365-246x.1998.00531.x, 40–86.

10.3 Tonga–Kermadec

10.3.1 Regional tectonics

More deep earthquakes ($h \geq 60$ km) occur within the Tonga–Kermadec region (Fig. 10.6) than in any other geographic region; moreover, there are more deep-focus ($h \geq 300$ km) earthquakes here than in all the remaining regions combined. The apparent cause of this remarkable activity is the subduction of the 70–100 Ma-old Pacific plate beneath the Fiji Plateau. The convergence rate estimated from GPS measurements is 16–24 cm/yr (Pelletier *et al.*, 1998) but only about 7 cm/yr in the NUVEL plate model; the discrepancy is due to vigorous spreading in the back arc basin (see Zellmer and Taylor, 2001). In the southern part of this region the convergence is more oblique and the rate is lower, or about 5 cm/yr. Thus,

I thank Doug Wiens for reviewing an earlier draft of this section.

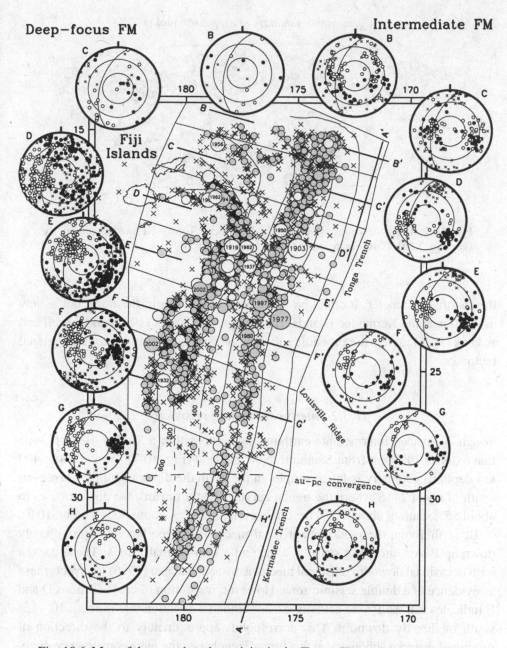

Fig. 10.6 Map of deep earthquake activity in the Tonga–Kermadec region. Filled circles, open circles, and crosses indicate selected deep earthquakes in the Harvard CMT, EV Centennial, and EHB catalogs (see description of selection process in the introduction to this chapter). Solid lines labeled A–A′ to H–H′ indicate orientations of cross sections A to H; lines show events included in each section; equal-area plots summarize T (solid circles), B (crosses), and P (open circles) axes orientations of selected Harvard CMT for intermediate-depth (right side) and deep-focus (left side) earthquakes. Other lines show locations of the Tonga and Kermadec trenches and 100-km depth contours for the Wadati–Benioff zone. The dashed line indicates the direction of Australia–Pacific (au–pc) plate convergence.

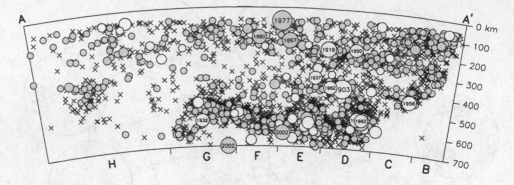

Fig. 10.7 Along-strike cross section A–A′ of deep Tonga–Kermadec seismicity (see Fig. 10.6). Earthquake symbols are as in Fig. 10.6.

the thermal parameter Φ could be as high as 12,000 km in the north but as low as 4000 near the Kermadec Islands. Hamburger and Isacks (1987) and Ansell and Adams (1986) review the probable subduction history and summarize the regional tectonics.

10.3.2 Intermediate-focus seismicity

Tonga–Kermadec intermediate earthquakes occur within a Wadati–Benioff zone that extends 2000 km from Samoa to New Zealand (Fig. 10.7). Above about 400 km depth its geometry is fairly simple: in the north the dip is about 50°; however, south of about 24° S where the trench curves slightly inward, the dip increases to about 55°; south of about 30° S the dip is even greater, or about 60° (see Fig. 10.8).

Beneath Tonga the predominant focal mechanism type has an approximately downdip P axis, an along-strike B axis, and a slab-normal P axis. These coexist with occasional downdip tensional mechanisms; Kawakatsu (1986) interprets these as evidence of a double seismic zone. However, inspection of cross sections D and E indicates that the P axes actually cluster around a direction that is about 10°–15° south of directly downdip. This corresponds approximately to the direction of motion of material subducted at the Tonga Trench; i.e., the plate convergence direction is approximately east–west but the strike of the trench is more nearly 16° east of north, so that subducting material moves downward along a path somewhat south of downdip.

Beneath Kermadec the focal mechanism pattern is more complicated. Mechanisms with approximately downdip P axes are again the most common type (see Fig. 10.6, sections G and H). However, here the T and B axes are not strongly clustered.

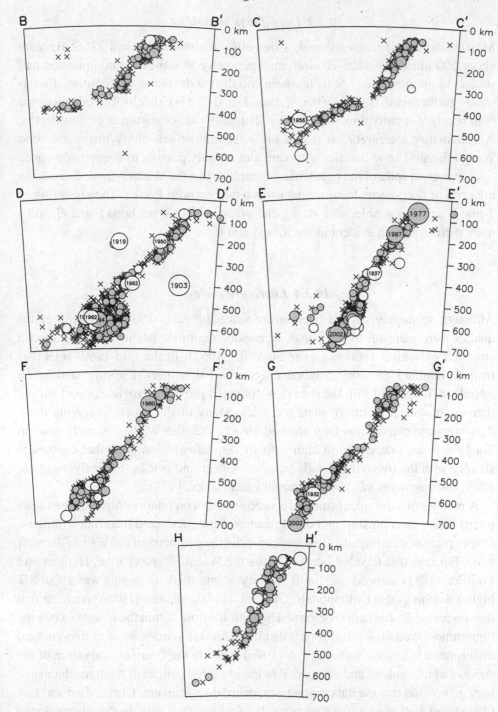

Fig. 10.8 Trench-normal cross sections B–B′ to H–H′ of deep Tonga–Kermadec seismicity (see Fig. 10.6). Earthquake symbols are as in Fig. 10.6.

10.3.3 Deep-focus seismicity

Most of the deep-focus earthquakes lie between about 17° S and 27° S. Beneath about 500 km the Wadati–Benioff zone geometry is somewhat complicated and subject to interpretation. At its northern end the subducted slab is contorted westward – further south (see Fig. 10.8, sections D and E) it is either broken into pieces or bent nearly vertically downward before flattening out somewhat at greater depths. An alternative interpretation is that the westernmost seismicity forms a second Wadati–Benioff zone that lies westward and roughly parallel to the primary zone.

However, in spite of the geometrical complexity of the Wadati–Benioff zone, the majority of deep-focus focal mechanisms have downdip P axes. The clustering of T and B axes is variable, with strong clustering evident in sections F and G, and a more diffuse pattern evident in sections D and E.

10.3.4 Literature review

Although seismologists had been aware since the early 1930s that deep earthquakes were plentiful in the Tonga–Kermadec region, it began receiving special attention only after 1960 (e.g., see Suyehiro, 1962). In the mid-1960s scientists from Columbia University installed a network of short-period seismic stations on islands in Tonga and Fiji; the data they collected provided crucial support for the then-controversial theory of plate tectonics. Many of the papers analyzing these data are now classics, as they showed clearly that the Wadati–Benioff zone in Tonga was associated with a high-velocity, low-attenuation zone that contrasted sharply with the overlying mantle (e.g., see Oliver and Isacks, 1967; Sykes *et al.*, 1969; Mitronovas *et al.*, 1969; Barazangi and Isacks, 1971).

A number of subsequent studies focused attention on relationships between seismicity, focal mechanisms, and the geometry of the subducted material. Billington (1980) relocated earthquakes to better constrain the geometry of the Wadati–Benioff zone. For rays that traveled laterally along the Wadati–Benioff zone, Huppert and Frohlich (1981) showed that the P velocity at intermediate depths was about 8% higher than in global earth models. Giardini and Woodhouse (1986) speculate that the subducted lithosphere is deformed by regional flow within the mantle so that the bottommost Wadati–Benioff zone is displaced 500 km southward of its hypothetical undeformed location. Nothard *et al.* (1996) evaluate the Gaussian curvature of the subducted lithosphere and compare it to the observed pattern of focal mechanisms; they do not find that the data require southward displacement. Christensen and Lay (1988) and Jiao *et al.* (2000) propose that features that exist in the plate prior to subduction control the observable features of deep earthquake focal mechanisms;

e.g., several of the larger intermediate-depth earthquakes including the 22 June 1977 event (see below) lie along the trajectory of the subducting Louisville Ridge, a well-defined bathymetric feature which currently meets the Tonga Trench at about 25° S.

Tomographic methods have begun to provide a more detailed picture of the velocity and attenuation structure within and above the subducted lithosphere. Zhou (1990) performed travel-time tomography to evaluate velocity structure using reported ISC arrivals; Fischer *et al.* (1991) applied residual-sphere analysis to study the dip and depth extent of high-velocity material. The most detailed tomographic results have come from a program which began in December 1993 and included deployments of 12 broad-band island stations as well as 30 ocean-bottom seismographs (Xu and Wiens, 1997; Zhao *et al.*, 1997; Roth *et al.*, 1999). Tomography results are contradictory concerning velocities in the lower mantle beneath Tonga; Hall and Spakman (2002) find high velocities extending to 1500 km depth, presumably attributable to slab penetration. However, Gurnis *et al.* (2000) find near-normal velocities and no evidence for penetration.

The most intriguing recent publications concern the nature of the possible "outboard" Wadati–Benioff zone that lies west of the main deep-focus activity at depths of 300–700 km. On the basis of relocated hypocenters and patterns in focal mechanism P axes, Chen and Brudzinski (2001) propose that the outboard earthquakes form a distinct structure (see also Section 10.4). In subsequent investigations they conclude that the outboard zone is highly anisotropic but that P and S velocities are not higher than in the surrounding mantle (Brudzinski and Chen, 2003; 2005; Chen and Brudzinski, 2003). Their interpretation is that the outboard region consists of cold, metastable olivine subducted prior to that in the main Wadati–Benioff zone above and to the east. Because the subduction rate is so fast it is cold and thus permits earthquake activity, but its low temperature makes it highly metastable and too buoyant to penetrate the lower mantle because it hasn't yet transformed to spinel.

The extraordinarily high rate of deep-focus activity in Tonga–Kermadec suggests this region occupies one end-member of conditions where deep seismicity is possible; thus, the seismicity places constraints on possible mechanisms for deep earthquakes (see Chapter 7). For example, Giardini and Woodhouse (1984) observed that many deep-focus earthquakes appeared to occur on localized planar features oriented parallel to nodal planes of focal mechanisms. Similarly, Wiens and Snider (2001) analyzed data from a temporary network deployed between 1993 and 1995 and identified several clusters of 10–30 "repeating" deep-focus earthquakes, some with nearly identical locations and focal mechanisms. They suggested that runaway shear melting was more likely to permit repeating ruptures than was

transformational faulting, a conclusion also consistent with the large spatial extent
of the region occupied by aftershocks of the 9 March 1994 earthquake (see below).

10.3.5 Significant earthquakes

Considering that overall rates of deep earthquake activity are so high within the
Tonga–Kermadec region, there are surprisingly few very large earthquakes (i.e.,
with $M_W \geq 7.5$). Nevertheless, within the depth range 375–475 km, Huang and
Okal (1998) report that the Tonga–Fiji quakes that occurred on 23 May 1956
($M_{W(\text{CMT-Hist})} = 7.6$) and 21 May 1962 ($M_{W(\text{CMT-Hist})} = 7.5$) are the largest known.
Several other quakes are of special interest.

26 May 1932: $h_{\text{EVC}} = 569$ km; $m_{B(\text{Abe})} = 7.5$. This earthquake is historically
important because, when it took place, it was the largest deep earthquake to occur
since Wadati's (1928) paper appeared. It thus received some attention as a "representative" deep-focus event. For example, Brunner (1938) concentrated on the
numerous body-wave phases that it generated, and in his paper included pho-
tographic copies of no fewer than 51 seismograms. Since it had seven reported
aftershocks, Gutenberg and Richter (1938) concluded (wrongly) that deep earthquake aftershock behavior was little different from that of shallow earthquakes.
Subsequently, Okal (1997) has studied this event thoroughly (the inset shows his
focal mechanism). He concluded that the focal depth was probably about 560 km,
that at least five aftershocks did occur, and that its moment was 3.4×10^{20} N-m
(i.e., $M_W = 7.7$). This makes it comparable in size to the 9 March 1994 and
19 August 2002 Tonga earthquakes, but a factor of ten smaller than the 1970
Colombian or 1994 Bolivian deep-focus events. Huang and Okal (1998) indicate
the 1932 event is the largest historical earthquake worldwide in the depth range
525–575 km.

22 June 1977: $h_{\text{CMT}} = 61$ km; $M_{W(\text{CMT})} = 8.1$. This is the largest Tongan earthquake to occur since the WWSSN was installed in the early 1960s, and possibly
the largest ever. It is tectonically interesting because its location places it approximately along the subducted extension of the Louisville ridge; also, unlike most
Tonga intermediate earthquakes, it had a normal-faulting, downdip T-type focal
mechanism (see inset). Even though it received considerable attention, there is

controversy about almost every aspect of its basic properties (Christensen and Lay, 1988; Lundgren and Okal, 1988; Silver and Jordan, 1983). The various investigations report focal depths ranging between about 60 km and more than 100 km, source-time functions ranging between 50 and 165 seconds, and rupture occurring along both the near-horizontal and the near-vertical nodal planes. The source was highly complex and there were aftershocks with depths spanning a depth range of 150 km. Zhang and Lay (1989a; 1989b) suggest this source complexity is responsible for the various controversies; their interpretation is that the rupture nucleated near a depth of 65 km, followed 25 s later by a second subevent with moment concentrated in the depth range 70–80 km. A coincident slip that radiated only long-period energy had a centroid depth near 100 km and occurred along a steeply dipping fault plane. Because of its size and depth, and because it occurred when digital seismograms were first becoming available, this earthquake provided the first unambiguous recordings of so-called core modes (Masters and Gilbert, 1981), i.e., spheroidal modes of elastic-gravitational oscillation dominated by shear energy in the inner core. This earthquake generated a tsunami with an amplitude of 40 cm at Suva, Fiji, and T waves that were felt on Tahiti, a distance of 2800 km from its epicenter (Talandier and Okal, 1979).

26 January 1983: $h_{\text{CMT}} = 224$ km; $M_{\text{W(CMT)}} = 7.0$. Riedesel's (1985) quite thorough study of this earthquake in ten different frequency bands indicated that its CLVD (non-double-couple) component is anomalously large – on the order of 75–90% (see inset). Whereas systematic errors in determining moment tensors may be responsible for the great majority of non-double-couple mechanisms reported by the Harvard group, this particular non-double-couple mechanism seems to be undeniably genuine, and may be put forth as early "proof" that non-double-couple earthquakes do actually occur.

9 March 1994: $h_{\text{CMT}} = 568$ km; $M_{\text{W(CMT)}} = 7.6$. Fortuitously, this earthquake occurred within an eight-station temporarily-deployed digital seismic network, and thus has been studied even more thoroughly than most large deep-focus events. The most significant result concerns the extent of co-seismic rupture. Wiens *et al.* (1994) and McGuire *et al.* (1997) found that rupture occurred along a near-vertical plane and extended at least 10–20 km beyond the bounds of previously known seismic activity; they concluded that this would be in material at least 200° C warmer than that which usually limits the occurrence of smaller earthquakes. Tibi *et al.* (1999) studied the rupture process using teleseismic body waves, and concluded that rupture occurred along a nearly horizontal plane, in disagreement with McGuire *et al.* (1997). However, both studies agreed that the moment release must have extended outside of the cold, previously seismic active slab core. In the 41 days following this earthquake, Wiens and McGuire (2000) identified no fewer than 144 aftershocks with magnitudes ranging from 3.2 to 6.0; although aftershocks this numerous are not remarkable for comparable-sized shallow-focus earthquakes, they are unprecedented for a deep-focus event. Because this earthquake occurred only three months before the great 9 June 1994 Bolivian earthquake, there are several papers comparing the two events in a special section of *Geophysical Research Letters* (e.g., Goes and Ritsema, 1995; Lundgren and Giardini, 1995).

19 August 2002 (two events): $h_{CMT} = 631$ km; $M_{W(CMT)} = 7.6$; $h_{CMT} = 699$ km; $M_{W(CMT)} = 7.7$. The first of these events was followed two minutes later by an earthquake with m_b 5.9 located 290 km away. Then, an additional five minutes later, the M_W 7.7 earthquake occurred at the second location; this second shock was reportedly felt in Auckland, New Zealand. The two later earthquakes occurred in a region where no activity had occurred previously. Several regional stations recorded these earthquakes; Tibi *et al.* (2002) found one foreshock and 21 aftershocks accompanying the first event; the second sequence included the aforementioned foreshock and one aftershock. Although it seems plausible that the first sequence triggered the second, the second sequence occurred well after the arrival of P or S and at a distance where static stress changes should be negligible. Thus Tibi and others concluded that they were triggered "by transient effects in regions near criticality, but where earthquakes have difficulty nucleating without external influences." The second large event is extraordinarily deep, even though its depth as reported in the EHB catalog (675 km) and by Tibi and others (664 km) is somewhat less than that reported by Harvard. Although there are earthquakes in the CMT Historical and EV Centennial catalogs with $M > 7$ and depths of 660–680 km occurring in 1922, 1940, 1963 and 1969, this 2002 earthquake is undoubtedly the deepest large earthquake since 1977. Persh and Houston (2002) report that that the 2002 earthquakes had unusually short rupture durations (15.5 and 11.6 s, respectively) for such large events.

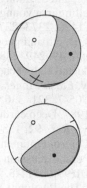

10.3.5 Tonga–Kermadec references

Ansell, J. and D. Adams, 1986. Unfolding the Wadati–Benioff zone in the Kermadec–New Zealand region, *Phys. Earth Planet. Int.*, **44**, 274–280.

Barazangi, M. and B. Isacks, 1971. Lateral variations of seismic-wave attenuation in the upper mantle above the inclined earthquake zone of the Tonga island arc: deep anomaly in the upper mantle, *J. Geophys. Res.*, **76**, 8493–8516.

Billington, S., 1980. *The Morphology and Tectonics of the Subducted Lithosphere in the Tonga–Fiji–Kermadec Region from Seismicity and Focal Mechanism Solutions*, Ph.D. dissertation, Cornell University, 273 pp.

• Brudzinski, M. R. and W.-P. Chen, 2003. A petrologic anomaly accompanying outboard earthquakes beneath Fiji–Tonga: corresponding evidence from broad-band P and S waveforms, *J. Geophys. Res.*, **108**, doi:10.1029/2002JB002012, 2299.

• 2005. Earthquakes and strain in subhorizontal slabs, *J. Geophys. Res.*, **110**, doi:10.1029/2004JB003470, B08303.

• Brunner, G. J., 1938. The deep earthquake of May 26, 1932 near the Kermadec Islands, *Gerlands Beitr. z. Geophysik*, **53**, 1–64.

Chen, W.-P. and M. R. Brudzinski, 2001. Evidence for a large-scale remnant of subducted lithosphere beneath Fiji, *Science*, **292**, doi:10.1126/science.292.5526.2475, 2475–2479.

2003. Seismic anisotropy in the mantle transition zone beneath Fiji–Tonga, *Geophys. Res. Lett.*, **30**, doi:10.1029/2002GL016330, 1682.

Christensen, D. H. and T. Lay, 1988. Large earthquakes in the Tonga region associated with subduction of the Louisville Ridge, *J. Geophys. Res.*, **93**, 13367–13389.

Fischer, K. M., K. C. Creager, and T. H. Jordan, 1991. Mapping the Tonga slab, *J. Geophys. Res.*, **96**, 14403–14427.

Giardini, D. and J. H. Woodhouse, 1984. Deep seismicity and modes of deformation in Tonga subduction zone, *Nature*, **307**, doi:10.1038/307505a0, 505–509.

1986. Horizontal shear flow in the mantle beneath the Tonga arc, *Nature*, **319**, doi:10.1038/319551a0, 551–555.

Goes, S. and J. Ritsema, 1995. A broadband P wave analysis of the large deep Fiji Island and Bolivia earthquakes of 1994, *Geophys. Res. Lett.*, **22**, doi:10.1029/95GL02011, 2249–2252.

Gurnis, M., J. Ritsema, H.-J. van Heijst, and S. Zhong, 2000. Tonga slab deformation: the influence of a lower mantle upwelling in a young subduction zone, *Geophys. Res. Lett.*, **27**, doi:10.1029/2000GL011420, 2373–2376.

Gutenberg, B. and C. F. Richter, 1938. Depth and geographical distribution of deep-focus earthquakes, *Geol. Soc. Amer. Bull.*, **49**, 249–288.

Hall, R. and W. Spakman, 2002. Subducted slabs beneath the eastern Indonesia–Tonga region: insights from tomography, *Earth Planet. Sci. Lett.*, **201**, doi:10.1016/S0012-821X(02)00705-7, 321–336.

Hamburger, M. W. and B. L. Isacks, 1987. Deep earthquakes in the southwest Pacific: a tectonic interpretation, *J. Geophys. Res.*, **92**, 13841–13854.

Huang, W.-C. and E. A. Okal, 1998. Centroid moment tensor solutions for deep earthquakes predating the digital era: discussion and inferences, *Phys. Earth Planet. Int.*, **106**, doi:10.1016/S0031-9201(97)00111-8, 191–218.

Huppert, L. N. and C. Frohlich, 1981. The P velocity within the Tonga Benioff zone determined from traced rays and observations, *J. Geophys. Res.*, **86**, 3773–3782.

Jiao, W., P. G. Silver, Y. Fei, and C. T. Prewitt, 2000. Do intermediate- and deep-focus earthquakes occur on preexisting weak zones? An examination of the Tonga subduction zone, *J. Geophys. Res.*, **105**, doi:10.11029/2000JB900314, 28125–18138.

Kawakatsu, H., 1986. Downdip tensional earthquakes beneath the Tonga arc: a double seismic zone? *J. Geophys. Res.*, **91**, 6432–6440.

Lundgren, P. and D. Giardini, 1995. The June 9 Bolivia and March 9 Fiji deep earthquakes of 1994: I. Source processes, *Geophys. Res. Lett.*, **22**, doi:10.1029/95GL02233, 2241–2244.

Lundgren, P. R. and E. A. Okal, 1988. Slab decoupling in the Tonga arc: the June 22, 1977 earthquake, *J. Geophys. Res.*, **93**, 13355–13366.

Masters, G. and F. Gilbert, 1981. Structure of the inner core inferred from observations of its spheroidal shear modes, *Geophys. Res. Lett.*, **8**, 569–571.

McGuire, J. J., D. A. Wiens, P. J. Shore, and M. G. Bevis, 1997. The March 9, 1994 (M_W 7.6), deep Tonga earthquake: rupture outside the seismically active slab, *J. Geophys. Res.*, **102**, doi:10.1029/96JB03185, 15163–15182.

Mitronovas, W., B. Isacks, and L. Seeber, 1969. Earthquake locations and seismic wave propagation in the upper 250 km of the Tonga island arc, *Bull. Seismol. Soc. Amer.* **59**, 1115–1135.

Nothard, S., D. McKenzie, J. Haines, and J. Jackson, 1996. Gaussian curvature and the relationship between the shape and the deformation of the Tonga slab, *Geophys. J. Int.*, **127**, 311–327.

Okal, E. A., 1997. A reassessment of the deep Fiji earthquake of 26 May 1932, *Tectonophysics*, **275**, doi:10.1016/0040-1951(97)00024-3, 313–329.

Oliver, J. and B. Isacks, 1967. Deep earthquake zones, anomalous structures in the upper mantle, and the lithosphere, *J. Geophys. Res.*, **72**, 4259–4275.

Pelletier, B., S. Calmant, and R. Pillett, 1998. Current tectonics of the Tonga–New Hebrides region, *Earth Planet. Sci. Lett.*, **164**, doi:10.1016/S0012-821X(98)00212-X, 263–276.

Persh, S. E. and H. Houston, 2002. Source time functions of the August 19, 2002 Tonga deep earthquakes compared with a global deep earthquake catalog, *EOS, Trans. Amer. Geophys. Un.*, **83** (**47**), Fall Meet. Suppl., Abstract 562C-1206.

Riedesel, M. A., 1985. *Seismic Moment Tensor Recovery at Low Frequencies*, Ph.D. dissertation, University of California San Diego., 245 pp.

Roth, E. G., D. A. Wiens, L. M. Dorman, J. Hildebrand, and S. C. Webb, 1999. Seismic attenuation tomography of the Tonga–Fiji region using pair phase methods, *J. Geophys. Res.*, **104**, doi:10.1029/1998JB900052, 4795–4809.

Silver, P. G. and T. H. Jordan, 1983. Total moment spectra of fourteen large earthquakes, *J. Geophys. Res.*, **88**, 3273–3293.

Suyehiro, S., 1962. Deep earthquakes in the Fiji region, *Papers Meterol. Geophys.*, **13**, 216–238.

Sykes, L. R., B. L. Isacks, and J. Oliver, 1969. Spatial distribution of deep and shallow earthquakes of small magnitudes in the Fiji–Tonga region, *Bull. Seismol. Soc. Amer.*, **59**, 1093–1113.

Talandier, J. and E. A. Okal, 1979. Human perception of T waves: the June 22, 1977 Tonga earthquake felt on Tahiti, *Bull. Seismol. Soc. Amer.*, **69**, 1475–1486.

Tibi, R., C. H. Estabrook, and G. Bock, 1999. The 1996 June 17 Flores Sea and 1994 March 9 Fiji–Tonga earthquakes: source processes and deep earthquake mechanisms, *Geophys. J. Int.*, **138**, doi:10.1046/j.1365-246x.1999.00879.x, 625–642.

Tibi, R., D. A. Wiens, and H. Inoue, 2003. Remote triggering of deep earthquakes in the 2002 Tonga sequences, *Nature*, **424**, doi:10.1038/nature01903, 921–925.

Wadati, K., 1928. Shallow and deep earthquakes, *Geophys. Mag.*, **1**, 162–202.

Wiens, D. A. and J. J. McGuire, 2000. Aftershocks of the March 9, 1994, Tonga earthquake: the strongest known deep aftershock sequence, *J. Geophys. Res.*, **105**, doi:10.1029/2000JB900097, 19067–19083.

Wiens, D. A. and N. O. Snider, 2001. Repeating deep earthquakes: evidence for fault reactivation at great depth, *Science*, **293**, doi:10.1126/science.1063042, 1463–1466.

Wiens, D. A., J. J. McGuire, P. J. Shore, M. G. Bevis, K. Draunidalo, G. Prasad, and S. P. Helu, 1994. A deep earthquake aftershock sequence and implications for the rupture mechanism of deep earthquakes, *Nature*, **372**, doi:10.1038/372540a0, 540–543.

Xu, Y. and D. A. Wiens, 1997. Upper mantle structure of the southwest Pacific from regional waveform inversion, *J. Geophys. Res.*, **102**, doi:10.1029/97JB02564, 27439–27451.

Zellmer, K. and B. Taylor, 2001. A three-plate kinematic model for Lau Basin opening, *Geochem. Geophys. Geosyst.*, **2**, doi:10.1029/2000GC000106.

Zhang, J. and T. Lay, 1989a. Duration and depth of faulting of the 22 June 1977 Tonga earthquake, *Bull. Seismol. Soc. Amer.*, **79**, 51–66.

  1989b. A new method for determining the long-period component of the source time function of large earthquakes, *Geophys., Res. Lett.*, **16**, 275–278.

Zhao, D., Y. Xu, D. A. Wiens, L. Dorman, J. Hildebrand, and S. Webb, 1997. Depth extent of Lau backarc spreading center and its relationship to subduction process, *Science*, **278**, doi:10.1126/science.278.5336.254, 254–257.

Zhou, H., 1990. Mapping of P-wave slab anomalies beneath the Tonga, Kermadec and New Hebrides arcs, *Phys. Earth Planet. Int.*, **61**, 199–229.

10.4 Vitiaz Cluster and South Fiji Basin "outboard" earthquakes

10.4.1 Deep seismicity

There are a significant number of deep-focus earthquakes in the mantle between the Tonga and Vanuatu trenches that occur singly or in groups separate from, and possibly not associated with, the Wadati–Benioff zones in either arc (see Fig. 10.9). The largest group is the so-called Vitiaz Cluster situated somewhat southward of the Vitiaz[3] Trench. This group includes about 160 earthquakes recorded teleseismically between 1949 and 1996 with relocated depths ranging between 580 and 683 km. They occupy an approximately horizontal zone with an E–W dimension of about 800 km extending eastward from the northern Vanuatu Trench. There are also about 40 deep-focus hypocenters scattered beneath the Fiji Basin; most of these occur in two "fingers" of activity that are either "detached" or extend "outboard" from most Tonga–Kermadec activity.

10.4.2 Literature review

Sykes (1964) identified eight earthquakes in the Vitiaz Cluster and was the first to recognize that they might be significant. None appear in the GR catalog, probably because regional station coverage was poor and these earthquakes rarely have magnitudes m_b exceeding 5.6. Subsequently, Pascal *et al.* (1973) and Okal and Kirby (1998) identified and relocated the remaining known hypocenters. They also evaluated available focal mechanisms and found no obvious pattern as there is considerable variation in the orientations of P, T and B axes (Fig. 10.9). Only one Vitiaz-region earthquake has a reported magnitude exceeding 7.0; this occurred in the south finger on 26 May 1986 ($h_{CMT} = 568\,km$; $M_{W(CMT)} = 7.1$). Brudzinski and Chen (2005) evaluated waveforms of this event to determine focal mechanisms for two subevents, and noted that over a two-week period it generated six aftershocks that appear in the EHB catalog.

There has been considerable speculation concerning the origin of the Vitiaz earthquakes. Pascal *et al.* (1973) and Barazangi *et al.* (1973) evaluated their travel times and frequency content at regional stations and concluded that there probably was no high-velocity slab beneath 300 km connecting the intermediate Vanuatu and deep-focus Vitiaz activity; thus, it did not occur in lithosphere subducted beneath the Vanuatu arc. What is undoubtedly true is that the presence of very fast subduction of old lithosphere produces colder-than-usual temperature conditions in the mantle that make both the Vitiaz and South Fiji Basin activity possible.

Because they occur approximately parallel to the Vitiaz Trench and because the northernmost section of the Tonga Wadati–Benioff zone curves westward, Hamburger and Isacks (1987) and Okal and Kirby (1998) suggest these earthquakes may occur in lithosphere subducted about 10 Ma at the Vitiaz Trench. In some plate models, prior to the formation of the Fiji Basin the Vitiaz Trench formed a boundary connecting the southern Vanuatu and northern Tonga trenches and there

---

[3] Sometimes this is spelled Vityaz.

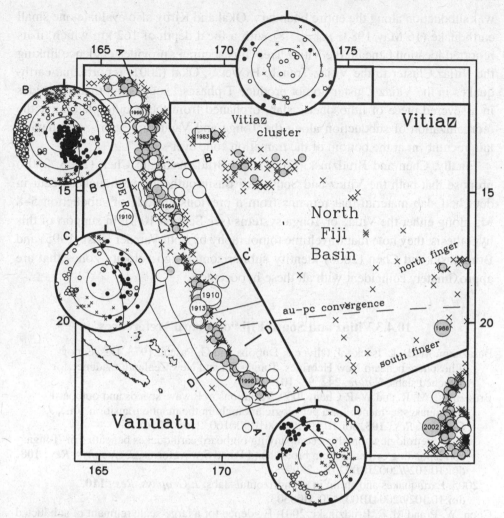

Fig. 10.9 Map of deep earthquake activity in the Vanuatu (New Hebrides) and Vitiaz regions. Filled circles, open circles, and crosses indicate selected deep earthquakes in the Harvard CMT, EV Centennial, and EHB catalogs (see description of selection process in the introduction to this chapter). The square is the m_b 3.8 earthquake of 16 May 1983 discussed by Okal and Kirby (1998). Solid lines labeled A–A' to D–D' indicate the extent and orientation of the cross sections (Fig. 10.10), lines surround included events, and equal-area plots summarize T (solid circles), B (crosses), and P (open circles) axis orientations of Harvard CMT. The equal area plot for the Vitiaz deep-focus earthquakes (top, center) includes all Harvard data; other plots include selected mechanisms only. The thick solid line labeled 'DE R' indicates where the d'Entrecasteaux Ridge impinges on the Vanuatu arc; other lines show the direction of Australia–Pacific (au–pc) plate convergence and the axis of the Vanuatu Trench.

was subduction along the entire boundary. Okal and Kirby also evaluate one small earthquake (16 May 1983, $m_b = 3.8$) with a focal depth of 162 km which, if its reported location beneath the Vitiaz Trench is accurate, provides evidence linking the Vitiaz Cluster to the Vitiaz Trench. However, Okal (2001) reports that earthquakes in the Vitiaz Cluster do not produce T phases,[4] suggesting that it "resides in a severed piece of lithospheric slab, orphaned from the Pacific plate after the reorganization of subduction along the Tonga and Vanuatu systems, and having lain recumbent at the bottom of the transition zone ever since."

Finally, Chen and Brudzinski (2001) and Brudzinski and Chen (2003; 2005) propose that both the Vitiaz and South Fiji Basin outboard earthquakes occur in detached slab material that remains from a previous episode of subduction 5–8 Ma along either the Vitiaz or Tonga systems (see Section 10.3). In support of this hypothesis, they note that traveltime tomography by both Van der Hilst (1995) and Brudzinski and Chen (2000) identify subhorizontal high-velocity zones that are approximately coincident with all these hypocenters.

10.4.3 Vitiaz and South Fiji "outboard" references

Barazangi, M., B. L. Isacks, J. Oliver, J. Dubois, and G. Pascal, 1973. Descent of lithosphere beneath New Hebrides, Tonga-Fiji and New Zealand: evidence for detached slabs, *Nature*, **242**, 98–101.

Brudzinski, M. R. and W.-P. Chen, 2000. Variations in P wave speeds and outboard earthquakes: evidence for a petrologic anomaly in the mantle transition zone, *J. Geophys. Res.*, **105**, doi:10.1029/2000JB900160, 21666–21682.

2003. A petrologic anomaly accompanying outboard earthquakes beneath Fiji–Tonga: corresponding evidence from broad-band P and S waveforms, *J. Geophys. Res.*, **108**, doi:10.1029/2002JB002012, 2299.

2005. Earthquakes and strain in subhorizontal slabs, *J. Geophys. Res.*, **110**, doi:10.1029/2004JB003470, B08303.

Chen, W.-P. and M. R. Brudzinski, 2001. Evidence for a large-scale remnant of subducted lithosphere beneath Fiji, *Science*, **292**, doi:10.1126/science.292.5526.2475, 2475–2479.

Hamburger, M. W. and B. L. Isacks, 1987. Deep earthquakes in the southwest Pacific: a tectonic interpretation, *J. Geophys. Res.*, **92**, 13841–13854.

Okal, E. A., 2001. 'Detached' deep earthquakes: are they really? *Phys. Earth Planet. Int.*, **127**, doi:10.016/S0031-9201(01)00224-2, 109–143.

Okal, E. A. and S. H. Kirby, 1998. Deep earthquakes beneath the Fiji Basin, SW Pacific: Earth's most intense deep seismicity in stagnant slabs, *Phys. Earth Planet. Int.*, **109**, doi:10.1016/S0031-9201(98)001116-2, 25–63.

Pascal, G., J. Dubois, M. Barazangi, B. L. Isacks, and J. Oliver, 1973. Seismic velocity anomalies beneath the New Hebrides Island arc: evidence for a detached slab in the upper mantle, *J. Geophys. Res.*, **78**, 6998–7004.

---

[4] T phases are high-frequency, highly dispersed body waves that travel great distances within the ocean's water column. In 1940 Daniel Linehan of Weston Observatory first assigned this name to these tertiary arrivals, thus distinguishing them from P (primary) and S (secondary) arrivals.

Sykes, L. R., 1964. Deep-focus earthquakes in the New Hebrides region, *J. Geophys. Res.*, **69**, 5353–5355.

Van der Hilst, R., 1995. Complex morphology of subducted lithosphere in the mantle beneath the Tonga Trench, *Nature*, **374**, doi:10.1038/374154a0, 154–157.

10.5 Vanuatu (New Hebrides)

10.5.1 Regional tectonics and deep seismicity

The islands that form the nation of Vanuatu achieved independence in 1980; previously they had been called the New Hebrides and were administered under a peculiar joint British-and-French colonial government. Although scientists still use both names interchangeably, nowadays Vanuatu is preferred, even though the Vanuatu government doesn't administer some of the northern- and southernmost islands in the arc.

The Vanuatu island arc occurs in response to convergence between the India–Australia and Pacific plates. At the plate boundary the crustal age is ~40 Ma and the NUVEL-1 rate is ~8 cm/yr, consistent with a thermal parameter Φ of about 3000 km. However, as with the Tonga arc, the rate of lithospheric consumption is uncertain because of spreading in the North Fiji Basin (Pelletier *et al.*, 1998). The d'Entrecasteaux Ridge, a broad bathymetric feature on the India–Australia plate, complicates the subduction process, and where it impinges on the arc it is undoubtedly responsible for the disappearance of the trench, the high relief on the islands of Santo and Malekula, and the existence of a secondary chain of islands in the backarc (e.g., Maewo and Pentecost).

There is intermediate-depth seismicity along most of the arc (Fig. 10.9); and there are a few earthquakes with reported focal depths at about 12° N between 300 and 350 km; however, none have magnitudes m_b larger than 5.0. Throughout most of the arc maximum focal depths are between 150 and 300 km and the dip of the Wadati–Benioff zone is about 70° (Fig. 10.10).

Beneath northern Vanuatu the majority of focal mechanisms have downdip T axes and along-strike B axes (Fig. 10.9). However, south of 18° S the orientations become more scattered, and where the arc bends eastward south of 20° S the T axes become more horizontal, possibly indicating that the arc is under along-strike tension. After correcting for variability in the orientation of the trench, Christova and Scholz (2003) and Christova *et al.* (2004) observed that there was strong clustering of slab-normal P axes along the entire arc, and thus concluded that the dominant forces here were normal to the subducting slab.

I thank Shamita Das for reviewing an earlier draft of this section.

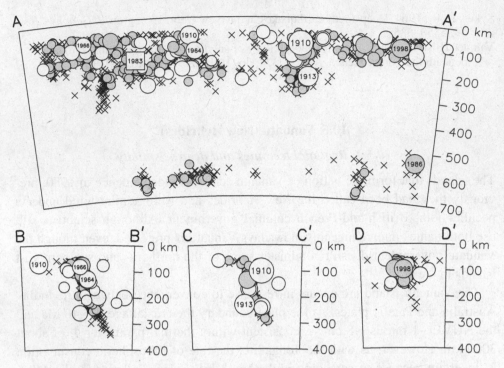

Fig. 10.10 Along-strike cross section A–A′ and trench-normal sections B–B′ to D–D′ of seismicity in the Vanuatu–Vitiaz regions. Earthquake symbols are as in Fig. 10.9.

10.5.2 Literature review

In the discussion of New Hebrides seismicity in the GR catalog, Gutenberg and Richter noted that earthquakes were difficult to locate here but that "the region is one of the most active sources for intermediate shocks." Except for Dubois (1971), between 1964 and 1975 much of the literature focused on the Vitiaz deep-focus cluster, and whether there was a high-velocity slab connecting the Vanuatu and Vitiaz activity (see Section 10.4.2). Subsequently, Pascal *et al.* (1978) and Hănus and Vănek (1983) presented numerous cross sections of teleseismically recorded hypocenters delineating the Wadati–Benioff zone.

In central Vanuatu where the d'Entrecasteaux Ridge impinges on the arc, there is an intriguing gap in seismicity beneath about 50 km depth. Marthelot *et al.* (1985) and Prevot *et al.* (1991) investigated this using teleseismic observations augmented with data from a 19-station regional network. These data delineate the gap clearly, demonstrate that shear waves attenuate significantly as they pass through it, and indicate that mantle velocity is lower in the gap than in adjacent seismically active areas. Both Hănus and Vănek (1983) and Chatelaine *et al.* (1992) suggest that the lithosphere is detached above where the gap occurs.

Coudert *et al.* (1981) investigated the Wadati–Benioff zone in southern Vanuatu using data from OBSs and a temporary local network. Coudert found that the maximum depths shallowed from about 300 km at 20° S to about 150 km at 21° S, and that hypocenters determined from teleseismic data were remarkably similar to those determined using local network observations. Chiu *et al.* (1985) used ray tracing to interpret multiple P and P-to-S converted phases at the regional network; they interpret these as P traveling in the inner and outer portions of the slab, and a conversion at the slab surface. If this correct, it implies that intermediate-depth activity doesn't occur in the coldest interior of the slab but in an outer zone where the velocity is somewhat lower.

10.5.3 Significant earthquakes

Between 1909 and 2001, the Harvard and EV Centennial catalogs list 58 intermediate-depth earthquakes in Vanuatu with magnitudes of 7 or greater; this is more than in any other geographic region. Surprisingly, there have been few detailed investigations of any of these events: Blum's (1936) investigation of travel times from the earthquake of 24 June 1935 ($h_{EVC} = 142\,km$; $m_{B(Abe)} = 7.1$) was essentially only an abstract. The earthquake of 9 July 1964 ($h_{EVC} = 131\,km$; $m_{B(Abe)} = 7.4$) occurred at almost the same location; Gupta and Hamada (1975) used its surface waves to evaluate upper mantle structure. Tajima *et al.* (1990) published a detailed analysis of the source process of the Santa Cruz Islands earthquake of 31 December 1966 ($h_{EVC} = 83\,km$; $M_{S(Abe)} = 7.9$), but concluded that its focal depth was only 40 km. The reported m_B of 7.9 for the earthquake of 16 June 1910 ($h_{Abe} = 100\,km$) makes it one of the ten largest intermediate-depth earthquakes in the EV Centennial catalog. Several Vanuatu earthquakes in the Harvard catalog have M_W of about 7.3–7.4.

10.5.4 Vanuatu (New Hebrides) references

Blum, V. J., 1936. The intermediate earthquake of 24 June, 1935, *Bull. Seismol. Soc. Amer.*, **26**, 195–196.

Chatelain, J.-L., P. Molnar, R. Prevot, and B. Isacks, 1992. Detachment of part of the downgoing slab and uplift of the New Hebrides (Vanuatu) islands, *Geophys. Res. Lett.*, **19**, 1507–1510.

Chiu, J.-M., B. L. Isacks, and R. K. Cardwell, 1985. Propagation of high-frequency seismic waves inside the subducted lithosphere from intermediate-depth earthquakes recorded in the Vanuatu arc, *J. Geophys. Res.*, **90**, 12741–12754.

Coudert, E., B. L. Isacks, M. Barazangi, R. Louat, R. Cardwell, A. Chen, J. Dubois, G. Latham, and B. Pontoise, 1981. Spatial distribution and mechanisms of earthquakes in the southern New Hebrides arc from a temporary land and ocean bottom seismic network and from worldwide observations, *J. Geophys. Res.*, **86**, 5905–5925.

Christova, C. and C. H. Scholz, 2003. Stresses in the Vanuatu subducting slab: a test of two hypotheses, *Geophys. Res. Lett.*, **30**, doi:10.1029/2003GL017701, 1790.

Christova, C., C. H. Scholz, and H. Kao, 2004. Stress fields in the Vanuatu (New Hebrides) Wadati–Benioff zone inferred by the inversion of earthquake focal mechanisms, *J. Geodynamics*, **37**, doi:10.1016/j.jog.2003.11.001, 125–137.

Dubois, J., 1971. Propagation of P waves and Rayleigh waves in Melanesia: structural implications, *J. Geophys. Res.*, **76**, 7217–7240.

Gupta, H. K. and K. Hamada, 1975. Rayleigh- and Love-wave dispersion up to 140-second-period range in the Indonesia–Philippine region, *Bull. Seismol. Soc. Amer.*, **65**, 507–521.

Hănus, V. and J. Vănek, 1983. Deep structure of the Vanuatu (New Hebrides) island arc: intermediate depth collision of subducted lithospheric plates, *New Zealand J. Geol. Geophys.*, **26**, 133–154.

Marthelot, J.-M., J.-L. Chatelain, B. L. Isacks, R. K. Cardwell, and E. Coudert, 1985. Seismicity and attenuation in the central Vanuatu (New Hebrides) islands: a new interpretation of the effect of subduction of the d'Entrecasteaux fracture zone, *J. Geophys. Res.*, **90**, 8641–8650.

Pascal, G., B. L. Isacks, M. Barazangi, and J. Dubois, 1978. Precise relocations of earthquakes and seismotectonics of the New Hebrides island arc, *J. Geophys. Res.*, **83**, 4957–4973.

Pelletier, B., S. Calmant, and R. Pillet, 1998. Current tectonics of the Tonga–New Hebrides region, *Earth Planet. Sci. Lett.*, **164**, doi:10.1016/S0012-821X(98)00212-X, 263–276.

Prevot, R., S. W. Roecker, B. L. Isacks, and J.-L. Chatelain, 1991. Mapping of low P wave velocity structures in the subducting plate of the central New Hebrides, southwest Pacific, *J. Geophys. Res.*, **96**, 19825–19842.

Tajima, F., L. J. Ruff, H. Kanamori, J. Zhang, and K. Mogi, 1990. Earthquake source processes and subduction regime in the Santa Cruz Islands region, *Phys. Earth Planet. Int.*, **61**, 269–290.

10.6 New Guinea–Solomons

10.6.1 Regional tectonics and deep seismicity

Intermediate-depth earthquakes occur along a nearly continuous band stretching from New Guinea eastward through New Britain, New Ireland and the Solomon Islands (Figs. 10.11 and 10.12). Overall, these occur in response to oblique convergence of ~10.5 cm/yr between the Australian and Pacific plates. However, the lithosphere beneath New Guinea is continental, and the Ontong–Java plateau (north of the Solmon Islands) is a large igneous province that resists subduction because of its thickened crust. To accommodate the Australia–Pacific motion, several small plates, including the Solomon and South Bismark plates, have formed between New Guinea and the Solomons.[5] All these factors make it impossible to settle on a single value for the thermal parameter. However, in some parts of this region

I thank Geoff Abers and Patricia Cooper for reviewing an earlier draft of this section.

[5] For a discussion of the various proposed plates, see Johnson and Molnar (1972), Weissel and Anderson (1978), and Wallace *et al.* (2004).

there are deep-focus earthquakes and Φ may be as great as 6000 km; elsewhere it is considerably less.

Thus the geometry of the Wadati–Benioff zone is complicated. The Solomon Islands form an arc-trench system with the Australian plate subducting northward along the arc's southern boundary. However, before the Ontong–Java plateau reached the plate boundary about 2 Ma, the trench lay along the north boundary of the arc and the subduction direction was southward. Today there are indications that convergence is still proceding in both directions. Although there are a few isolated earthquakes at depths of 540 km, most have depths shallower than about 180 km and focal mechanisms exhibiting arc-normal P axes (cross section F, Fig. 10.12). Towards the west, the subduction of the Woodlark spreading center coincides with a distinct gap in intermediate teleseismically reported seismicity (cross section A). Then, west of the spreading center the activity resumes vigorously as the Solomon plate subducts northward under a pronounced bend in the New Britain Trench. East of the bend (cross section E), most intermediate-depth quakes have focal mechanisms with vertical T and arc-normal P axes; to the west (cross section D) the T and P orientations are highly scattered. Hypocenters as deep as 600 km lie on both sides of the bend.

Still further west, numerous earthquakes with depths as great as 235 km occur just beyond the western margin of the Solomon plate where the New Britain zone reaches New Guinea (cross section C). Careful relocations indicate that the Wadati–Benioff zone here is saddle-shaped, with limbs dipping both to the north and to the south. East of 144° E there is no active trench north of New Guinea, and so these earthquakes probably originate within lithosphere attached to the Solomon plate rather than the South Bismark plate. Again, the orientations of P and T axes are highly scattered, although T axes for events at depths greater than about 90 km seem to conform to the curvature of the saddle. There are a few earthquakes with depths as great as 170 km in a separate group beneath the Papuan Peninsula, southwest of the Solomon plate. And across central New Guinea there is a diffuse band of intermediate earthquakes with depths as great as 100 km, apparently produced by subduction beneath New Guinea's north coast (cross section B).

10.6.2 Literature review

GR noted that the central New Guinea, New Britain, and Solomons regions were tectonically distinct, and commented on the presence of intermediate- and deep-focus earthquakes there. However, there was little analysis of the deep seismicity until about 1970; then Denham (1969), Santo (1970), and Johnson *et al.* (1971) all presented cross sections across the region and discussed how Australian–Pacific convergence influences regional tectonics. Ripper (1970) is historically interesting

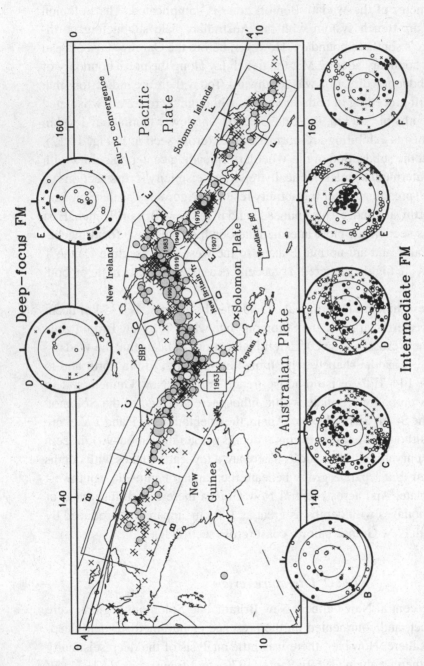

Fig. 10.11 Map of deep earthquake activity in New Guinea–Solomons region. Filled circles, open circles, and crosses indicate selected deep earthquakes in the Harvard CMT, EV Centennial, and EHB catalogs (see description of selection process in the introduction to this chapter). Solid lines labeled A–A' to F–F' indicate locations and orientations of cross sections A to F, thin lines indicate geographic boundaries, and equal-area plots summarize T (solid circles), B (crosses), and P (open circles) axis orientations of selected Harvard CMT for intermediate-depth (plots B, C, D, E and F below) and deep-focus (plots D and E above) earthquakes. Other lines indicate locations of plate boundaries. The dashed line indicates the direction of Australia–Pacific (au–pc) plate convergence. Labels "SBP" and "Woodlark SC" indicate the South Bismark plate and Woodlark spreading center, respectively.

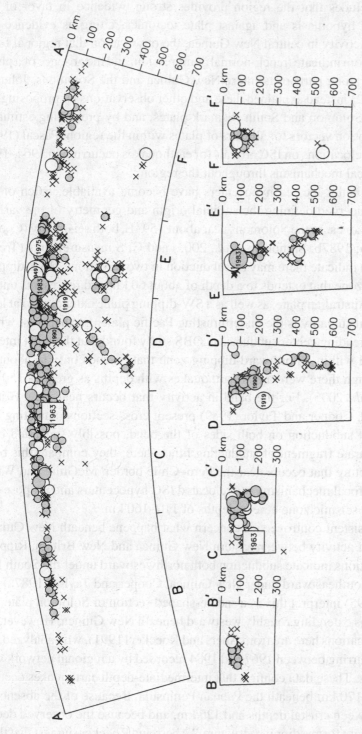

Fig. 10.12 Cross sections A–A' to F–F' of deep seismic activity in New Guinea–Solomons region (see Fig. 10.11). Earthquake symbols are as in Fig. 10.11.

because he concludes that the region provides strong evidence in favor of the expanding Earth hypothesis and against plate tectonics, citing as evidence the paucity of deep activity in central New Guinea, the observation that regional focal mechanisms seldom indicate trench-normal compression, and the absence of a physiographic trench along parts of northern New Guinea and the Solomons. Johnson and Molnar (1972) instead explained these and other observations by proposing the existence of the Solomon and South Bismark plates, and by presenting estimates of the relative motion vectors for all pairs of plates within the region. Pascal (1979) performed JHD relocations on ISC arrivals for earthquakes occurring in 1964–1973 and evaluated focal mechanisms throughout the region.

As more and better-located hypocenters have become available, much of the literature has concerned defining the surficial origin and geometry of the various Wadati–Benioff zones. In the Solomons near about 159° E, both seismicity (Cooper and Taylor, 1985; 1987b; Shinohara *et al.*, 2003) and GPS measurements (Tregoning *et al.*, 1998) indicate there may be subduction in two directions, a NE-dipping Wadati–Benioff zone that extends to a depth of about 60 km and represents underthrusting of the Australian plate, as well as a SW-dipping plane extending to at least 150 km which presumably is the underthrusting Pacific plate. Further west where the Woodlark spreading center subducts, an OBS study found that the most intense activity occurred within a northward-dipping zone that extends only to about 60 km depth, although there were a few earthquakes with depths as great as 150 km (Yoneshima *et al.*, 2005). For the bend in activity that occurs near New Britain and New Ireland, Cooper and Taylor (1989) present cross sections indicating that there is coherent subduction on both sides of the bend, possibly separated by a tear or by a separate fragment of subducting lithosphere; they compare this bend to a similar structure that occurs near the Peru–Chile border. McGuire and Wiens (1995) interpret focal mechanisms and relocated ISC hypocenters and propose that there is a double seismic zone here at depths of 110–160 km.

The most persistent controversies concern what happens beneath New Guinea. For the cluster of activity between eastern New Guinea and New Britain, Ripper's (1982) cross sections indicate subduction both northwestward under the South Bismark plate and southeastward under New Guinea. Cooper and Taylor (1987a) and Pegler *et al.* (1995) interpret this as a saddle-shaped section of Solomon plate material with an axis extending roughly westward beneath New Guinea. However, the most accurate locations here are from Abers and Roecker (1991), who analyzed 635 earthquakes occurring between 1967 and 1984 recorded by a regional network with about 20 stations. These data confirm that intermediate-depth earthquakes occur to depths of about 170 km beneath the Papuan Peninsula. Because of the absence of hypocenters between crustal depths and 125 km, and because the observed deeper earthquakes "do not form a dipping structure," Abers and Roecker suggest that these may not be caused by ordinary subduction; rather, they may correspond to deep crustal and upper mantle earthquakes similar to those reported beneath mountainous regions in Morocco, the Karakorum (Hindu Kush), and Taiwan.

Like the Papuan Peninsula earthquakes, the intermediate hypocenters in western New Guinea west of about 141° E also are diffuse and do not form an obvious planar zone. However, Seno and Kaplan (1988) interpret these as a Wadati–Benioff zone extending to depths of 100–150 km originating from southwestern-dipping subduction along northern New Guinea.

10.6.3 Significant earthquakes

There are 49 intermediate-depth earthquakes with magnitudes of 7.0 or greater in the EV Centennial and CMT catalogs. Only two (18 March 1983 and 30 December 1990) have occurred since 1950, had depths clearly exceeding 60 km, and magnitudes exceeding 7.4.

26 February 1963: $h_{EVC} = 181$ km; $m_{B(Abe)} = 7.3$. After Aki (1966) proposed the concept of scalar moment, it was determined initially by comparing frequency-domain observations and synthetics for surface waves from large shallow earthquakes. Unlike earthquakes with surface foci, deep earthquakes generate higher-mode Love waves; thus, to demonstrate how to find scalar moments for deep earthquakes, Fukao and Abe (1971) calculated synthetics in the time domain for this 1963 New Guinea earthquake (the inset presents their focal mechanism). They obtained a scalar moment of 2.5×10^{20} N-m, corresponding to an M_W of 7.6.[6]

18 March 1983: $h_{CMT} = 70$ km; $M_{W(CMT)} = 7.7$. This earthquake is the largest known in the New Guinea–Solomons region. It produced landslides and minor damage along the southeast coast of New Ireland. Silverstein *et al.* (1986) present processed strong-motion records as recorded at three stations on Bouganville Island situated about 270 km from the epicenter. The observed peak accelerations ranged from 0.02–0.30 g.

[6] Fukao and Abe's (1971) mechanism is a dip slip with vertical fault plane, nodal plane N5° W (T axis in western hemisphere), other nodal plane horizontal. They claimed Isacks and Molnar's (1971) mechanism (second plane strikes N5° W with 20° dip) is inconsistent with the Love wave data.

10.6.4 New Guinea–Solomons references

Abers, G. A. and S. W. Roecker, 1991. Deep structure of an arc-continent collision: earthquake relocation and inversion for upper mantle P and S wave velocities beneath Papua New Guinea, *J. Geophys. Res.*, **96**, 6379–6401.

Aki, K., 1966, Generation and propagation of G waves from the Niigata earthquake of June 16, 1964. 2. Estimation of earthquake movement, released energy, and stress–strain drop from G wave spectrum, *Bull. Earthquake Res. Inst., Tokyo Univ.*, **44**, 73–88.

Cooper, P. and B. Taylor, 1985. Polarity reversal in the Solomon Islands arc, *Nature*, **314**, doi:10.1038/314428a0, 428–430.

   1987a. Seismotectonics of New Guinea: a model for arc reversal following arc–continent collision, *Tectonics*, **6**, 53–67.

 • 1987b. The spatial distribution of earthquakes, focal mechanisms and subducted lithosphere in the Solomon Islands. In *Marine Geology and Geophysics in the Woodlark Basin–Solomon Islands Region*, eds. B. Taylor and N. Exon, Houson, TX, Circum-Pacific Council for Energy and Mineral Resources, 67–88.

 • 1989. Seismicity and focal mechanisms at the New Britain Trench related to deformation of the lithosphere, *Tectonophysics*, **164**, doi:10.1016/0040-1951(89)90231-X, 25–40.

Denham, D., 1969. Distribution of earthquakes in the New-Guinea–Solomon Islands region, *J. Geophys. Res.*, **74**, 4290–4299.

Fukao, Y. and K. Abe, 1971. Multi-mode Love waves excited by shallow and deep earthquakes, *Bull. Earthquake Res. Inst, Tokyo Univ.*, **49**, 1–12.

Isacks, B. L. and P. Molnar, 1971. Distribution of stresses in the descending lithosphere from a global survey of focal-mechanism solutions of mantle earthquakes, *Rev. Geophys. Space Phys.*, **9**, 103–174.

Johnson, T. and P. Molnar, 1972. Focal mechanisms and plate tectonics of the southwest Pacific, *J. Geophys. Res.*, **77**, 5000–5032.

Johnson, R. W., D. E. Mackenzie, and L. E. Smith, 1971. Seismicity and late Cenozoic volcanism of parts of Papua-New Guinea, *Tectonophysics*, **12**, doi:10.1016/0040-1951(71)90064-3, 15–22.

McGuire, J. J. and D. A. Wiens, 1995. A double seismic zone in New Britain and the morphology of the Solomon plate at intermediate depths, *Geophys. Res. Lett.*, **22**, doi:10.1029/95GL01806, 1965–1968.

Pascal, G., 1979. Seismotectonics of the Papua New Guinea–Solomon Islands region, *Tectonophysics*, **57**, doi:10.1016/0040-1951(79)90099-4, 7–34.

 • Pegler, G., S. Das, and J. H. Woodhouse, 1995. A seismological study of the eastern New Guinea and the western Solomon Sea regions and its tectonic implications, *Geophys. J. Int.*, **122**, 961–981.

Ripper, I. D., 1970. Global tectonics and the New Guinea–Solomon Islands region, *Search*, **1**, 226–232.

   1982. Seismicity of the Indo-Australian/Solomon Sea plate boundary in the southwest Papua region, *Tectonophysics*, **87**, doi:10.1016/0040-1951(82)90233-5, 355–369.

Santo, T., 1970 Regional study on the characteristic seismicity of the world. Part IV. New Britain Island region, *Bull. Earthquake Res. Inst., Tokyo Univ.*, **48**, 127–144.

Seno, T. and D. E. Kaplan, 1988. Seismotectonics of western New Guinea, *J. Phys. Earth*, **36**, 107–124.

Shinohara, M., K. Suyehiro, and T. Murayama, 2003. Microearthquake seismicity in relation to double convergence around the Solomon Islands arc by ocean-bottom seismometer observation, *Geophys. J. Int.*, **153**, doi:10.1046/j.1365-246X.2003. 01940.x, 691–698.

Silverstein, B. L., A. G. Brady, and P. N. Mork, 1986. *Processed Strong-Motion Records Recorded on Bouganville Island, Papua New Guinea; Earthquakes of December 13, 1981 and March 18, 1983*, United States Geological Survey, Open File Rept. 86–0264.

Tregoning, P., F. Tan, J. Gilliland, H. McQueen, and K. Lambeck, 1998. Present-day crustal motion in the Solomon Islands from GPS observations, *Geophys. Res. Lett.*, **25**, doi:10.1029/*98GL52761*, 3627–3630.

Wallace, L. M., C. Stevens, E. Silver, R. McCaffrey, W. Loratung, S. Hasiata, R. Stanaway, R. Curley, R. Rosa, and J. Taugaloidi, 2004. GPS and seismological constraints on active tectonics and arc–continent collision in Papua New Guinea: implications for mechanics of microplate rotations in a plate boundary zone, *J. Geophys. Res.*, **109**, doi:10.1029/2003JB002481, B05404.

Weissel, J. K. and R. N. Anderson, 1978. Is there a Caroline plate? *Earth Planet. Sci. Lett.*, **41**, 143–158.

Yoneshima, S., K. Mochizuki, E. Araki, R. Hino, M. Shinohara, and K. Suyehiro, 2005. Subduction of the Woodlark Basin at New Britain Trench, Solomon Islands region, *Tectonophysics*, **397**, doi:10.1016/j.tecto.2004.12.008, 225–239.

10.7 Indonesia (Sunda–Banda Arc)

10.7.1 Regional tectonics and deep seismicity

The deep earthquakes of Indonesia are numerous, various, and important historically; yet the literature describing them can be confusing. Much of this is for linguistic reasons – there are two or more names used commonly for many of the region's geographic features (Fig. 10.13). This is partly attributable to political turmoil (Indonesia was a Dutch colony until achieving independence in 1949), partly due to changes in conventions concerning the transliteration of names originating in local languages, and partly because there seems to be no common policy about whether to translate regional names given by European explorers. Thus, Indonesia's capital city, Djakarta, is also Jakarta and (formerly) Batavia. Seram and Ceram are the names for one of the Spice Islands (also known as the Moluccas); it lies beween Sulawesi (or Suluwesi; formerly, Celebes) and Irian Jaya (formerly, Netherlands New Guinea and now West Papua). On some maps, Vogelkop, the westernmost region of Irian Jaya, is labeled Bird's Head. Because Indonesia (formerly, the Netherlands East Indies) is a political entity, for tectonic discussion the entire region is sometimes called the Sunda arc, the Sunda–Banda arc, and even the Java arc. And of course, Java can be Djava or Jawa.

I thank Shamita Das and Rob McCaffrey for reviewing an earlier draft of this section.

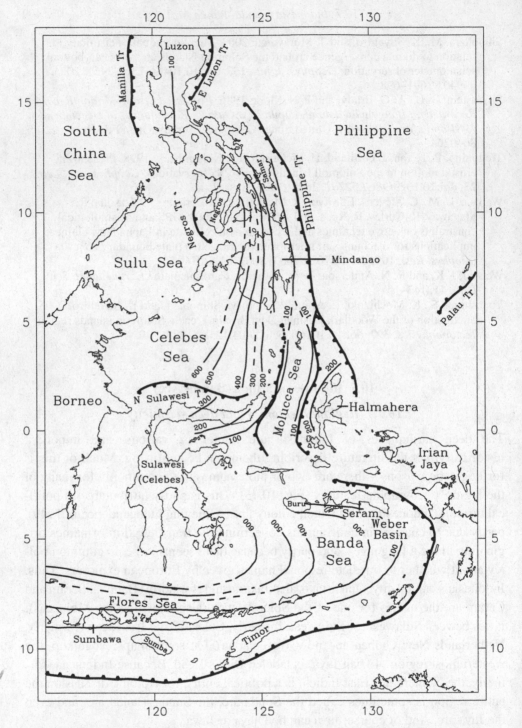

Fig. 10.13 Wadati–Benioff zone geometry and geographic place names for significant features in the Indonesia and Philippine regions. Subducting boundaries are
from Hamilton (1979), Lee and Lawver (1995) and Milsom (2001); teeth are on
the overriding plate. Contours of intermediate- and deep-focus activity are from
Burbach and Frohlich (1986).

If this wasn't enough, the subduction tectonics in eastern Indonesia and the southern Philippines are more complex than anywhere else in the world.[7] Moreover, earthquakes are numerous (Fig. 10.14) and some of the world's largest events occur here. From the south, the Australian plate converges on the Eurasian plate. East of 107° E the Wadati–Benioff zone is steep but mostly regular (Fig. 10.15), although beneath Timor there is a paucity of intermediate-depth seismicity. At about 130° E the Wadati–Benioff zone curves sharply around to form a "shoehorn" or "scoop" that impinges on Irian Jaya. The controversy here concerns whether:

(1) the shoehorn merges continuously with the Philippine Wadati–Benioff zone to the north;
(2) the southwestern-dipping intermediate seismicity near Seram and beneath the Weber Basin represents subduction of one limb of the Philippine plate, and thus is separate from the rest of the shoehorn; or
(3) the Molucca Sea/Seram/Weber Basin region is situated at the intersection of the Java, Philippine, and New Guinea subduction zones where fragments of continental crust complicate subduction, and thus the activity comes from the subduction of several different microplates.

Earthquake focal mechanisms reflect the tectonic complexity in eastern Indonesia, as both T and P axes show considerable scatter (Fig. 10.14), especially for deep-focus earthquakes. There is also scatter in the T axes for intermediate-depth earthquakes, although the majority seem to be of the downdip T variety.

Beneath the Banda Sea a few hypocenters with depths exceeding 500 km are displaced northward of activity in the main Wadati–Benioff zone. Proponents of the various positions above interpret the displaced deep hypocenters as:

(1) attached to the shoehorn, representing a flattening out of the northward subduction of the Australian plate at depths greater than 500 km;
(2) attached to the subducting Philippine plate or even a subducting microplate, representing the deep-focus extension of the southwestern-dipping zone of intermediate-depth seismicity beneath the Weber Basin; or
(3) "Outboard" earthquakes (see Section 5.1.4), not intimately connected to any of the recognized or alleged Wadati–Benioff zones.

In western Indonesia, the convergence of the Australian plate along the Java Trench and the Andaman Sea becomes increasingly oblique[8] and deep-focus earthquakes are entirely absent west of about 107° E (see Fig. 10.15, section A–A′). There is a long-simmering controversy about whether this absence is attributable to the

[7] Hamilton (1979) puts this rather poetically in the first sentence of his opus on Indonesian tectonics, stating: "The tangled patterns of land and sea, and of deep-sea trenches, volcanic chains, and zones of mantle earthquakes, have long fascinated structural geologists and geophysicists."

[8] Like taxonomists, among plate modelers there are both "lumpers" and "splitters". Thus, east and north of Sumatra the splitters divide the Australian plate; the subducting plate is the Indian plate. Moreover, between the Sunda trench and mainland Eurasia they identify two smaller plates, the Burma and the Sunda plates.

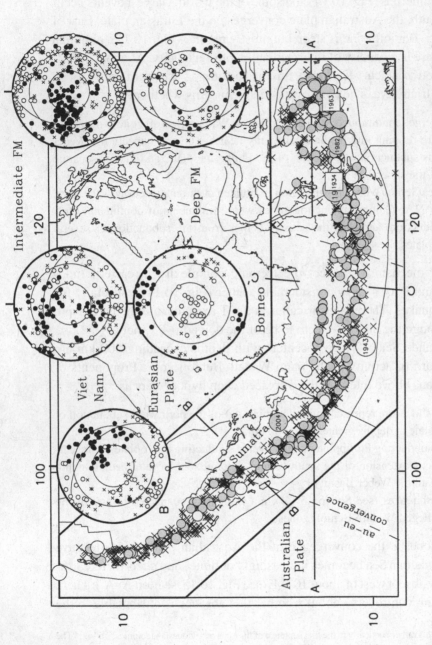

Fig. 10.14 Map of deep earthquake activity in Indonesia. Filled circles, open circles, and crosses indicate selected deep earthquakes in the Harvard CMT, EV Centennial, and EHB catalogs; open squares are other significant events (see description of selection process in the introduction to this chapter). Solid lines labeled A–A' to C–C' indicate locations and orientations of cross sections A to C, thin lines surround events included in each section, and equal-area plots summarize T (solid circles), B (crosses), and P (open circles) axis orientations of selected Harvard CMT for intermediate-depth (above) and deep-focus (below) earthquakes; the rightmost two equal-area plots summarize orientations only for the eastern portion of section A not included in sections B and C. Other lines indicate locations of plate boundaries. The dashed line indicates the direction of Australia–Eurasian (au–eu) plate convergence. This map does not show numerous deep earthquakes that occur beneath the Philippines and northern Sulawesi (see Fig. 10.16).

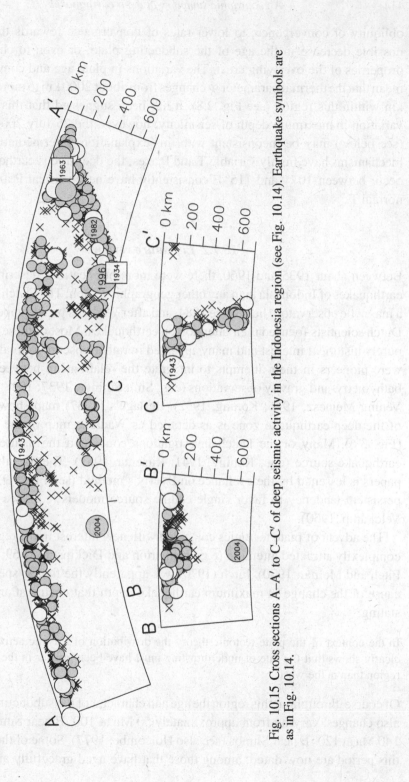

Fig. 10.15 Cross sections A–A′ to C–C′ of deep seismic activity in the Indonesia region (see Fig. 10.14). Earthquake symbols are as in Fig. 10.14.

obliquity of convergence, to lower rates of convergence towards the west, to a possible decrease in the age of the subducting plate, or even to changes in the properties of the overlying crust. The variations in plate age and convergence rate mean that the thermal parameter Φ changes from about 2000 km to more than 10,000 km within this region (see Fig. 2.8). It has been suggested that this explains the variation in maximum depth of seismicity; however, the 25 July 2004 earthquake (see below) may be inconsistent with this explanation. Intermediate-depth focal mechanisms have highly variable T and P axes; the deep-focus earthquakes which occur between 107° and 115° E consistently have near-vertical P axes and slab-normal T.

10.7.2 Literature review

Between about 1935 and 1960, there were more publications describing the deep earthquakes of Indonesia than any other geographic region. The Dutch had operated a magnetic observatory here since 1884, and after Wadati's papers appeared, several Dutch scientists focused their attention on earthquakes. Most of these papers are of purely historical interest and many appeared in rather obscure journals. The Dutch were pioneers in their attempts to integrate the relationship between seismicity, bathymetry, and gravity observations (e.g., Smit Sibinga, 1937; 1938; Visser 1938; Vening Meinesz, 1946; Koning, 1952). Berlage's (1937) map showing contours of the deep earthquake zone is as detailed as Wadati's map of the Japanese arc (Fig. 3.6). Many of the Dutch investigations concerned the nature of the deep earthquake source (e.g., Koning, 1941; Ritsema, 1957). The usefulness of these papers is lessened by their reliance on Byerly's method (see Fig. 8.9) and by their persistent tendency to favor single-couple source models (e.g., see Ritsema and Veldkamp, 1960).

The advent of plate tectonics coincided with new interest in the region, when its complexity attracted attention (e.g., Hatherton and Dickinson, 1969; Santo, 1969; Fitch and Molnar, 1970). Fitch (1970) was apparently the first to speculate on the cause of the change in maximum earthquake depth that occurs at around 107° E, stating:

In the context of the plate tectonic theory the distribution of seismic activity with depth clearly shows that the rate of underthrusting must have been greater in the eastern Sunda region than in the west.

Of course, throughout this region the age and character of the subducting lithosphere also changes, varying from approximately 50 Ma at 100° E near Sumatra to about 140 Ma at 120° E near Sumba (see also Holcombe, 1977). Some of the papers from this period are now dated; among those that have aged gracefully are Hamilton's

(1979) extraordinary monograph on the regional tectonics and Cardwell *et al.*'s (1978) summary of the seismicity.

A still-unresolved question concerns the continuity of the eastern Indonesian Wadati–Benioff zone; i.e., how many independent fragments of subducting lithosphere do the earthquakes represent? Richardson (1993) shows a gap in seismicity beneath Timor. However, McCaffrey *et al.* (1985) operated a temporary network of stations in western Timor above the gap boundary; McCaffrey (1989) shows that high-frequency S waves travel through the gap and concludes that the lithosphere is continuous there. The main controversy concerns whether the Wadati–Benioff zone extending southwest of Seram is attached to the main northward-dipping zone beneath Sumbawa. McCaffrey (1989) concludes that these are separate fragments, while Milsom (2001) argues the lithosphere is in one continuous piece. Cardwell *et al.* (1978) show the lithosphere as continuous and with the northward subduction beneath the Banda Sea flattening laterally out beneath 500 km. However, the best data and relocations are from Schöffel and Das (1999) and Das (2004), who favor the two-slab hypothesis. They also conclude that the deep "flat" Banda hypocenters lie along the direct extension of the westward dipping Wadati–Benioff zone beneath the Weber Basin. Near 120° E and beneath about 500 km depth the near-vertical Wadati–Benioff zone bends backwards so that the Australian plate subduction direction turns slightly southward. Tomographic analyses do not resolve these controversies; however, both Puspito and Shimazki (1995) and Widiyantoro and Van der Hilst (1997) do find that the high-velocity region extends into the lower mantle ($\sim$1500 km depth) beneath the central parts of the arc in Sumbawa, even though it reaches only to about 500 km beneath Sumatra and 700 km beneath the Banda Sea.

There have been some impressive recent compilations of seismicity and focal mechanisms in the central portion of Indonesia between Timor and Sumatra. Schöffel and Das (1999) hand-picked 1790 arrivals from P, S, pP, sP, PcP and ScP phases to relocate Indonesian earthquakes and present detailed cross sections of the Wadati–Benioff zone, while Slancová *et al.* (2000) summarize regional focal mechanisms. The accuracy of these relocations allowed Das *et al.* (2000) to conclude that the deep-focus mechanisms are consistent with thickening of the subducting lithosphere, i.e., the focal mechanisms are systematically different for events on the lower and upper sides of the subducting plate.

There are fewer investigations focusing on the western Sumatra and Andaman Sea regions where subduction becomes oblique and there are no deep-focus earthquakes. However, Ghose and Oike (1988), Ghose *et al.* (1990) and Guzmán-Speziale and Ni (1996) do evaluate the seismicity and stress within the subducted lithosphere here, while both Dasgupta and Mukhopadhyay (1993) and Slancová et al. (2000) summarize focal mechanisms. Fauzi *et al.* (1996) analyzed 30 months of data from a local network in northwestern Sumatra. Their observations confirm that earthquakes extend to depths of 200–250 km, and they conclude that the subducted lithosphere is broadly bent rather than separated into distinct lateral segments.

10.7.3 Significant earthquakes

The Indonesian arc is extraordinarily active, and among its earthquakes are some of the largest deep and shallow[9] events on the planet. Some, like the great Banda Sea earthquake of 1 February 1938 ($h_{EVC} = 35$ km; $h_{Okal\&Reymond} = 60$ km; $M_{Okal\&Reymond} = 8.6$) are shallow in most catalogs but have centroid depths suggesting an intermediate focus (Okal and Reymond, 2003). In any case, the EV Centennial and Harvard catalogs report about 40 earthquakes with depths of 60 km or more and magnitudes of 7 or greater. Only a few have received concentrated attention.

29 June 1934: $h_{Reese–Okal} = 627$ km; $m_{B(Abe)} = 7.0$. This earthquake is chiefly significant because GR assigned it a depth of 720 km, gave this depth an "A" grade for quality, and subsequently called it the "deepest of all known large earthquakes" (e.g., Richter, 1958). Although numerous subsequent publications have repeated this assertion, it almost certainly was shallower. Although Berlage (1937) and Koning (1941; 1942) listed the depth as 700 km, Berlage's (1935) depth from (pP-P) intervals and the published ISS depth were both 680 km. And the GR depth provoked Jeffreys (1942) to publish a rather thorough analysis of all available arrival times, arguing that the maximum credible depth was 650 km. Curiously, apparently neither Gutenberg nor Richter ever referenced Jeffreys' paper. Finally, Rees and Okal (1987) relocated the earthquake using modern methods and the arrivals reported to the ISS. They obtained a depth of 627 km; the EV Centennial catalog relocation placed it at 648 km; these results thus agree substantially with Jeffreys. Koning (1942) reports first motions for this earthquake at 18 stations and discusses possible focal mechanisms; however, apparently there has been no modern waveform analysis to determine the mechanism.

4 November 1963: $h_{Abe} = 108$ km; $m_{B(Abe)} = 7.8$; $M_{W(Osada \& Abe)} = 8.3$. If we accept Osada and Abe's (1981) moment of 3.1×10^{21} N-m for this earthquake, it is the largest earthquake ever with a depth of 100 km or greater, including the 1994 Bolivia deep-focus earthquake. It occurred at the eastern end of the Indonesian Wadati–Benioff zone, just where the zone begins to bend sharply northward. Hearn and Webb's (1984) intensity map indicates it was widely felt throughout northern

---

[9] The shallow-focus earthquake of 26 December 2004 had M_W 9.0 and produced a tsunami that killed more than 250,000 people.

Australia, even though its epicenter was about 500 km north of the coastline; intensities were MMI V in Darwin. Both Osada and Abe (1981) and Welc and Lay (1987) (see inset focal mechanism) investigated the source process. Although Welc and Lay found the rupture lasted 80 s or more, the bulk of the moment release occurred within 50 s, initiating at a depth of about 120 km and expanding laterally and downdip over a vertical extent of about 50 km. The along-strike rupture length was about 100 km. Osada and Abe studied the source by evaluating surface waves; they determined a fault dimension of $90 \times 70\,\text{km}^2$, a stress drop of 120 bars, a slip of 7 m, and suggested that this event ruptured the entire thickness of the elastic plate.

24 November 1983: $h_{\text{CMT}} = 157\,\text{km}$; $M_{\text{W(CMT)}} = 7.4$. This earthquake, with an epicenter about a degree west of the 1963 earthquake, is unusual because it had a significant non-double-couple component. Riedesel (1985) studied it in the frequency domain and found non-double-couple components of 62–78% at frequencies of 3.5–6.5 mHz; this compares to 60 per cent for the Harvard CMT. Although Michael-Leiba (1984) used P-wave first motions to determine a pure double-couple mechanism, these observations were generally consistent with the non-double-couple solution as well. Michael-Leiba also mentioned that there were about ten aftershocks recorded teleseismically within a month of this event.

17 June 1996: $h_{\text{CMT}} = 584\,\text{km}$; $M_{\text{W(CMT)}} = 7.9$ (see Fig. 2.3). This is the fourth largest deep-focus earthquake known (see Table 1.2) and the second largest to occur in the era of digital recording. There were several investigations of its source process. Goes *et al.* (1997) and Tibi *et al.* (1999) evaluated subevents within the rupture; both found that rupture was bilateral with two to four subevents occurring over about 20–25 seconds and extending over a distance of 70–105 km. Antolik *et al.* (1999) obtained similar results by inverting waveforms to find the slip on cells with dimensions of 5 km (see Fig. 6.11). The various analyses found somewhat differing results concerning rupture velocity and stress drop. Tinker *et al.* (1998) were able to locate 17 teleseismically recorded aftershocks which occurred within five days. These hypocenters spanned a slab-normal distance of 57 km, and thus Wiens

(1998) suggested that some must have occurred outside of the previously active seismic zone. The mainshock was well-recorded in the distance range 120°–150°, allowing Okal and Cansi (1998) and Deuss *et al.* (2000) to make the first (tentative) identification of PKJKP, the shear phase traveling within the inner core.

25 July 2004: $h_{CMT} = 601$ km; $M_{W(CMT)} = 7.3$. This earthquake was highly isolated, occurring about 300 km to the east of previous deep-focus activity. With respect to its relatively large size and isolation it is similar to the 1954 Spanish and 1970 Colombia deep-focus events. Also, it occurred approximately at the eastern boundary of a region where it has been suggested that the thermal parameter Φ was too low to permit deep-focus seismicity (see Fig. 2.8).

10.7.4 Indonesia (Sunda–Banda Arc) references

Antolik, M., D. Dreger, and B. Romanowicz, 1999. Rupture processes of large deep-focus earthquakes from inversion of moment rate functions, *J. Geophys. Res.*, **104**, doi:10.1029/1998JB00042, 863–894.

Berlage, H. P., 1935. The earthquake in the Flores Sea on June 29, 1934, with a focal depth of nearly 700 km, *Handelingen Zevende Nederlandsch-Indisch Natuurwetenschappelijk Congres Batavia*, 658–665.

1937. A provisional catalogue of deep-focus earthquakes in the Netherlands East Indies 1918–1936, *Gerlands Beitr. z. Geophysik*, **50**, 7–17.

Burbach, G. V. and C. Frohlich, 1986. Intermediate and deep seismicity and lateral structure of subducted lithosphere in the circum-Pacific region, *Rev. Geophys.*, **24**, 833–874.

•Cardwell, R. K., B. L. Isacks, and D. E. Karig, 1978. Geometry of the subducted lithosphere beneath the Banda Sea in eastern Indonesia from seismicity and fault plane solutions, *J. Geophys. Res.*, **83**, 2825–2838.

•Das, S., 2004. Seismicity gaps and the shape of seismic zones in the Banda Sea region from relocated hypocenters, *J. Geophys. Res.*, **109**, doi:10.1029/2004JG003192, B12303.

Das, S., H.-J. Schöffel, and F. Gilbert, 2000. Mechanism of slab thickening near 670 km under Indonesia, *Geophys. Res. Lett.*, **27**, doi:10.1029/1999GL010865, 831–834.

Dasgupta, S. and M. Mukhopadhyay, 1993. Seismicity and plate deformation below the Andaman arc, northeastern Indian Ocean, *Tectonophysics*, **225**, doi:10.1016/0040-1951(93)90314-A, 529–542.

Deuss, A., J. H. Woodhouse, H. Paulssen, and J. Trampert, 2000. The observation of inner core shear waves, *Geophys. J. Int.*, **142**, doi:10.1046/j.1365-246x.2000.00147.x, 67–73.

Fauzi, R. McCaffrey, D. Wark, Sunaryo and P. Y. Prih Haryadi, 1996. Lateral variation in slab orientation beneath Toba Caldera, northern Sumatra, *Geophys. Res. Lett.*, **23**, doi:10.1029/96GL00381, 443–446.

Fitch, T. J., 1970. Earthquake mechanisms and island arc tectonics in the Indonesian-Philippine region, *Bull. Seismol. Soc. Amer.*, **60**, 565–591.

Fitch, T. J. and P. Molnar, 1970. Focal mechanisms along inclined earthquake zones in the Indonesia–Philippine region, *J. Geophys. Res.*, **75**, 1431–1444.

Ghose, R. and K. Oike, 1988. Characteristics of seismicity distribution along the Sunda arc: some new observations, *Bull. Disaster Prevention Res. Inst.*, **38**, 29–48.

Ghose, R., S. Yoshioka, and K. Oike, 1990. Three-dimensional numerical simulation of the subduction dynamics in the Sunda arc region, southeast Asia, *Tectonophysics*, **181**, doi:10.1016/0040-1951(90)90018-4, 223–255.

Goes, S., L. Ruff, and N. Winslow, 1997. The complex rupture process of the 1996 deep Flores, Indonesia earthquake (M_W 7.9) from teleseismic P-waves, *Geophys. Res. Lett.*, **24**, doi:10.1029/97GL01245, 1295–1298.

• Guzmán-Speziale, M. and J. Ni, 1996. Seismicity and active tectonics of the western Sunda arc. In *The Tectonic Evolution of Asia*, eds. A. Yin and T. M. Harrison, Cambridge, U. K., Cambridge University Press, 63–84.

• Hamilton, W., 1979. *Tectonics of the Indonesian Region*, Washington, DC, Geological Survey Professional Paper 1078, 345 pp.

Hatherton, T. and W. R. Dickinson, 1969. The relationship between andesitic volcanism and seismicity in Indonesia, the Lesser Antilles, and other island arcs, *J. Geophys. Res.*, **74**, 5301–5310.

Hearn, S. J. and J. P. Webb, 1984. Continental-scale felt effects of the large Banda Sea earthquake of 4 November 1963, *Bull. Seismol. Soc. Amer.*, **74**, 349–351.

Holcombe, C. J., 1977. Earthquake foci distribution in the Sunda arc and the rotation of the backarc area, *Tectonophysics*, **43**, doi:10.1016/0040-1951(77)90115-9, 169–180.

Jeffreys, H., 1942. The deep earthquake of 1934 June 29, *Mon. Not. Roy. Astron. Soc. Geophys. Supp.*, **5**, 33–36.

Koning, L. P. G., 1941. On the mechanism of deep-focus earthquakes, *Gerlands Beitr. z. Geophysik*, **58**, 159–197.

    1942. On the determination of the faultplanes in the hypocenter of the deep-focus earthquake of June 29, 1934 in the Netherlands East Indies, *Proc. Koninklijke Nederlandse Akad. v. Wetenschappen*, **45**, 636–642.

    1952. Earthquakes in relation to their geographical distribution, depth and magnitude, *Proc. Koninklijke Nederlandse Akad. v. Wetenschappen*, **B55**, 60–77, 174–206.

Lee, T.-Y. and L. A. Lawver, 1995. Cenozoic plate reconstruction of southeast Asia, *Tectonophysics*, **251**, doi:10.1016/0040-1951(95)00023-2, 85–138.

Michael-Leiba, M. O., 1984. The Banda Sea earthquake of 24 November 1983: evidence for intermediate depth thrust faulting in the Benioff zone, *Phys. Earth Planet. Int.*, **36**, 95–98.

McCaffrey, R., 1989. Seismological constraints and speculations on Banda arc tectonics, *Netherlands J. Sea Research*, **24**, 141–152.

McCaffrey, R., P. Molnar, and S. W. Roecker, 1985. Microearthquake seismicity and fault plane solutions related to arc–continent collision in the eastern Sunda arc, Indonesia, *J. Geophys. Res.*, **90**, 4511–4528.

Milsom, J., 2001. Subduction in eastern Indonesia: how many slabs? *Tectonophysics*, **338**, doi:10.1016/0040-1951(01)00137-8, 167–178.

Okal, E. A. and Y. Cansi, 1998. Detection of PKJKP at intermediate periods by progressive multi-channel correlation, *Earth Planet. Sci. Lett.*, **164**, doi:10.1016/S0012-821X(98)00210-6, 23–30.

Osada, M. and K. Abe, 1981. Mechanism and tectonic implications of the great Banda Sea earthquake of November 4, 1963, *Phys. Earth Planet. Int.*, **25**, 129–139.

Puspito, N. T. and K. Shimazki, 1995. Mantle structure and seismotectonics of the Sunda and Banda arcs, *Tectonophysics*, **251**, doi:10.1016/0040-1951(95)00063-1, 215–228.

Rees, B. A. and E. A. Okal, 1987. The depth of the deepest historical earthquakes, *Pure Appl. Geophys.*, **125**, 699–715.

Richardson, A., 1993. Lithosphere structure and dynamics of the Banda arc collision zone: eastern Indonesia, *Bull. Geol. Soc. Malaysia*, **33**, 105–118.

Richter, C. F., 1958. *Elementary Seismology*, San Francisco, W. H. Freeman, 768 pp.

Riedesel, M. A., 1985. *Seismic Moment Tensor Recovery at Low Frequencies*, Ph.D. dissertation, University of California at San Diego, 245 pp.

Ritsema, A. R., 1957. Earthquake-generating stress systems in southeast Asia, *Bull. Seismol. Soc. Amer.*, **47**, 267–279.

Ritsema, A. R. and J. Veldkamp, 1960. *Fault Plane Mechanisms of Southeast Asian Earthquakes*, Mededelingen en Verhandelingen 76, Koninklijke Nederlandse Met. Inst., 63 pp.

Santo, T., 1969. Regional study on the characteristic seismicity of the world. Part II. From Burma down to Java, *Bull. Earthquake Res. Inst., Tokyo Univ.*, **47**, 1049–1061.

• Schöffel, H.-J. and S. Das, 1999. Fine details of the Wadati–Benioff zone under Indonesia and its geodynamic implications, *J. Geophys. Res.*, **104**, doi:10.1029/1999JB900091, 13101–13114.

• Slancová, A., A. Špičák, V. Hănus, and J. Vănek, 2000. How the state of stress varies in the Wadati–Benioff zone: indications from focal mechanisms in the Wadati–Benioff zone beneath Sumatra and Java, *Geophys. J. Int.*, **143**, doi:10.1046/j.1365-246X.2000.01304.x, 909–943.

Smit Sibinga, G. L., 1937. On the relation between deep-focus earthquakes, gravity, and morphology in the Netherlands East Indies, *Gerlands Beitr. z. Geophysik*, **51**, 402–409.

   1938. Additional note on the relation between deep-focus earthquakes, gravity, and morphology in the Netherlands East Indies, *Gerlands Beitr. z. Geophysik*, **53**, 392–394.

Tibi, R., C. H. Estabrook, and G. Bock, 1999. The 1996 June 17 Flores Sea and 1994 March 9 Fiji–Tonga earthquakes: source processes and deep earthquake mechanisms, *Geophys. J. Int.*, **138**, doi:10.1046/j.1365-246x.1999.00879.x, 625–642.

Tinker, M. A., S. L. Beck, W. Jiao, and T. C. Wallace, 1998. Mainshock and aftershock analysis of the June 17, 1996, deep Flores Sea earthquake sequence: implications for the mechanism of deep earthquakes and the tectonics of the Banda Sea, *J. Geophys. Res.*, **103**, doi:10.1029/97JB03533, 9987–10001.

Vening Meinesz, F. A., 1946. Deep-focus and intermediate earthquakes in the East Indies, *Proc. Koninklijke Nederlandse Akad. v. Wetenschappen*, **49**, 855–865.

Visser, S. W., 1938. Seismic isobaths in the East Indian archipelago, *Gerlands Beitr. z. Geophysik*, **53**, 389–391.

Welc, J. L. and T. Lay, 1987. The source rupture process of the great Banda Sea earthquake of November 4, 1963, *Phys. Earth Planet. Int.*, **45**, 242–254.

Widiyantoro, S. and R. van der Hilst, 1997. Mantle structure beneath Indonesia inferred from high-resolution tomographic imaging, *Geophys. J. Int.*, **130**, 167–182.

Wiens, D. A., 1998. Source and aftershock properties of the 1996 Flores Sea earthquake, *Geophys. Res. Lett.*, **25**, doi:10.1029/98GL00417, 781–784.

10.8 Philippines

10.8.1 Regional tectonics and deep seismicity

At first glance, the subduction tectonics of the Philippines seem simple enough: the Philippines plate converges westward towards the Eurasian plate, producing deep earthquake activity that extends northward from Indonesia to Taiwan (Figs. 10.16 and 10.17). A peculiarity is that the relative plate motions are poorly known because the Philippine plate is nearly devoid of islands and surrounded by consuming plate boundaries. The usual consensus is that the Philippines–Eurasia convergence is somewhat oblique and the rate is about 10 cm/yr; at the Philippine Trench the age of subducting lithosphere is in the 33–66 Ma range; this suggests the thermal parameter is about 2000–4000 km.

A closer look reveals that complexity is rampant as there is consumption of lithosphere along at least five independent subduction zones, each producing a separate Wadati–Benioff zone (see Fig. 10.13 for place names).

- *Mindanao*: The longest and most active zone lies beneath Mindanao (sections C–C′ and E–E′; Figs. 10.16 and 10.18). Subduction occurs along the Philippine Trench, producing a west-dipping Wadati–Benioff zone that produces both intermediate- and deep-focus earthquakes. This zone extends from the Celebes Sea north to the island of Samar. It includes large earthquakes with depths as great as 644 km, although there is a pronounced gap in seismic activity between depths of about 200–500 km beneath Mindanao.

- *Halmahera*: Subduction in the eastern Molucca Sea produces an east-dipping Wadati– Benioff zone beneath Halmahera (section D–D′). This extends to a depth of about 230 km.

- *North Sulawesi*: There are large earthquakes with depths as great as 250 km beneath the north arm of Sulawesi (section B–B′). One interpretation is that these originate from westward subduction in the western Molucca Sea.

- *Negros*: There is ongoing subduction along the Negros Trench, producing an east-dipping Wadati–Benioff zone. Although this zone extends to a depth of 100 km or more, the activity rate is low and there are few well-located hypocenters; thus, its presence isn't obvious from teleseismic locations (e.g., section E–E′, left).

I thank Rob McCaffrey for reviewing an earlier draft of this section.

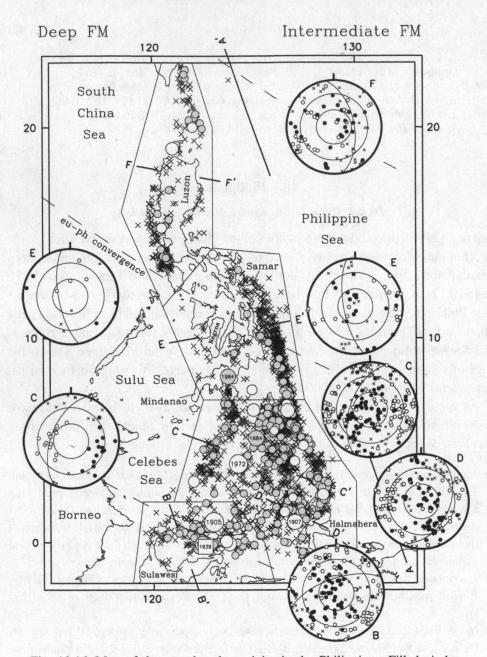

Fig. 10.16 Map of deep earthquake activity in the Philippines. Filled circles, open circles, and crosses indicate selected deep earthquakes in the Harvard CMT, EV Centennial, and EHB catalogs; open squares are other significant events (see description of selection process in the introduction to this chapter). Solid lines labeled A–A′ to F–F′ indicate locations and orientations of cross sections A to F, thin lines surround events included in each section, and equal-area plots summarize T (solid circles), B (crosses), and P (open circles) axis orientations of selected Harvard CMT for intermediate-focus (B, C, D, E, and F at right) and deep-focus (C and E at left) earthquakes. The dashed line indicates the direction of Eurasian–Philippine (eu–ph) plate convergence as reported by Seno et al. (1993).

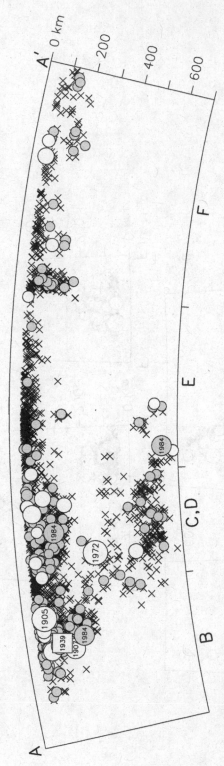

Fig. 10.17 Summary cross section A–A' for Philippine region. Earthquake symbols are as in Fig. 10.16.

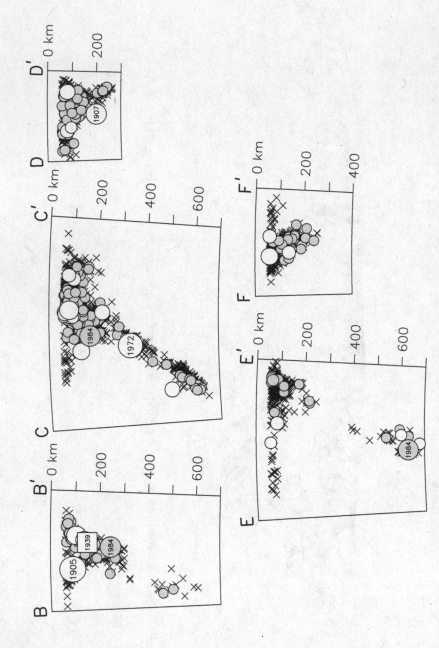

Fig. 10.18 Cross sections B–B′ to F–F′ of deep seismic activity in the Philippine region. Earthquake symbols as in Fig. 10.16.

- *West Luzon*: There is active subduction along the Manilla Trench on the western side of Luzon (section F–F′). This produces an east-dipping Wadati–Benioff zone that extends to a depth of about 250 km.

As suggested above, there is uncertainty about the geometry of some of these zones and the surface origin of the subducted material. For example, for the north Sulawesi and Halmahera zones the usual interpretation is that the lithosphere here is saddle-shaped as it is being consumed on both sides of the Molucca Sea. And some interpretations also have the lithosphere subducting from the north along the North Sulawesi Trench with the associated Wadati–Benioff zone beneath the Sulawesi's north arm (e.g., see Hamilton, 1979). Finally, there is subduction along both sides of Luzon – in the east along the East Luzon Trough as well as along the Manilla Trench in the west. However, for the westward-dipping earthquake activity connected to the East Luzon Trough, there are no large-magnitude and few well-determined earthquakes with depths exceeding 60 km.

Apparently because of the tectonic complexity, throughout this region focal mechanism orientations are quite scattered. However, there is an overall tendency for intermediate-focus earthquakes to have downdip T axes and for deep-focus earthquakes to have downdip P axes.

10.8.2 Literature review

Both Willis (1944) and GR mentioned the presence of deep earthquakes in the Philippines and commented on the complexity of the tectonics. GR stated that "there is great complication, with intersecting structures of different types and trends." Fitch and Molnar (1970) presented nine focal mechanisms from Philippine Wadati–Benioff zones, including four from the north Sulawesi–Halmahera zones. They called the west Luzon zone "a proposed arc" and noted that "seismic activity is too sparse to yield well defined profiles [but] bathymetric data for the region along the west coast of Luzon suggest that the arc is convex toward Asia and imply that a seismic zone dips towards the east."

Subsequently there have been several investigations of the Philippine region that evaluate the Wadati–Benioff zone structure using teleseismic locations and focal mechanisms. Cardwell *et al.* (1980) studied the entire region, and presented 12 cross sections of selected hypocenters, mostly using ISC data. These clearly show the presence of the east-dipping west Luzon zone extending from the Manilla Trench to a depth of about 200 km. Hypocenters determined using data from stations on Taiwan suggest that this east-dipping zone may extend to southern or even central Taiwan (Roecker *et al.*, 1987; Lin *et al.*, 1999). On the east side of Luzon Hamburger *et al.* (1983) also indicate that there is subduction along the East Luzon Trough;

however, their cross section and those of Lin and Tsai (1981) show a diffuse cloud of hypocenters with only scattered events beneath 60 km. Cardwell *et al.* (1980) include two cross sections beneath the north arm of Sulawesi which show a near-vertical zone of rather intense activity extending from the surface to about 200 km depth. For the Negros zone they state that "earthquakes do not clearly define a seismic zone that dips east from the Negros Trench, but earthquakes with depths in the mantle beneath the islands indicate that subduction of lithosphere has occurred". However, Acharya and Aggarwal's (1980) cross section shows a Wadati–Benioff zone here more clearly. For the region north of Mindanao, Yang *et al.* (1996) and Bautista *et al.* (2001) include more recent hypocenters and present EW cross sections at 1°intervals.

For brief periods there has been a temporary network of short-period stations operating on various islands around the Molucca Sea (McCaffrey, 1982; McCaffrey and Sutardjo, 1982). McCaffrey (1983) evaluated these data and concluded that the north Sulawesi zone represented westward subduction extending to depths of 400 km, and the Halmahera zone was an expression of eastward subduction extending to about 200 km. Because there was considerable variability in the frequency content of body phases from deep quakes recorded at different island stations, he concluded that the subducted lithosphere beneath Halmahera may not be continuous.

10.8.3 Significant earthquakes

There are about 30 deep earthquakes in the EV Centennial and Harvard catalogs with magnitudes of 7 or greater. Two in the north Sulawesi zone have m_B of 7.8 (22 January 1905, $h_{Abe} = 90$ km; 21 December 1939, $h_{Abe} = 150$ km); GR stated that these were "among the largest [intermediate shocks] known." The Harvard catalog has three earthquakes with M_W greater than 7, all occurring in 1984 (Mindanao: 5 March 1984; $h_{CMT} = 644$ km, $M_{W(CMT)} = 7.3$; north Sulawesi: 6 August 1984, $h_{CMT} = 253$ km, $M_{W(CMT)} = 7.4$; Mindanao: 20 November 1984, $h_{CMT} = 181$ km, $M_{W(CMT)} = 7.5$). Surprisingly, none of these earthquakes has been the focus of special investigation.

10.8.4 Philippines references

Acharya, H. K. and Y. P. Aggarwal, 1980. Seismicity and tectonics of the Philippine Islands, *J. Geophys. Res.*, **85**, 3239–3250.

Bautista, B. C., M. L. P. Bautista, K. Oike, F. T. Wu, and R. S. Punongbayan, 2001. A new insight on the geometry of subducting slabs in northern Luzon, Philippines, *Tectonophysics*, **339**, doi:10.1016/0040-1951(01)00120-2, 279–310 .

•Cardwell, R. K., B. L. Isacks, and D. E. Karig, 1980. The spatial distribution of earthquakes, focal mechanisms solutions, and subducted lithosphere in the Philippine and northeastern Indonesian islands. In *The Tectonic and Geologic Evolution of Southeast Asian Seas and Islands: Part 1*, ed. D. E. Hayes, Geophys. Mon. 23, Washington, DC, American Geophysical Union, 1–35.

Fitch, T. J. and P. Molnar, 1970. Focal mechanisms along inclined earthquake zones in the Indonesia–Philippine region, *J. Geophys. Res.*, **75**, 1431–1444.

Hamburger, M. W., R. K. Cardwell, and B. L. Isacks, 1983. Seismotectonics of the northern Philippine island arc. In *The Tectonic and Geologic Evolution of Southeast Asian Seas and Islands: Part 2*, ed. D. E. Hayes, Geophys. Mon. 27, Washington, DC, American Geophysical Union, 1–22.

Hamilton, W., 1979. *Tectonics of the Indonesian Region*, Washington, DC, Geological Survey Professional Paper 1078, 345 pp.

Lin, C. H., B. S. Huang, and R. J. Rau, 1999. Seismological evidence for a low-velocity layer within the subducted slab of southern Taiwan, *Earth Planet. Sci. Lett.*, **174**, doi:10.1016/S0012-821X(99)00255-1, 231–240.

Lin, M. T. and Y. B. Tsai, 1981. Seismotectonics in Taiwan–Luzon area, *Bull. Inst. Earth Sci., Acad. Sin.*, **1**, 51–82.

McCaffrey, R., 1982 Lithospheric deformation within the Molucca Sea arc–arc collision: evidence from shallow and intermediate earthquake activity, *J. Geophys. Res.*, **87**, 3663–3678.

1983. Seismic wave propagation beneath the Molucca Sea arc–arc collision zone, Indonesia, *Tectonophysics*, **96**, doi:10.1016/0040-1951(83)90243-3, 45–57.

McCaffrey, R. and R. Sutardjo, 1982. Reconnaissance microearthquake survey of Sulawesi, Indonesia, *Geophys. Res. Lett.*, **9**, 793–796.

Okal, E. A. and D. Reymond, 2003. The mechanism of great Banda Sea earthquake of 1 February 1938: applying the method of preliminary determination of focal mechanism to a historical event, *Earth Planet. Sci. Lett.*, **216**, doi:10.1016/S0012-821X(03)00475-8, 1–15.

Roecker, S. W., Y. H. Yeh, and Y. B. Tsai, 1987. Three-dimensional P and S wave velocity structures beneath Taiwan: deep structure beneath an arc–continent collision, *J. Geophys. Res.*, **92**, 10547–10570.

Seno, T., S. Stein, and A. E. Gripp, 1993. A model for the motion of the Philippine Sea plate consistent with NUVEL-1 and geological data, *J. Geophys. Res.*, **98**, 17941–17948.

Willis, B., 1944. Philippine earthquakes and structure, *Bull. Seismol. Soc. Amer.*, **34**, 69–81.

Yang, T. F., T. Lee, C.-H. Chen, S.-H. Cheng, U. Knittel, R. S. Punongbayan, and A. R. Rasdas, 1996. A double island arc between Taiwan and Luzon: consequence of ridge subduction, *Tectonophysics*, **258**, doi:10.1016/0040-1951(95)00180-82, 85–101.

10.9 Ryukyu–Taiwan

10.9.1 Regional tectonics and deep seismicity

The Ryukyu arc stretches between Taiwan and Kyushu, the southernmost large island of Japan (Fig. 10.19). Here the Philippine and Eurasian plates converge at

I thank Honn Kao for reviewing an earlier draft of this section.

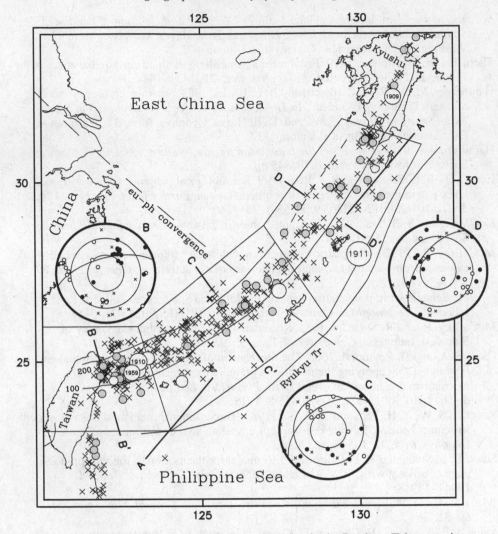

Fig. 10.19 Map of intermediate-depth earthquakes in the Ryukyu–Taiwan region. Filled circles, open circles, and crosses indicate selected earthquakes in the Harvard CMT, EV Centennial, and EHB catalogs (see description of selection process in the introduction to this chapter). Solid lines labeled A–A' to D–D' indicate locations and orientations of cross sections A to D, thin lines surround events included in each section, and equal-area plots summarize T (solid circles), B (crosses), and P (open circles) axis orientations of selected Harvard CMT for intermediate-depth earthquakes. Other lines indicate locations of plate boundaries. The dashed line indicates the direction of Eurasian–Philippine (eu–ph) plate convergence.

6.5–8.5 cm/yr and the age of the subducting lithosphere is about 40–50 Ma. The Ryukyu Wadati–Benioff zone extends to depths of 200–300 km with a north- and westward dip that is about 60° near Taiwan but closer to 45–50° along most of the arc (Fig. 10.20). Thus the thermal parameter Φ is about 3000 km near Kyushu, and decreases to about 2000 km near Taiwan. There is a distinct change in the tectonics

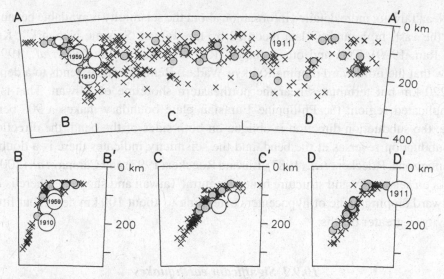

Fig. 10.20 Cross sections A–A′ to D–D′ of deep seismicity in the Ryukyu–Taiwan region. Earthquake symbols as in Fig. 10.19.

where the arc meets Taiwan; beneath the island there is a gap in seismicity at depths exceeding 100 km, and then to the south the Wadati–Benioff zone dips eastward.

10.9.2 Literature review

GR say little about the deep seismicity here, and it didn't receive much attention until Katsumata and Sykes (1969) reviewed the regional seismicity using teleseismic observations. The two focal mechanisms available to them indicated that the Ryukyu slab underwent downdip compression. However, Shiono *et al.* (1980) and Kao and Chen (1991) concur that earthquake mechanisms indicate "downdip extension along the northern end of the arc near Kyushu but abruptly change to downdip compression along the rest of the arc." Earthquakes that have occurred more recently indicate that both varieties of mechanism occur occasionally in the northern section (see Fig. 10.19). Christova (2004) evaluated the stress field and demonstrated that it was compatible with downdip extension despite the existence of a few exceptional events.

There has been considerable recent interest in the geometry of the subducted lithosphere in this region as inferred from the distribution of hypocenters and from tomographic analysis. Lallemand *et al.* (2001) evaluated teleseismic observations, while Nakamura *et al.* (2003) also used recordings from ocean-bottom seismograph and regional stations. Along the Ryukyu arc the tomographic images clearly indicate that the subducting slab is detached beneath the depths of the deepest hypocenters.

Near Taiwan, more detailed information about the seismicity is available because of "the vast, high-quality data recorded by the Taiwan Seismic Network" (Kao and Rau, 1999). This and other studies (Rau and Wu, 1995; Kao *et al.*, 1998) show that the northward dipping Ryukyu Wadati–Benioff zone extends to a depth of 220 km but terminates near the northeastern shoreline of Taiwan. This is a complicated region; the Philippine–Eurasian plate boundary makes a 90° bend here, the subduction direction is oblique on both sides of the bend, the direction of subduction reverses at the bend, and the seismicity indicates there is a double seismic zone. Roecker *et al.* (1987), Lin and Roecker (1993) and Carena *et al.* (2002) focus on seismicity and structure beneath central Taiwan and show that there is an eastward dipping plane of hypocenters extending to about 100 km depth, but little activity at greater depths.

10.9.3 Significant earthquakes

In this region the EV Centennial catalog reports 10 earthquakes with magnitudes of 7.0 or more. Three have reported m_B of 7.5 or more – the 1911 shock discussed below, and two which occurred just east of Taiwan in 1910 and 1959. The 1959 earthquake occurred when the controversy about single- and double-couple sources was at its peak. Ritsema (1962) and Scholte and Ritsema (1962) used it as evidence favoring the single-couple source model, a model now no longer in favor. The largest earthquake in the Harvard catalog occurred on 2 January 1981 ($M_{W(CMT)} = 6.9$, $h_{CMT} = 217$ km) and had an epicenter near that of the 1911 event.

15 June 1911: $h_{Abe} = 160$ km; $m_{B(Abe)} = 8.1$. GR asserted that this was one of the largest intermediate-depth earthquakes known anywhere. From inspection of a published seismogram and because of the distribution of intensity, Utsu (1979) concludes its depth was about 100 km; however, it had many features that suggest a shallower focus is more likely. According to Japanese historical compilations (Utsu, 1979; Usami, 1996), it produced a tsunami, it had several foreshocks, and there were aftershocks that lasted for more than two months. In any case, it caused several fatalities, collapsed hundreds of dwellings on islands near the epicenter, and also caused damage on Okinawa, 300 km distant. No focal mechanism is available.

10.9.4 Ryukyu–Taiwan references

Carena, S., J. Suppe, and H. Kao, 2002. Active detachment of Taiwan illuminated by small earthquakes and its control of first-order topography, *Geology*, **30**, doi:10.1130/0091–7613(2002)030, 935–938.

Christova, C., 2004. Stress field in the Ryukyu–Kyushu Wadati–Benioff zone by inversion of focal mechanism data, *Tectonophysics*, **384**, doi:10.1016/j.tecto.2004.03.010, 175–189.

Kao, H. and W.-P. Chen, 1991. Earthquakes along the Ryukyu–Kyushu arc: strain segmentation, lateral compression, and the thermomechanical state of the plate interface, *J. Geophys. Res.*, **96**, 21443–21485.

• Kao, H. and R.-J. Rau, 1991. Detailed structures of the subducted Philippine Sea plate beneath northeast Taiwan: a new type of double seismic zone, *J. Geophys. Res.*, **104**, doi:10.1029/1998JB/900010, 1015–1033.

Kao, H., S. J. Shen, and K.-F. Ma, 1998. Transition from oblique subduction to collision: earthquakes in the southernmost Ryukyu arc–Taiwan region, *J. Geophys. Res.*, **103**, doi:10.1029/97JB03510, 7211–7229.

Katsumata, M. and L. R. Sykes, 1969. Seismicity and tectonics of the western Pacific: Izu–Mariana–Caroline and Ryukyu–Taiwan regions, *J. Geophys. Res.*, **74**, 5923–5948.

• Lallemand, S., Y. Font, H. Bijwaard, and H. Kao, 2001. New insights on 3-D interactions near Taiwan from tomography and tectonic implications, *Tectonophysics*, **335**, doi:10.1016/S0040–1951(01)00071–3, 229–253.

Lin, C. H. and S. W. Roecker, 1993. Deep earthquakes beneath central Taiwan: mantle shearing in an arc–continent collision, *Tectonics*, **12**, 745–755.

Nakamura, M., Y. Yoshida, D. Zhao, H. Katao, and S. Nishimura, 2003. Three-dimensional P- and S-wave velocity structures beneath the Ryukyu arc, *Tectonophysics*, **369**, 121–143, doi:10.1016/S0040–1951(03)00172–0.

Rau, R.-J. and F. T. Wu, 1995. Tomographic imaging of lithospheric structures under Taiwan, *Earth Planet. Sci. Lett.*, **133**, doi:10.1016/0012–821X(95)00076-O, 517–532.

Ritsema, A. R., 1962. P and S amplitudes of two earthquakes of the single force couple type, *Bull. Seismol. Soc. Amer.*, **52**, 723–746.

Roecker, S. W., Y. H. Yeh, and Y. B. Tsai, 1987. Three-dimensional P and S wave velocity structures beneath Taiwan: deep structure beneath an arc–continent collision, *J. Geophys. Res.*, **92**, 10547–10570.

Scholte, J. G. J. and A. R. Ritsema, 1962. Generation of earthquakes by a volume source with moment, *Bull. Seismol. Soc. Amer.*, **52**, 747–765.

Shiono, K., T. Mikumo, and Y. Ishikawa, 1980. Tectonics of the Kyushu–Ryukyu arc as evidenced from seismicity and focal mechanisms of shallow to intermediate-depth earthquakes, *J. Phys. Earth*, **28**, 17–43.

Usami, T., 1996. *Materials for Comprehensive Listing of Destructive Earthquakes in Japan* (revised and enlarged edition), Tokyo, University of Tokyo Press, 493 pp.

Utsu, T., 1979. Seismicity of Japan from 1885 through 1925: a new catalog of earthquakes of *M*>6 felt in Japan and smaller earthquakes which caused damage in Japan (in Japanese), *Bull. Earthquake Res. Inst., Tokyo Univ.*, **54**, 253–308.

10.10 Mariana–Bonin

10.10.1 Regional tectonics and deep seismicity

A series of archipelagos extends 3000 km southward from central Japan; these are: Izu, Bonin, Mariana, Yap, and Palau (Fig. 10.21). For historical reasons, this section treats the seismicity beneath the Izu islands in the section on Japan, although tectonically it is not distinct from the Bonin activity. Three of the southernmost islands in the Bonin chain are also known as the Volcano Islands.

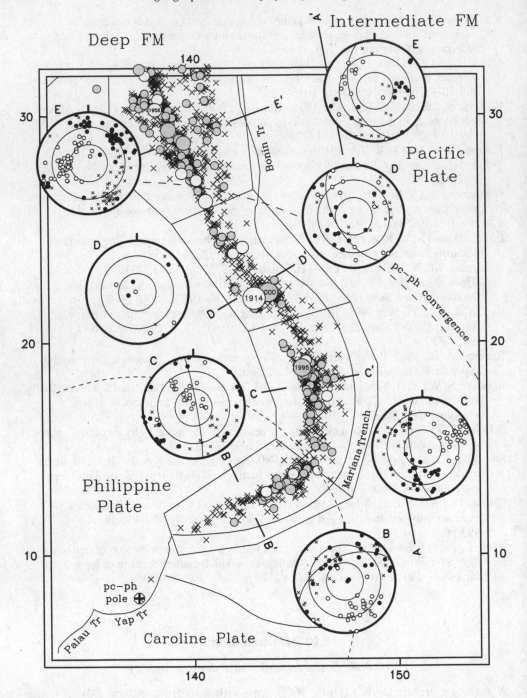

Along the Izu–Bonin and Mariana arcs, the Pacific plate subducts westward beneath the Philippine plate. The rate is 4.3 cm/yr at 30° N and then decreases regularly southward, as the pole of Pacific–Philippine convergence lies near the Yap Trench (see Fig. 10.21). The age of the lithosphere subducting in the Mariana–Bonin region is $\sim$ 130–150 Ma, making it the oldest lithosphere subducting anywhere. Intermediate-depth earthquakes form a nearly continuous chain beneath both arcs (Fig. 10.22). Both arcs also have numerous deep-focus earthquakes, although between the active regions deep-focus seismicity is sparse or absent. Beneath the Bonin islands the Wadati–Benioff zone dips steeply and then appears to bend and flatten out horizontally at about 500 km depth (cross section E). Focal mechanisms of both intermediate- and deep-focus earthquakes exhibit downdip compression. Variation in both the rate and obliquity of subduction cause the thermal parameter Φ to vary, but generally it is relatively high, or about 5000–9000 km.

In contrast to Izu–Bonin, beneath the central Marianas activity extends to depths exceeding 650 km, and there is back-arc spreading so that the rate of lithospheric consumption may be as high as 6.4 cm/yr (Furukawa, 1994). Here the Wadati–Benioff zone is nearly vertical and even appears to bend backwards somewhat (cross sections C and D). Focal mechanisms for deep-focus activity show downdip compression. However, at intermediate depths the dominant direction is not downdip. Rather, most of the mechanisms exhibit compression along a direction normal to the Wadati–Benioff zone, possibly because the convergence direction is highly oblique, or perhaps because the lithosphere is undergoing bending.

The maximum depth, size, and rate of seismic activity all decrease south of the central Marianas. Along the Yap Trench the convergence rate is extraordinarily slow, $\sim$0.6 cm/yr or less, and the age of the subducting lithosphere is 30–35 Ma. Here at the edge of the Pacific plate an additional small plate, the Caroline plate, subducts slowly beneath the Philippine plate. This forms the Yap Trench and is responsible for a few earthquakes with magnitudes of 4–5 which the ISC catalog assigns depths of 60–110 km.

Fig. 10.21 Map of deep earthquake activity in the Mariana–Izu–Bonin region. Filled circles, open circles, and crosses indicate selected deep earthquakes in the Harvard CMT, EV Centennial, and EHB catalogs (see description of selection process in the introduction to this chapter). Solid lines labeled A–A' to E–E' indicate locations and orientations of cross sections A to E, and thin lines surround events included in each section. Equal-area plots summarize T (solid circles), B (crosses), and P (open circles) axis orientations of selected Harvard CMT for intermediate-focus (B, C, D, and E at right) and deep-focus earthquakes (C, D, and E at left). Other lines indicate locations of plate boundaries. The dashed line indicates the direction of Eurasian–Philippine (eu–ph) plate convergence; note the pole of plate motion from Seno *et al.* (1993) at lower left.

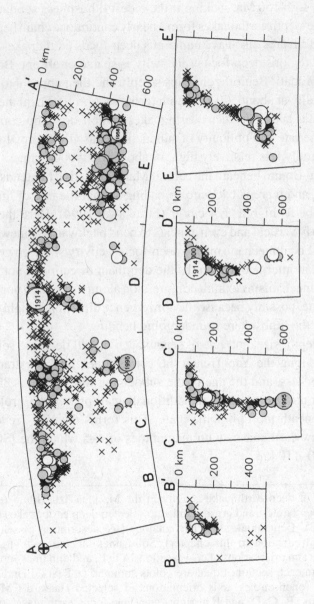

Fig. 10.22 Cross sections A–A' to E–E' of deep seismic activity in the Mariana–Izu–Bonin region. Earthquake symbols as in Fig. 10.21.

10.10.2 Literature review

Even though there is a nearly continuous band of intermediate-depth earthquakes connecting the Izu–Bonin and Mariana arcs, many papers discussing their seismicity and structure focus on the contrasts between them (e.g., Katsumata and Sykes, 1969; Stein *et al.*, 2003). Whereas deep-focus activity extends continuously from Japan southward beneath Izu–Bonin, beneath the Mariana arc there are significant gaps. Beneath Izu–Bonin, both seismicity and tomographic inversions indicate a change from near-vertical to near-horizontal structure at about 500 km depth (e.g., Fukao *et al.*, 1992). Moreover, Widiyantoro *et al.* (1999) show that the slab–mantle velocity contrast is smaller for S than for P in the horizontal Izu–Bonin slab, and less than both S or P at comparable depths beneath the Marianas. However, in the central Marianas beneath 500 km the Wadati–Benioff zone is vertical and tomography indicates that high-velocity material continues well beneath the deepest seismicity. Samowitz and Forsyth (1981) report a double seismic zone at depths of 80–120 km beneath the central Mariana arc; however, the evidence is weak as they observe exactly three hypocenters in the lower zone, all with locations determined using only teleseismic data.

A number of investigations use Izu–Bonin and Mariana earthquakes as sources to resolve deep structure in the neighboring mantle. Two studies investigate patterns in travel-time residuals at Japanese stations to evaluate whether there is high-velocity lithospheric material filling gaps in seismic activity. Okino *et al.* (1989) focus on the "outboard" earthquake 200 km west of the Bonin deep high-activity region (4 July 1982; $h_{CMT} = 545$ km, $M_{W(CMT)} = 6.7$) and conclude a horizontal high-velocity region connects these events. Similarly, Ohtaki and Kaneshima (1994) find a high-velocity region extending south from the Bonin deep cluster towards the Mariana zone; and Brudzinski *et al.* (1997) conclude that velocity anomalies above the 660 km discontinuity suggest there is a horizontal remnant of subducted slab there. Collier and Helffrich (1997) and Castle and Creager (1998) evaluate arrivals between P and S to map regional topography on mantle discontinuities near 410 km and 660 km. Kaneshima and Helffrich (1998), Castle and Creager (1999) and Kruger *et al.* (2001) stack records at American stations and elsewhere to image localized scatterers in the mantle well below the transition zone.

Along the Yap Trench, most studies agree that the so-called Caroline plate here moves independently of the Pacific plate and subducts beneath the Philippine plate (e.g., see Weissel and Anderson, 1978; Fujiwara *et al.*, 2000; Kobayashi, 2004). No earthquakes with focal depths exceeding 40 km near Yap were detected by seismic stations operating during either a one-week ocean-bottom or a seven-month island deployment (Sato *et al.*, 1997).

10.10.3 Significant earthquakes

The Harvard and EV Centennial catalogs report 25 intermediate- and deep-focus earthquakes in this region with magnitudes m_B or M_W of 7.0 or greater. Only the 1914 and 2000 earthquakes discussed below have magnitudes exceeding 7.5.

24 November 1914: $h_{Abe} = 110\,km$; $m_{B(Abe)} = 7.9$. This large earthquake and the event of 28 March 2000 occurred at nearly the same location. I have found no publications describing either event, no reports in the standard catalogs indicating that either generated a tsunami, and no focal mechanism for the 1914 event.

18 February 1956: $h_{CMT\text{-}Hist} = 480\,km$; $M_{W(CMT\text{-}Hist)} = 7.3$. When this earthquake occurred the seismological community was not in agreement concerning whether the earthquake source was a single or double couple, and whether shallow and deep earthquakes might satisfy different source models. Thus, Kasahara (1963), Hirasawa and Stauder (1964) and Honda *et al.* (1965) all studied this event and concluded its body waves clearly supported the double-couple, and not the single-couple model. Their published mechanism agreed closely with that in CMT Historical catalog (see inset).[10] This earthquake's hypocenter was close to the bend in the Bonin Wadati–Benioff zone where its orientation changes from near-vertical to near-horizontal; several large earthquakes have been associated with this feature, notably in 1984 (Yoshida, 1988) and 2000.

23 August 1995: $h_{CMT} = 599\,km$; $M_{W(CMT)} = 7.1$. This Mariana deep-focus quake is unusual because it generated numerous aftershocks. Tibi *et al.* (2001) were able to relocate 17, and concluded that most occupied an area about 20 km × 25 km at 486–603 km depth. They and Antolik *et al.* (1999) studied the source and found that its duration was about 5 sec and the rupture had a spatial extent of about 15 km both vertically and laterally. Wu and Chen (1999) noted that the combined scalar moment of the aftershocks was large, about one-sixth of the moment of the mainshock. Indeed, four aftershocks that occurred the following day were large enough to be in the CMT catalog; there are also two possible foreshocks within ten days prior to the event. Both Kaneshima and Helffrich (1998) and Kruger *et al.* (2001) used this sequence as a source to image scatterers in the lower mantle. The

---

[10] However, the figures in the earlier studies may confuse modern readers as they display the mechanism on an upper-hemisphere projection.

1995 earthquake is close spatially and has a similar moment and focal mechanism to the earthquake of 17 October 1979 ($h_{CMT} = 584$ km; $M_{W(CMT)} = 7.0$); Lundgren and Giardini (1992) found that most hypocenters near the 1979 event occurred along one of its nodal planes. Wiens and Gilbert (1996) evaluated these and other deep-focus earthquakes with numerous aftershocks and concluded they tend to occur in subduction zones where the mantle is cold, i.e. where the thermal parameter is large.

28 March 2000: $h_{CMT} = 100$ km; $M_{W(CMT)} = 7.6$. The Harvard mechanism for this earthquake had a large non-double-couple component (54%, see inset). See also the discussion of the 24 November 1914 event (above).

10.10.4 Mariana–Bonin references

Antolik, M., D. Dreger, and B. Romanowicz, 1999. Rupture processes of large deep-focus earthquakes from inversion of moment rate functions, *J. Geophys. Res.*, **104**, doi:10.1029/1998JB00042, 863–894.

Brudzinski, M. R., W.-P. Chen, R. L. Nowack, and B.-S. Huang, 1997. Variations of P wave speeds in the mantle transition zone beneath the northern Philippine Sea, *J. Geophys. Res.*, **102**, doi:10.1029/97/JB00212, 11815–11828.

Castle, J. C. and K. C. Creager, 1998. Topography of the 660-km seismic discontinuity beneath Izu–Bonin: implications for tectonic history and slab deformation, *J. Geophys. Res.*, **103**, doi:10.1029/98JB00503, 12511–12527.

Castle, J. C. and K. C. Creager, 1999. A steeply dipping discontinuity in the lower mantle beneath Izu–Bonin, *J. Geophys. Res.*, **104**, doi:10.1029/1999JB900011, 7279–7292.

Collier, J. D. and G. R. Helffrich, 1997. Topography of the '410' and '660' km seismic discontinuities in the Izu–Bonin subduction zone, *Geophys. Res. Lett.*, **24**, doi:10.1029/97GL01383, 1535–1538.

Fujiwara, T., T. Chiori, A. Nishizawa, K. Fujioka, K. Kobayashi, and Y. Iwabuchi, 2000. Morphology and tectonics of the Yap Trench, *Mar. Geophys. Res.*, **21**, 69–86.

Fukao, Y., M. Obayashi, H. Inoue, and M. Nenbai, 1992. Subducting slabs in the mantle transition zone, *J. Geophys. Res.*, **97**, 4809–4822.

Furukawa, Y., 1994. Two types of deep seismicity in subducting slabs, *Geophys. Res. Lett.*, **21**, doi:10.1029/94GL01083, 1181–1184.

Hirasawa, T. and W. Stauder, 1964. Spectral analysis of body waves from the earthquake of February 18, 1956, *Bull. Seismol. Soc. Amer.*, **54**, 2017–2035.

Honda, H., T. Hirasawa, and M. Ichikawa, 1965. The mechanism of the deep earthquake that occurred south of Honshu, Japan on February 18, 1956, *Bull. Earthquake Res. Inst., Tokyo Univ.*, **43**, 661–669.

Kaneshima, S. and G. Helffrich, 1998. Detection of lower mantle scatterers northeast of Mariana subduction zone using short-period array data, *J. Geophys. Res.*, **103**, doi:10.1029/97JB02565, 4825–4838.

Kasahara, K., 1963. Radiation mode of S waves from a deep-focus earthquake as derived from observations, *Bull. Seismol. Soc. Amer.*, **53**, 643–659.

Katsumata, M. and L. R. Sykes, 1969. Seismicity and tectonics of the western Pacific: Izu–Mariana–Carolone and Ryukyu–Taiwan regions, *J. Geophys. Res.*, **74**, 5923–5948.

Kobayashi, K., 2004. Origin of the Palau and Yap trench-arc systems, *Geophys. J. Int.*, **157**, doi:10.1111/j.1365-246X.2003.02244.x, 1303.

Kruger, F., M. Baumann, F. Scherbaum, and W. Weber, 2001. Mid mantle scatterers near the Mariana slab detected with a double array method, *Geophys. Res. Lett.*, **28**, doi:10.1029/2000GL011570, 667–670.

Lundgren, P. R. and D. Giardini, 1992. Seismicity, shear failure and modes of deformation in subduction zones, *Phys. Earth Planet. Int.*, **74**, 63–74.

Ohtaki, T. and S. Kaneshima, 1994. Continuous high velocity aseismic zone beneath the Izu–Bonin arc, *Geophys. Res. Lett.*, **21**, doi:10.1029/93GL03050, 1–4.

Okino, K., M. Ando, S. Kaneshima, and K. Hirahara, 1989. The horizontally lying slab, *Geophys. Res. Lett.*, **15**, 1059–1062.

Samowitz, I. R. and D. W. Forsyth, 1981. Double seismic zone beneath the Mariana island arc, *J. Geophys. Res.*, **86**, 7013–7021.

Sato, T., J. Kasahara, H. Katao, N. Tomiyama, K. Mochizuki, and S. Koresawa, 1997. Seismic observations at the Yap Islands and the northern Yap Trench, *Tectonophysics*, **271**, doi:10.1016/S0040-1951(96)00251-X, 285–294.

Seno, T., S. Stein, and A. E. Gripp, 1993. A model for the motion of the Philippine Sea plate consistent with NUVEL-1 and geophysical data, *J. Geophys. Res.*, **98**, 17941–17948.

Stein, R. J., M. J. Fouch, and S. L. Klemperer, 2003. An overview of the Izu–Bonin–Mariana subduction factory. In *Inside the Subduction Factory*, ed. J. Eiler, Washington, DC, American Geophysical Union Monograph **138**, 175–222.

Tibi, R., D. A. Wiens, and J. A. Hildebrand, 2001. Aftershock locations and rupture characteristics of the 1995 Mariana deep earthquake, *Geophys. Res. Lett.*, **28**, doi:10.1029/2001GL013059, 4311–4314.

Weissel, J. K. and R. N. Anderson, 1978. Is there a Caroline plate? *Earth Planet. Sci. Lett.*, **41**, 143–158.

Widiyantoro, S., B. L. N. Kennett, and R. D. van der Hilst, 1999. Seismic tomography with P and S data reveals lateral variations in the rigidity of deep slabs, *Earth Planet. Sci. Lett.*, **173**, doi:10.1016/S0012-821X(99)00216-2, 91–100.

Wiens, D. A. and H. J. Gilbert, 1996. Effect of slab temperature on deep-earthquake aftershock productivity and magnitude–frequency relations, *Nature*, **384**, doi:10.1028/384153a01 53-156.

Wu, L.-R. and W.-P. Chen, 1999. Anomalous aftershocks of deep earthquakes in Mariana, *Geophys. Res. Lett.*, **26**, doi:10.1029/1999GL900389, 1977–1980.

Yoshida, S., 1988. Waveform inversion for rupture processes for two deep earthquakes in the Izu–Bonin region, *Phys. Earth Planet. Int.*, **52**, 85–101.

10.11 Japan

10.11.1 Regional tectonics and deep seismicity

In early plate models Japan was a fragment of buoyant lithosphere that straddled the Eurasian and North American plates; in subsequent models it sits instead on the more recently proposed Amuria and Okhotsk plates (see Seno *et al.*, 1996; Heki *et al.*, 1999). The location of some of the plate boundaries is a matter of controvery; however, in all the models three plates converge just east of Japan to form a triple junction (Fig. 10.23). Along Japan's northeast boundary the Pacific plate converges along the Japan Trench at about 9 cm/yr; it forms a distinctly planar Wadati–Benioff zone that dips at 35° and extends to a depth of about 580 km (Figs. 10.24 and 10.25). Along this boundary the thermal parameter Φ is unusually high, or about 7000 km. To the north, the Wadati–Benioff zone undergoes a distinct bend at the "Hokkaido corner" and continues along the Kuril arc. Focal mechanisms of intermediate-depth earthquakes show considerable variation, apparently because of the influence of the Hokkaido corner and also because there is a well-developed double-Wadati–Benioff zone here.

The second arm of the triple junction lies offshore of east-central Japan where the Pacific and Philippine plates converge at about 5 cm/yr; to the southeast this boundary becomes the Izu–Bonin arc. Here the Wadati–Benioff zone dips at about 50° and extends to depths of about 500 km (Fig. 10.25). The subducting lithosphere is old ($\sim$150 Ma) and thus the thermal parameter Φ is about 5000 km. Focal mechanisms of both intermediate- and deep-focus earthquakes mostly exhibit downdip compression.

The third arm of the triple junction – the Philippine–North America (Okhotsk) boundary – is more complicated. Most plate models present this as a relatively short and inactive segment that joins a second triple junction located in central Japan, the Philippine–North America (Okhotsk)–Eurasia (Amuria) triple junction. In any case, along the southeastern coast of Japan the Philippine plate converges at about 5 cm/yr. South of Japan this forms an intermediate-depth Wadati–Benioff zone that becomes the Ryukyu arc (see section 10.9). The age of the subducting material is $\sim$60 Ma or less and thus the thermal parameter Φ is lower, or about 2000 km. Beneath south-central Japan some regional reconstructions show two

I thank Junji Koyama for reviewing an earlier draft of this section.

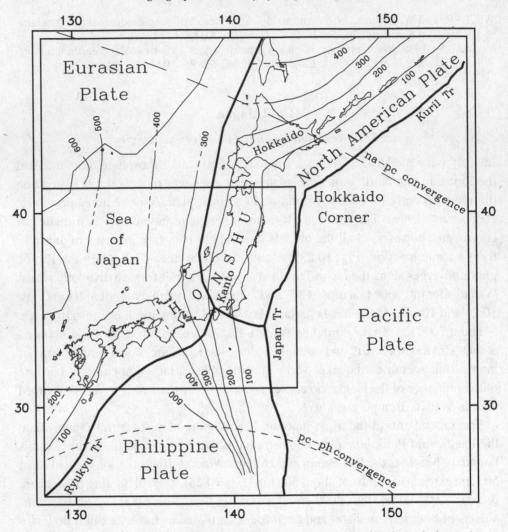

Fig. 10.23 Map showing the relationship of Japan to major plate boundaries and 100 km depth contours of Wadati–Benioff zones. The rectangle in the left-center of the map indicates the geographic extent of the region discussed in this section. Dashed lines indicate relative-motion directions between the North America–Pacific (na–pc) and Pacific–Philippine (pc–ph) plates.

Wadati–Benioff zones, one atop another, produced respectively by the subduction of the Philippine and the Pacific plates (see below).

10.11.2 *Literature review*

The literature on the deep earthquakes of Japan is enormous, larger than for any other geographic area. This is partly because deep earthquakes and damaging earthquakes

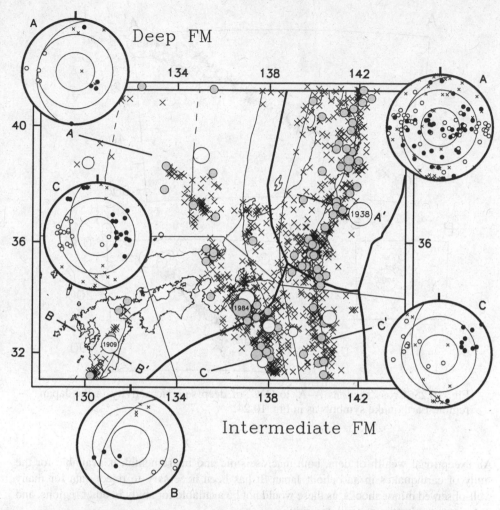

Fig. 10.24 Map of deep earthquake activity in the Japan region. Filled circles, open circles, and crosses indicate selected deep earthquakes in the Harvard CMT, EV Centennial, and EHB catalogs (see description of selection process in the introduction to this chapter). Solid lines labeled A–A′ to C–C′ indicate locations and orientations of cross sections A to C, and thin lines surround events included in each section. Equal-area plots summarize T (solid circles), B (crosses), and P (open circles) axis orientations of selected Harvard CMT for intermediate-focus (plot B below map and plots A and C at right) and deep-focus earthquakes (plots A and C at left). Other lines are depth contours of Wadati–Benioff zones, locations of plate boundaries, and depth contours of the Wadati–Benioff zone (see Fig. 10.23).

are numerous here, partly because Japanese seismologists are themselves numerous, and partly because both seismograph networks and seismology as a science became well-established earlier in Japan than elsewhere. Indeed, GR noted that there were more than 100 stations operating in Japan in 1936, and observed that:

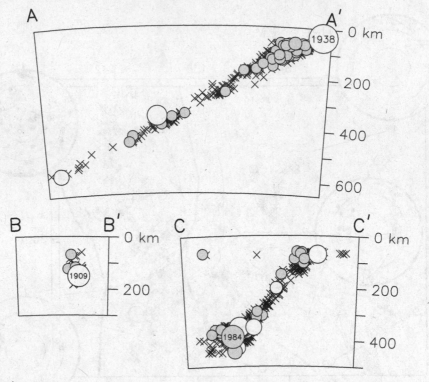

Fig. 10.25 Cross sections A–A′ to C–C′ of deep seismic activity in the Japan region. Earthquake symbols as in Fig. 10.24.

An exceptional wealth of data, both macroseismic and instrumental, is available for the study of earthquakes in and about Japan. It has been necessary to reject data for many well-observed minor shocks, as these would not be available for study in other regions, and only serve to blur the general description.

Similarly, this review will be less comprehensive than for other regions, and will ignore much of the historical literature.

However, it is worth emphasizing that Japanese seismologists have been responsible for many of the most significant observational discoveries concerning deep earthquakes. For example, Omori (1907) first called attention to the anomalous polarization of body waves for what we now know was a large deep-focus earthquake (see the Preface). Papers by Kiyoo Wadati convinced the scientific community of the existence of deep earthquakes (Wadati, 1928; 1929; see section 3.3) and first described what we now call Wadati–Benioff zones (Wadati, 1935; see section 3.4). Honda (1932; 1934) first displayed the pattern of P-wave first motions around a deep earthquake as a "beach ball" and described the radiation pattern of P and S that we now call a double couple. More recently, Hasegawa *et al.* (1978b) provided evidence for the existence of double-Wadati–Benioff zones; and Kawakatsu

(1991) has measured the isotropic component of deep earthquakes and found it to be insignificant. It is also noteworthy that except for the 1940–1955 period during World War II and the reconstruction, much of the Japanese seismological research (including all referenced above) has been published in English. Often, however, papers in Japanese preceded these contributions; e.g., see Tsumura (1973) concerning double Wadati–Benioff zones. Utsu (1971) thoroughly reviews the rather considerable body of pre-1970 literature in both English and Japanese that describes the geometry of the Wadati–Benioff zone and the evidence that it is embedded in a high-Q, high velocity zone underlying a low-Q mantle.

Considering this record of discovery, it is perhaps surprising that Japanese seismologists didn't author the initial papers credited for providing the seismological evidence supporting plate tectonics. However, few of the pre-1970 Japanese papers focus on comparisons between Japan and other geographic regions. Thus, for the early seismological evidence for plate tectonics most people cite publications written by seismologists at Lamont-Doherty in New York (e.g., Isacks *et al.*, 1968), even though their essential data concerning earthquakes and structure in the Tonga–Kermadec region was vastly inferior to the then-available information for Japan.

Because the Japanese seismograph network possesses numerous stations in operation for many decades, since at least the 1930s there has been more detailed information available concerning the seismicity, velocity structure, and attenuation beneath Japan than anywhere else in the world. For example, beneath Kanto, Ishida (1992) used arrivals from 244 earthquakes recorded at 66 stations to determine a 3-D velocity model, and then used this model to relocate 47,500 earthquakes. The hypocenters suggest that the subduction of the Philippine Sea plate beneath the Eurasian plate is more extensive than shown in Fig. 10.23. Ishida concludes that the Philippine Sea plate subducts northwestward beneath Japan's southern coast and extends to a depth of about 90 km beneath Kanto. It thus overlies the Pacific plate, which subducts westward, as shown in Fig. 10.23. Beneath northern Honshu, Igarashi *et al.* (2001) also relocated hypocenters and summarized information from 1106 focal mechanisms determined using first motions from the regional network.

Because of its abundant earthquakes and favorable event-station geometry, Japan is a favorite location to apply tomographic inversion methods to determine P-velocity, S-velocity, and attenuation. Initially researchers applied Aki's method to determine velocity within a grid of rectangular cells (e.g., Hirahara, 1977; 1981; Hirahara and Mikumo, 1980; Hori and Aki, 1982; Ishida and Hasemi, 1988; Ishida, 1992). Hirahara (1980; 1981) interpreted the results in terms of P and S contrasts between slab and mantle, finding P and S contrasts of 5–6% and 6–8% between 50–250 km depth range, 3% and 5% at at 350 km, and 2% and 2–3% at 450 km depth. More recent comprehensive tomographic inversions include Fukao *et al.* (1992), Van der Hilst *et al.* (1993), Zhao and Hasegawa (1993), and Zhao *et al.* (1994). Sekiguchi (2001) investigated the region between Kanto and concludes that the Philippine Sea plate does subduct northwestward there, but he obtains a slightly different configuration than Ishida. Both Sekiguchi (1991) and Tsumura *et al.* (2000) have performed inversions to determine attenuation structure within and above the Wadati–Benioff zone.

The plethora of earthquakes and stations has encouraged investigators to analyze observations of shear-wave polarization anisotropy in an effort to localize its source. The initial studies analyzed S waves from intermediate-depth earthquakes at Japanese stations (e.g., Ando *et al.*, 1980; 1983) and concluded that there was polarized material in localized regions of the upper mantle overlying the Wadati–Benioff zone. However, more recent investigations also evaluated teleseismic SKS phases and ScS phases from deep earthquakes in the Japan, Ryukyu, and Izu–Bonin zones (Fouch and Fischer, 1996; Hiramatsu and Ando, 1996; Hiramatsu *et al.*, 1997). These studies suggest that the polarized material occurs near the sources, as the fast directions determined for different earthquakes vary by as much as 50° from one another.

Perhaps the most remarkable aspect of Japanese intermediate seismicity is the pervasive "double Wadati–Benioff zone" which consists of two separate planes of earthquakes separated by 30–40 km, beginning at about 50 km depth and extending to about 190 km, where the two planes appear to merge (see Fig. 5.7). This feature does not occur in all subduction zones but is well-defined beneath northern Honshu, where upper-zone earthquakes tend to have downdip compressional mechanisms, while lower-zone earthquakes exhibit downdip extension. After Tsumura (1973) and Hasegawa *et al.* (1978b) announced the presence of double zones beneath Kanto and northern Honshu, several studies confirmed that it extends along the entire northeastern Japan Wadati–Benioff zone (Hasegawa *et al.*, 1979; Kawakatsu and Seno, 1983; Hasegawa *et al.*, 1994). There have been various explanations offered to explain the occurrence of double zones (see Section 5.1.2), including thermal-chemical explanations involving dehydration reactions of subducting mineral assemblages.

A second extraordinary feature is a high-impedance interface that coincides with, or lies slightly above, the double zone, extends from about 40 km to 150 km depth, and produces high-amplitude S-to-P and P-to-S converted phases. Evidence that the feature is associated with the upper zone comes from the observation that S-to-P converted phases are never observed from hypocenters in the upper zone. The interface was first detected because it produced precursors to ScS, or ScSp phases (e.g., Hasegawa *et al.*, 1978a, Okada, 1979; Nakanishi, 1980; Nakanishi *et al.*, 1981; Iidaka and Obara, 1993). Subsequently Matsuzawa *et al.* (1987; 1990), Obara and Sato (1988) and Obara (1989) have investigated SP and PS arrivals. Zhao *et al.* (1992; 1997) performed inversions of SP data to fix the location and extent of the interface. These studies, as well as Ohmi and Hori (2000), indicate that this feature extends from 34.5° N to 42° N beneath northeastern Japan, and Oda and

Douzen (2001) observe it beneath southwestern Japan. Modeling studies are unable to explain the high amplitudes observed if the interface is a simple velocity change, and suggest it must be some kind of relatively thin, layered structure (Matsuzawa *et al.*, 1987; Helffrich and Stein, 1993; Ohmi and Hori, 2000).

The Japanese network is also favorably situated for investigating structure within the northernmost Izu–Bonin subduction zone, although the resolution is less than for beneath Japan itself. Iidaka *et al.* (1992) evaluate the velocity structure within the subducting slab beneath the Izu region. Nakamura *et al.* (1998) evaluate S-to-P converted phases from earthquakes with depths of 350–460 km, and conclude they arise from a surface lying about 20 km above the hypocenters, presumably the upper surface of the subducting lithosphere. Iidaka and Furukawa (1994) and Iidaka and Obara (1997) report that there is an unusual double seismic zone situated between the depths of about 310–410 km.

10.11.3 Significant earthquakes

There are 13 intermediate- and deep-focus earthquakes with magnitudes m_B or M_W 7.0 or greater in the EV Centennial and CMT catalogs. In addition, four earthquakes occurred on 5–6 November 1938 for which Abe and GR assigned m_B or M_S between 7.0 and 7.7 and depths of 60 km. Subsequent investigations and the occurrence of tsunamis indicate that these were probably shallower (e.g., Abe, 1977; Kawakatsu and Seno, 1983).

21 January 1906: $h_{Abe} = 340$ km; $m_{B(Abe)} = 7.7$. This earthquake is important because it was large, and because both Omori (1907) (see the Preface) and Zoeppritz (1912) (see Section 3.2.3) commented on its unusual properties, which in retrospect are attributable to its depth. From an observational perspective this was in some sense the "first" deep earthquake. Abe (1985) reexamined seismograms published by Omori and Szirtes (1909), determined a focal mechanism (see inset), and concluded that its moment was 1.5×10^{20} N-m, corresponding to M_W 7.4. Utsu (1979) presents an intensity map for this event.

29 March 1928: $h_{CMT-Hist} = 428$ km; $M_{W(CMT-Hist)} = 7.1$. Many of the earthquakes Kiyoo Wadati analyzed to "discover" deep earthquakes were too small to be widely recorded teleseismically. However, Wadati (1929) assigned this event a depth of "about 500" km, published a table of 56 S-P intervals for Japanese stations, and noted that it was "the largest deep earthquake in the recent years." Stechschulte

(1932) then collected teleseismic records from 61 stations; in an exceedingly thorough paper he analyzed travel times, presented reproductions of 11 seismograms, and concluded that the teleseismic observations supported a depth of 410±30 km. This demonstrated that regional and teleseismic observations were in substantial agreement. The published hypocenters for this earthquake lie between 31.35° N (CMT-Hist) and 32.9° (Wadati), straddling the border of what this book designates as the Japan and the Mariana–Bonin regions.

1 January 1984: $h_{CMT} = 384$ km; $M_{W(CMT)} = 7.2$. This earthquake's hypocenter is nearly the same as that of the 1906 earthquake (see above). Ekström *et al.* (1986) reported that it had a significant non-double-couple component, and to understand its origin Kuge and Kawakatsu (1990) evaluated the source process. They concluded that the non-double-couple source was real; the earthquake consisted of two subevents with different mechanisms, one a thrust and one strike-slip, each with approximately equal scalar moments and separated in time by four seconds (see Fig. 6.24). This underscored the limitations of subevent analyses of source process that assume identical mechanisms for the subevents (e.g., Sugi *et al.*, 1989).

10.11.4 Japan references

Abe, K., 1977. Tectonic implications of the large Shioya-Oki earthquakes of 1938, *Tectonophysics*, **41**, doi:10.1016/0040-1951(77)90135-6, 269–289.

    1985. Re-evaluation of the large deep earthquake of Jan. 21, 1906, *Phys. Earth Planet. Int.*, **39**, 157–166.

Ando, M., Y. Ishikawa, and H. Wada, 1980. S-wave anisotropy in the upper mantle under a volcanic area in Japan, *Nature*, **286**, doi:10.1038/286043a0, 43–46.

Ando, M., Y. Ishikawa, and F. Yamazaki, 1983. Shear wave polarization anisotropy in the upper mantle beneath Honshu, Japan, *J. Geophys. Res.*, **88**, 5850–5864.

Ekström, G., A. M. Dziewonski, and J. M. Steim, 1986. Single station CMT; application to the Michoacan, Mexico, earthquake of September 19, 1985, *Geophys. Res. Lett.*, **13**, 173–176.

Fouch, M. J. and K. M. Fischer, 1996. Mantle anisotropy beneath northwest Pacific subduction zones, *J. Geophys. Res.*, **101**, doi:10.1029/96JB00881, 15987–16002.

Fukao, Y., M. Obayashi, H. Inoue, and M. Nenbai, 1992. Subducting slabs in the mantle transition zone, *J. Geophys. Res.*, **97**, 4809–4822.

Hasegawa, A., N. Umino, and A. Takagi, 1978a. Double-planed deep seismic zone and upper mantle structure in the northeastern Japan arc, *Geophys. J. Roy. Astron. Soc.*, **54**, 281–296.

1978b. Double-planed structure of the deep seismic zone in the northeastern Japan arc, *Tectonophysics*, **47**, doi:10.1016/0040-1951(78)90150-6, 43–58.

Hasegawa, A., N. Umino, A. Takagi, and Z. Suzuki, 1979. Double-planed deep seismic zone and the anomalous structure in the upper mantle beneath northeastern Honshu (Japan), *Tectonophysics*, **57**, doi:10.1016/0040-1951(79)90098-2,1–6.

Hasegawa, A., S. Horiuchi, and N. Umino, 1994. Seismic structure of the northeastern Japan convergent margin: a synthesis, *J. Geophys. Res.*, **99**, 22295–22311.

Helffrich, G. and S. Stein, 1993. Study of the structure of the slab–mantle interface using reflected and converted seismic waves, *Geophys. J. Int.*, **115**, 14–40.

Heki, K., S. Miyazaki, H. Takahashi, M. Kasahara, F. Kimata, S. Miura, N. F. Vasilenko, A. Ivashchenko, and K.-D. An, 1999. The Amurian plate motion and current plate kinematics in eastern Asia, *J. Geophys. Res.*, **104**, doi:10.1029/1999JB900295, 29147–29155.

Hirahara, K., 1977. A large-scale three-dimensional seismic structure under the Japan Islands and the Sea of Japan, *J. Phys. Earth*, **25**, 393–417.

1980. Three-dimensional shear velocity structure beneath the Japan Islands and its tectonic implications, *J. Phys. Earth*, **28**, 221–241.

1981. Three-dimensional seismic structure beneath southeast Japan: the subducting Philippine Sea plate, *Tectonophysics*, **79**, doi:10.1016/0040-1951(81)90231-6, 1–44.

Hirahara, K. and T. Mikumo, 1980. Three-dimensional seismic structure of subducting lithospheric plates under the Japan Islands, *Phys. Earth Planet. Int.*, **21**, 109–119.

Hiramatsu, Y. and M. Ando, 1996. Seismic anisotropy near source region in subduction zones around Japan, *Phys. Earth Planet Int.*, **95**, doi:10.1016/0031-9201(95)03119-7, 237–250.

Hiramatsu, Y., M. Ando, and Y. Ishikawa, 1997. ScS wave splitting of deep earthquakes around Japan, *Geophys. J. Int.*, **128**, 409–424.

Honda, H., 1932. On the types of the seismograms and the mechanism of deep earthquakes, *Geophys. Mag.*, **5**, 301–326.

1934. On the amplitude of the P and the S waves of deep earthquakes, *Geophys. Mag.*, **8**, 153–164.

Hori, A. and K. Aki, 1982. Three-dimensional velocity structure beneath the Kanto district, Japan, *J. Phys. Earth*, **30**, 255–281.

Igarashi, T., T. Matsuzawa, N. Umino, and A. Hasegawa, 2001. Spatial distribution of focal mechanisms for interplate and intraplate earthquakes associated with the subducting Pacific plate beneath the northeastern Japan arc: a triple-planed deep seismic zone, *J. Geophys. Res.*, **106**, doi:10.1029/2000JB900386, 2177–2191.

Iidaka, T. and Y. Furukawa, 1994. Double seismic zone for deep earthquakes in the Izu–Bonin subduction zone, *Science*, **263**, 1116–1118.

Iidaka, T. and K. Obara, 1993. The upper boundary of the subducting Pacific plate estimated from ScSp waves beneath the Kanto region, Japan, *J. Phys. Earth*, **41**, 103–108.

1997. Seismological evidence for the existence of anisotropic zone in the metastable wedge inside the subducting Izu–Bonin slab, *Geophys. Res. Lett.*, **24**, doi:10.1029/97GL03277, 3305–3308.

Iidaka, T., M. Mizoue, and K. Suyehiro, 1992. Seismic velocity structure of the subducting Pacific plate in the Izu–Bonin region, *J. Geophys. Res.*, **97**, 15307–15319.

Isacks, B. L., J. Oliver, and L. R. Sykes, 1968. Seismology and the new global tectonics, *J. Geophys. Res.*, **73**, 5855–5899.

Ishida, M., 1992. Geometry and relative motion of the Philippine Sea plate and Pacific plate beneath the Kanto–Tokai District, Japan, *J. Geophys. Res.*, **97**, 489–513.

Ishida, M. and A. H. Hasemi, 1988. Three-dimensional fine velocity structure and hypocentral distribution of earthquakes beneath the Kanto–Tokai District, Japan, *J. Geophys. Res.*, **93**, 2076–2094.

Kawakatsu, H., 1991. Insignificant isotropic component in the moment tensor of deep earthquakes, *Nature*, **351**, doi:10.1038/351050a0, 50–53.

Kawakatsu, H. and T. Seno, 1983. Triple seismic zone and the regional variation of seismicity along the northern Honshu arc, *J. Geophys. Res.*, **88**, 4215–4230.

Kuge, K. and H. Kawakatsu, 1990. Analysis of a deep "non-double-couple" earthquake using very broadband data, *Geophys. Res. Lett.*, **17**, 227–230.

Matsuzawa, T., N. Umino, A. Hasegawa, and A. Takagi, 1987. Estimation of thickness of a low-velocity layer at the surface of the descending oceanic plate beneath the northeastern Japan arc by using synthesized PS-wave, *Tohoku Geophys. J.*, **31**, 19–28.

Matsuzawa, T., T. Kono, A. Hasegawa, and A. Takagi, 1990. Subducting plate boundary beneath the northeastern Japan arc estimated from SP converted waves, *Tectonophysics*, **181**, doi:10.1016/0040-1951(90)90012-W, 123–133.

Nakamura, M., M. Ando, and T. Ohkura, 1998. Fine structure of deep Wadati–Benioff zone in the Izu–Bonin region estimated from S-to-P converted phase, *Phys. Earth Planet. Int.*, **106**, doi:10.1016/S0031-9201(97)00109-X, 63–74.

Nakanishi, I., 1980. Precursors to ScS phases and dipping interface in the upper mantle beneath southwestern Japan, *Tectonophysics*, **69**, doi:10.1016/0040-1951(80)90125-0, 1–35.

Nakanishi, I., K. Suyehiro, and T. Yokota, 1981. Regional variations of amplitudes of ScSp phases observed in the Japanese Islands, *Geophys. J. Roy. Astron. Soc.*, **67**, 615–634.

Obara, K., 1989. Regional extent of the S wave reflector beneath the Kanto District, Japan, *Geophys. Res. Lett.*, **16**, 839–842.

Obara, K. and H. Sato, 1988. Existence of an S wave reflector near the upper plane of the double seismic zone beneath the southern Kanto District, Japan, *J. Geophys. Res.*, **93**, 15037–15045.

Oda, H. and T. Douzen, 2001. New evidence for a low-velocity layer on the subducting Philippine Sea plate in southwest Japan, *Tectonophysics*, **332**, doi:10.1016/0040-1951(00)00157-8, 347–358.

Ohmi, S. and S. Hori, 2000. Seismic wave conversion near the upper boundary of the Pacific plate beneath the Kanto District, Japan, *Geophys. J. Int.*, **141**, doi:10.1046/j.1365-246X.2000.00086.x, 136–148.

Okada, H., 1979. New evidence of the discontinuous structure of the descending lithosphere as revealed by ScSp phase, *J. Phys. Earth*, **27** Supp., S53–S64.

Omori, F.,1907. Seismograms showing no preliminary tremor, *Bull. Imperial Earthquake Investigation Committee*, **1**, No. 3, 145–154.

Sekiguchi, S., 1991. Three-dimensional Q structure beneath the Kanto–Tokai district, Japan, *Tectonophysics*, **195**, doi:10.1016/0040-1951(91)90145-I, 83–104.

2001. A new configuration of an aseismic slab of the descending Philippine Sea plate revealed by seismic tomography, *Tectonophysics*, **341**, doi:10.1016/S0040-1951(01)00182-2, 19–32.

Seno, T., T. Sakurai and S. Stein, 1996. Can the Okhotsk plate be discriminated from the North American plate? *J. Geophys. Res.*, **101**, doi:10.1029/*96JB00532*, 11305–11315.

Stechschulte, V. C., 1932. The Japanese earthquake of March 29, 1928, and the problem of depth of focus, *Bull. Seismol. Soc. Amer.*, **22**, 81–137.

Sugi, N., M. Kikuchi, and Y. Fukao, 1989. Mode of stress release within a subducting slab of lithosphere: implication of source mechanism of deep and intermediate-depth earthquakes, *Phys. Earth Planet. Int.*, **55**, 106–125.

Szirtes, S., 1909. *Seismogramme des Japanischen Erdbebens am 21 Januar 1906*, Strassburg, Assoc. Int. Sismol., Publ. Bur. Centrl., 50 pp., 7 plates.

Tsumura, K., 1973. Microearthquake activity in the Kanto District (in Japanese), *Publ. 50th Anniv. Great Kanto Earthquake*, **1923**, Tokyo, Earthquake Research Institute, Tokyo University, 67–87.

Tsumura, N., S. Matsumoto, S. Horiuchi, and A. Hasegawa, 2000. Three-dimensional attenuation structure beneath the northeastern Japan arc estimated from spectra of small earthquakes, *Tectonophysics*, **319**, doi:10.1016/S0040-1951(99)00297-8, 241–260.

• Utsu, T., 1971. Seismological evidence for anomalous structure of island arcs with special reference to the Japanese region, *Rev. Geophys. Space Phys.*, **9**, 839–890.

1979. Seismicity of Japan from 1885 through 1925: a new catalog of earthquakes of $M > 6$ felt in Japan and smaller earthquakes which caused damage in Japan (in Japanese), *Bull. Earthquake Res. Inst., Tokyo Univ.*, **54**, 253–308.

Van der Hilst, R. D., E. R. Engdahl, and W. Spakman, 1993. Tomographic inversion of P and pP data for aspherical mantle structure below the northwest Pacific region, *Geophys. J. Int.*, **115**, 264–302.

Wadati, K., 1928. Shallow and deep earthquakes, *Geophys. Mag.*, **1**, 161–202.

1929. Shallow and deep earthquakes (2nd paper), *Geophys. Mag.*, **2**, 1–36.

1935. On the activity of deep-focus earthquakes in the Japan Islands and neighbourhoods, *Geophys. Mag.*, **8**, 305–325.

Zhao, D. and A. Hasegawa, 1993. P wave tomographic imaging of the crust and upper mantle beneath the Japan Islands, *J. Geophys. Res.*, **98**, 4333–4353.

Zhao, D., A. Hasegawa, and S. Horiuchi, 1992. Tomographic imaging of P and S wave velocity structure beneath northeastern Japan, *J. Geophys. Res*, **97**, 19909–19928.

Zhao, D., A. Hasegawa, and H. Kanamori, 1994. Deep structure of Japan subduction zone as derived from local, regional, and teleseismic events, *J. Geophys. Res.*, **99**, 22313–22329.

Zhao, D., T. Matsuzawa, and A. Hasegawa, 1997. Morphology of the subducting slab boundary in the northeastern Japan arc, *Phys. Earth Planet. Int.*, **102**, doi:10.1016/S0031-9201(96)03258-X, 89–104.

Zoeppritz, K., 1912. Über Erdbebenwellen V.A.5, zwei verschiedene Arten von Erdbeben, *Nachr. Akad. Wiss., Göttingen Math.-Phys. Klasse*, **K1**, 132–134.

10.12 Kuril–Hokkaido

10.12.1 Regional tectonics and deep seismicity

Along the Kuril arc the Pacific plate converges towards Eurasia; however, the overriding plate is usually considered to be the North American plate (or the Okhotsk

I thank Alexei Gorbatov for reviewing an earlier draft of this section.

plate – see Section 10.11.1). At the Kuril Trench the age of the subducting plate is poorly known but on the order of 100 Ma. The convergence rate is about 8 cm/yr and approximately perpendicular to the Kuril arc; thus the thermal parameter Φ is about 6000 km. The subduction is mostly regular except near two distinct "corners" along the Pacific plate boundary (Figs. 10.23 and 10.26). To the northeast the Aleutian–Kamchatka corner accommodates the Pacific plate's abrupt shift from the arc-parallel motion along the western Aleutians to the arc-normal motion along the Kuril Trench. Then, just south of Hokkaido there is a second distinct bend; there is abundant deep-focus seismicity on both sides of this "Hokkaido corner" but the inclination angles of the Wadati–Benioff zones are different. There is also an absence of deep-focus activity at the corner (Fig. 10.27), possibly indicating a tear in the subducting lithosphere. This Kuril–Hokkaido section will discuss the deep seismicity north of $40.2°$ N, which includes the Kuril Wadati–Benioff zone as well as the activity surrounding the Hokkaido corner.

The characteristics of the Kuril Wadati–Benioff zone vary somewhat between the two corners. The region just northeast of the Hokkaido corner is remarkable for generating extraordinarily large intermediate-depth earthquakes, including several of the largest known anywhere. At intermediate depths just northwest of the corner, focal mechanisms are irregular – rather than being either downdip tension or compression they exhibit a component of along-strike tension (see Fig. 10.27 section B). In the central Kurils the dip of the Wadati–Benioff zone is about $45°$ and seismicity extends to depths exceeding 650 km; an earthquake occurring on 9 February 1998 had a depth of 680–690 km. Here mechanisms with both downdip T and downdip P are common at intermediate depth (see section C). Beneath northern Kamchatka the maximum depth of seismicity begins to decrease, reaching about 100 km near the Aleutian–Kamchatka corner (see section D). Along the entire Kuril zone mechanisms of deep-focus earthquakes are regular, indicating downdip compression.

10.12.2 Literature review

Prior to about 1970 there are few publications concerning earthquakes within the Kuril Wadati–Benioff zone. However, GR and other early seismologists do mention the region and comment about its high activity rate and extraordinarily large earthquakes. An unusual feature of the literature on the Kurils is that there are a relatively large number of publications concerned with the properties of individual large earthquakes.

There have also been numerous publications focused on the seismicity and structure of the Hokkaido corner. Most conclude that the bend at the corner causes the subducting lithosphere to separate or "tear" (e.g., Aoki, 1974; Moriya, 1979;

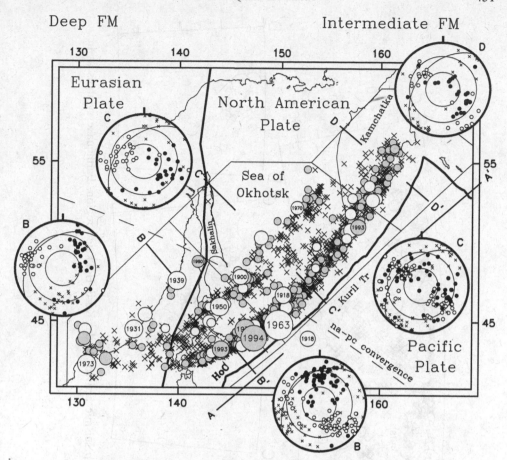

Fig. 10.26 Map of deep earthquake activity in the Kuril–Hokkaido region. Filled circles, open circles, and crosses indicate selected deep earthquakes in the Harvard CMT, EV Centennial, and EHB catalogs (see description of selection process in the introduction to this chapter). Solid lines labeled A–A' to D–D' indicate locations and orientations of cross sections A to D, and thin lines surround events included in each section. Equal-area plots summarize T (solid circles), B (crosses), and P (open circles) axis orientations of selected Harvard CMT for intermediate-depth (plots B, C and D at right) and deep-focus earthquakes (plots B and C at left). Other thick lines indicate locations of plate boundaries; dashed lines show the direction of North America–Pacific (na–pc) plate convergence.

Katsumata *et al.*, 2003). However, because many focal mechanisms near the corner have a component of along-slab tension, an alternative interpretation is that the slab yields aseismically as it stretches and thins around the corner (see Sasatani, 1976a; Lundgren and Giardini, 1990). There is a double Wadati–Benioff zone beneath Hokkaido; however, unlike in central Japan, earthquakes occur more frequently in the lower plane (Suzuki *et al.*, 1983; Ozel and Moriya, 2003).

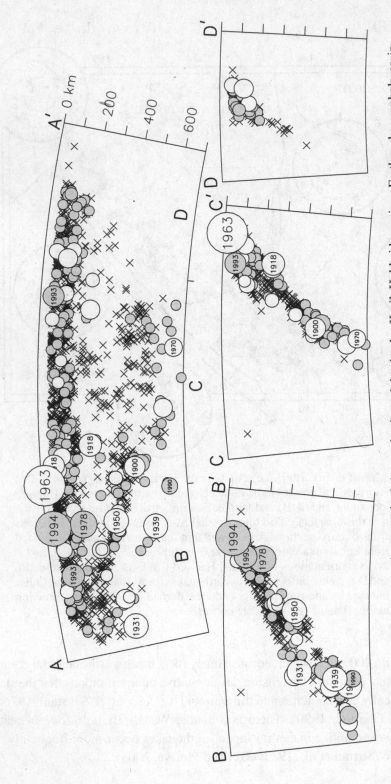

Fig. 10.27 Cross sections A–A' to D–D' of deep seismic activity in the Kuril–Hokkaido region. Earthquake symbols as in Fig. 10.26.

The Wadati–Benioff zone that extends from Hokkaido to Kamchatka is generally planar (Hănus and Vănek, 1984a; 1984b; 1984c; 1988). Gorbatov *et al.* (1997) show that variations in the thermal parameter Φ may control the decrease in maximum depth beneath Kamchatka. There is a reduced level of activity between depths of about 250–350 km; Hănus and Vănek claim to identify another smaller gap near 100 km. At depths of about 50–175 km there is a well-developed double Wadati–Benioff zone that extends to at least 54° N (Kao and Chen, 1994; 1995; Gorbatov *et al.*, 1994). There are also several large "outboard" earthquakes with depths of 440–640 km that occur about 200 km west of the main Wadati–Benioff zone (Glennon and Chen, 1993; and see Fig. 10.27 section B). There is ongoing controversy (see below) about whether these occur within a vertical remnant slab from a previous episode of subduction, or within a deflected extension of the main Kuril zone that extends horizontally westward of the deepest seismicity there. The EHB catalog also reports several small earthquakes with depths between 60 km and 120 km beneath Sakhalin Island.

Because Kuril earthquakes are often large, deep, and well recorded both teleseismically and by regional stations in Japan, Siberia, and China, their travel-time residuals are useful for assessing what happens to subducted material after it reaches depths of 650 km. Jordan (1977) initiated these investigations by evaluating travel time residuals from an earthquake that occurred on 29 January 1971, concluding that subducted material reached depths of at least 1000 km beneath Okhotsk. This result has been controversial and thus there have been numerous detailed studies of residuals for Kuril deep earthquakes (e.g., Creager and Jordan, 1984; Zhou *et al.*, 1990; Gaherty *et al.*, 1991; Schwartz *et al.*, 1991; Ding and Grand, 1994; Glennon and Chen, 1995b; Pankow and Lay, 1999). The overall conclusion is that part of the alleged slab-penetration signal may arise from velocity variations within the lower mantle; nevertheless, the subducting lithosphere does appear to penetrate into the lower mantle along some, but not all, parts of the arc.

The Kuril region is similarly attractive as a region for travel-time tomography, and these efforts produce a similar lack of agreement about mantle structure. Although Zhou and Clayton (1990) found evidence for fast velocities extending beneath 650 km in the Kurils, Spakman *et al.* (1989) analyzed synthetic data and demonstrated that even when there is no velocity anomaly beneath 650 km, tomography produces "resolution artifacts" or "ghosts" which look very similar to the slab-like structures penetrating beneath 650 km observed in inversions of real data. Other tomographic investigations find little or no evidence that the slab penetrates beneath 650 km; rather, they conclude that the lithosphere flattens out and extends horizontally at that depth (e.g., Fukao *et al.*, 1992; Van der Hilst *et al.*, 1993; Takenaka *et al.*, 1999). However, Gorbatov *et al.* (2000) incorporate both teleseismic and regional data in their inversion and find a broad velocity anomaly that extends well below 650 km in the southern Kurils. They also find a region of high velocity extending from 300 km to 600 km depth beneath easternmost Asia, which they interpret as a remnant fragment of a past episode of subduction. Inversions providing information about specific areas include Kamiya *et al.* (1989), who find the subducting lithosphere around the Hokkaido corner is continuous to a depth of 200 km, but split into two fragments at greater depths. In the northern part of the Kuril region, Gorbatov *et al.* (1999) use P phases from the Russian regional network to image the subducted slab and overlying upper mantle beneath Kamchatka.

10.12.3 Significant earthqakes

This region has about 50 intermediate- and deep-focus earthquakes with magnitudes m_B or M_W of 7.0 or greater in the EV Centennial and CMT catalogs, including nine with magnitudes of 7.5 or more. Abe *et al.* (1970) evaluated free oscillations and determined a depth of 60 km and magnitude M_W of 8.5 for the 13 October 1963 earthquake. However, this is uncertain; other sources report a shallow depth and Balakina (1989) notes that it caused a tsunami with maximum heights of 4–5 m over approximately 200 km of coastline in the Kurils.

20 February 1931: $h_{EVC} = 357$ km, $m_{B(Abe)} = 7.4$. This was among the first large deep-focus earthquakes to occur following the appearance of Wadati's papers which used regional observations to infer that deep quakes existed. It was well-recorded both regionally and teleseismically; it thus received attention because seismologists wished to confirm that teleseismic observations were in accord with Wadati's assertions and because they wished to understand the earthquake source process. Scrase (1933) made a thorough study of travel times for numerous phases including pP, sP, and sS, which he had been the first to identify (Scrase, 1931a; 1931b). He also noted that surface waves were completely absent at 60% of the stations, and that a phase some catalogs listed as the commencement of the Love wave was actually sSSS and SSSS. Wadati and Isikawa (1933) investigated regional travel times of P and S; they also evaluated first motions and showed that two nodal lines separated zones where horizontal components of P pointed either towards or away from the epicenter. Kawasumi and Yosiyama (1934) evaluated first motions at 74 teleseismic and 17 Japanese stations and concluded that the radiation pattern was consistent with what we now call a double-couple source, having two perpendicular nodal planes (the inset shows mechanism determined by Honda, 1962, from first motions). Today these papers seem unremarkable unless one realizes that when this earthquake occurred, the existence of deep earthquakes was controversial, the paucity of their surface waves had not been confirmed, the origin and significance of phases such as pP and sP had not been recognized, and the near-ubiquity of the double-couple radiation pattern had not been established.

30 August 1970: $h_{CMT-Hist} = 650$ km, $M_{W(CMT-Hist)} = 7.3$. This is the largest of the deep-focus earthquakes beneath the Sea of Okhotsk whose travel-time residuals have been studied in excruciating detail (see above) to determine the fate of subducting lithosphere. Two other well-studied events in this group occurred in 5 September 1970 and 29 January 1971; Balakina and Kislovskaya (1975), Sasatani (1974; 1976b) and Strelitz (1975; 1980) investigated the mechanisms of all three earthquakes. For the 30 August event Sasatani and Strelitz found a source with three or four subevents having onset times separated by 5–10 s.

29 September 1973: $h_{CMT-Hist} = 593$ km, $M_{W(CMT-Hist)} = 7.8$. This earthquake attracted attention because of its size and because it generated well-recorded multiple ScS phases and higher-mode Rayleigh waves (Yoshida and Tsujiura, 1975; Okal, 1979). Several researchers investigated its source process – Furumoto and Fukao (1976) using Rayleigh waves and Koyama (1978), Fukao and Kikuchi (1987), and Wu and Chen (2001) using body waves. All these studies are consistent with a moment of about 5–7 $\times$ 10^{20} N-m and a source having several subevents occurring over 25–30 s; however, in contrast with the previous studies, Wu and Chen favor a subhorizontal rupture surface. To evaluate the orientation and depth extent of subducted lithosphere beneath the Sea of Japan, Creager and Jordan (1984; 1986) investigated this earthquake's traveltime residuals.

6 December 1978: $h_{CMT} = 181$ km, $M_{W(CMT)} = 7.8$. This earthquake occurred just northwest of the Hokkaido corner and is unusual because it had 19 teleseismically reported aftershocks in the ISC catalog (see Willemann and Frohlich, 1987), and because both its rupture and aftershocks had a horizontal and a vertical extent on the order of 100 km. The source was highly complex and there are discrepancies concerning its reported focal depth (100–181 km), scalar moment ($3.3–10 \times 10^{20}$ N-m) and source duration (45–70 s). Subevent analyses by both Brüstle and Müller (1987) and Lundgren *et al.* (1988) find three principal subevents, the shallowest with a depth near 100 km and the deepest near 200 km. Analyses of longer-period data by Silver and Jordan (1983), Romanowicz and Guillemant (1984), and Kasahara and Sasatani (1985) obtain focal depths midway between these values. These studies are generally consistent with a focus initiating near 100 km depth and a rupture that traveled vertically downward to about 200 km along a fault striking perpendicular to the trench (e.g., see Lundgren and Giardini (1990)).

12 May 1990: $h_{CMT} = 613\,$km, $M_{W(CMT)} = 7.2$. This earthquake, often called the Sakhalin Island earthquake, attracted attention because it is an unusual "outboard" event situated about 200 km west of other regional deep-focus seismicity. It is also the largest such event since a similar earthquake with a depth of 543 km occurred on 21 April 1939 (see Glennon and Chen, 1993). Kuge (1994) investigated the source rupture process and found that the 1990 earthquake had two distinct subevents separated in space by about 15 km but occurring less than 2 seconds apart; thus the apparent rupture speed exceeded the shear velocity. However, Glennon and Chen (1995a) disagree, concluding that the two subevents occurred 7 km and 2.4 s apart, with somewhat different focal mechanisms. Both Iidaka and Obara (1994) and Hiramatsu and Ando (1995) evaluated ScS-wave polarizations for this earthquake at stations across Japan.

15 January 1993: $h_{CMT} = 100\,$km, $M_{W(CMT)} = 7.6$. This large intermediate-depth earthquake, known as the Kushio-Oki earthquake, occurred just east of Hokkaido, produced "landslides and liquifaction (that) caused extensive damage to roads, housings, harbors, and lifelines" (Yamamura, 1996; Koseki *et al.*, 2000), and was extraordinarily well-recorded by the Hokkaido University network. It had important implications for the origin of double seismic zones, as its aftershocks and rupture both formed a nearly horizontal surface joining the lower and upper planes of the Hokkaido double zone. Takeo *et al.* (1993) studied the source process and concluded that the rupture was confined to an area of about 40 km × 20 km, and Kosuga *et al.* (1996) and Ozel (1999) evaluated the pattern of focal mechanisms in both the upper and lower zones. Suzuki and Kasahara (1996) report that there were 432 aftershocks with *M* of 1.7 or greater in the 45 days following the mainshock; these extended over an area of 50 km × 40 km at depths of 100–110 km. From this they concluded that

the descending slab of about 100 km thickness constitutes two layers: a seismic, brittle upper layer and an aseismic, ductile lower layer. Small ruptures may tend to occur easily along both boundary surfaces of the upper layer and the action of forces, especially unbending, and form the doubled seismic zone.

4 October 1994: $h_{CMT} = 68$ km, $M_{W(CMT)} = 8.3$. In the Harvard CMT catalog, this is the largest earthquake with a centroid depth exceeding 60 km and the third largest of any depth prior to 2004. Because of their depth this and the 9 June 1994 Bolivia earthquake produce strong higher-mode free oscillations useful for constraining models of deep-Earth structure (e.g., see Tromp and Zanzerkia, 1995; Durek and Romanowicz, 1999).

All other catalogs assign this earthquake a shallow focal depth; it produced subsidence of about 60 cm on nearby Shikotan Island (Yeh *et al.*, 1995; Ozawa, 1996), and a tsunami that extended along the entire coast of Japan (e.g., McGehee and McKinney, 1997). Both depths may be correct if the rupture began at shallow depths but ultimately tore the entire slab (e.g., see Tanioka *et al.*, 1995); however, there are controversies about its source mechanism (see Piatanesi *et al.*, 1999). Katsumata *et al.* (1995) investigated this earthquake's aftershocks; Kikuchi and Kanamori (1995), Tanioka *et al.* (1995), and Morikawa and Sasatani (2003) evaluated strong motion records it generated; and Takahashi and Hirata (2003) discuss its relation to other nearby large events, such as the 6 December 1978 earthquake (see above).

10.12.4 Kuril–Hokkaido references

Abe, K., Y. Sato, and J. Frez, 1970. Free oscillations of the Earth excited by the Kurile Islands earthquake of 1963, *Bull. Earthquake Res. Inst., Tokyo Univ.*, **48**, 87–114.

Aoki, H., 1974. Plate tectonics of arc-junction at central Japan, *J. Phys. Earth*, **22**, 141–161.

Balakina, L. M., 1989. The Urup earthquake of October 13, 1963 ($M \sim 8.4$–8.5) in the lithosphere beneath the Kurile island arc, *Izv., Earth Phys.*, **25**, 781–792.

Balakina, L. M. and V. V. Kislovskaya, 1975. Characteristic features of the focal mechanism of some deep earthquakes in the Okhotsk Sea, *Izv., Earth Physics*, **8**, 491–497.

Brüstle, W. and G. Müller, 1987. Stopping phases in seismograms and the spatio-temporal extent of earthquakes, *Bull. Seismol. Soc. Amer.*, **77**, 47–68.

Creager, K. C. and T. H. Jordan, 1984. Slab penetration into the lower mantle, *J. Geophys. Res.*, **89**, 3031–3049.

  1986. Slab penetration into the lower mantle beneath the Mariana and other island arcs of the northwest Pacific, *J. Geophys. Res.*, **91**, 3573–3589.

Ding, X.-Y. and S. P. Grand, 1994. Seismic structure of the deep Kurile subduction zone, *J. Geophys. Res.*, **99**, 23767–23786.

Durek, J. J. and B. Romanowicz, 1999. Inner core anisotropy inferred by direct inversion of normal mode spectra, *Geophys. J. Int.*, **139**, doi:10.1046/j.1365-246x.1999.00961.x, 599–622.

Fukao, Y. and M. Kikuchi, 1987. Source retrieval for mantle earthquakes by interactive deconvolution of long-period P-waves, *Tectonophysics*, **144**, doi:10.1016/0040-1951(87)90021-7, 249–269.

Fukao, Y., M. Obayashi, H. Inoue, and M. Nenbai, 1992. Subducting slabs in the mantle transition zone, *J. Geophys. Res.*, **97**, 4809–4822.

Furumoto, M. and Y. Fukao, 1976. Seismic moment of great deep shocks, *Phys. Earth Planet. Int.*, **11**, 352–357.

Gaherty, J. B., T. Lay, and J. E. Vidale, 1991. Investigations of deep slab structure using long-period S waves, *J. Geophys. Res.*, **96**, 16349–16367.

Glennon, M. A. and W.-P. Chen, 1993. Systematics of deep-focus earthquakes along the Kuril–Kamchatka arc and their implications on mantle dynamics, *J. Geophys. Res.*, **98**, 735–769.

  1995a. Ruptures of deep-focus earthquakes in the northwestern Pacific and their implications on seismogenesis, *Geophys. J. Int.*, **120**, 706–720.

  1995b. Travel times from earthquakes near southern Kuril and their implications for the fate of subducted lithosphere, *Phys. Earth Planet. Int.*, **88**, doi:10.1016/0031-9201(94)02991-J, 177–191.

Gorbatov, A., G. Suárez, V. Kostoglodov, and E. Gordeev, 1994. A double-planed seismic zone in Kamchatka from local and teleseismic data, *Geophys. Res. Lett.*, **21**, doi:10.1029/94GL01593, 1675–1678.

• Gorbatov, A., V. Kostoglodov, G. Suárez, and E. Gordeev, 1997. Seismicity and structure of the Kamchatka subduction zone, *J. Geophys. Res.*, **102**, doi:10.1029/96JB03491, 17883–17898.

Gorbatov, A., J. Dominguez, G. Suárez, V. Kostoglodov, D. Zhao, and E. Gordeev, 1999. Tomographic imaging of the P-wave velocity structure beneath the Kamchatka peninsula, *Geophys. J. Int.*, **137**, doi:10.1046/j.1365-246X.1999.t01-1-00801.x, 269–279.

Gorbatov, A., S. Widiyantoro, Y. Fukao, and E. Gordeev, 2000. Signature of remnant slabs in the north Pacific from P-wave tomography, *Geophys. J. Int.*, **142**, doi:10.1046/j.1365-246x.2000.00122.x, 27–36.

Hănus, V. and J. Vănek, 1984a. Earthquake distribution and volcanism in Kamchatka, Kurile Islands, and Hokkaido. Part 1. Kamchatka and northern Kuriles, *Studia Geoph. et Geod.*, **28**, 36–55.

1984b. Earthquake distribution and volcanism in Kamchatka, Kurile Islands, and Hokkaido. Part 2. Central Kurile Islands, *Studia Geoph. et Geod.*, **28**, 129–148.

1984c. Earthquake distribution and volcanism in Kamchatka, Kurile Islands, and Hokkaido. Part 3. Southern Kuriles and Hokkaido, *Studia Geoph. et Geod.*, **28**, 248–271.

1988. Deep structure, volcanism and Wadati–Benioff zone of the northwestern Pacific convergent margin, *J. Geodynamics*, **10**, doi:10.1016/0264-3707(88)90004-X, 25–41.

Hiramatsu, Y. and M. Ando, 1995. Attenuation anisotropy beneath the subduction zones in Japan, *Geophys. Res. Lett.*, **22**, doi:10.1029/95GL01606, 1653–1656.

Honda, H., 1962. Earthquake mechanism and seismic waves, *Geophys. Notes, Tokyo Univ.*, **15**, *Supp.*, 1–97.

Iidaka, T. and K. Obara, 1994. Shear-wave polarization anisotropy in the upper mantle from a deep earthquake, *Phys. Earth Planet. Int.*, **82**, 19–25.

Jordan, T. H., 1977. Lithospheric slab penetration into the lower mantle beneath the Sea of Okhotsk, *J. Geophys.*, **43**, 473–496.

Kamiya, S., T. Miyatake, and K. Hirahara, 1989. Three-dimensional P-wave velocity structure beneath the Japanese Islands, *Bull. Earthquake Res. Inst., Tokyo Univ.*, **64**, 457–485.

Kao, H. and W.-P. Chen, 1994. The double seismic zone in Kuril–Kamchatka: the tale of two overlapping single zones, *J. Geophys. Res.*, **99**, doi:10.1029/93JB03409, 6913–6930.

1995. Transition from interplate slip to double seismic zone along the Kuril–Kamchatka arc, *J. Geophys. Res.*, **100**, doi:10.1029/95JB00239, 9881–9903.

Kasahara, M. and T. Sasatani, 1985. Source characteristics of the Kunashiri Strait earthquake of December 6, 1978 as deduced from strain seismographs, *Phys. Earth Planet. Int.*, **37**, 124–134.

Katsumata, K., M. Ichiyanagi, M. Miwa, M. Kasahara, and H. Miyamachi, 1995. Aftershock distribution of the October 4, 1994 M_W 8.3 Kurile Island earthquake determined by a local seismic network in Hokkaido, Japan, *Geophys. Res. Lett.*, **22**, doi:10.1029/95GL01316, 1321–1324.

Katsumata, K., N. Wada, and M. Kasahara, 2003. Newly imaged shape of the deep seismic zone within the subducting plate beneath the Hokkaido corner, Japan–Kurile arc–arc junction, *J. Geophys. Res.*, **108**, 2565, doi:10.1029/2002JB002175.

Kawasumi, H. and R. Yosiyama, 1934. On the mechanism of a deep-seated earthquake as revealed by the distribution of the initial motions at stations throughout the world, *Proc. Imperial Acad. Japan*, **10**, 345–348.

Kikuchi, M. and H. Kanamori, 1995. The Shikotan earthquake of October 4, 1994: lithospheric earthquake, *Geophys. Res. Lett.*, **22**, doi:10.1029/95GL00883, 1025–1028.

Koseki, J., O. Matsuo, T. Sasaki, K. Saito, and M. Yamashita, 2000. Damage to sewer pipes during the 1993 Kushiro-Oki and the 1994 Toho-Oki earthquakes, *Soils and Foundations*, **40**, 99–111.

Kosuga, M., T. Sato, A. Hasegawa, T. Matsuzawa, S. Suzuki, and Y. Motoya, 1996. Spatial distribution of intermediate-depth earthquakes with horizontal or vertical planes beneath northeastern Japan, *Phys. Earth Planet. Int.*, **93**, doi:10.1016/0031-9201(95)03089-163-89.

Koyama, J., 1978 Seismic moment of the Vladivostok deep-focus earthquake of September 29, 1973, deduced from P waves and mantle Rayleigh waves, *Phys. Earth Planet. Int.*, **16**, 307–317.

Kuge, K., 1994. Rapid rupture and complex faulting of the May 12, 1990, Sakhalin deep earthquake: analysis of regional and teleseismic broadband data, *J. Geophys. Res.*, **99**, 2671–2685.

Lundgren, P. R. and D. Giardini, 1990. Lateral structure of the subducting Pacific plate beneath the Hokkaido corner from intermediate and deep earthquakes, *Pure Appl. Geophys.*, **134**, 385–404.

Lundgren, P. R., E. A. Okal, and S. Stein, 1988. Body-wave deconvolution for variable source parameters; application to the 1978 December 6 Kuriles earthquake, *Geophys. J. Int.*, **94**, 171–180.

McGehee, D. and J. P. McKinney, 1997. Tsunami detection and warning capability using nearshore submerged pressure transducers: case study of 4 October 1994 Shikotan tsunami, *Advances in Natural and Techological Hazards Research*, **9**, 133–143.

Morikawa, N. and T. Sasatani, 2003. Source spectral characteristics of two large intra-slab earthquakes along the southern Kurile–Hokkaido arc, *Phys. Earth Planet. Int.*, **137**, doi:10.1016/S0031-9201(03)00008-6, 67–80.

Moriya, T., 1979. Seismic studies of the upper mantle beneath the arc-junction at Hokkaido: folded structure of intermediate-depth seismic zone and attenuation of seismic waves, *Adv. Earth Planet. Sci.*, **6**, 467–475.

Okal, E. A., 1979. Higher-mode Rayleigh waves as individual seismic phases, *Earth Planet. Sci. Lett.*, **43**, 162–167.

Ozawa, S., 1996. Geodetic inversion for the fault model of the 1994 Shikotan earthquake, *Geophys. Res. Lett.*, **23**, doi:10.1029/96GL02049, 2009–2012.

Ozel, N., 1999. Different stress directions in the aftershock focal mechanisms of the Kushiro-Oki earthquake of Jan. 15, 1993, SE Hokkaido, Japan, and horizontal rupture in the double seismic zone, *Tectonophysics*, **313**, doi:10.1016/S0040-1951(99)00207-3, 307–327.

Ozel, N. and T. Moriya, 2003. Focal mechanisms of intermediate-depth earthquakes beneath southeastern Hokkaido, Japan: implications of the double seismic zone, *Pure Appl. Geophys.*, **160**, 2279–2399, doi:10.1007/s00024-003-2396-y.

Pankow, K. L. and T. Lay, 1999. Constraints on the Kurile slab from shear wave residuals sphere analysis, *J. Geophys. Res.*, **104**, doi:10.1029/1999JB900039, 7255–7278.

Piatanesi, A., P. Heinrich, and S. Tinti, 1999. The October 4, 1994 Shikotan (Kurile Islands) tsunamigenic earthquake: an open problem on the source mechanism, *Pure Appl. Geophys.*, **154**, doi:10.1007/s000240050244, 555–574.

Romanowicz, B. and P. Guillemant, 1984. An experiment in the retrieval of depth and source mechanism of large earthquakes using very long period Rayleigh-wave data, *Bull. Seismol. Soc. Amer.*, **74**, 417–437.

Sasatani, T., 1974. Source process of a deep-focus earthquake in the Sea of Okhotsk as deduced from long-period P and SH waves, *J. Phys. Earth*, **22**, 279–297.

1976a. Mechanism of mantle earthquakes near the junction of the Kurile and northern Honshu arcs, *J. Phys. Earth*, **24**, 341–354.

1976b. Source process of a large deep-focus earthquake of 1970 in the Sea of Okhotsk, *J. Phys. Earth*, **24**, 27–42.

Scrase, F. J., 1931a. Deep focus earthquakes, *Nature*, **127**, 486.

1931b. The reflected waves from deep focus earthquakes, *Proc. Roy. Soc. London*, **A132**, 213–235.

• 1933. The characteristics of a deep focus earthquake: a study of the disturbance of February 20, 1931, *Phil. Trans. Roy. Soc., London*, **A231**, 207–234.

Schwartz, S. Y., T. Lay, and S. L. Beck, 1991. Shear wave travel time, amplitude, and waveform analysis for earthquakes in the Kurile slab: constraints on deep slab structure and mantle heterogeneity, *J. Geophys. Res.*, **96**, 14445–14460.

Silver, P. G. and T. H. Jordan, 1983. Total-moment spectra of fourteen large earthquakes, *J. Geophys. Res.*, **88**, 3273–3293.

Spakman, W., S. Stein, R. van der Hilst, and R. Wortel, 1989. Resolution experiments for NW Pacific subduction zone tomography, *Geophys. Res. Lett.*, **16**, 1097–1100.

Strelitz, R., 1975. The September 5, 1970 Sea of Okhotsk earthquake: a multiple event with evidence of triggering, *Geophys. Res. Lett.*, **2**, 124–127.

1980. The fate of the downgoing slab: a study of the moment tensors from body waves of complex deep-focus earthquakes, *Phys. Earth Planet. Int.*, **21**, 83–96.

Suzuki, S. and M. Kasahara, 1996. Unbending and horizontal fracture of the subducting Pacific plate, as evidenced by the 1993 Koshiro-Oki and the 1981 and 1987 intermediate-depth earthquakes in Hokkaido, *Phys. Earth Planet. Int.*, **93**, doi:10.1016/0031-9201(95)03090-5, 91–104.

Suzuki, S., T. Sasatani, and Y. Motoya, 1983. Double seismic zone beneath the middle of Hokkaido, Japan, in the southwestern side of the Kurile arc, *Tectonophysics*, **96**, doi:10.1016/0040-1951(83)90244-5, 59–76.

Takahashi, H. and H. Hirata, 2003. The 2000 Nemuro-Hanto-Oki earthquake, off eastern Hokkaido, Japan, and the high intraslab activity in the southwestern Kuril Trench, *J. Geophys. Res.*, **108**, doi:10.1029/2002JB001813, 2178.

Takenaka, S., H. Sanshadokoro, and S. Yoshioka, 1999. Velocity anomalies and spatial distributions of physical properties in horizontally lying slabs beneath the northwestern Pacific region, *Phys. Earth Planet. Int.*, **112**, doi:10.1016/S0031-9201(99)00038-2, 137–157.

Takeo, M., S. Ide, and Y. Yoshida, 1993. The 1993 Kushiro-Oki, Japan, earthquake: a high stress-drop event in a subducting slab, *Geophys. Res. Lett.*, **20**, 2607–2610.

Tanioka, Y., L. Ruff, and K. Satake, 1995. The great Kurile earthquake of October 4, 1994 tore the slab, *Geophys. Res. Lett.*, **22**, doi:10.1029/95GL01656, 1661–1664.

Tromp, J. and E. Zanzerkia, 1995. Toroidal splitting observations from the great 1994 Bolivia and Kuril Islands earthquakes, *Geophys. Res. Lett.*, **22**, doi:10.1029/95GL01810, 2297–2300.

Van der Hilst, R. D., E. R. Engdahl, and W. Spakman, 1993. Tomographic inversion of P and pP data for aspherical mantle structure below the northwest Pacific region, *Geophys. J. Int.*, **115**, 264–302.

Wadati, K. and T. Isikawa, 1933. On deep-focus earthquakes in the northern part of the Japan Sea, *Geophys. Mag.*, **7**, 291–305.

Willemann, R. J. and C. Frohlich, 1987. Spatial patterns of aftershocks of deep focus earthquakes, *J. Geophys. Res.*, **92**, 13927–13943.

Wu, L.-R. and W.-P. Chen, 2001. Rupture of the large (M_W 7.8), deep earthquake of 1973 beneath the Japan Sea with implications for seismogenesis, *Bull. Seismol. Soc. Amer.*, **91**, 102–111.

Yamamura, E., 1996. Land hazards of the Kushiro-Oki earthquake of 15 January 1993, *GeoJournal*, **38**, 345–348.

Yeh, H., V. Titov, V. Gusiakov, E. Pelinovsky, V. Khramushin, and V. Kaistrenko, 1995. The 1994 Shikotan earthquake tsunamis, *Pure Appl. Geophys.*, **144**, 855–874.

Yoshida, M. and M. Tsujiura, 1975. Spectrum and attenuation of multiply reflected core
    phases, *J. Phys. Earth*, **23**, 31–42.
Zhou, H.-W. and R. W. Clayton, 1990. P and S wave travel time inversions for subducting
    slab under the island arcs of the northwest Pacific, *J. Geophys. Res.*, **95**, 6829–6851.
Zhou, H.-W., D. L. Anderson, and R. W. Clayton, 1990. Modeling of residual spheres for
    subduction zone earthquakes. 1. Apparent slab penetration signatures in the NW
    Pacific caused by deep diffuse mantle anomalies, *J. Geophys. Res.*, **95**, 6799–6827.

10.13 Alaska and the Aleutians

10.13.1 Regional tectonics and deep seismicity

The Aleutian Trench marks the boundary between the converging Pacific and North American plates and extends more than 2500 km from southern Alaska to Kamchatka, northeast of Japan (Fig. 10.28). The eastern portion of this boundary is remarkable for producing the world's second and third largest shallow earthquakes of the twentieth century (28 March 1964, $M_W = 9.2$; 9 March 1957, $M_W = 9.1$). In southern Alaska near Cook Inlet and along the Alaskan Peninsula, the convergence rate is about 6 cm/yr and the direction is nearly perpendicular to the trench. Here the thermal parameter Φ is about 2000 km. However, the convergence becomes progressively oblique and the ocean floor becomes older as one moves westward along the Aleutians; west of about 175° E the motion is nearly trench-parallel (Fig. 10.29) and there are no active volcanoes.

Beneath the Alaskan Peninsula and much of the Aleutian arc intermediate-depth earthquakes form a well-defined Wadati–Benioff zone that has a dip of 50°–60°. However, there is a pronounced gap in intermediate activity west of 175° E, presumably because the obliquity of the convergence affects subduction (Fig. 10.30). Focal mechanisms in the Aleutians do not exhibit any simple pattern, as their P- and T-axis orientations are scattered and definitely not clustered in the downdip direction. For events with magnitude of 5 or greater the maximum depths are about 280 km, although there are a few deeper small-magnitude earthquakes in the ISC catalog (Fig. 10.31).

In southern Alaska the Wadati–Benioff zone geometry is more complicated. Near Cook Inlet it bends northward and, although it may be segmented, it reaches to Mt. Denali where it bends again to the northeast. Most earthquakes in the Cook–Denali zone have depths of 150 km or less, although the regional network detects a few with depths as great as 200 km. The majority of focal mechanisms here have principal axes with approximately downdip T and along-strike P orientations. The regional network also records small-magnitude earthquakes with depths as great as 100 km in a southeast-trending zone about 200 km southeast of Denali beneath the Wrangell mountains.

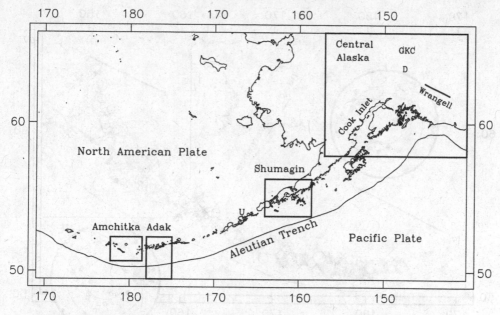

Fig. 10.28 Principal geographic features of Alaska–Aleutian region. Labeled rectangles indicate approximate coverage of regional seismic networks. Solid line
labeled "Wrangell" shows approximate location of 100-km contour of the Wrangell
Wadati–Benioff zone. Abbreviations: U – Unalaska Island; D – Denali; GKC –
Gold King cluster.

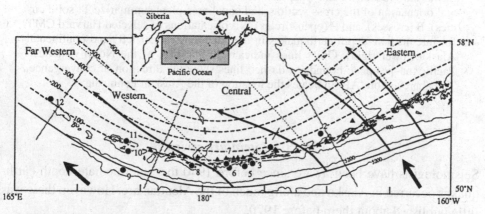

Fig. 10.29 Boyd and Creager's (1991) model of the geometry and kinematics of
subducted lithosphere in the Aleutian arc. Thick solid arrows are relative plate
motions; thin solid arrows are paths taken by subducted material, projected onto
surface of the Earth; and dashed lines are isodepths in km of the descending slab.
Note that the path takes on a significant along-strike component for subduction in
the central and western Aleutians. The medium solid line is the smoothed trench
profile; light lines show bathymetry in meters; and light dashed lines show the path
material would travel if there was no subduction. Triangles are active volcanoes,
and numbered black dots are locations of earthquakes used to construct the model.
Figure reproduced from Boyd and Creager (1991).

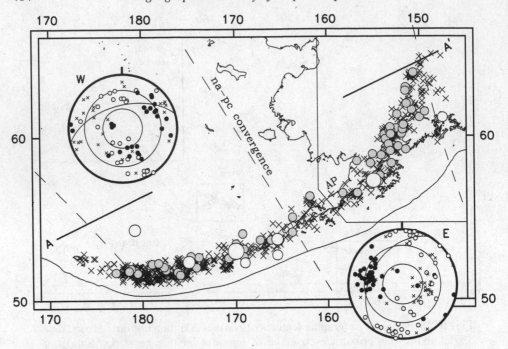

Fig. 10.30 Map of intermediate-depth earthquakes in the Alaska–Aleutian region. Filled circles, open circles, and crosses indicate selected earthquakes in the Harvard CMT, EV Centennial, and EHB catalogs (see description of selection process in the introduction to this chapter). The solid line A–A′ indicates the location and orientation of the cross section, and equal-area plots summarize T (solid circles), B (crosses), and P (open circles) axis orientations of selected Harvard CMT for intermediate-depth earthquakes in the Aleutians (upper left) and continental Alaska (lower right). Other lines indicate the location of the North America–Pacific (na–pc) plate boundary; dashed lines show the direction of convergence. The abbreviation 'AP' indicates the location of the Alaskan Peninsula.

10.13.2 Literature review

Seismologists have been aware since before 1940 that intermediate-depth earthquakes occurred in Alaska and along much of the Aleutian arc. However, there was little published about them before 1970.

Subsequently, three developments took place that provided support for the installation of several regional networks in Alaska. First, the great 1964 Alaskan earthquake emphasized the obvious – that the Alaskan mainland and its emerging oil and minerals industry were subject to considerable seismic hazard. This led to the installation of a rather extensive network (ultimately, more than 40 stations) in Central Alaska in the region surrounding Cook Inlet. Second, the U.S. decision to detonate a series of underground nuclear explosions on Amchitka Island

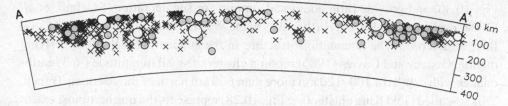

Fig. 10.31 Cross section A–A' of deep seismicity in the Alaska–Aleutian region. Earthquake symbols as in Fig. 10.30.

provided the impetus to operate a series of eight seismic stations there between 1970 and 1973. When the nuclear program ended the scientists involved installed a similar 14-station network centered on Adak Island, where there was a permanent naval station and logistics were simpler. This network operated from 1974 to 1990. Finally, hoping to record the seismic activity prior to a great earthquake, beginning in 1973 seismologists installed a network in the Shumagin Islands to monitor the "gap" between the rupture zones of the 1964 and 1957 Alaskan earthquakes. This network operated until 1991, but its most stable period, with about 18 stations, was from 1982 to 1989 (Abers, 1994). There was also a five-station network on Unalaska Island that monitored seismicity from 1980–1982 (Boyd and Jacob, 1986).

Analyses of the data collected by these networks dominates the literature on Alaskan and Aleutian seismicity. Most papers that review the geometry of the Alaska–Aleutian Wadati–Benioff zone de-emphasize teleseismic observations and instead focus on cross sections of data from several of these regional networks (e.g., Jacob *et al.*, 1977; Davies and House, 1979; Taber *et al.*, 1991). With respect to sophistication of analysis applied to collected data and number of papers published, these regional networks have been as scientifically productive as those in any subduction zone other than Japan. This is remarkable, especially considering that Alaska and the Aleutians are sparsely populated, transportation is difficult, and weather is often poor. At the writing of this book, only the Central Alaska network still operates.

Much of the Central Alaska literature focuses on the structure of the highly active Cook Inlet–Denali Wadati–Benioff zone and evaluates how it relates to the Wrangell activity to the southeast. For the Cook–Denali zone, both Pulpan and Frohlich (1985) and Ratchkovski and Hansen (2002) performed careful relocations and concluded that distinct variations in the strike of the Wadati–Benioff zone imply that there are at least three segments of subducted lithosphere. Several investigators (Li *et al.*, 1995; Lu, 1997; Ratchkovsky *et al.*, 1997b) have analyzed focal mechanisms in this region; generally they are consistent with the Harvard CMT (see

Fig. 10.30) and provide little additional support for segmentation. Stachnik *et al.* (2004) evaluate signals from earthquakes in the Wadati–Benioff zone north of Cook Inlet to determine the attenuation structure in the subducting slab and overlying mantle. Gedney and Davies (1986) report a cluster of small magnitude (< 3) earthquakes with depths of 100–120 km more than 600 km north of the Aleutian Trench. This so-called Gold King cluster (see Fig. 10.28) represents the northernmost extent of the Aleutian–Alaskan Wadati–Benioff zone.

Page *et al.* (1989) provide the best description of the Wrangell Wadati–Benioff zone southeast of Denali. They conclude that the Cook–Denali zone and this Wrangell zone "define adjacent limbs of a north-dipping buckle, or cuspidal ridge, in the contorted, but continuous subducting lithosphere." There are gaps in the seismicity between the two zones at depths greater than about 45 km, however, and it is possible these indicate that there are tears in the subducted lithosphere.

Along the Aleutian arc, early cross sections of locations from the Amchitka, Adak, and Shumagin networks indicated a distinct bend or steepening in the Wadati–Benioff zone beneath about 100 km depth (Engdahl, 1977; Reyners and Coles, 1982; Hauksson *et al.*, 1984). This bend is not evident in locations determined from teleseismic data (Barazangi and Isacks, 1979; Ekström and Engdahl 1989; Boyd and Creager, 1991). Moreover, relocations of the regional network data using ray tracing and velocity models incorporating a higher-velocity subducting slab demonstrated that the bend is an artifact of the presence of the slab (Engdahl *et al.*, 1977; Spencer and Engdahl, 1983; Hauksson, 1985; McLaren and Frohlich, 1985; see Fig. 5.6). Moreover, the bend also disappears if one augments the regional network data with readings from ocean bottom seismograph stations situated so that arriving phases do not travel up the slab (Frohlich *et al.*, 1982).

There have been reports that the earthquakes form a double Wadati–Benioff zone beneath Adak (Engdahl and Scholz, 1977), the Shumagins (Hudnut and Taber, 1987) and Central Alaska (Ratchkovsky *et al.*, 1997a; 1997b). The evidence supporting this conclusion isn't very strong. Beneath Adak the double zone reported above is almost certainly a product of the location artifact described above. In the Shumagins the two zones are poorly defined when the hypocenters are separated into southwest and northeast groups (Abers, 1992); there are also P-to-S converted phases arriving before S that complicate location (Helffrich and Abers, 1997). Apparent double zones can also be artifacts caused when different combinations of P and S readings are used to locate events in the supposed upper and lower zones. House and Jacob (1983) evaluated teleseismic focal mechanisms for 11 intermediate-depth earthquakes in the Aleutians and concluded that they fell into downdip-P and downdip-T groups. However, presently available focal mechanism data for the Aleutians indicates that the pattern of focal mechanisms is more complex (see Fig. 10.30).

Several investigations use data from these networks to evaluate the detailed structure and depth extent of the subducting lithospheric slab. Engdahl and Gubbins (1987) evaluated mantle structure beneath the Adak network by inverting travel times for 151 earthquakes and incorporating pP and sP readings from teleseismic stations to constrain focal depths. They found slab velocities up to 10% faster than ordinary mantle, and concluded that the slab extended well below the deepest seismic activity at 270 km depth (see also Boyd and Creager, 1991). Gorbatov *et al.* (2000) have constructed tomographic images of the deeper mantle beneath the Aleutians using both teleseismic data and regional phases recorded by Russian stations.

There have been additional analyses in the Shumagins and Central Alaska. Kissling and Lahr (1991) performed a tomographic inversion of 110,000 P times from 4928 earthquakes recorded by the Central Alaska network to evaluate the upper 200 km of velocity structure in and above the Wadati–Benioff zone. Similarly, Zhao *et al.* (1995) inverted 142,908 P waves from 12,237 shallow and intermediate earthquakes recorded at 177 stations in Alaska and the Shumagins, and concluded the subducting Pacific plate has a thickness of 45–55 km and a P wave velocity 3–6% higher than that of the surrounding mantle. Abers and Sarker (1996) suggest that a waveguide at the top of the slab is responsible for observed dispersion of P waveforms from earthquakes with depths exceeding 100 km recorded by stations in the Shumagin network. Helffrich and Abers (1997) analyze slab structure by modeling travel times of P-to-S converted phases that arrive between P and S. They interpret these as arising from the top of a low velocity layer which extends to 150 km depth and is 8±2% slower than the overlying mantle, while the slab is 7% faster than the mantle. Finally, Yang *et al.* (1995) and Wiemer *et al.* (1999) have evaluated anisotropy in the mantle above the deep seismic zone, and Wiemer and Benoit (1996) have evaluated how *b* values vary within the Wadati–Benioff zone.

10.13.3 Significant earthquakes

The largest intermediate-depth Alaska–Aleutian earthquakes in the CMT catalog have magnitudes of 6.6; however, the EV Centennial catalog lists eight with m_B between 7.0 and 7.4. None of these earthquakes has been the subject of special study.

10.13.4 Alaska and Aleutian references

Abers, G. A., 1992. Relationship between shallow- and intermediate-depth seismicity in the eastern Aleutian subduction zone, *Geophys. Res. Lett.*, **20**, 2019–2022.

1994. Three-dimensional inversion of regional P and S arrival times in the east
Aleutians and sources of subduction zone gravity highs, *J. Geophys. Res.*, **99**,
4395–4412.

Abers, G. A. and G. Sarker, 1996. Dispersion of regional body waves at 100–150 km
depth beneath Alaska: in situ constraints on metamorphism of subducted crust,
Geophys. Res. Lett., **23**, doi:10.1029/96GL00974, 1171–1174.

Barazangi, M. and B. L. Isacks, 1979. A comparison of the spatial distribution of mantle
earthquakes determined from data produced by local and by teleseismic networks for
the Japan and Aleutian arcs, *Bull. Seismol. Soc. Amer.*, **69**, 1763–1770.

Boyd, T. M. and K. C. Creager, 1991. The geometry of Aleutian subduction:
three-dimensional seismic imaging, *J. Geophys. Res.*, **96**, 2267–2291.

Boyd, T. M. and K. Jacob, 1986. Seismicity of the Unalaska region, Alaska, *Bull. Seismol.
Soc. Amer.*, **76**, 463–481.

Davies, J. N. and L. House, 1979. Aleutian subduction zone seismicity, volcano-trench
separation, and their relation to great thrust-type earthquakes, *J. Geophys. Res.*, **84**,
4583–4591.

Ekström, G. and E. R. Engdahl, 1989. Earthquake source parameters and stress
distribution in the Adak Island region of the central Aleutian Islands, Alaska, *J.
Geophys. Res.*, **94**, 15499–15519.

Engdahl, E. R., 1977. Seismicity and plate subduction in the central Aleutians. In *Island
Arcs, Deep Sea Trenches and Back-Arc Basins*, eds. M. Talwani and W. C. Pitman,
Washington, DC, American Geophysical Union, 259–271.

Engdahl, E. R. and D. Gubbins, 1987. Simultaneous travel time inversion for earthquake
locations and subduction zone structure in the central Aleutian islands, *J. Geophys.
Res.*, **92**, 13855–13862.

Engdahl, E. R. and C. H. Scholz, 1977. A double Benioff zone beneath the central
Aleutians: an unbending of the lithosphere, *Geophys. Res. Lett.*, **4**, 473–476.

Engdahl, E. R., N. H. Sleep, and M.-T. Lin, 1977. Plate effects in north Pacific subduction
zones, *Tectonophysics*, **37**, doi:10.1016/0040-1951(77)90041-5, 95–116.

Gedney, L. and J. N. Davies, 1986. Additional evidence for downdip tension in the Pacific
plate beneath central Alaska, *Bull. Seismol. Soc. Amer.*, **76**, 1207–1214.

Frohlich, C., S. Billington, E. R. Engdahl, and A. Malahoff, 1982. Detection and location
of earthquakes in the central Aleutian subduction zone using island and ocean bottom
seismograph stations, *J. Geophys. Res.*, **87**, 6853–6864.

Gorbatov, A., S. Widiyantoro, Y. Fukao, and E. Gordeev, 2000. Signature of remnant
slabs in the north Pacific from P-wave tomography, *Geophys. J. Int.*, **142**,
doi:10.1046/j.1365–246x.2000.00122.x, 27–36.

Hauksson, E., 1985. Structure of the Benioff zone beneath the Shumagin Islands, Alaska:
relocation of local earthquakes using three-dimensional ray tracing, *J. Geophys. Res.*,
90, 635–649.

Hauksson, E., J. Armbruster, and S. Dobbs, 1984. Seismicity patterns (1963–1983) as
stress indicators in the Shumagin seismic gap, Alaska, *Bull. Seismol. Soc. Amer.*, **74**,
2541–2558.

Helffrich, G. and G. A. Abers, 1997. Slab low-velocity layer in the eastern Aleutian
subduction zone, *Geophys. J. Int.*, **130**, 640–648.

House, L. S. and K. H. Jacob, 1983. Earthquakes, plate subduction, and stress reversals in
the eastern Aleutian arc, *J. Geophys. Res.*, **88**, 9347–9373.

Hudnut, K. W. and J. J. Taber, 1987. Transition from double to single Wadati–Benioff
seismic zone in the Shumagin Islands, Alaska, *Geophys. Res. Lett.*, **14**,
143–146.

Jacob, K. H., K. Nakamura, and J. N. Davies, 1977. Trench–volcano gap along the Alaska–Aleutian arc: facts, and speculations on the role of terrigenous sediments. In *Island Arcs, Deep Sea Trenches and Back-Arc Basins*, eds. M. Talwani and W. C. Pitman, Washington, DC, American Geophysical Union, 243–258.

Kissling, E. and J. C. Lahr, 1991. Tomographic image of the Pacific slab under southern Alaska, *Eclogae Geologicae Helvetiae*, **84**, 297–315.

Li, Z., N. Biswas, G. Tytgat, H. Pulpan, and M. Wyss, 1995. Stress directions along the Alaska Wadati–Benioff zone from the inversion of focal mechanism data, *Tectonophysics*, **246**, doi:10.1016/0040-1951(94)00257-A, 163–170.

Lu, Z., M. Wyss, and H. Pulpan, 1997. Details of stress directions in the Alaska subduction zone from fault plane solutions, *J. Geophys. Res.*, **102**, doi:10.1029/96JB03666, 5385–5402.

McLaren, J. P. and C. Frohlich, 1985. Model calculations of regional network locations for earthquakes in subduction zones, *Bull. Seismol. Soc. Amer.*, **75**, 397–413.

Page, R. A., C. D. Stephens, and J. C. Lahr, 1989. Seismicity of the Wrangell and Aleutian Wadati–Benioff zones and the North American plate along the trans-Alaska crustal transect, Chugach mountains and Copper River Basin, southern Alaska, *J. Geophys. Res.*, **94**, 16059–16082.

Pulpan, H. and C. Frohlich, 1985. Geometry of the subducted plate near Kodiak Island and Lower Cook Inlet, Alaska, determined from relocated earthquake hypocenters, *Bull. Seismol. Soc. Amer.*, **75**, 791–810.

•Ratchkovski, N. A. and R. A. Hansen, 2002. New evidence for segmentation of the Alaska subduction zone, *Bull. Seismol. Soc. Amer.*, **92**, 1754–1765.

Ratchkovsky, N. A., J. Pujol, and N. N. Biswas, 1997a. Relocation of earthquakes in the Cook Inlet area, south central Alaska, using the joint hypocenter determination method, *Bull. Seismol. Soc. Amer.*, **87**, 620–636.

1997b. Stress pattern in the double seismic zone beneath Cook Inlet, south-central Alaska, *Tectonophysics*, **281**, doi:10.1016/0040-1951(97)00042-5, 163–171.

Reyners, M. and K. S. Coles, 1982. Fine structure of the dipping seismic zone and subduction mechanics in the Shumagin Islands, Alaska, *J. Geophys. Res.*, **87**, 356–366.

Spencer, C. P. and E. R. Engdahl, 1983. A joint hypocenter location and velocity inversion technique applied to the central Aleutians, *Geophys. J. Roy. Astron. Soc.*, **72**, 399–415.

Stachnik, J. C., G. A. Abers, and D. H. Christensen, 2004. Seismic attenuation and mantle wedge temperatures in the Alaska subduction zone, *J. Geophys. Res.*, **109**, doi:10.1029/2004JB003018.

Taber, J. J., S. Billington, and E. R. Engdahl, 1991. Seismicity of the Aleutian arc. In *Neotectonics of North America*, eds. by D. B. Slemmons, E. R. Engdahl, M. D. Zoback, and D. D. Blackwell, Boulder, CO, Geological Society of America, 29–46.

Wiemer, S. and J. P. Benoit, 1996. Mapping the *b*-value anomaly at 100 km depth in the Alaska and New Zealand subduction zones, *Geophys. Res. Lett.*, **23**, doi:10.1029/96GL01233, 1557–1560.

Wiemer, S., G. Tytgat, M. Wyss, and U. Duenkel, 1999. Evidence for shear-wave anisotropy in the mantle wedge beneath south central Alaska, *Bull. Seismol. Soc. Amer.*, **89**, 1313–1322.

Yang, X., K. M. Fischer, and G. A. Abers, 1995. Seismic anisotropy beneath the Shumagin Islands segment of the Aleutian–Alaska subduction zone, *J. Geophys. Res.*, **100**, doi:10.1029/95JB01425, 18165–18177.

Zhao, D., D. Christensen, and H. Pulpan, 1995. Tomographic imaging of the Alaska subduction zone, *J. Geophys. Res.*, **100**, doi:10.1029/95JB00046, 6487–6504.

10.14 Western North America

10.14.1 Regional tectonics and deep seismicity

Along western North America the Juan de Fuca and Gorda plates subduct beneath the North American continent. This produces hazardous volcanoes (notably Rainier and Mt. St. Helens) and an inclined zone of earthquakes extending along the western Washington/Canada border (Fig. 10.32). Some of these earthquakes with depths of 40–70 km are large and pose a significant hazard to the cities of Seattle and Vancouver.[11] The Juan de Fuca and Gorda plates are very young, everywhere less than 10 Ma, the convergence rate is only about 4 cm/yr, and the Wadati–Benioff zone dip angle is shallow, about 20° or less. Thus the thermal parameter Φ is ~150 km, making it among the lowest in the world.

In northern California there is an eastward-dipping Wadati–Benioff zone that is fairly active to depths of about 40 km; and there have been a few small and reasonably well-recorded intermediate-depth earthquakes with depths of 60–90 km reported beneath west-central California and northern Utah. As in Washington, these deeper earthquakes are recorded by regional networks that have operated for decades and which routinely record earthquakes with magnitudes as small as 2 or less. The reported depths exceeding 80 km may or may not be accurate.

10.14.2 Literature review

Nuttli's (1952) analysis of the 13 April 1949 earthquake (see below) demonstrated that intermediate-depth earthquakes do occur in western North America. However, because such events are rare and their magnitudes are mostly small, only after dense regional networks were in place did investigators plot cross sections that suggested they formed a Wadati–Benioff zone (for Washington see Taber and Smith, 1985, Weaver and Baker, 1988, and Wahlstrom, 1993; for California, see Cockerham, 1984, Walter, 1986; and Smith *et al.*, 1993). Beneath Oregon–Washington, a tomographic inversion using teleseismic arrivals recorded during a 1993 IRIS-PASCAL deployment demonstrates that the deep seismicity occurs within an inclined high-velocity region extending to depths as great as 100 km in the vicinity of Mt. Baker (Rondenay *et al.*, 2001). Beneath British Columbia, Cassidy and Waldhauser (2003)

I thank John Cassidy for reviewing an earlier draft of this section.

[11] Truly great shallow earthquakes can occur in this region, although none have occurred in historic times. There is strong evidence that a Cascadia earthquake with M_W of about 9.0 produced the damaging tsunami which struck the coast of Japan on 27 January 1700 (e.g., see Satake *et al.*, 2003).

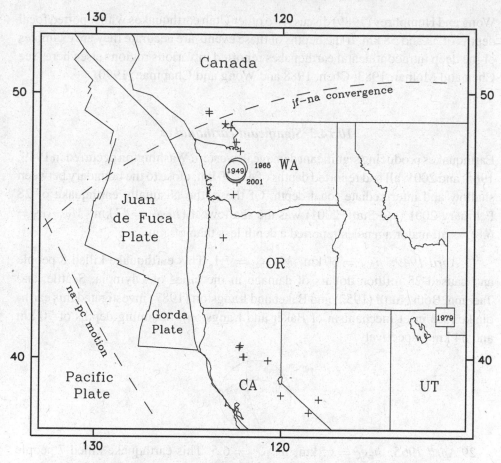

Fig. 10.32 Map of intermediate-depth earthquake activity in the western North America. Circles labeled 1949, 1965, and 2001 are damaging earthquakes mentioned in the text that have struck western Washington; the open square labeled 1979 is the magnitude 3.8 Utah quake reported by Zandt and Richins (1979); '+' symbols are better-located intermediate-depth earthquakes from Walter (1986), Wahlstrom (1993), or other regional network locations. Other lines indicate plate boundaries and directions of relative plate motion.

located hypocenters as deep as 65 km and present evidence that they form a double seismic zone; Hacker *et al.* (2003) conclude that the pattern of seismicity observed here may be attributable to dehydration reactions occurring within the young, hot subducting lithosphere.

The Utah earthquake of 24 February 1979 ($h = 80$–$90\,\text{km}$, $M_L = 3.8$) is puzzling. Its depth is suspicious as it is an outlier among huge numbers of shallower events located by the Utah network. However, Zandt and Richins (1979) note that its location incorporated data from 39 stations with an azimuthal gap of 124°. And

Wong and Humphrey (1989) discuss two other Utah earthquakes with reported focal depths of 53 and 58 km. If the depths of these events are accurate they are examples of the deep intracontinental earthquakes reported in various regions elsewhere (see Chen and Molnar, 1983; Chen, 1988 and Wong and Chapman, 1990).

10.14.3 Significant earthquakes

Earthquakes producing significant damage in western Washington occurred in 1949, 1965, and 2001; all had reported depths of 45–70 km, close to the boundary between shallow and intermediate focal depth. Of these, the Nisqually earthquake of 28 February 2001 (see Staff, 2001) was the shallowest ($h_{\text{CMT}} = 47$ km, $M_{\text{W(CMT)}} = 6.8$), as all major agencies reported a depth less than 60 km.

13 April 1949: $h_{\text{Abe}} = 60$ km; $M_{\text{S(Abe)}} = 7.1$. This earthquake killed 8 people and caused 25 million dollars of damage in the cities of Olympia, Seattle, and Tacoma. Both Nuttli (1952) and Baker and Langston (1987) investigated this earthquake (see inset mechanism of Baker and Langston), obtaining depths of 70 km and 54 km respectively.

29 April 1965: $h_{\text{EVC}} = 65$ km; $m_{\text{b(ISC)}} = 6.5$. This earthquake killed 7 people and caused 12.5 million dollars of damage, producing intensities of MMI VII to VIII in and around Seattle (Algermissen and Harding, 1965). The ISC, EHB, and EV Centennial catalogs, as well as Langston and Blum (1977), all report a depth between 60 and 66 km. The inset shows the source obtained by Langston and Blum from waveform modeling.

10.14.4 Western North America references

Algermissen, S. T. and S. T. Harding, 1965. Preliminary seismological report. In *The Puget Sound, Washington Earthquake of April 29, 1965*, U.S. Dept. Commerce, Coast and Geodetic Surv., Washington, U.S. Government Printing Office, 51 pp.

Baker, G. E. and C. A. Langston, 1987. Source parameters of the 1949 magnitude 7.1 south Puget Sound, Washington, earthquake as determined from long-period body waves and strong ground motions, *Bull. Seismol. Soc. Amer.*, **77**, 1530–1557.

•Cassidy, J. F. and F. Waldhauser, 2003. Evidence for both crustal and mantle earthquakes in the subducting Juan de Fuca plate, *Geophys. Res. Lett.*, **30**, 1095, doi:10.1029/2002GL015511.

Chen, W.-P., 1988. A brief update on the focal depths of intracontinental earthquakes and their correlations with heat flow and tectonic age, *Seismol. Res. Lett.*, **59**, 263–272.

Chen, W.-P. and P. Molnar, 1983. Focal depths of intracontinental and intraplate earthquakes and their implications for the thermal and mechanical properties of the lithosphere, *J. Geophys. Res.*, **88**, 4183–4214.

Cockerham, R. S., 1984. Evidence for a 180-km-long subducted slab beneath northern California, *Bull. Seismol. Soc. Amer.*, **74**, 569–576.

Hacker, B. R., S. M. Peacock, G. A. Abers, and S. D. Holloway, 2003. Subduction factory 2. Are intermediate-depth earthquakes in subducting slabs linked to metamorphic dehydration reactions? *J. Geophys. Res.*, **108**, doi:10.1029/2001JB001129.

Nuttli, O. W., 1952. The western Washington earthquake of April 13, 1949, *Bull. Seismol. Soc. Amer.*, **42**, 21–28.

Langston, C. A. and D. E. Blum, 1977. The April 29, 1965, Puget Sound earthquake and the crustal and upper mantle structure of western Washington, *Bull. Seismol. Soc. Amer.*, **67**, 693–711.

Rondenay, S., M. G. Bostock, and J. Shragge, 2001. Multiparameter two-dimensional inversion of scattered teleseismic body waves. 3. Application to the Cascadia 1993 data set, *J. Geophys. Res.*, **106**, doi:10.1029/2000JB000039, 30795–30807.

Satake, K., K. Wang, and B. F. Atwater, 2003. Fault slip and seismic moment of the 1700 Cascadia earthquake inferred from Japanese tsunami descriptions, *J. Geophys. Res.*, **108**, doi:10.1029/2003JB002521, 2535.

•Smith, S. W., J. S. Knapp, and R. C. McPherson, 1993. Seismicity of the Gorda plate, structure of the continental margin, and an eastward jump of the Mendocino triple junction, *J. Geophys. Res.*, **98**, 8153–8171.

Staff of the Pacific Northwest Seismograph Network, 2001. Preliminary report on the $M_W = 6.8$ Nisqually, Washington earthquake of 28 February 2001, *Seismol. Res. Lett.*, **72**, 352–361.

Taber, J. J. and S. W. Smith, 1985. Seismicity and focal mechanisms associated with the subduction of the Juan de Fuca plate beneath the Olympic Peninsula, Washington, *Bull. Seismol. Soc. Amer.*, **75**, 237–249.

Wahlstrom, R., 1993. Comparison of dynamic source parameters for earthquakes in different tectonic regions of the northern Cascadia subduction zone, *Tectonophysics*, **217**, doi:10.1016/0040-1951(93)90004-4, 205–215.

Walter, S. R., 1986. Intermediate-focus earthquakes associated with Gorda plate subduction in northern California, *Bull. Seismol. Soc. Amer.*, **76**, 583–588.

Weaver, C. S. and G. E. Baker, 1988. Geometry of the Juan de Fuca plate beneath Washington and northern Oregon from seismicity, *Bull. Seismol. Soc. Amer.*, **78**, 264–275.

Wong, I. G. and D. S. Chapman, 1990. Deep intraplate earthquakes in the western United States and their relationship to lithospheric temperatures, *Bull. Seismol. Soc. Amer.*, **80**, 589–599.

Wong, I. G. and J. R. Humphrey, 1989. Contemporary seismicity, faulting and the state of stress in the Colorado plateau, *Geol. Soc. Amer. Bull.*, **101**, 1127–1146.

Zandt, G. and W. D. Richins, 1979. An upper mantle earthquake beneath the middle Rocky Mountains in NE Utah, *Earthquake Notes*, **50**, 69–70.

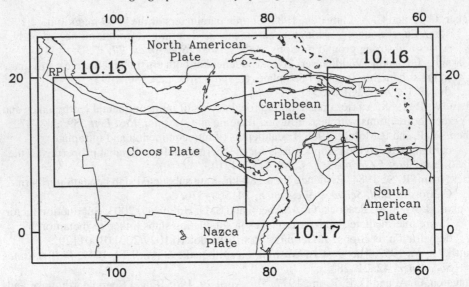

Fig. 10.33 Map showing the six plates that affect deep earthquake activity in the Central American (Section 10.15), Caribbean (Section 10.16), and Colombia–Venezuela–Panama (Section 10.17) regions. Label RP indicates the location of the Rivera plate.

10.15 Central America

10.15.1 Regional tectonics and deep seismicity

Intermediate-depth earthquakes occur between the southern Mexico and the Panama–Costa Rica border in response to convergence of the Rivera and Cocos plates with Central America (Figs. 10.33 and 10.34). In Guatemala, Belize, and Mexico the overlying plate is the North American plate and the convergence rate is about 6 cm/yr; to the southeast it is the Caribbean plate and the rate is slightly higher. The triple junction between the three plates occurs approximately where the Tehuantepec Ridge intersects the Pacific coastline. The regional tectonics are complicated by the fact that the subducting lithosphere is quite young; some of the subducting material is small platelets formed as the ridge/transform system that is the Rivera/Cocos boundary converged on Central America. Beneath Costa Rica and Nicaragua the thermal parameter Φ is about 1500 km, and less elsewhere where the subducting ocean floor is younger.

The Central American Wadati–Benioff zone has three major sections. The northwest section is beneath Mexico and has few earthquakes with focal depths exceeding 120 km (Fig. 10.35). Here the Rivera plate is being subducted and the dip is

I thank Geoff Abers for reviewing an earlier draft of this section.

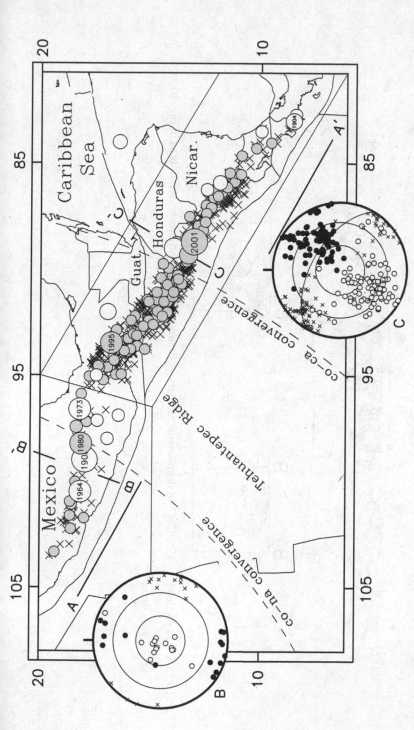

Fig. 10.34 Map of intermediate-depth earthquakes in Central America. Filled circles, open circles, and crosses indicate selected earthquakes in the Harvard CMT, EV Centennial, and EHB catalogs (see description of selection process in the introduction to this chapter). Solid lines A–A′ to C–C′ indicate locations and orientations of cross sections, and equal-area plots summarize T (solid circles), B (crosses), and P (open circles) axis orientations of selected Harvard CMT for intermediate-depth earthquakes within the indicated regions. Dashed lines show the direction of convergence for the Cocos–North America (co–na) and Cocos–Caribbean Co–Ca plates; other lines are plate boundaries and political boundaries.

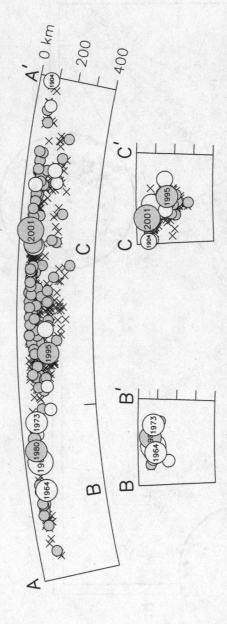

Fig. 10.35 Cross sections A–A′ to C–C′ of deep seismic activity in Central America. Earthquake symbols as in Fig. 10.34.

about 45°. The central section, southeast of the Rivera fracture zone, has a shallow dip and becomes nearly flat-lying in some areas. The southeastern section extends from Chiapas in southern Mexico to northwestern Costa Rica; it has dips as great as 80°–85° and maximum focal depths which reach 200 km and more. Indeed, beneath Honduras and Nicaragua the ISC catalog reports a few hypocenters with depths as great as 345 km. A peculiarity of the southeastern section is that there is little volcanism where the overlying plate is within Mexico, but there is quite active volcanism on the overlying plate from Guatemala to Costa Rica. Beneath Costa Rica the Wadati–Beniofff zone becomes systematically shallower, and the deep activity terminates to the southeast.

Focal mechanisms in Central America are quite consistent and reflect the geometry of the Wadati–Benioff zone. In the shallow-dipping section beneath Mexico the majority of earthquakes have normal-faulting mechanisms with horizontal T axes oriented approximately perpendicular to the Mexican coastline (Fig. 10.34, section B). However, in the southeastern Chiapas-to-Costa Rica section the T axes dip more steeply, but still consistently exhibit downdip tension (see section C).

10.15.2 Literature review

There have been numerous investigations which utilized teleseismically recorded hypocenters and focal mechanisms determined from first motions to investigate the geometry and segmentation of the Central American Wadati–Benioff zone (e.g., Molnar and Sykes, 1969; Hănus and Vănek, 1977–78; Burbach *et al.*, 1984; LeFevre and McNally, 1985; Wolters, 1986; and Pardo and Suárez, 1995). These studies are in general agreement, although there are considerable differences in interpretation concerning the extent of segmentation within the subducted lithosphere; e.g., Stoiber and Carr (1973) delineated no fewer than seven segments on the basis of an analysis of volcanism on the overlying plate, whereas Burbach *et al.* (1984) found only three on the basis of seismicity.

In several areas there are local networks or JHD studies that constrain the geometry of the Wadati–Benioff zone. At its northwestern edge beneath Jalisco, Mexico, Pardo and Suárez (1993) inferred from both locally recorded and relocated teleseismic data that the slab reaches a depth of about 150 km and has a dip of 50°. For a local network situated to the southeast beneath Guererro (~17° S, 101° W), Suárez *et al.* (1990) found no intermediate-depth earthquakes. However, this network is situated near the coast above the shallower part of the subducting plate, and thus places little constraint on the depths of the deeper earthquakes that occur further inland. Singh and Pardo (1993) also evaluated the seismicity in Guerrero and concluded that the slab flattens out at a depth of about 50–60 km and extends inland for a distance of about 175 km. They suggest that it must go deeper to the east where there are volcanoes, but there are no hypocenters there to confirm this. Beneath Chiapas near the Gulf of Tehuantepec, Havskov *et al.* (1982), Rebollar *et al.* (1999a), and Bravo *et al.* (2004) evaluate local network data and find a planar slab extending to 200 km depth with a dip of 45°. Ponce *et al.* (1992) analyzed data from six permanent and 15 temporary stations and concluded that the transition between the flat-lying slab in Guererro and the dipping slab and Chiapas was continuous, i.e., it occurred as a broad flexure rather than as a discrete tear. Gutscher *et al.* (2000) review various factors that control whether the subducting slab is flat-lying or inclined. Pardo and Suárez (1995) present a thorough overall review of the Wadati–Benioff zone geometry beneath Mexico.

As we move southeast along the Pacific coast the maximum depth of seismicity gradually gets deeper.[12] Near Managua, Nicaragua, Dewey and Algermisson (1974) performed JHD relocations of earthquakes and found a dip of about 60° extending to 200 km depth. For Nicaragua and northern Costa Rica, Protti *et al.* (1995) combined data from several local networks and found that the dip was 80°–85° and earthquakes occurred down to 200 km depth. Beneath southern Costa Rica both the dip and the maximum extent of activity gradually become somewhat shallower. Neither Protti *et al.* (1995) or Quintero and Güendel (2000) observed any earthquakes with focal depths exceeding 50 km east of 83.85° W, which is approximately where the Cocos ridge subducts.

Several recent investigations suggest that dehydration of subducting materials explains the pattern of seismicity beneath Nicaragua and Costa Rica. Ranero *et al.* (2003) show that faulting in the lithosphere beneath the outer rise of Costa Rica extends to depths of at least 20 km; at greater depths Abers *et al.* (2003) find a low-velocity layer coincident with the top of the subducting slab. Husen *et al.* (2003) modeled metamorphic dehydration and suggested that their model explains various features of the seismicity including the decrease in depth beneath southeastern Costa Rica. However, Peacock *et al.* (2005) concluded that the along-strike variation in calculated thermal structure is relatively minor and inadequate by itself to explain variations in the distribution of Wadati–Benioff earthquakes and arc geochemistry. They thus suggested that variations in seismicity were attributable to other factors such as regional variations in slab stresses, crustal thickness, incoming sediment load, and the distribution of hydrous minerals in the incoming lithosphere.

Sometimes the literature creates confusion concerning the depths of Central American earthquakes because several authors use "intermediate" to apply to earthquakes with focal depths clearly shallower than 60 km (e.g., Cocco *et al.*, 1997; Quintanar *et al.*, 1999). This is perhaps natural because this region regularly experiences large and destructive earthquakes with focal depths in the 40–80 km

---

[12] I have been unable to find publications presenting detailed results from the local networks in Guatemala or Nicaragua.

range; thus it is unlikely that Wadati or Gutenberg would have chosen the shallow-intermediate boundary at 60 km if they had been residents of Central America (see discussion in Section 2.1). In any case, the risk associated with these earthquakes has provoked a number of investigations of their mechanisms and surface ground motions (e.g., Singh and Mortera, 1991; Pacheco and Singh, 1995; Rebollar *et al.*, 1999a; Vănek *et al.*, 2000; Iglesias *et al.*, 2002).

10.15.3 Significant earthquakes

In the twentieth century the Abe, EV Centennial and Harvard catalogs list about 20 intermediate-depth earthquakes in Central America with magnitudes exceeding 7.0. Two very early events (20 December 1904, $h_{Abe} = 60\,km$, $m_{B(Abe)} = 7.8$; 26 March 1908, $h_{Abe} = 80\,km$, $m_{B(Abe)} = 7.7$) and the El Salvador earthquake of 13 January 2001 (see below) have the largest reported magnitudes. No other earthquakes with depths that may exceed 60 km have magnitudes exceeding 7.4.

There is considerable hazard associated with Central American intermediate-depth earthquakes. The National Geophysical Data Center A (1992) indicates that four intermediate Central American earthquakes have killed 50 people or more. Of these, recent research indicates that the El Salvador earthquake of 6 May 1951 was actually shallow (Ambraseys *et al.*, 2001). However, the earthquakes in Guerrero on 6 July 1964 ($h_{pP-P} = 94$; $m_{B(Abe)} = 7.2$), Veracruz on 28 August 1973 (see below), and Oaxaca on 24 October 1980 ($h_{CMT} = 63\,km$; $M_{W(CMT)} = 7.2$) were genuine intermediate events.

28 August 1973: $h_{pP-P} = 83\,km$; $m_{B(Abe)} = 7.3$. This earthquake produced extensive damage in the Mexican state of Veracruz and is called the Veracruz earthquake. It killed 212 people in Ciudad Serdan, which had a population of 12,000; altogether it reportedly killed about 600 people. It produced intensities of MMI VI as far away as Mexico City. The articles by Irvine (1973) and Meehan (1974) describe the damage it caused to buildings, with each presenting more than 20 photographs. Singh and Wyss (1976) evaluated the source in the frequency domain, and concluded that the earthquake had a scalar moment of 4.8×10^{19} N-m and a stress drop of 95 bars. The inset focal mechanism is from Jiménez and Ponce (1978).

21 October 1995: $h_{CMT} = 164\,km$; $M_{W(CMT)} = 7.2$. This earthquake occurred in Chiapas, Mexico, near the bottom edge of the Wadati–Benioff zone as determined from local network data. Rebollar *et al.* (1999) investigated its strong motion properties and noted that the local network recorded no aftershocks.

13 January 2001: h_{CMT} = 56 km; M_W = 7.7. This earthquake occurred offshore of El Salvador near the shallow-intermediate boundary. Both the Harvard CMT and the EHB catalogs reported a shallow focus but the ISC listed a pP-P depth of 73 km. Choy and Kirby (2004) discuss its source, which is unusual both because of its normal-faulting mechanism and also because its apparent stress was somewhat high. It is an important earthquake, as it caused about 1200 fatalities, mostly from landslides (Salazar and Seo, 2003; Valée *et al.*, 2003; Rose *et al.*, 2004).

10.15.4 Central America references

Abers, G. A., T. Plank, and B. R. Hacker, 2003. The wet Nicaraguan slab, *Geophys. Res. Lett.*, **30**, doi:10.1029/2002GL015649, 1098.

Ambraseys, N. N., J. J. Bommer, E. Buforn, and A. Udías, 2001. The earthquake sequence of May 1951 at Jucuapa, El Salvador, *J. Seismology*, **5**, doi:10.1023/A:1009883313414, 23–39.

Bravo, H., C. J. Rebollar, A. Uribe, and O. Jiménez, 2004. Geometry and state of stress of the Wadati–Benioff zone in the Gulf of Tehuantepec, Mexico, *J. Geophys. Res.*, **109**, doi:10.1029/2003JB002854.

Burbach, G. V., C. Frohlich, W. D. Pennington, and T. Matumoto, 1984. Seismicity and tectonics of the subducted Cocos plate, *J. Geophys. Res.*, **89**, 7719–7735.

Choy, G. L. and S. H. Kirby, 2004. Apparent stress, fault maturity and seismic hazard for normal-fault earthquakes at subduction zones, *Geophys. J. Int.*, **159**, doi:10.1111/j.1365-246X.2004.02449.x, 991.

Cocco, M., J. Pacheco, S. K. Singh, and F. Courboulex, 1997. The Zihuatanejo, Mexico, earthquake of 1994 December 10 (M = 6.6): source characteristics and tectonic implications, *Geophys. J. Int.*, **131**, 135–145.

Dewey, J. W. and S. T. Algermissen, 1974. Seismicity of the Middle America arc–trench system near Managua, Nicaragua, *Bull. Seismol. Soc. Amer.*, **64**, 1033–1048.

• Gutscher, M.-A., W. Spakman, H. Bijwaard, and E. R. Engdahl, 2000. Geodynamics of flat subduction: seismicity and tomographic constraints from the Andean margin, *Tectonics*, **19**, doi:10.1029/1999TC001152, 814–833.

Hănus, V. and J. Vănek, 1977–8. Subduction of the Cocos plate and deep active fracture zones of Mexico, *Geofísica Internacional*, **17**, 14–53.

Havskov, J., S. K. Singh, and D. Novolo, 1982. Geometry of the Benioff zone in the Tehuantepec area in southern Mexico, *Geofísica Internacional*, **21**, 325–330.

• Husen, S., R. Quintero, E. Kissling, and B. Hacker, 2003. Subduction-zone structure and magmatic processes beneath Costa Rica constrained by local earthquake tomography and petrological modeling, *Geophys. J. Int.*, **155**, doi:10.1046/j.1365–246X.2003.01984.x, 11–32.

Iglesias, A., S. K. Singh, J. F. Pacheco, and M. Ordaz, 2002. A source and wave propagation study of the Copalillo, Mexico, earthquake of 21 July 2000 (M_W 5.9): implications for seismic hazard in Mexico City from inslab earthquakes, *Bull. Seismol. Soc. Amer.*, **92**, 1060–1071.

Irvine, H. M., 1973. The Veracruz earthquake of 28 August 1973, *Bull. New Zealand Soc. Earthquake Eng.*, **7**, 2–13.

Jiménez, Z. and L. Ponce, 1978. Focal mechanism of six large earthquakes in northern Oaxaca, Mexico, for the period 1928–1973, *Geofísica Internacional*, **17**, 379–386.

LeFevre, L. V. and K. C. McNally, 1985. Stress distribution and subduction of aseismic ridges in the Middle America subduction zone, *J. Geophys. Res.*, **90**, 4495–4510.

McCann, W. R. and W. D. Pennington, 1990. Seismicity, large earthquakes, and the margin of the Caribbean plate. In *The Caribbean Region: the Geology of North America, vol. H.*, eds. G. Dengo and J. E. Case, Boulder, CO, Geological Society America, 291–306.

Meehan, J. F., 1974. Reconnaissance report of the Veracruz earthquake of August 28, 1973, *Bull. Seismol. Soc. Amer.*, **64**, 2011–2025.

Molnar, P. and L. R. Sykes, 1969. Tectonics of the Caribbean and Middle America regions from focal mechanisms and seismicity, *Geol. Soc. Amer. Bull.*, **80**, 1639–1684.

National Geophysical Data Center A. *Catalog of Significant Earthquakes 2150 BC – AD 1991*, Boulder, CO US Dept. of Commerce.

Pacheco, J. F. and S. K. Singh, 1995. Estimation of ground motions in the Valley of Mexico from normal-faulting, intermediate-depth earthquakes in the subducted Cocos plate, *Earthquake Spectra*, **11**, 233–247.

Pardo, M. and G. Suárez, 1993. Steep subduction geometry of the Rivera plate beneath the Jalisco block in western Mexico, *Geophys. Res. Lett.*, **20**, 2391–2394.

• 1995. Shape of the subducted Rivera and Cocos plates in southern Mexico: seismic and tectonic implications, *J. Geophys. Res.*, **100**, doi:10.1029/95JB00919, 12357–12373.

Peacock, S. E., P. E. van Keken, S. D. Holloway, B. R. Hacker, G. Abers, and R. L. Fergason, 2005. Thermal structure of the Costa Rica–Nicaragua subduction zone: slab metamorphism, seismicity and arc magmatism, *Phys. Earth Planet. Int.*, **149**, doi:10.1016/j.pepi.2004.08.030, 187–200.

Ponce, L., R. Gaulon, G. Suárez, and E. Lomas, 1992. Geometry and state of stress of the downgoing Cocos plate in the isthmus of Tehuantepec, *Geophys. Res. Lett.*, **19**, 773–776.

• Protti, M., F. Güendel, and K. McNally, 1995. Correlation between the age of the subducting Cocos Plate and the geometry of the Wadati–Benioff zone under Nicaragua and Costa Rica. In *Geologic and Tectonic Development of the Caribbean Plate Boundary in Southern Central America*, ed. P. Mann, Boulder, CO, Geological Society of America, Special Paper 295, 309–325.

Quintanar, L., J. Yamamoto, and Z. Jiménez, 1999. Source mechanism of two 1994 intermediate-depth-focus earthquakes in Guerrero, Mexico, *Bull. Seismol. Soc. Amer.*, **89**, 1004–1018.

Quintero, R. and F. Güendel, 2000. Stress field in Costa Rica, Central America, *J. Seismology*, **4**, do:10.1023/A:1009867405248, 297–319.

Ranero, C. R., J. P. Morgan, K. McIntosh, and C. Reichert, 2003. Bending-related faulting and mantle serpentinization at the Middle America Trench, *Nature*, **425**, doi:10.1038/nature01961, 367–373.

Rebollar, C. J., V. H. Espindola, A. Uribe, A. Mendoza, and A. Perez-Vertti, 1999a. Distribution of stresses and geometry of the Wadati–Benioff zone under Chiapas, Mexico, *Geofísica Internacional*, **38**, 95–106.

Rebollar, C. J., L. Quinanar, J. Yamamoto, and A. Uribe, 1999. Source process of the Chiapas, Mexico, intermediate-depth earthquake ($M_W = 7.2$) of 21 October 1995, *Bull. Seismol. Soc. Amer.*, **89**, 348–358.

Rose, W. I., J. J. Bommer, D. L López, M. J. Carr, and J. J. Major, eds., 2004. *Natural Hazards in El Salvador*. Boulder, CO, Geological Society of America Special Paper 375.

Salazar, W. and K. Seo, 2003. Earthquake disasters of 13 January and 13 February, 2001, El Salvador, *Seismol. Res. Lett.*, **74**, 420–439.

Singh, S. K. and F. Mortera, 1991. Source time functions of large Mexican subduction earthquakes, morphology of the Benioff zone, age of the plate, and their tectonic implications, *J. Geophys. Res.*, **96**, 21487–21502.

Singh, S. K. and M. Pardo, 1993. Geometry of the Benioff zone and state of stress in the overriding plate in central Mexico, *Geophys. Res. Lett.*, **20**, 1483–1486.

Singh, S. K. and M. Wyss, 1976. Source parameters of the Orizba earthquake of August 28, 1973, *Geofísica Internacional*, **16**, 165–184.

Stoiber, R. E. and M. J. Carr, 1973. Quaternary volcanic and tectonic segmentation of central America, *Bull. Volcanol.*, **37**, 304–325.

Suárez, G., T. Monfret, G. Wittlinger, and C. David, 1990. Geometry of subduction and depth of the seismogenic zone in the Guerrero gap, *Nature*, **345**, doi:10.1038/345336a0, 336–338.

Valée, M., M. Bouchon, and S. Y. Schwartz, 2003. The 13 January 2001 El Salvador earthquake: a multidata analysis, *J. Geophys. Res.*, **108**, doi:10.1029/2002JB001922, 2203.

Vǎnek, J., A. Špičák, and V. Hǎnus, 2000. Position of the disastrous 1999 Puebla earthquake in the seismotectonic pattern of Mexico, *Bull. Seismol. Soc. Amer.*, **90**, 786–789.

Wolters, B., 1986. Seismicity and tectonics of southern Central America and adjacent regions with special attention to the surroundings of Panama, *Tectonophysics*, **128**, doi:10.1016/0040-1951(86)90306-9, 21–46.

10.16 The Caribbean

10.16.1 Regional tectonics and deep seismicity

Where the oceanic lithosphere of the North American and Caribbean plates con-verges there is a series of volcanic islands, the Lesser Antilles, and intermediate-focus earthquakes forming a steeply dipping Wadati–Benioff zone that extends to a depth of about 200 km (See Fig. 10.36). The eastern Caribbean is distinctive in

I thank Ray Russo for reviewing an earlier draft of this section.

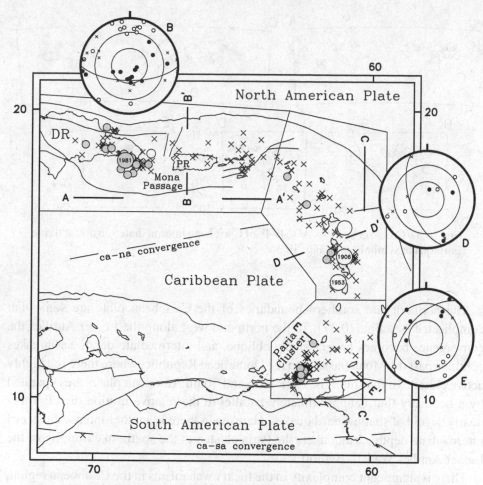

Fig. 10.36 Map of Caribbean intermediate-depth earthquakes. Filled circles, open circles, and crosses indicate intermediate-depth earthquakes in the Harvard CMT, EV Centennial, and EHB catalogs. Solid lines A–A′ to E–E′ indicate locations and orientations of cross sections, and equal-area plots summarize T (solid circles), B (crosses), and P (open circles) axis orientations of selected Harvard CMT for intermediate-depth earthquakes within the indicated regions. Dashed lines show direction of convergence for Caribbean–North American (ca–na) and Caribbean–South American (ca–sa) plates; other lines indicate plate boundaries. DR is the Dominican Republic.

comparison with other island arcs because the lithosphere of the subducting North American plate is quite old (~80–100 Ma) while the convergence rate is quite slow (~2 cm/yr). These factors trade off so that the maximum thermal parameter Φ nowhere exceeds about 1500 km. The Caribbean is also unusual in that its seismic history is continuous back to the sixteenth century, when, following exploration by Columbus, colonization began.

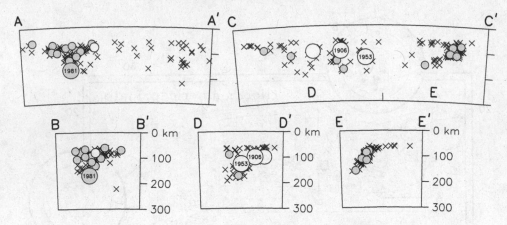

Fig. 10.37 Cross sections A–A' to E–E' of Caribbean intermediate seismic activity. Earthquake symbols as in Fig. 10.36.

The northern and southern boundaries of the Caribbean plate are somewhat complicated (see Fig. 10.37). To the north and west along the Lesser Antilles the convergence becomes progressively oblique, and intermediate-depth earthquakes continue past the Mona Passage to the Dominican Republic, where there is a highly active cluster. To the south the Caribbean and South American plates are separated by a boundary that is approximately parallel to the relative motion direction but nearly devoid of shallow earthquakes. Here again there is another intense cluster of intermediate-depth earthquakes, the Paria cluster, at the southern extremity of the Lesser Antilles Wadati–Benioff zone.

There is significant complexity in the focal mechanisms in the Caribbean region, possibly because relative plate motions are highly oblique where activity rates are highest. Mechanisms with approximately downdip T axes are the most common, but this is by no means universal.

10.16.2 Literature review

Sykes and Ewing (1965) undertook the earliest thorough investigation of the regional seismicity, relocating about 500 teleseismically-recorded earthquakes in the Caribbean, northern South America, and Central America. They noted the absence of shallow seismicity in northern South America between 64.5°–69° W, and also observed that "a zone of intermediate-depth earthquakes can be traced along the entire length of the Lesser Antilles." Their relocated hypocenters occurred within a zone about 50 km wide dipping about 60° to the west, and reaching depths as great as 200 km. They stated that "an intense source of intermediate earthquakes is located beneath the eastern end of Hispanola." The Sykes and Ewing study

was highly influential, as several subsequent investigations of regional plate geometry and seismicity reproduced maps and cross sections of their hypocenters (e.g.,
Molnar and Sykes, 1969; Stein *et al.*, 1982; Sykes *et al.*, 1982).

Various organizations have operated regional seismic networks in the eastern
and northeastern Caribbean (see McCann, 2002). Beneath Puerto Rico and eastern
Hispanola the subduction is highly oblique. For Puerto Rico and adjacent islands
to the east, Fischer and McCann (1984) evaluated data from an 18-station network
and determined hypocenters with depths as great as 150 km in a Wadati–Benioff
zone that dipped southward at an angle of 60°–70°. The Wadati–Benioff zone
beneath Hispanola is more complex. McCann and Pennington (1984) suggest that
it is fragmented, with one fragment being an extension of the activity in Puerto
Rico, and a second nearly vertical fragment situated to the south and extending to
a depth of about 200 km.

In the eastern Caribbean, Wadge and Shepherd (1984) performed JHD relocations using data collected by three short-period networks operating between 1973
and 1983 on various islands in the Lesser Antilles and the Virgin Islands. Their
relocations suggest the Wadati–Benioff zone consists of distinct segments separated by regions where activity is infrequent or absent. There is an active segment
north of 14° N where the Wadati–Benioff zone dips about 50°–60° to the WSW
and reaches a depth of 210 km. For the segment south of 14° N the maximum depth
is 170 km and the dip is 50°–60° to the WNW. Because of the low activity between
segments, Wadge and Shepherd would not speculate whether the subducting plate
was torn or deformed, or whether two separate plates subduct with different velocities. McCann and Sykes (1984) and McCann and Pennington (1990) suggest that
the observed segmentation and clustering are related to the subduction of aseismic
ridges on the downgoing plate.

In the southeastern Caribbean, there is a low-activity gap between 12°–13° N,
and then between 11°–12° N beneath the Venezuelan continental shelf there is
a group of earthquakes that Perez and Aggarwal (1981) called the Paria cluster that forms a nearly vertical zone extending to 165 km depth. Russo *et al.*
(1993) investigated earthquakes in the Paria cluster and argued that "intermediate
($165\,\text{km} > h > 70\,\text{km}$) depth thrust and dip slip events within the NW dipping slab
indicate that oceanic lithosphere, probably originally attached to South America,
subducts to the NW beneath the Caribbean plate." VanDecar *et al.* (2003) perform
tomographic analysis of data recorded by a temporary local array in Venezuela
and neighboring islands. They conclude that here the Caribbean plate overruns
the South American plate, and that subducted South American lithosphere lies
beneath South America along an aseismic continuation of the Lesser Antilles island
arc.

Several other investigators have evaluated focal mechanisms to constrain the
plate dynamics and investigate the nature of segment boundaries (Molnar and Sykes,

1969; Stein *et al.*, 1982). Johnston and Langston (1984) made a very careful study of the earthquake of 14 September 1981 ($h_{CMT} = 160$ km; $M_{W(CMT)} = 6.1$) and concluded that it possessed a downdip tensional mechanism. For the eastern Caribbean near 14° N, Russo *et al.* (1992) studied the 19 March 1953 earthquake ($h_{Russo} = 133$ km; $m_{B(Abe)} = 7.1$) and concluded it had a downdip tensional mechanism with an EW-trending B axis. This mechanism and the low rate of seismic activity immediately to the south may indicate that there is a tear or segment boundary in the subducted slab here.

10.16.3 Significant earthquakes

Several intermediate earthquakes in this region have reported magnitudes of 7 or greater, although none since 1964. From a review of felt reports Dorel (1981) indicates that the Lesser Antilles earthquakes occurring on 30 August 1844 and 9 January 1888 had depths of 150 km and 100 km and magnitudes of 7.0 and 7.5, respectively. And in the GR catalog the earthquake of 3 December 1906 at 100 km depth has magnitude M_S 7.5.[13] Two earthquakes listed by the National Geophysical Data Center A (1992) as having intermediate focal depth and causing a significant number of deaths occurred on 4 August 1946 and 30 July 1967. However, in both cases other sources (GR, Abe, ISC, EV Centennial) indicate that their depths were shallow.

10.16.4 Caribbean references

Dorel, J., 1981. Seismicity and seismic gap in the Lesser Antilles arc and earthquake hazard in Guadeloupe, *Geophys. J. Roy. Astron. Soc.*, **67**, 679–695.

Fischer, K. M. and W. R. McCann, 1984. Velocity modeling and earthquake relocation in the northeast Caribbean, *Bull. Seismol. Soc. Amer.*, **74**, 1249–1262.

Johnston, D. E. and C. A. Langston, 1984. The effect of assumed source structure on inversion of earthquake source parameters: the eastern Hispaniola earthquake of 14 September 1981, *Bull. Seismol. Soc. Amer.*, **74**, 2115–2134.

McCann, W. R., 2002. Microearthquake data elucidate details of Caribbean subduction zone, *Seismol. Res. Lett.*, **73**, 25–32.

• McCann, W. R. and W. D. Pennington, 1990. Seismicity, large earthquakes, and the margin of the Caribbean plate. In *The Caribbean Region: the Geology of North America, vol. H.*, eds. G. Dengo and J. E. Case, Boulder, CO, Geological Society of America, 291–306.

McCann, W. R. and L. R. Sykes, 1984. Subduction of aseismic ridges beneath the Caribbean plate: implications for the tectonics and seismic potential of the northeastern Caribbean, *J. Geophys. Res.*, **89**, 4493–4519.

Molnar, P. and L. R. Sykes, 1969. Tectonics of the Caribbean and Middle America regions from focal mechanisms and seismicity, *Geol. Soc. Amer. Bull.*, **80**, 1639–1684.

---

[13] Dorel (1981) lists this earthquake as occurring on 16 February 1906; this is apparently erroneous as there is no large earthquake listed on that date in the GR, Abe, ISC, or EV Centennial catalogs.

National Geophysical Data Center A. *Catalog of Significant Earthquakes 2150 BC–AD 1991*, Boulder, CO, U.S. Dept. of Commerce.

Perez, O. J. and Y. P. Aggarwal, 1981. Present-day tectonics of the southeastern Caribbean and northeastern Venezuela, *J. Geophys. Res.*, **86**, 10791–10804.

Russo, R. M., E. A. Okal, and K. C. Rowley, 1992. Historical seismicity of the southeastern Caribbean and tectonic implications, *Pure Appl. Geophys.*, **139**, 87–120.

Russo, R. M., R. C. Speed, E. A. Okal, J. B. Shepherd, and K. C. Rowley, 1993. Seismicity and tectonics of the southeastern Caribbean, *J. Geophys. Res.*, **98**, 14299–14319.

Stein, S., J. F. Engeln, D. A. Wiens, K. Fujita, and R. C. Speed, 1982. Subduction seismicity and tectonics in the Lesser Antilles arc, *J. Geophys. Res.*, **87**, 8642–8664.

Sykes, L. R. and M. Ewing, 1965. The seismicity of the Caribbean region, *J. Geophys. Res.*, **70**, 5065–5074.

Sykes, L. R., W. R. McCann, and A. L. Kafka, 1982. Motion of Caribbean plate during last 7 million years and implications for earlier Cenozoic movements, *J. Geophys. Res.*, **87**, 10656–10676.

VanDecar, J. C., R. M. Russo, D. E. James, W. B. Ambeh, and M. Franke, 2003. Aseismic continuation of the Lesser Antilles slab beneath continental South America, *J. Geophys. Res.*, **108**, doi:10.1029/2001JB000884, 2043.

• Wadge, G. and J. B. Shepherd, 1984. Segmentation of the Lesser Antilles subduction zone, *Earth Planet. Sci. Lett.*, **71**, 297–304.

10.17 Colombia–Venezuela–Panama

10.17.1 Regional tectonics and deep seismicity

While the presence of intermediate-depth earthquakes in northern Colombia and Venezuela is indisputable, their tectonic origins are a matter for ongoing speculation. The basic problem is that the geology in the region is complex, as both the Nazca and Caribbean plates are presently subducting beneath Colombia (Figs. 10.33 and 10.38). Moreover, the plate geometry within the region has undergone considerable change, as the formation of the Cocos and Nazca plates from the former Farallon plate only took place within the past 20 Ma.

About two-thirds of the intermediate-depth earthquakes in this region originate from a highly localized focus known as the Bucaramanga nest (see Fig. 5.9). The nest lies at a depth of about 160 km beneath the town of Bucaramanga, Colombia, has dimensions of only a few km, and produces about eight earthquakes per year with m_b of 4.7 or greater.

10.17.2 Literature review

Although several investigators have carefully examined hypocenters, focal mechanisms, and geology in this region, there is little agreement concerning the origin and geometry of the subducted lithosphere responsible for the intermediate-depth seismicity. The earliest thorough study was by Dewey (1972), who stated that:

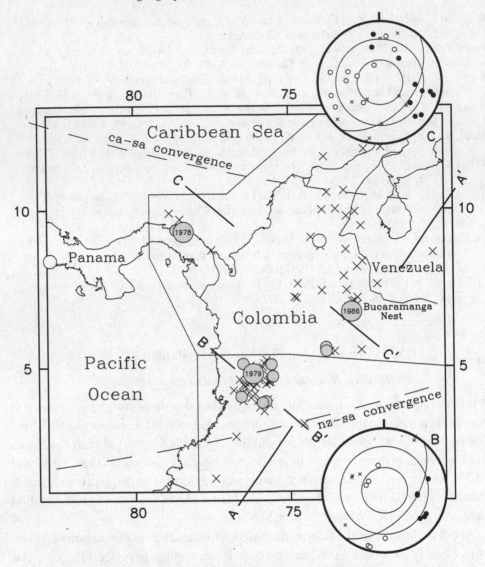

Fig. 10.38 Map of intermediate earthquakes in the Colombia–Venezuela–Panama region. Filled circles, open circles, and crosses indicate selected intermediate earthquakes in the Harvard CMT, EV Centennial, and EHB catalogs (see description of selection process in the introduction to this chapter). Solid lines A–A′ to C–C′ indicate locations and orientations of cross sections, and equal-area plots summarize T (solid circles), B (crosses), and P (open circles) axis orientations of selected Harvard CMT for intermediate-depth earthquakes within the indicated regions. Dashed lines show the direction of convergence for Caribbean–South America (ca–sa) and Nazca–South American (nz–sa) plates.

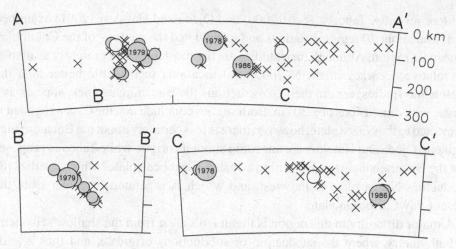

Fig. 10.39 Cross sections A–A′ to C–C′ of Colombia–Venezuela–Panama inter-mediate seismic activity. Earthquake symbols as in Fig. 10.38.

the tectonics of northwestern Venezuela are a result of eastward motion of the Caribbean plate with respect to the South American plate [although] the principal interface between these two plates may have changed within the last 5 Ma. . . . Evidence from seismicity for a continuous thin sinking slab of lithosphere beneath northeastern Colombia is not as compelling as evidence for such slabs beneath some island arc regions, because the intermediate-depth hypocenters outside of the Bucaramanga source are not numerous enough to unambiguously 'require' a continuous slab . . . If these earthquakes do occur in the same lithospheric slab as that containing the Bucaramanga earthquakes, that slab must have an approximately northerly strike to 10° N and dip to the east . . . There is not, however, an obvious zone of shallow seismicity to mark the site where the lithospheric slab containing the intermediate-depth earthquakes is initially thrust into the mantle.

However, Pennington (1981) favored a somewhat different geometry. In Colombia–Venezuela he identified two separate slab segments (Figs. 10.38 and 10.39): an ESE dipping "Cauca" segment (cross section B), which presumably represents a subducted fragment of the Nazca plate; and, further to the north, an ESE dipping Bucaramanga segment (section C) which "is apparently continuous with the Caribbean seafloor northwest of Colombia." Subsequently, Malavé and Suárez (1995) concluded that Pennington's Cauca and Bucaramanga segments comprise a single piece of subducted lithosphere that is still attached to the Caribbean plate. However:

towards the southwestern part of the study area, understanding the geometry of the slab is more complicated. The dip of the slab appears to change near the suture zone due to the collision of the Panamanian arc. In this region it is more difficult to ascertain whether the slab is still attached to the Caribbean plate, since it cannot be established with certainty to which tectonic plate the earthquakes are related: the Caribbean or Nazca plates.

More recently, Taboada *et al.* (2000) and Ojeda and Havskov (2001) examined data from about 20 regional stations and reevaluated the geology of the Colombia–Venezuela region. Although their studies are the most thorough presently available, the relative accuracy of their hypocentral locations seems little better than the teleseismic studies; e.g., in their cross sections the Bucaramanga nest appears as a cluster with dimensions of ~30 km. Both studies conclude that the Cauca segment is connected to the Nazca plate; however, there is less certainty about the Bucaramanga segment. Ojeda and Havskov are uncertain about its origin, while Taboada proposes that the Bucaramanga nest lies within a "paleo-Caribbean plate" which overlies the subducting Nazca plate to the west, and which is separated by a tear from the present-day Caribbean plate.

A major difficulty in this region is that it isn't clear from the shallow seismicity or bathymetry where the subduction or subductions originate, and thus a wide range of models is possible (e.g., see Corredor, 2003). A travel-time tomography study by Van der Hilst and Mann (1994) found a high-velocity shallow region extending to the northwest of the Bucaramanga intermediate seismicity. However, in the Caribbean Sea offshore of Colombia there is an absence of shallow activity, and there are even two isolated earthquakes with reported focal depths of about 60 km (12 March 1968 and 28 April 1978).

The unique character of the Bucaramanga nest (see Section 5.1.3 and Fig. 5.9) has made it a major topic of many of the regional reviews (Dewey, 1972; Pennington, 1981; Malavé and Suárez, 1995) and the sole focus of several other investigations. The nest wasn't widely recognized as unusual before the publications of Tryggvason and Lawson (1970) and Santo (1970). Frohlich *et al.* (1995) performed JHD relocations on teleseismically recorded nest events and found that they were confined to a region with dimensions of 13 km × 18 km × 12 km; the locations determined by Schneider *et al.* (1987) from 16 days of local network observations were within a region having dimensions of 5.1 km × 3.4 km × 3.8 km. In spite of the nest's small size, focal mechanisms of nest earthquakes are somewhat variable (e.g. Cortés and Angelier, 2005). However, probably because the nest is so localized in space, it has never generated earthquakes with magnitudes larger than about 6.0; thus there are no felt reports from Bucaramanga mentioned in Ramirez (1933), no obvious Bucaramanga hypocenters in the GR catalog,[14] and none occurring prior to 1958 in the Rothé catalog (see also discussion in Section 5.1.3). The β value for earthquakes produced by the nest exceeds 1.0 (see Fig. 5.10), which is atypical for intermediate depths.

---

[14] However, GR list the earthquake of 29 May 1930 at 7° N 74 1/2° W, 220 km depth, magnitude 6, quality CCC. This is quite possibly a Bucaramanga earthquake, especially since the relocations of Russo *et al.* (1992) placed other contemporary events in the region such as the earthquake of 1 February 1926 some 300 km to the east of the GR locations.

According to the EHB catalog there is sporadic intermediate-depth activity beneath and just offshore the Caribbean margin of Panama, but nearly all these events have m_b less than 5.0. Adamek *et al.* (1988) compared observed and synthetic waveforms for the largest of these events, the earthquake of 1 July 1978 with m_b 5.2, and concluded it had a focal depth of 67 km. The Harvard CMT catalog reports a centroid depth of 59 km and M_W of 5.8.

10.17.3 Significant earthquakes

Some intermediate-depth earthquakes in this region are destructive. The National Geophysical Data Center A (1992) lists six in Colombia and Venezuela that killed 20 or more people. However, some of these, such as the earthquakes of 31 January 1906 and 9 February 1967, probably actually had shallow foci. Note that the next section, 10.18, discusses seismicity in southern Colombia, including the celebrated 1922 and 1970 deep-focus earthquakes.

23 November 1979: $h_{pP-P} = 110\,\text{km}$; $M_{W(CMT)} = 7.2$. This is the largest intermediate-depth earthquake listed in the Harvard and Abe catalogs in the Colombia–Venezuela region. It killed more than 50 people, including many buried by earthquake-related mudslides. There was heavy structural damage in the towns of Cali and Medellin (Ramirez and Goberna, 1980).

29 June 1986: $h_{CMT} = 160\,\text{km}$; $M_{W(CMT)} = 5.8$. This earthquake is representative of the larger known Bucaramanga nest events, and was felt strongly in Bucaramanga. It was also felt in many parts of northern and central Colombia and the San Cristobal–Maricaibo region of Venezuela. Frohlich *et al.* (1995) estimated that its moment of 5.1×10^{17} N-m corresponds to a fault dimension of about 4 km, or about the same as the dimensions of the nest as determined by Schneider *et al.* (1987). The Harvard CMT for this earthquake (see inset) had an unusually large CLVD component of 79%.[15]

[15] For this earthquake the ISC reports an m_b of only 5.2. Several of the larger Bucaramanga earthquakes in the CMT catalog have large CLVD components; it would be interesting to investigate whether these earthquakes have complex source-time functions. This might explain both the CLVD sources and the disparity in the reported magnitudes.

10.17.4 Colombia–Venezuela–Panama references

Adamek, S., C. Frohlich, and W. D. Pennington, 1988. Seismicity of the Caribbean-Nazca boundary: constraints on microplate tectonics of the Panama region, *J. Geophys. Res.*, **93**, 2053–2075.

Corredor, F., 2003. Seismic strain rates and distributed continental deformation in the northern Andes and three-dimensional seismotectonics of northwestern South America, *Tectonophysics*, **372**, doi:10.1016/S0040-1951(03)00276-2, 147–166.

Cortés, M. and J., Angelier, 2005. Current states of stress in the northern Andes as indicated by focal mechanisms of earthquakes, *Tectonophysics*, **403**, 29–58. doi:10.1016/j.tecto.2005.03.020.

Dewey, J. W., 1972. Seismicity and tectonics of western Venezuela, *Bull. Seismol. Soc. Amer.*, **62**, 1711–1751.

Frohlich, C., K. Kadinsky-Cade, and S. D. Davis, 1995. A reexamination of the Bucaramanga, Colombia, earthquake nest, *Bull. Seismol. Soc. Amer*, **85**, 1622–1634.

• Malavé, G. and G. Suárez, 1995. Intermediate-depth seismicity in northern Colombia and western Venezuela and its relationship to Caribbean plate subduction, *Tectonics*, **14**, doi:10.1029/95TC00334, 617–628.

National Geophysical Data Center A, 1992. *Catalog of Significant Earthquakes 2150 BC–AD 1991*, Boulder, CO, U.S. Dept. of Commerce.

Ojeda, A. and J. Havskov, 2001. Crustal structure and local seismicity in Colombia, *J. Seismology*, **5**, doi:10.1023/A:1012053206408, 575–593.

• Pennington, W. D., 1981. Subduction of the eastern Panama Basin and seismotectonics of northwestern South America, *J. Geophys. Res.*, **86**, 10753–10770.

Ramirez, J., 1933. Earthquake history of Colombia, *Bull. Seismol. Soc. Amer.*, **23**, 13–22.

Ramirez, J. E. and J. R. Goberna, 1980. *Terremotos Colombianos Noviembre 23 y Deciembre 12 de 1979*, Publicacion del Instituto Geofisico de los Andes Colombianos, Universidad Janeriana, Ser. A: Sismologia No. 45, 95 pp.

Russo, R. M., E. A. Okal, and K. C. Rowley, 1992. Historical seismicity of the southeastern Caribbean and tectonic implications, *Pure Appl. Geophys.*, **139**, 87–120.

Santo, T., 1970. Regional study on the characteristic seismicity of the world, Part VI. Colombia, Rumania and South Sandwich Islands, *Bull. Earthquake Res. Inst., Tokyo Univ.*, **48**, 1089–1105.

• Schneider, J. F., W. D. Pennington, and R. P. Meyer, 1987. Microseismicity and focal mechanisms of the intermediate-depth Bucaramanga nest, Colombia, *J. Geophys. Res.*, **92**, 13913–13926.

• Taboada, A., L. A. Rivera, A. Fuenzalida, A. Cisternas, H. Philip, H. Bijwaard, J. Olaya, and C. Rivera, 2000. Geodynamics of the northern Andes: subductions and intracontinental deformation (Colombia), *Tectonics*, **19**, doi:10.1029/2000TC900004, 787–813.

Tryggvason, E. and J. E. Lawson, 1970. The intermediate earthquake source near Bucaramanga, Colombia, *Bull. Seismol. Soc. Amer.*, **60**, 269–276.

Van der Hilst, R. and P. Mann, 1994. Tectonic implications of tomographic images of subducted lithosphere beneath northwestern South America, *Geology*, **22**, doi:10.1130/0091-7613(1994)022, 451–454.

10.18 South America

10.18.1 Regional tectonics and important issues

This section summarizes the characteristics of South American deep earthquakes produced by the convergence of the Nazca plate with the continental lithosphere of South America. The previous section (10.17) considers earthquakes in northern Colombia and Venezuela, a tectonically complex area where the subduction of both the South American and the Caribbean plates affect intermediate-depth seismicity.

The subduction of the Nazca plate produces one of the world's great mountain chains, the Andes, and a well-developed intermediate-depth Wadati–Benioff zone that extends from Ecuador to Chile, a distance of more than 4000 km. This makes it Earth's laterally most extensive Wadati–Benioff zone lying completely beneath continental lithosphere. From a deep earthquake perspective South America is remarkable for several reasons. First, during the twentieth century it experienced 15 intermediate- and deep-focus earthquakes with magnitudes of 7.5 or greater – almost twice as many as any other geographic region. Also, the three largest deep-focus earthquakes occurring in the twentieth century were all South American events – these were the earthquakes of 17 January 1922, 31 July 1970 and 9 June 1994. Partly because of interest in the 1970 and 1994 earthquakes, there have been more papers published concerning deep-focus earthquakes in South America than for any other geographic region. Finally, some South American earthquakes with reported focal depths of 60–200 km are very destructive; e.g., the earthquake of 25 January 1939 in Chillan, Chile, killed approximately 30,000 people, making it the most deadly earthquake of any depth in Chilean history, and the most deadly deep earthquake anywhere (see Section 1.2).

For South American deep-focus earthquakes, the plethora of very large events is not accompanied by numerous smaller-magnitude events (e.g., see Suyehiro, 1967). In effect, this means that b values in South America are significantly lower than in a geographic region such as Tonga–Kermadec that produces more earthquakes altogether, but has fewer very large earthquakes (see Section 4.4.3 and Fig. 1.14).

The oldest seafloor on the Nazca plate is offshore of northern Chile and has an age of about 50 Ma; the subducting lithosphere becomes progressively younger in both directions as one approaches the ridge-transforms that form the northern and southern boundary of the Nazca plate. The convergence rate for subduction is about 7–8 cm/yr, and thus the thermal parameter Φ is no more than about 2500 km at present. However, for deep-focus seismicity a somewhat larger value seems likely, as it is plausible that past subduction at steeper dip angles consumed older seafloor.

10.18.2 Intermediate-focus seismicity

The geometry of the Wadati–Benioff zone beneath South America is more complicated than that observed in typical oceanic subduction zones. The tectonic segmentation of the Andes coincides with the segmentation of the subducted Nazca plate,
which has nearly horizontal segments and 30° east-dipping segments (Figs. 10.40–
10.42). Different investigators divide the plate into different numbers of segments;
however, all agree that there are at least five important divisions:

(1) Between about 2° S and 18° S in Ecuador–Peru the intermediate-depth Wadati–Benioff
    zone is nearly horizontal with a dip of 5° or less (see Fig. 10.42, cross sections B and
    C), and there is no volcanism.
(2) Between about 18° S and 24° S in Peru–Chile the dip is about 30° (see sections D
    and E), and there are active volcanoes.
(3) Between 24° S and 27° S in Chile–Argentina there is a gap in activity at depths of
    125–300 km; the mantle there appears to be virtually aseismic.
(4) Between 27° S and 33° S in Chile–Argentina there is another flat-lying segment (see
    section F), again with no volcanism.
(5) South of 33° S the dip is again about 30° (see section G) and there is again volcanism.

A commonly-stated assertion concerning South American intermediate focal
mechanisms is that they exhibit downdip tension. This is certainly true for the
highly active 30° -dipping segment between 18° S and 24° S (section E) and mostly
true for the 30° -dipping segment south of 33° S. In the northern flat-lying segment
the T axes are approximately in the plane of the Wadati–Benioff zone, and oriented
roughly along the convergence direction (sections B and C). However, the T-axis
directions are highly scattered in the transition region near the Peru–Chile corner
(section D) and in the southern flat-lying section (section F).

10.18.3 Deep-focus seismicity

A remarkable feature of South American seismicity is a virtual absence of reliably-
located earthquakes between depths of about 325 km and 500 km. Then, at greater
depths deep-focus earthquakes occur in four separate regions (Fig. 10.40). In two
of these regions the activity is fairly regular and forms distinctly linear patterns that
approximately parallel the South American Trench.

The laterally longest zone occurs between depths of about 500 km and 620 km
beneath Argentina and Bolivia and extends from 15° S to 29° S. Since 1977 this
region has generated about one earthquake per year with M_W 5.5 or greater; the
highest rates of activity occur near its southern extremity. There have been several
earthquakes in this zone with magnitudes of 7 or greater, the most recent occurring
on 23 April 2000. At the northern end, between 15° S–19° S, the geometry is defined

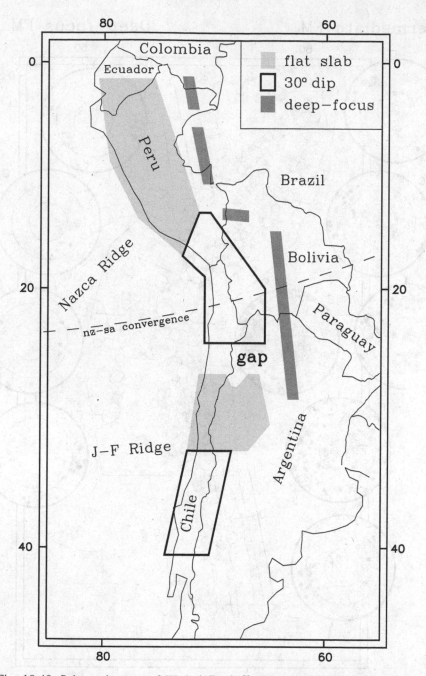

Fig. 10.40 Schematic map of Wadati–Benioff geometry in South America. At depths of 60–300 km the Wadati–Benioff zone is nearly horizontal in two regions; elsewhere it has a dip of about 30° except in central Chile–Argentina where there is a pronounced gap in activity at depths of 125–300 km ("gap" in figure). Most of the deep-focus earthquakes occur within two decidedly linear zones in Argentina–Bolivia and along the Peru–Brazil border; however, rare large deep-focus quakes also occur in western Bolivia and along the Colombia–Peru border. The dashed line shows the direction of Nazca–South American (nz–sa) plate convergence.

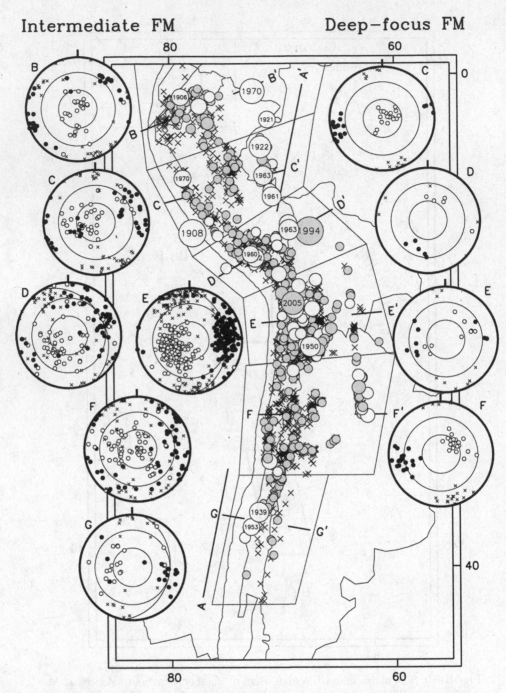

Fig. 10.41 Map of deep earthquake activity in South America. Filled circles, open circles, and crosses indicate selected intermediate earthquakes in the Harvard CMT, EV Centennial, and EHB catalogs (see description of selection process in the introduction to this chapter). Solid lines A–A′ to G–G′ indicate locations and orientations of cross sections, and equal-area plots summarize T (solid circles), B (crosses), and P (open circles) axis orientations of selected Harvard CMT for intermediate-depth (left side) and deep-focus (right side) earthquakes. Other lines are political boundaries and the South American Trench.

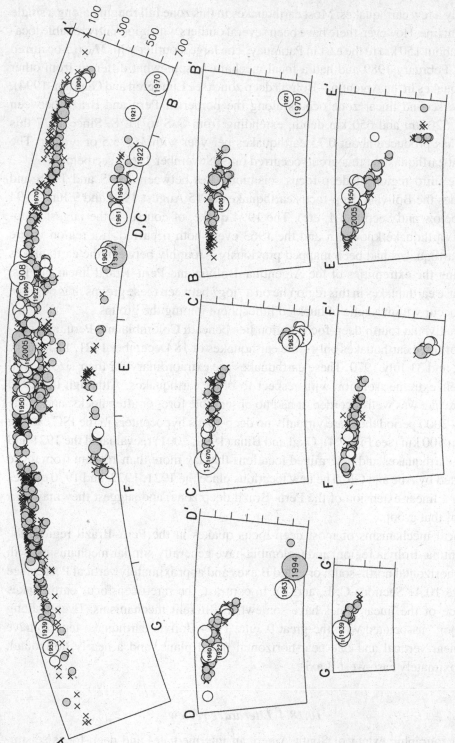

Fig. 10.42 Cross sections A–A′ to G–G′ of South American deep seismic activity. Earthquake symbols are as in Fig. 10.41.

by only a few earthquakes. Most earthquakes in this zone fall roughly along a single straight line. However, there have been several outliers with apparently reliable locations about 150 km to the east in Paraguay. The largest outlier with M_W 6.5 occurred on 28 February 1989 and had a focal mechanism somewhat different from other earthquakes in the Argentina–Bolivia deep zone (see Lundgren and Giardini, 1994).

The second linear zone occurs along the border of Peru and Brazil between about 520 km and 650 km depth, extending from 6° S to 11° S. Since 1977 this zone has produced about 0.75 earthquakes per year with M_W 5.5 or greater. The largest earthquake in this group occurred on 9 November 1963 (see below).

The third region of deep-focus seismicity lies between 13° S and 14° S and includes the Bolivian deep-focus earthquakes of 15 August 1963 and 9 June 1994 (see below and Section 1.1, etc). The 1994 quake, of course, is the largest deep-focus earthquake known; it and the 1963 event both occurred in a region where few earthquakes had been mapped previously – roughly between the earthquakes defining the extremities of the Argentina–Bolivia and Peru–Brazil linear groups. Because earthquakes in this region lie on a "jog" between these groups, it is possible they occur within a zone of buckled lithosphere joining the groups.

Finally, the fourth deep-focus region lies beneath Colombia and Peru, and consists of three earthquakes only – the earthquakes of 18 December 1921, 17 January 1922, and 31 July 1970. These earthquakes are extraordinary for their size as well as their extreme isolation with respect to other earthquakes. Although the 1970 earthquake was well-recorded, it had no observable fore- or aftershocks, and in the 1964–2000 period there are virtually no deep-focus hypocenters in the ISC catalog within 500 km (see Fig. 2.4). Okal and Bina (1994; 2001) reevaluated the 1921 and 1922 earthquakes and determined locations that lay more than 160 km from those reported by Abe and GR. The new locations place the 1921, 1922, and 1970 quakes along a linear extension of the Peru–Brazil deep zone, and suggest they may be a part of that group.

Focal mechanisms of most deep-focus quakes in the Peru–Brazil region, the Argentina–Bolivia region, and Colombia have generally similar mechanisms with near-horizontal north–south oriented B axes and approximately vertical P axes (see Figure 10.41 Sections C, E, and F). In contrast, the rare deep-focus earthquakes outside of the linear zones have somewhat different mechanisms. Events along the "jog" associated with the great 9 June 1994 Bolivia earthquake tend to have one near-vertical and one near-horizontal nodal plane, and a nearly horizontal, approximately east–west B axis.

10.18.4 Literature review

The geographic extent of South American intermediate- and deep-focus seismicity was evident to Gutenberg and Richter (1940) during the preparation of early versions of the GR catalog; however, except for studies of individual events the region got surprisingly little attention prior to the advent of plate tectonics in the late 1960s. Subsequently there were a number of reviews that utilized teleseismically determined locations and/or focal mechanisms to evaluate the geometry and mechanics of the Wadati–Benioff zone (e.g., Santo, 1969; Stauder, 1973; Swift and Carr, 1974; Hănus and Vănek, 1976; 1978).

Several more recent reviews of teleseismic data have focused attention on the intermediate-depth seismicity. Jordan *et al.* (1983) present an especially succinct summary of South American tectonics and its relationship to the subduction of the Nazca plate. Cahill and Isacks (1992) and Chen *et al.* (2001) evaluate selected teleseismic earthquake locations to assess subducted plate geometry and present a compilation of available focal mechanisms. Araujo and Suárez (1994), Comte and Suárez (1995) and Tavera and Buforn (2001) concentrate on Wadati–Benioff geometry and focal mechanisms in Chile and Peru. Finally, Gutscher *et al.* (2000) present an impressive and highly useful review of flat and low-angle subduction regions world-wide, focusing on South America where most of them occur. Their paper includes 3-D views of the Wadati–Benioff zone, tomographic sections of the slab and overlying mantle, and a thorough discussion of the various proposed causes of flat subduction.

Compared to many other subduction zones, there have been an extraordinary number of temporary local seismic networks operated in various locations within South America. These include: for the flat-slab region between 2° S and 18° S in Peru – Suárez *et al.* (1990) and Norabuena *et al.* (1994); for the 30° -dipping and seismic gap regions between 18° S and 27° S in Peru–Chile–Argentina – Comte and Suárez (1994), Comte *et al.* (1994; 1999), Delouis *et al.* (1996), Dorbath (1997), Myers *et al.* (1998), Graeber and Asch (1999), Schurr *et al.* (1999), and Haberland and Rietbrock (2001); for the flat-slab segment between 27° S–33° S in Chile–Argentina – Smalley and Isacks (1987), Pujol *et al.* (1991), Smalley *et al.* (1993), Comte *et al.* (2002), and Pardo *et al.* (2002a); and, for the region of transition to the 30° dipping segment south of 33° S – Fuenzalida *et al.* (1992), Bohm *et al.* (2002), and Wagner *et al.* (2005).

For intermediate-depth earthquakes, much of the literature focuses on four questions. First, how many independent segments do the focal mechanisms and domains of seismic activity define? This is largely a matter of perspective, as some researchers identify as few as five (Barazangi and Isacks, 1976; Jordan *et al.*, 1983) and others as many as sixteen (Slancová *et al.*, 2000).

Second, what is mechanically responsible for the occurrence of the flat-lying, volcano-free sections, which are so extensive in South America and seldomly observed elsewhere? Bevis and Isacks (1984) and Creager *et al.* (1995) suggest that the concave-oceanward bend of the South American Trench in the Peru–Chile region may cause lateral membrane stresses in the subducted lithosphere that are reduced by the presence of flat-lying sections. However, Sacks (1983) and Gutscher *et al.* (2000) favor the argument that the subduction of buoyant oceanic lithosphere produces the flatness, noting that the flat regions occur approximately where distinct topographic features such as the Nazca Ridge and the Juan Fernandez Ridge intersect with the trench.

Third, do the regions separating major flat- and 30° -dipping segments represent tears in the subducted slab or regions of continuous but deformed lithosphere? Beneath Ecuador, Pennington (1981) concludes that there is a northwest-dipping segment of lithosphere extending to about 200 km depth that is connected to the flat-lying Peru segment. Further south, an analysis of teleseismic data led Barazangi and Isacks (1979) to propose that there was a tear in the slab between the Peru and the Peru–Chile segments. However, Hasegawa and Sacks (1981) and Schneider and Sacks (1987) evaluated local network data and found hypocenters in the intervening region, indicating that it was continuous. Muñoz (2005) goes so far as to suggest that the flat-slab seismicity does not take place in subducted lithosphere; rather, it represents brittle deformation in colder mantle material while the subducted lithosphere "lies not horizontally and deeper in the tectonosphere." Although it is often difficult for readers such as myself to evaluate the reliability of the locations of critical events, most publications reporting the results from local networks favor the presence of continuous, deformed lithosphere (e.g., Boyd *et al.*, 1984; Grange *et al.*, 1984a; 1984b; Fuenzalida *et al.*, 1992).

Finally, is there convincing evidence that there is a double seismic zone beneath northern Chile? Kono *et al.* (1985) noted that the focal mechanism for the earthquake of 17 January 1977 at about 25° S and 134 km depth was nearly opposite that of other, downdip tensional events in this region, and suggested that teleseismically determined hypocenters confirmed the presence of a double seismic zone. However, Cahill and Isacks (1986) argued that this was only an apparent double zone caused by the misidentification of sP as pP. An analysis of data from two different local networks led Comte and Suárez (1994) and Comte *et al.* (1999) to conclude that there was a double zone, with the upper group showing along-slab tension and the lower group along-slab compression. But Graeber and Asch (1999) evaluated 100 days of data from 31 three-component stations deployed in this region and found no clear evidence for a double seismic zone in the distribution of hypocenters. Somewhat further north, at about 22° S, Rietbrock and Waldhauser (2004) show that there is a double seismic zone at depths of 90–120 km, with the two planes separated by about 10–15 km; focal mechanisms in both zones are of the downdip T type.

In the literature describing observations of deep-focus events, an important issue concerns whether the lithospheric slab is continuous or absent between depths of about 325 km and 500 km. That is, are the Argentina–Bolivia and Peru–Brazil deep-focus earthquake zones connected to or separate from the intermediate activity to the west? Isacks and Barazangi (1973) note that the deep-focus earthquakes in northern Chile produce peculiar high-frequency late-arriving S phases at some stations along the South American coast. They interpret these as evidence that the lithospheric slab is continuous within the distinct seismic gap that separates the deep-focus and intermediate-focus activity. However, Snoke *et al.* (1974a; 1974b) argue that these phases might instead be underside reflections from an interface at about 200 km depth. Wortel (1984) notes that these deep-focus earthquakes probably occur within lithosphere that was older than 70 million years when it subducted, and argue that this implies that the slab is probably detached. However, James and Snoke (1990) conclude that the slab is probably continuous through the aseismic region. They observed that deep-focus earthquakes in the Peru–Brazil group produced an anomalous high-amplitude arrival about 1.5 s after the direct P at a station in Peru, which they interpreted as a phase trapped within the subducting lithosphere. From a tomographic analysis of teleseismic travel times, Engdahl *et al.* (1995) conclude that the slab is "probably continuous"; however, the cross sections they show suggest it is continuous in the Peru–Chile area but are not convincing elsewhere. Gorbatov *et al.* (1996) evaluate the thermal parameter Φ for South American deep-focus earthquakes, and argue that variations in Φ indicate detachment for deep-focus earthquakes south of 26° S in Chile but not elsewhere.

Probably because very large deep-focus events are so common in South America, there are numerous papers that focus on the mechanics of the earthquake source. Although a few papers investigate properties of groups of earthquakes having a range of magnitudes (e.g., Wyss, 1970), most concern individual large events (see below).

10.18.5 Significant earthquakes

During the twentieth century there were altogether about 80 South American deep earthquakes with magnitudes of 7.0 or greater, including about 15 greater than 7.5. Intermediate-depth events can be quite destructive, especially in areas subject to landslides or where construction is poor (e.g., see Lemoine *et al.*, 2002; Pardo *et al.*, 2002b); there have been about ten intermediate-depth earthquakes that killed 50 or more people.

17 January 1922: $h_{\text{CMT-Hist}} = 664$ km, $M_{\text{W(CMT-Hist)}} = 7.9$. Okal and Bina (1994) investigated this earthquake and a large ($h_{\text{Okal\&Bina}} = 630$ km; $M_{\text{W(Okal\&Bina)}} = 7.4$) companion that occurred one month earlier on 18 December 1921. They concluded that the 1922 earthquake had a moment M_0 of 6×10^{20} N-m, making it the third largest deep-focus earthquake known. Its location has a problematic history; rather than assign it a deep focus, Byerly (1924) concluded that it was actually three separate nearly simultaneous shallow earthquakes (see Section 3.3.2). The GR locations of the 1921 and 1922 earthquakes place them about 200 km from the 31 July 1970 Colombia earthquake. Okal and Bina's relocations put them at a slightly greater distance, but suggest that all three earthquakes lie along a low-activity roughly linear extension of the Peru–Brazil deep-focus zone (Okal and Bina, 2001). For unknown reasons, the EV Centennial catalog assigned the 1922 earthquake a depth of 359 km, and located it about 400 km south of all other known locations.

25 January 1939: $h_{Abe} = 60$ km, $m_{B(Abe)} = 7.6$. This earthquake, with an instrumental epicenter slightly northwest of Chillan, Chile ($\sim$36° S), was felt throughout southern South America. The National Geophysical Data Center A (1992) estimates it killed 28,000 people and caused \$100,000,000 in damage. According to Lomnitz (1970):

The earthquake destroyed Chillan and caused great damage in the surrounding towns of the Central Valley and in the Concepcion area. There were about 30,000 dead, chiefly in Chillan and Concepcion. The area of greatest damage was of the order of 45,000 square km, and included the Central Chilean region west of the Andean foothills between the towns of Linares and Los Angeles. The amount of damage and the number of casualties make this the most destructive earthquake in Chilean history. The high number of deaths may be partly attributed to its occurring close to midnight. A survey of damage in Chillan showed that adobe houses were mostly demolished, but other types of construction also suffered important damage; this must be connected with the virtual absence of engineering design or provisions against lateral forces. The Coast Range batholith was uplifted by more than five feet near the Central Valley margin; but . . . no tsunami and no coastal uplift were described.

Contemporary accounts describe the damage but avoid mentioning deaths (Bastiancig, 1939; Del Canto *et al.*, 1940; Komischke, 1939; Saita, 1940), apparently because of government censorship (see Section 1.2). Jara (1968) presents a highly personalized, autobiographical and non-technical account of this earthquake from his perspective as a young man who experienced it in Concepcion. Kelleher (1972) indicates that the rupture zone extended from 35.5° to 37.3° S.

Beck *et al.* (1998) present the most thorough discussion of the source properties of this earthquake. Beck and her Chilean coauthors performed waveform modeling and determined a focal mechanism (see inset). There is some uncertainty about the focal depth; Beck's group strongly favors a depth of 80–100 km but note that a very shallow ($\sim$10 km) depth is a possibility. Komischke (1939) presents one eyewitness account that describes the water table in Chillan changing several hours prior to the earthquake, and both Del Canto *et al.* (1940) and Saita (1940) report that a leveling survey after the earthquake showed that Chillan had uplifted about two meters with respect to Concepcion. However, none of these accounts report any surface faulting as might be expected for a shallow hypocenter of this magnitude.

9 December 1950: $h_{Abe} = 100$ km, $m_{B(Abe)} = 7.7$. This earthquake occurred on the Chile–Argentina border somewhat inland of Antofagasta. Huang and Okal (1998) present it as the world's largest in the depth range 100–275 km, mentioning Okal's (1992) scalar moment of 2.6×10^{21} N-m. Kausel and Campos (1992) determined a scalar moment of 1×10^{21} N-m. Although both values are less than those reported for the Banda Sea earthquake of 4 November 1963, the 1950 earthquake was undeniably huge. Bolt (1970) and Kausel and Campos (1992) both redetermined the focal depth and determined 76 km and 96 km, respectively. Kausel and Campos also studied the source process and determined a focal mechanism (see inset).

15 August 1963: $h_{CMT-Hist} = 573$ km, $M_{W(CMT-Hist)} = 7.7$. This earthquake occurred along the Peru–Bolivia border in a region which, as noted by Alsop and Brune (1965), "had not been very active – at least for deep earthquakes – before or after this earthquake." Its source elicited comment from several authors, including Chandra (1970; 1975), who noted that its focal mechanism differed from those of deep-focus events in either the Peru–Brazil region to the north or the Chile–Argentina region to the south. He also stated that

the earthquake is interpreted to have occurred in a series of jerks during the course of fault propagation, or in other words, it is composed of multiple events.

This apparently is the first reference in the literature to a deep earthquake having a complex source-time function. Its complexity is partly responsible for it being the subject of various controversies in the literature (e.g., see Chandra, 1975; Dziewonski and Gilbert, 1974; Geller, 1974). This earthquake was also the first deep event for which free oscillations were reported (Alsop and Brune, 1965) and the first deep earthquake analyzed to determine rupture velocities. It and the 9 November 1963 earthquake are regularly included in analyses of significant, large, or complex deep earthquakes (e.g., Furumoto and Fukao, 1976; Kikuchi and Fukao, 1988). This 1963 earthquake received new attention after the 6 June 1994 Bolivia earthquake occurred with a hypocenter about 160 km to the east of it and 100 km deeper. Both Brüstle and Müller (1987) and Estabrook (1999) reevaluated the source process for the 1963 quake; Estabrook determined that the rupture front propagated over a distance of about 70 km, and that its complexity produced a mechanism with a large non-double-couple component.

9 November 1963, western Brazil: $h_{CMT\text{-}Hist} = 596$ km, $M_{W(CMT\text{-}Hist)} = 7.7$. This earthquake occurred within the Peru–Brazil deep-focus zone and provided important data for early investigations of the deep earthquake source. Bollinger (1968) performed a directivity analysis and determined a rupture velocity of 3 km/s and a fault length of 60 km. The method he used – evaluating the azimuthal variation of pulse widths of long-period P waves – is now considered standard, but his publication was the first to apply it to a deep earthquake. Other, similar studies followed and obtained varying results; Davies and Smith (1968) found a rupture velocity of 3.8 km/s and a length of 27 km, while Khattri's (1972) results agreed with Bollinger except concerning which nodal plane was the fault plane. The most thorough study was by Fukao (1972; 1973) who determined a moment M_0 of 2.2×10^{20} N-m, a rupture velocity of 2.1–2.5 km/s, and a fault area of 510–680 km^2. The body waves from this earthquake also temporarily provoked a controversy about whether fault rupture could allow "return motion" or "backswings of fault displacement" (Kikuchi and Fukao, 1976; Strelitz, 1977). P-wave observations from this earthquake and the nearby event of 3 November 1965 provided data for one of the earlier studies to determine the attenuation structure of the lower mantle (Teng, 1968).

31 July 1970, Colombia: $h_{\text{CMT-Hist}} = 623\,\text{km}$, $M_{W(\text{CMT-Hist})} = 8.1$. This earthquake is remarkable both for its size and for its isolation from other seismic activity. When it occurred it was the largest deep-focus earthquake in history, a distinction that it held for 24 years (see the 9 June 1994 Bolivia earthquake). In spite of its size it had no distinct fore- or aftershocks, and it took place in a location where neither the ISC nor the historical catalogs showed any previous seismicity. The nearest prior activity was the 18 December 1921 and 17 January 1922 earthquakes that occurred about 200 km to the south (see Fig. 2.4).

The 1970 earthquake provided the principal data for about 20 publications. Because of its depth and because it occurred a few years after the confirmation that large earthquakes could stimulate the Earth's free oscillations, it encouraged seismologists to develop improved methods for retrieving normal mode periods and constructing synthetic seismograms (see Dratler *et al.*, 1971; Nowroozi, 1972; Mendiguren, 1973; Gilbert and Dziewonski, 1975; Luh and Dziewonski, 1975). Mode periods determined from observations of the 1970 Colombia earthquake have been essential for constraining gross earth models; especially the models 1066A[16] and PREM (see Section 8.2.2).

There have also been persistent investigations of the nature of the source of the 1970 Colombian earthquake. Analysis of body wave subevents by Furumoto (1977), Strelitz (1980), Fukao and Kikuchi (1987) and Estabrook (1999) all generally agree that the rupture traveled roughly southward for about 60–100 seconds, along a high-angle fault with a dimension of about 60–100 km. There has been a spirited controversy about whether the 1970 earthquake source had an implosive isotropic component and whether it had a low-frequency precursor (Dziewonski and Gilbert, 1974; Hart and Kanamori, 1975; Kennett and Simons, 1976; Mendiguren and Aki, 1978; Okal and Geller, 1979; Mendiguren, 1980). However, a more recent investigation that incorporates various methodological improvements in normal mode theory (Russakoff *et al.*, 1997) indicates that there was no resolvable isotropic component or low-frequency precursor.

---

[16] Gilbert and Dziewonski (1975) proposed gross earth model 1066A, named because it was a best fit to 1066 different observable parameters, mostly the periods of normal modes, and because (as stated by one of its proponents at an AGU presentation) "it is now used by more earth scientists than any gross earth model since the time of William the Conqueror."

9 June 1994, Bolivia: $h_{CMT} = 647$ km; $M_{W(CMT)} = 8.2$. This is the largest deep-focus earthquake of the twentieth century. It was felt with Modified Mercalli Intensities of MMI III or greater over a large part of South America, and felt by some in tall buildings in the United States and southern Canada, 6000–8600 km from the epicenter (see Fig. 1.1 and Anderson *et al.*, 1995). It reportedly caused five deaths in Peru, making it the most deadly deep-focus earthquake in the National Geophysical Data Center A compilation. As discussed in Chapter 1, there were two broadband three-component digital seismograph networks operating nearby when the earthquake occurred – one with 21 stations in Bolivia, and the other with 9 stations in Brazil. More research papers have been published concerning this earthquake than any other deep earthquake, including 23 that appeared in a special issue of *Geophysical Research Letters* (Kirby *et al.*, 1995; Wallace, 1995).

As discussed in Chapters 1, 6 and 7, analysis of its rupture history has provided especially important data that constrain mechanical models to explain deep earthquakes. It occurred five months after a 10 January 1994 deep-focus earthquake with M_W 6.9 situated about 200 km to the west. Unlike the 1970 Colombia earthquake, the June 1994 quake had numerous aftershocks (Myers *et al.*, 1995; Tinker *et al.*, 1995), although many of these were only detectable because the temporary networks were in place. While the source subevents appear to lie on a sub-horizontal plane that corresponds to an east-then-northward rupture along one of the nodal planes (Beck *et al.*, 1995; Chen, 1995; Estabrook and Bock, 1995; Lundgren and Giardini, 1995; Silver *et al.*, 1995; Ihmlé, 1998), the aftershocks occurred to the east of the initial rupture along a northwest trending plane with a dip of roughly 45°. Creager *et al.* (1995) thus suggest that the 1994 earthquake sequence may have occurred in a bend in the subducted lithosphere. Moreover, Silver *et al.* (1995) and Ihmlé (1998) speculate that the published foci occupy too great a lateral extent to fit within the cold portion of the subducted lithospheric slab where most deep-focus earthquakes originate.

Like the 1970 Colombian earthquake, this earthquake's size and depth allowed it to stimulate seldom-observed overtone modes of the Earth's free oscillations (Resovsky and Ritzwoller, 1995; Ritzwoller and Resovsky, 1995; Tromp and Zanzerkia, 1995). For this reason it has provided essential data for investigations of deep earth structure (e.g., Durek and Romanowicz, 1999).

13 June 2005: $h_{USGS} = 119$ km, $M_{W(USGS)} = 7.8$. This intermediate-depth earthquake collapsed buildings and killed several people near Iquique in northern Chile. The preliminary focal mechanism reported by the USGS (see inset) possesses a downdip T axis and is very similar to previously reported mechanisms from earthquakes in this region.

10.18.6 South America references

Alsop, L. E. and J. N. Brune, 1965. Observation of free oscillations excited by a deep earthquake, *J. Geophys. Res.*, **70**, 6165–6174.

Anderson, J. G., M. Savage, and R. Quaas, 1995. 'Strong' ground motions in North America from the Bolivia earthquake of June 9, 1994 ($M_W = 8.3$), *Geophys. Res. Lett.*, **22**, doi:10.1029/95GL01808, 2293–2296.

Araujo, M. and G. Suárez, 1994. Geometry and state of stress of the subducted Nazca plate beneath central Chile and Argentina: evidence from teleseismic data, *Geophys. J. Int.*, **116**, 283–303.

Barazangi, M. and B. L. Isacks, 1976. Spatial distribution of earthquakes and subduction of the Nazca plate beneath South America, *Geology*, **4**, doi:10.1130/0091–7613, 686–69.

  1979. Subduction of the Nazca plate beneath Peru: evidence from spatial distribution of earthquakes, *Geophys. J. Roy. Astron. Soc.*, **57**, 537–555.

Bastiancig, A., 1939. El terremoto del 24 de enero de 1939 en Chile: observaciones y consideraciones relacionadas con la edificación, *Scientia*, **5**(20), 178–188.

Beck, S. L., P. Silver, T. C. Wallace, and D. James, 1995. Directivity analysis of the deep Bolivian earthquake of June 9, 1994, *Geophys. Res. Lett.*, **22**, doi:10.1029/95GL01089, 2257–2260.

Beck, S. L., S. Barrientos, E. Kausel, and M. Reyes, 1998. Source characteristics of historic earthquakes along the central Chile subduction zone, *J. South Amer. Earth Sci.*, **11**, doi:10.1016/S0895-9811(98)00005-4, 115–129.

Bevis, M. and B. L. Isacks, 1984. Hypocentral trend surface analysis: probing the geometry of Benioff zones, *J. Geophys. Res.*, **89**, 6153–6170.

Bohm, M., S. Luth, H. Ecktler, G. Asch, K. Bataille, C. Bruhn, A. Rietbrock, and P. Wigger, 2002. The southern Andes between 36° and 40° S latitude: seismicity and average seismic velocities, *Tectonophysics*, **356**, doi:10.1016/S0040-1951(02)00399-2, 275–289.

Bollinger, G. A., 1968. Determination of earthquake fault parameters from long-period P waves, *J. Geophys. Res.*, **73**, 785–807.

Bolt, B. A., 1970. PdP and PKiKP waves and diffracted PcP waves, *Geophys. J. Roy. Astron. Soc.*, **20**, 367–382.

Boyd, T. M., J. A. Snoke, I. S. Sacks, and A. Rodriguez, B., 1984. High-resolution determination of the Benioff zone geometry beneath southern Peru, *Bull. Seismol. Soc. Amer.*, **74**, 559–568.

Brüstle, W. and G. Müller, 1987. Stopping phases in seismograms and the spatio-temporal extent of earthquakes, *Bull. Seismol. Soc. Amer.*, **77**, 47–68.

Byerly, P., 1924. The South American earthquakes of January 17, 1922, *Bull. Seismograph Stations (Berkeley)*, **2**, 50–54.

Cahill, T. and B. L. Isacks, 1986. An apparent double-planed Benioff zone beneath northern Chile resulting from misidentification of reflected phases, *Geophys. Res. Lett.*, **13**, 333–336.

1992. Seismicity and shape of the subducted Nazca plate, *J. Geophys. Res.*, **97**, 17503–17529.

Chandra, U., 1970. The Peru–Bolivia border earthquake of August 15, 1963, *Bull. Seismol. Soc. Amer.*, **60**, 639–646.

1975. On the focal mechanism of the Peru–Bolivia border earthquake of August 15, 1963, *Bull. Seismol. Soc. Amer.*, **65**, 1033–1034.

Chen, W.-P., 1995. En echelon ruptures during the great Bolivian earthquake of 1994, *Geophys. Res. Lett.*, **22**, doi:10.1029/95GL01805, 2261–2264.

Chen, P.-F., C. R. Bina, and E. A. Okal, 2001. Variations in slab dip along the subducting Nazca plate, as related to stress patterns and moment release of intermediate-depth seismicity and to surface volcanism, *Geochem., Geophys., Geosyst.*, **2**, doi:10.1029/2001GC000153.

Comte, D. and G. Suárez, 1994. An inverted double seismic zone in Chile: evidence of phase transformation in the subducted slab, *Science*, **263**, 212–215.

1995. Stress distribution and geometry of the subducting Nazca plate in northern Chile using teleseismically recorded earthquakes, *Geophys. J. Int.*, **122**, 419–440.

Comte, D., M. Pardo, L. Dorbath, C. Dorbath, H. Haessler, L. Rivera, A. Cisternas, and L. Ponce, 1994. Determination of seismogenic interplate contact zone and crustal seismicity around Antofagasta, northern Chile using local data, *Geophys. J. Int.*, **116**, 553–561.

Comte, D., L. Dorbath, M. Pardo, T. Monfret, H. Haessler, L. Rivera, M. Frogneux, B. Glass, and C. Meneses, 1999. A double-layered seismic zone in Arica, Northern Chile, *Geophys. Res. Lett.*, **26**, doi:10.1029/1999GL900447, 1965–1968.

Comte, D., H. Haessler, L. Dorbath, M. Pardo, T. Monfret, A. Lavenu, B. Pontoise, and Y. Hello, 2002. Seismicity and stress distribution in the Copiapo, northern Chile subduction zone using combined on- and off-shore seismic observations, *Phys. Earth Planet. Int.*, **132**, doi:10.1016/S0031-9201(02)00052-3, 197–217.

Creager, K. C., L.-Y. Chiao, J. P. Winchester, and E. R. Engdahl, 1995. Membrane strain rates in the subducting plate beneath South America, *Geophys. Res. Lett*, **22**, doi:10.1029/95GL02321, 2321–2324.

Davies, J. B. and S. W. Smith, 1968. Source parameters of earthquakes, and discrimination between earthquakes and nuclear explosions, *Bull. Seismol. Soc. Amer.*, **58**, 1503–1517.

Del Canto, H., P. Godoy P., E. Aguirre S., J. Muñoz-Christi, and J. Ibáñez V., 1940. Informe de la comisión gubernativa sobre los efectos producidos por el terremoto de enero 1939, *Anales del Instituto de Ingenieros de Chile*, 376–395.

Delouis, B., A. Cisternas, L. Dorbath, L. Rivera, and E. Kausel, 1996. The Andean subduction zone between 22 and 25°S (northern Chile): precise geometry and state of stress, *Tectonophysics*, **259**, doi:10.1016/0040-1951(95)00065-8, 81–100.

Dorbath, C., 1997. Mapping the continuity of the Nazca plate through its aseismic part in the Arica Elbow (Central Andes), *Phys. Earth Planet. Int.*, **101**, doi:10.1016/S0031-9201(96)03206-2, 163-173.

Dratler, J., W. E. Farrell, B. Block, and F. Gilbert, 1971. High-Q overtone modes of the Earth, *Geophys. J. Roy. Astron. Soc.*, **23**, 399-410.

Durek, J. J. and B. Romanowicz, 1999. Inner core anisotropy inferred by direct inversion of normal mode spectra, *Geophys. J. Int.*, **139**, doi:10.1046/j.1365-246x.1999.00961.x, 599-622.

Dziewonski, A. M. and F. Gilbert, 1974. Temporal variation of the seismic moment tensor and the evidence of precursive compression for two deep earthquakes, *Nature*, **247**, 185-188.

Engdahl, E. R., R. D. van der Hilst, and J. Berrrocal, 1995. Imaging of subducted lithosphere beneath South America, *Geophys. Res. Lett.*, **22**, doi:10.1029/95GL02013, 2317-2320.

Estabrook, C. H., 1999. Body wave inversion of the 1970 and 1963 South American large deep-focus earthquakes, *J. Geophys. Res.*, **104**, doi:10.1029/1999JB900244, 28751-28767.

Estabrook, C. H. and G. Bock, 1995. Rupture history of the great Bolivian earthquake: slab interaction with the 660-km discontinuity? *Geophys. Res. Lett.*, **22**, doi:10.1029/95GL02234, 2277-2280.

Fuenzalida, A., M. Pardo, A. Cisternas, L. Dorbath, C. Dorbath, D. Comte, and E. Kausel, 1992. On the geometry of the Nazca plate subducted under central Chile (32-34.5° S) as inferred from microseismic data, *Tectonophysics*, **205**, doi:10.1016/0040-1951(92)90413-Z, 1-11.

Fukao, Y., 1972. Source process of a large deep-focus earthquake and its tectonic implications – the western Brazil earthquake of 1963, *Phys. Earth Planet. Int.*, **5**, 61-76.

1973. Source process of a large deep-focus earthquake and its tectonic implications – the western Brazil earthquake of 1963: reply, *Phys. Earth Planet. Int.*, **7**, 120-121.

Fukao, Y. and M. Kikuchi, 1987. Source retrieval for mantle earthquakes by interactive deconvolution of long-period P-waves, *Tectonophysics*, **144**, doi:10.1016/0040-1951(87)90021-7, 249-269.

• Furumoto, M., 1977. Spacio-temporal history of the deep Colombia earthquake of 1970, *Phys. Earth Planet. Int.*, **15**, 1-12.

Furumoto, M. and Y. Fukao, 1976. Seismic moment of great deep shocks, *Phys. Earth Planet. Int.*, **11**, 352-357.

Geller, R. J., 1974. Evidence of precursive compression for two deep earthquakes, *Nature*, **252**, 28-29.

• Gilbert, F. and A. M. Dziewonski, 1975. An application of normal mode theory to the retrieval of structural parameters and source mechanisms from seismic spectra, *Phil. Trans. Roy. Soc., London*, **A278**, 187-269.

Gorbatov, A., V. Kostoglodov, and E. Burov, 1996. Maximum seismic depth versus thermal parameter of subducted slab: application to deep earthquakes in Chile and Bolivia, *Geofísica Internacional*, **35**, 41-50.

Graeber, F. M. and G. Asch, 1999. Three-dimensional models of P wave velocity and P-to-S velocity ratio in the southern central Andes by simultaneous inversion of local earthquake data, *J. Geophys. Res.*, **104**, doi:10.1029/1999JB900037, 20237-20256.

Grange, F., P. Cunningham, J. Gagnepain, D. Hatzfeld, P. Molnar, L. Ocala, A. Rodrígues, S. W. Roecker, J. M. Stock, and G. Suárez, 1984a. The configuration of the seismic zone and the downgoing slab in southern Peru, *Geophys. Res. Lett.*, **11**, 38-41.

Grange, F., D. Hatzfeld, P. Cunningham, P. Molnar, S. W. Roecker, G. Suárez, A. Rodrígues, and L. Ocola, 1984b. Tectonic implications of the microearthquake seismicity and fault plane solutions in southern Peru, *J. Geophys. Res.*, **89**, 6139–6152.

Gutenberg, B. and C. F. Richter, 1940. Deep-focus earthquakes in America, *Proc. 6th Pacific Sci. Congr.*, **1**, 149–150.

• Gutscher, M.-A., W. Spakman, H. Bijwaard, and E. R. Engdahl, 2000. Geodynamics of flat subduction: seismicity and tomographic constraints from the Andean margin, *Tectonics*, **19**, doi:10.1029/1999TC001152, 814–833.

Haberland, C. and A. Rietbrock, 2001. Attenuation tomography in the western central Andes: a detailed insight into the structure of a magmatic arc, *J. Geophys. Res.*, **106**, doi:10.1029/2000JB900472, 11151–11167.

Hǎnus, V. and J. Vǎnek, 1976. Intermediate aseismicity of the Andean subduction zone and recent andesitic volcanism, *J. Geophys.*, **42**, 219–223.

1978. Morphology of the Andean Wadati–Benioff zone, andesitic volcanism, and tectonic features of the Nazca plate, *Tectonophysics*, **44**, doi:10.1016/0040-1951(78)90063-X, 65–77.

Hart, R. J. and H. Kanamori, 1975. Search for precursive compression for a deep earthquake, *Nature*, **253**, 333–335.

Hasegawa, A. and I. S. Sacks, 1981. Subduction of the Nazca plate beneath Peru as determined from seismic observations, *J. Geophys. Res.*, **86**, 4971–4980.

Huang, W.-C. and E. A. Okal, 1998. Centroid moment tensor solutions for deep earthquakes predating the digital era: discussion and inferences, *Phys. Earth Planet. Int.*, **106**, doi:10.1016/S0031-9201(97)00111-8, 191–218.

Ihmlé, P. F., 1998. On the interpretation of subevents in teleseismic waveforms: the 1994 Bolivia deep earthquake revisited, *J. Geophys. Res.*, **103**, doi:10.1029/98JB00603, 17919–17932.

Isacks, B. L. and M. Barazangi, 1973. High frequency shear waves guided by a continuous lithosphere descending beneath western South America, *Geophys. J. Roy. Astron. Soc.*, **33**, 129–139.

James, D. E. and J. A. Snoke, 1990. Seismic evidence for continuity of the deep slab beneath central and eastern Peru, *J. Geophys. Res.*, **95**, 4989–5001.

Jara, J. V., 1968. *El terremoto de Chillan de 1939 y otros recuerdos*, Santiago, Chile, Assoc. Chilena de Escritores, 94 pp.

Jordan, T. E., B. L. Isacks, R. W. Allmendinger, J. A. Brewer, V. A. Ramos, and C. J. Ando, 1983. Andean tectonics related to geometry of the subducted Nazca plate, *Geol. Soc. Amer. Bull.*, **94**, 341–361.

Kausel, E. and J. Campos, 1992. The $M_S = 8$ tensional earthquake of 9 December 1950 of northern Chile and its relation to the seismic potential of the region, *Phys. Earth Planet. Int.*, **72**, 220–235.

Kelleher, J. A., 1972. Rupture zones of large South American earthquakes and some predictions, *J. Geophys. Res.*, **77**, 2087–2103.

Kennett, B. L. N. and R. S. Simons, 1976. An implosive precursor to the Colombia earthquake 1970 July 31, *Geophys. J. Roy. Astron. Soc.*, **44**, 471–482.

Khattri, K., 1972. Body wave directivity functions for two-dimensional fault model and kinematic parameters of a deep focus earthquake, *J. Geophys. Res.*, **77**, 2062–2071.

Kikuchi, M. and Y. Fukao, 1976. Seismic return motion, *Phys. Earth Planet. Int.*, **12**, 343–349.

1988. Seismic wave energy inferred from long-period body wave inversion, *Bull. Seismol. Soc. Amer.*, **78**, 1707–1724.

• Kirby, S. H., E. A. Okal, and E. R. Engdahl, 1995. The 9 June 1994 Bolivian deep earthquake: an exceptional event in an extraordinary subduction zone, *Geophys. Res. Lett.*, **22**, doi:10.1029/95GL01802, 2233–2236.

Komischke, A., 1939. Observaciones sobre el terremoto del 24 de enero de 1939 en Chile central, *Scientia*, **5**(20), 163–175; also **6**(21), 2–7.

Kono, M., Y. Takahashi, and Y. Fukao, 1985. Earthquakes in the subducting slab beneath northern Chile: a double seismic zone, *Tectonophysics*, **112**, doi:10.1016/0040-1951(85)90180-5, 211–225.

Lemoine, A., R. Madariaga, and J. Campos, 2002. Slab-pull and slab-push earthquakes in the Mexican, Chilean and Peruvian subduction zones, *Phys. Earth Planet. Int.*, **132**, doi:10.1016/S0031-9201(02)00050-X, 157–175.

Lomnitz, C., 1970. Major earthquakes and tsunamis in Chile during the period 1535 to 1955, *Geol. Rundshau*, **59**, 938–960.

Luh, P. C. and A. M. Dziewonski, 1975. Theoretical seismograms for the Colombian earthquake of 1970 July 31, *Geophys. J. Roy. Astron. Soc.*, **43**, 679–695.

Lundgren, P. and D. Giardini, 1994. Isolated deep earthquakes and the fate of subduction in the mantle, *J. Geophys. Res.*, **99**, 15833–15842.

1995. The June 9 Bolivia and March 9 Fiji deep earthquakes of 1994: I. Source processes, *Geophys. Res. Lett.*, **22**, doi:10.1029/95GL02233, 2241–2244.

Mendiguren, J. A., 1973. Identification of free oscillation spectral peaks for 1970 July 31, Colombian deep shock using the excitation criterion, *Geophys. J. Roy. Astron. Soc.*, **33**, 281–321.

1980. Inversion of free oscillation data for an elliptical rotating earth in source moment tensor studies, *Pure Appl. Geophys.*, **118**, 1192–1208.

Mendiguren, J. A. and K. Aki, 1978. Source mechanism of the deep Colombian earthquake of 1970 July 31 from the free oscillation data, *Geophys. J. Roy. Astron. Soc.*, **55**, 539–556.

Muñoz, M., 2005. No flat Wadati–Benioff Zone in the central and southern central Andes, *Tectonophysics*, **395**, doi:10.1016/j.tecto.2004.09.002, 41–65.

Myers, S. C., T. C. Wallace, S. L. Beck, P. G. Silver, G. Zandt, J. Vandecar, and E. Minaya, 1995. Implications of spatial and temporal development of the aftershock sequence for the M_W 8.3 June 9, 1994 deep Bolivian earthquake, *Geophys. Res. Lett.*, **22**, doi:10.1029/95GL01600, 2269–2272.

Myers, S. C., S. Beck, G. Zandt, and T. Wallace, 1998. Lithospheric-scale structure across the Bolivian Andes from tomographic images of velocity and attenuation for P and S waves, *J. Geophys. Res.*, **103**, doi:10.1029/98JB00956, 21233–21252.

National Geophysical Data Center A, 1992. *Catalog of Significant Earthquakes 2150 BC–AD 1991*, Boulder, CO, U.S. Dept. of Commerce.

Norabuena, E. O., J. A. Snoke, and D. E. James, 1994. Structure of the subducting Nazca plate beneath Peru, *J. Geophys. Res.*, **99**, 9215–9226.

Nowroozi, A. A., 1972. Characteristic periods of fundamental and overtone oscillations of the Earth following a deep-focus earthquake, *Bull. Seismol. Soc. Amer.*, **62**, 247–274.

Okal, E. A., 1992. Use of the mantle magnitude M_m for the reassessment of the moment of historical earthquakes. II. Intermediate and deep events, *Pure Appl. Geophys.*, **139**, 59–85.

Okal, E. A. and C. R. Bina, 1994. The deep earthquakes of 1921–1922 in northern Peru, *Phys. Earth Planet. Int.*, **87**, 33–54.

2001. The deep earthquakes of 1997 in Western Brazil, *Bull. Seismol. Soc. Amer.*, **91**, 161–164.

Okal, E. A. and R. J. Geller, 1979. On the observability of isotropic seismic sources: the July 31, 1970 Colombian earthquake, *Phys. Earth Planet. Int.*, **18**, 176–196.

Pardo, M., D. Comte, and T. Monfret, 2002a. Seismotectonic and stress distribution in the central Chile subduction zone, *J. S. Amer. Earth Sci.*, **15**, doi:10.1016/S0895-9811(02)00003-2, 11–22.

Pardo, M., D. Comte, T. Monfret, R. Boroschek, and M. Astroza, 2002b. The October 15, 1997 Punitaqui earthquake ($M_W = 7.1$): a destructive event within the subducting Nazca plate in central Chile, *Tectonophysics*, **345**, doi:10.1016/S0040-1951(01)00213-X, 199–210.

Pennington, W. D., 1981. Subduction of the eastern Panama Basin and seismotectonics of northwestern South America, *J. Geophys. Res.*, **86**, 10753–10770.

Pujol, J., J. M. Chiu, R. Smalley, M. Regnier, B. Isacks, J. L. Chatelain, J. Vlasity, D. Vlasity, J. Castano, and N. Puebla, 1991. Lateral velocity variations in the Andean foreland in Argentina determined with the JHD method, *Bull. Seismol. Soc. Amer.*, **81**, 2441–2457.

Resovsky, J. S. and M. H. Ritzwoller, 1995. Constraining odd-degree earth structure with coupled free-oscillations, *Geophys. Res. Lett.*, **22**, doi:10.1029/95GL01996, 2301–2304.

Rietbrock, A. and F. Waldhauser, 2004. A narrowly spaced double-seismic zone in the subducting Nazca plate, *Geophys. Res. Lett.*, **31**, doi:10.1029/2004GL019610.

Ritzwoller, M. H. and J. S. Resovsky, 1995. The feasibility of normal mode constraints on higher degree structures, *Geophys. Res. Lett.*, **22**, doi:10.1029/95GL02231, 2305–2308.

Russakoff, D., G. Ekström, and J. Tromp, 1997. A new analysis of the great 1970 Colombia earthquake and its isotropic component, *J. Geophys. Res.*, **102**, doi:10.1029/97JB01645, 20423–20434.

Sacks, I. S., 1983. The subduction of young lithosphere, *J. Geophys. Res.*, **88**, 3355–3366.

Saita, T., 1940. The great Chilean earthquake of January 24, 1939 (in Japanese), *Bull. Earthquake Res. Inst., Tokyo Univ.*, **18**, 446–459.

Santo, T., 1969. Characteristics of seismicity in South America, *Bull. Earthquake Res. Inst., Tokyo Univ.*, **47**, 635–672.

Schneider, J. F. and I. S. Sacks, 1987. Stress in the contorted Nazca plate beneath southern Peru from local earthquakes, *J. Geophys. Res.*, **92**, 13887–13902.

Schurr, B., G. Asch, A. Rietbrock, R. Kind, M. Pardo, B. Heit, and T. Monfret, 1999. Seismicity and average velocities beneath the Argentine Puna Plateau, *Geophys. Res. Lett.*, **26**, doi:10.1029/1999GL005385, 3025–3028.

• Silver, P. G., S. L. Beck, T. C. Wallace, C. Meade, S. C. Myers, D. E. James, and R. Kuehnel, 1995. Rupture characteristics of the deep Bolivian earthquake of 9 June 1994 and the mechanism of deep-focus earthquakes, *Science*, **268**, 69–73.

Slancová, A., A. Špičák, V. Hănus, and J. Vănek, 2000. Delimitation of domains with uniform stress in the subducted Nazca plate, *Tectonophysics*, **319**, doi:10.1016/S0040-1951(99)00302-9, 339–364.

Smalley, R. F. and B. L. Isacks, 1987. A high-resolution local network study of the Nazca plate Wadati–Benioff zone under western Argentina, *J. Geophys. Res.*, **92**, 13903–13912.

Smalley, R., J. Pujol, M. Regnier, J.-M. Chiu, J.-L. Chatelain, B. L. Isacks, M. Araujo, and N. Puebla, 1993. Basement seismicity beneath the Andean precordillera thin-skinned thrust belt and implications for crustal and lithospheric behavior, *Tectonics*, **12**, 63–76.

Snoke, A. J., I. S. Sacks, and H. Okada, 1974a. A model not requiring continuous lithosphere for anomalous high-frequency arrivals from deep-focus South American earthquakes, *Phys. Earth Planet Int.*, **9**, 199–206.

1974b. Empirical models for anomalous high-frequency arrivals from deep-focus earthquakes in South America, *Geophys. J. Roy. Astron. Soc.*, **37**, 133–139.

Stauder, W., 1973. Mechanism and spatial distribution of Chilean earthquakes with relation to subduction of the oceanic plate, *J. Geophys. Res.*, **78**, 5033–5061.

Strelitz, R. A., 1977. Seismic return motion – comments, *Phys. Earth Planet. Int.*, **14**, 378–382.

1980. The fate of the downgoing slab: a study of the moment tensors from body waves of complex deep-focus earthquakes, *Phys. Earth Planet. Int.*, **21**, 83–96.

Suárez, G., J. Gagnepain, A. Cisternas, D. Hatzfeld, P. Molnar, L. Ocola, S. W. Roecker, and J. P. Viodé, 1990. Tectonic deformation of the Andes and the configuration of the subducted slab in central Peru: results from a microseismic experiment, *Geophys. J. Int.*, **103**, 1–12.

Suyehiro, S., 1967. A search for small, deep earthquakes using quadripartite stations in the Andes, *Bull. Seismol. Soc. Amer.*, **57**, 447–461.

Swift, S. A. and M. J. Carr, 1974. The segmented nature of the Chilean seismic zone, *Phys. Earth Planet Int.*, **9**, 183–191.

Tavera, H. and E. Buforn, 2001. Source mechanisms of earthquakes in Peru, *J. Seismology*, **5**, doi:10.1023/A:1012027430555, 519–539.

Teng, T., 1968. Attenuation of body waves and the Q structure of the mantle, *J. Geophys. Res.*, **73**, 2195–2208.

Tinker, M. A., T. C. Wallace, S. L. Beck, P. G. Silver, and G. Zandt, 1995. Aftershock source mechanisms from the June 9, 1994, deep Bolivian earthquake, *Geophys. Res. Lett.*, **22**, doi:10.1029/95GL01090, 2273–2276.

Tromp, J. and E. Zanzerkia, 1995. Toroidal splitting observations from the great 1994 Bolivia and Kuril Islands earthquakes, *Geophys. Res. Lett.*, **22**, doi:10.1029/95GL01810, 2297–2300.

Wagner, L. L., S. Beck, and G. Zandt, 2005. Upper mantle structure in the south central Chilean subduction zone (30° to 36° S), *J. Geophys. Res.*, **110**, doi:10.1029/2004JB003238, B01308.

Wallace, T. C., 1995. Introduction to the special issue on the great Bolivian earthquake of 1994, *Geophys. Res. Lett.*, **22**, doi:10.1029/95GL02070, 2231.

Wortel, M. J. R., 1984. Spatial and temporal variations in the Andean subduction zone, *J. Geol. Soc. London*, **141**, 783–791.

Wyss, M., 1970. Stress estimates for South American shallow and deep earthquakes, *J. Geophys. Res.*, **75**, 1529–1544.

10.19 South Shetland Islands

10.19.1 Regional tectonics and deep seismicity

There are occasional reports of intermediate-depth earthquakes occurring near the South Shetland Islands just north of the Antarctic Peninsula (see Fig. 10.43).

I thank Doug Wiens for reviewing an earlier draft of this section.

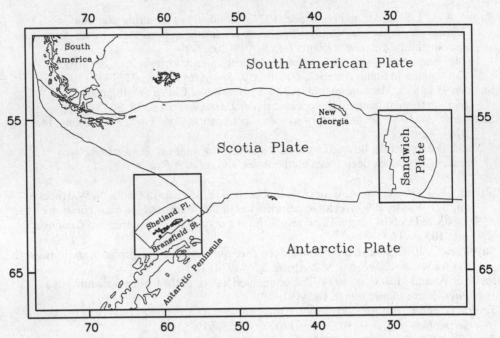

Fig. 10.43 Plate geometry in the South Shetland region (rectangle at left) and Scotia region (rectangle at right). The South Shetlands are the chain of islands north of the Bransfield Strait. According to Klepeis and Lawver (1996) they lie on the Shetland plate, which is bounded on the north by the Shetland Trench.

Between 1964 and 1999 there are exactly six such earthquakes in the ISC catalog between 60.5°–62.0° S and 56°–62.5° W; five have assigned focal depths between 60 and 84 km, and one (15 January 1975) at 167 km. All have magnitudes m_b of 4.9 or less except the earthquake of 15 September 1980, with M_W of 5.9. Although 131 stations reported this earthquake and the ISC reported a pP-P depth of 63 km based on six observations, Harvard and the EHB catalogs reported depths of 32 km and 20 km, respectively.

Some temporary local deployments indicate that there is deep seismicity here. From analysis of data recorded by a temporary small-aperture mini-array, Ibáñez *et al.* (1997) found 11 earthquakes with focal depths between 60 and 87 km and one earthquake at 130 km depth beneath Deception Island, one of the South Shetlands. These depths aren't convincing; the assessment of depth rests solely on the observation that arrivals had steep angles of incidence as indicated by the mini-array and by particle motions, and on S-P times fixing event-station distance. Robertson Maurice *et al.* (2003) analyzed data from seven land stations and 14 ocean bottom seismographs deployed on and around the South Shetlands. They list six earthquakes with focal depths exceeding 60 km, including one at 64 km which they assign an "A" rating for both epicenter and focal depth. Although they don't state either the number of S readings or the azimuthal gaps for these locations, the network geometry looks adequate and it is plausible the depths are accurate.

The tectonics in the southwestern part of the Scotia Sea are complex. Pelayo and Wiens (1989) conclude that the Scotia plate is distinct from the South American and Antarctic plates, and suggest that there is ongoing slow subduction along the northern boundary of the South Shetland Islands. For the earthquake of 30 March 1984 ($M_{W(CMT)} = 5.6$) they fit observed and synthetic waveforms and find a depth of 55 km, somewhat deeper than Harvard's reported depth of 40 km. Klepeis and Lawver (1996) interpret bathymetric data and seismic reflection profiles and present evidence that the South Shetland Islands lie within a microplate that they call the Shetland plate, bounded by the South Shetland Trench on the north and the Bransfield Strait to the south. Although, like Pelayo and Wiens, they believe there is ongoing plate convergence along the South Shetland Trench, they believe that relative plate motions may have changed significantly over the past 4 Ma and presently the lithosphere may be too buoyant for true subduction to occur there. The locations reported by Robertson Maurice *et al.* (2003) suggest that subduction does occur here and that there are earthquakes with depths exceeding 40 km. However, because observations of deep seismicity are sparse and contradictory, the presence of earthquakes here with focal depths exceeding 60 km isn't established unequivocally.

10.19.2 South Shetland Islands references

Ibáñez, J. M., J. Morales, G. Alguicil, J. Almedros, R. Ortiz, and E. Del Pezzo, 1997. Intermediate-focus earthquakes under South Shetland Islands (Antarctica), *Geophys. Res. Lett*, **24**, doi:10.1029/97GL00314, 531–534.

• Klepeis, K. A. and L. A. Lawver, 1996. Tectonics of the Antarctic–Scotia plate boundary near Elephant and Clarence Islands, Antarctica, *J. Geophys. Res.*, **101**, doi:10.1029/96JB01510, 20211–20231.

Pelayo, A. M. and D. A. Wiens, 1989. Seismotectonics and relative plate motions in the Scotia Sea region, *J. Geophys. Res.*, **94**, 7293–7320.

Robertson Maurice, S. D., D. A. Wiens, P. J. Shore, E. Vera, and L. M. Dorman, 2003. Seismicity and tectonics of the South Shetland Islands and Bransfield Strait from a regional broadband seismograph deployment, *J. Geophys. Res.*, **108**, doi:10.1029/2003JB002416, 2461.

10.20 Scotia arc

10.20.1 Regional tectonics and deep seismicity

The South Sandwich Islands – sometimes called the Southern Antilles or the Scotia arc – are the surface expression of an island arc produced by convergence of the

I thank Doug Wiens for reviewing an earlier draft of this section.

South American and Scotia plates at about 1 cm/yr (Pelayo and Wiens, 1989). However, there is vigorous back-arc spreading so that the islands themselves lie on a microplate, the Sandwich plate (see Fig. 10.43), thus altogether the lithosphere is consumed at about 7 cm/yr. Beneath the islands intermediate-depth earthquakes are numerous and some are quite large (Figs. 10.44 and 10.45). Yet because the islands are uninhabited, rarely visited, and remote, with truly awful weather, everything we know about these earthquakes is from teleseismic observations. Because of the paucity of nearby stations, epicenters are somewhat scattered; however, throughout the arc the dip of the Wadati–Benioff zone appears to be quite steep – 60° or more. The subducting lithosphere has an age of about 65 Ma at the north end of the arc and becomes progressively younger southward. Thus, the thermal parameter Φ is about 4000 km and less.

Although the arc geometry is fairly regular, north and south of 58° S there are differences in the character of the seismicity, the age of the subducting plate, and the depth of the trench. To the north the subducting plate is older and the trench is deeper – about 7 km. The most intense activity occurs near 56° S in a cluster that extends from the surface down to 180 km depth. It is possible that there is some deeper activity; the EHB catalog reports an m_b 4.3 hypocenter at 263 km occurring in 2001. This northern cluster has generated about six intermediate-depth events since 1910 with magnitudes of 7 and above, including two with magnitudes exceeding 7.5 (8 September 1961 and 26 May 1964). For the northern group the T axes cluster around a roughly southward axis with a dip of about 45° (see Fig. 10.44 section C).

South of 58° S the seismic activity is more diffuse and the trench is shallower – about 5.5 km. Although Isacks and Molnar (1971) and Forsyth (1975) report that mechanisms for this group exhibit downdip compression, there is no obvious pattern in the currently-available Harvard mechanisms (see section B). Although most of the earthquakes have focal depths of 180 km or less, there have been two well-recorded hypocenters with greater reported depths, both situated well apart from other activity. Brett (1977), Frankel and McCann (1979), the CMT Historical catalog, and the EHB all evaluated the earthquake of 7 October 1974 ($M_{W(CMT-Hist)} = 6.3$) and assigned it depths between 260 and 290 km. And both the Harvard and the EHB catalogs report depths in the 270–285 km range for the earthquake of 5 October 1997 ($M_{W(CMT)} = 6.3$).

10.20.2 Literature review

Although the GR catalog reported 10 earthquakes with focal depths of 60 km or greater in the South Sandwich arc, Santo's (1970) study is the first that investigated the seismicity in detail. Subsequently, Brett (1977) determined a set of station corrections by jointly relocating a set of selected hypocenters. Then he used the

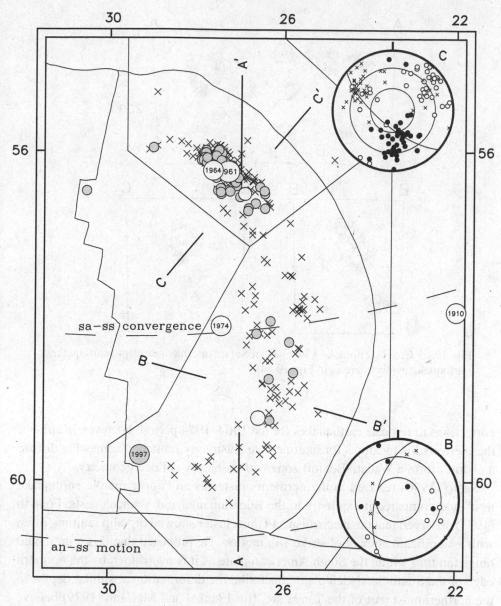

Fig. 10.44 Map of intermediate-depth earthquakes in the Scotia region. Filled circles, open circles, and crosses indicate selected earthquakes in the Harvard CMT, EV Centennial, and EHB catalogs (see description of selection process in the introduction to this chapter). Solid lines A–A′ to C–C′ indicate locations and orientations of cross sections, and equal-area plots summarize T (solid circles), B (crosses), and P (open circles) axis orientations of selected Harvard CMT for intermediate-depth earthquakes within the indicated regions. Other lines indicate locations of plate boundaries; dashed lines show the directions of South America–South Sandwich (sa–ss) and Antarctica–South Sandwich (an–ss) plate motions as reported by Pelayo and Wiens (1989).

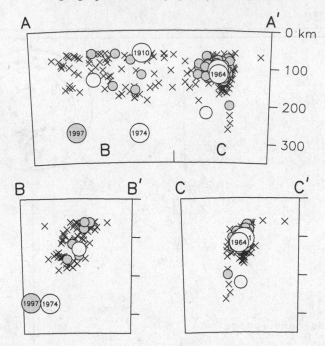

Fig. 10.45 Cross sections A–A′ to C–C′ of Scotia intermediate-depth earthquakes. Earthquake symbols are as in Fig. 10.44.

corrections to relocate earthquakes for the 1964–1974 period; however, in spite of these efforts his locations for intermediate seismicity remained somewhat diffuse and don't form a Wadati–Benioff zone with a sharply defined geometry.

Santo (1970) referred to the northern cluster as an "upper mantle earthquake nest" and compared it explicitly to the Bucaramanga and Vrancea nests. Forsyth (1975) observed that the mechanisms in this cluster showed dip-slip faulting, often with a significant amount of strike-slip motion. He proposed that this represented hinge faulting within the South American plate as it is overridden by the northern edge of the South Sandwich plate, much like the deformation at similar depths in the northernmost part of the Tonga arc. But Frankel and McCann (1979) believe the activity is more nearly comparable to the activity at the eastern end of the Banda arc, where the descending slab bends sharply and there is an intense, vertically oriented cluster of intermediate-depth earthquakes extending from 80 km to 200 km depth.

Several investigations report focal mechanisms determined from first-motions, including Isacks and Molnar (1971), Forsyth (1975), and Frankel and McCann (1979). Forsyth's (1975) compilation includes 15 deeper than 60 km. Harvard's mechanism for the 285 km-deep earthquake of 5 October 1997 is unusual because it had a CLVD component of 57% ($f_{\mathrm{CLVD}} = 0.285$).

Since about 1990, interest in the anisotropic structure of the inner core has focused attention on intermediate-depth South Sandwich Island earthquakes (e.g., see Creager, 1992). This is because their PKP phases recorded at stations in Alaska have ray paths that sample the inner core along a path that is roughly parallel to the Earth's rotation axis. Because these paths are especially sensitive to possible inner-core rotation, there is particular interest in evaluating residuals for earthquakes separated by long time periods, including those that occurred prior to 1970 (e.g., see Song and Richards, 1996).

10.20.3 Significant earthquakes

There are several inconsequential but inexplicable anomalies in the literature about large South Sandwich earthquakes. Undoubtedly the largest earthquake in the region occurred on 26 May 1964 ($M_{W(Abe)} = 7.8$; see below). However, a misprint in Abe's catalog assigns it a location of 27.7° E rather than 27.7° W, placing it 3000 km to the east of the South Sandwich Islands, south of Africa. Santo (1970) fails to mention that large earthquakes occurred here in 1961 and 1964. Instead, for an earthquake that occurred on 26 May 1965 exactly one year after the 1964 event he reports a magnitude of 7.5 and calls it a "great" earthquake, a conclusion not reached by others. Rothé's catalog includes two 1961 earthquakes with magnitudes larger than the 7.3 he reports for the 1964 event – one with a magnitude of 7.5 on 1 September 1961 at 131 km depth, and another with a magnitude of 7.7 on 8 September 1961 at 125 km depth.[17]

26 May 1964: $h_{EVC} = 116$ km; $M_{W(Abe)} = 7.8$. Abe (1972) analyzed the long period surface waves for this earthquake and determined a location, focal mechanism (see inset), and moment (6.2×10^{20} N-m); from the asymmetry of the surface waves he determined a rupture length (60–100 km) and rupture velocity (2–3 km/s). This was one of the earliest studies to evaluate earthquake source properties by comparing observed and synthetic seismograms, and possibly the first to do this using surface waves from a deep earthquake. Abe noted that there were six teleseismically locatable aftershocks within a day after the mainshock; since they occurred along a fan-shaped planar region corresponding to one of the nodal planes, he concluded that the source need not have any isotropic component.

---

[17] Rothé also orients his seismicity map for the South Sandwich region with south facing upward. To most viewers the Scotia arc thus appears backwards, a conclusion that is encouraged as his projection makes it easy to confuse the tips of South America and the Antarctic Peninsula, neither of which Rothé labels.

10.20.4 Scotia references

Abe, K., 1972. Focal process of the South Sandwich Islands earthquake of May 26, 1964, *Phys. Earth Planet. Int.*, **5**, 110–122.

Brett, G. P., 1977. Seismicity of the South Sandwich Islands region, *Geophys. J. Roy. Astron. Soc.*, **51**, 453–464.

Creager, K. C., 1992. Anisotropy of the inner core from differential travel times of the phases PKP and PKIKP, *Nature*, **356**, doi:10.1038/356309a0, 309–314.

Forsyth, D. W., 1975. Fault plane solutions and tectonics of the South Atlantic and Scotia Sea, *J. Geophys. Res.*, **80**, 1429–1443.

Frankel, A. and W. McCann, 1979. Moderate and large earthquakes in the South Sandwich arc: indicators of tectonic variation along a subduction zone, *J. Geophys. Res.*, **84**, 5571–5577.

Isacks, B. and P. Molnar, 1971. Distribution of stresses in the descending lithosphere from a global survey of focal-mechanism solutions to mantle earthquakes, *Rev. Geophys. Space Phys.*, **9**, 103–174.

• Pelayo, A. M. and D. A. Wiens, 1989. Seismotectonics and relative plate motions in the Scotia Sea region, *J. Geophys. Res.*, **94**, 7293–7320.

Santo, T., 1970. Regional study on the characteristic seismicity of the world. Part VI. Colombia, Rumania and South Sandwich Islands, *Bull. Earthquake Res. Inst., Tokyo Univ.*, **48**, 1089–1105.

Song, X. and P. G. Richards, 1996. Seismological evidence for differential rotation of the Earth's inner core, *Nature*, **382**, doi:10.1038/382221a0, 221–224.

10.21 Spain and Morocco

10.21.1 Regional tectonics and deep seismicity

Many geoscientists express surprise that there are intermediate- and deep-focus earthquakes beneath southern Spain and northern Africa (Fig. 10.46). Their incredulity is natural, as there is no trench or volcanic activity that would suggest that there is ongoing subduction, and – with one notable exception – most reported earthquakes are tiny, with magnitudes of 4 or less. The exception is the great, deep-focus Spanish earthquake of 29 March 1954 (see Section 1.3), with a reported magnitude M_W of 7.9 and a focal depth of ~640 km.

In addition to the 1954 deep-focus earthquake and three nearby events, in this region there has been intermediate-depth activity reported in three areas. These are:

(1) in the Western Alboran Sea at depths down to about 130 km, east of the Straits of Gibraltar and almost directly above the 1954 focus;

(2) in the Gulf of Cadiz at depths down to about 150 km, west of the Straits of Gibraltar; and

(3) beneath northern Morocco and also (possibly) beneath the Atlas Mountains at depths as great as 160 km.

I thank Agustín Udías for reviewing an earlier draft of this section.

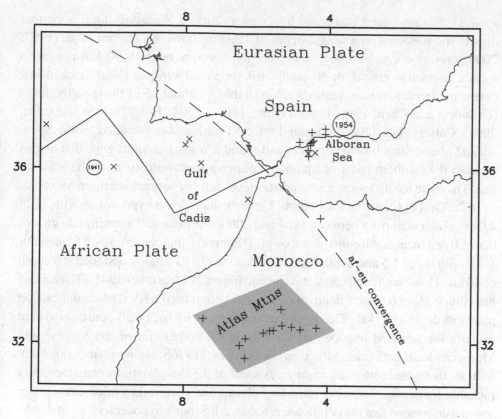

Fig. 10.46 Map of deep earthquake activity in the Spain–Morocco region. crosses are intermediate earthquakes with m_b of 4.0 or greater in the Gulf of Cadiz reported by the ISC. Plus symbols in Spain and the Alboran Sea are intermediate earthquakes investigated by Buforn *et al.* (1991a; 1997) and Grimison and Chen (1986). The gray area in Morocco indicates a region where existence of intermediate activity is uncertain; here plus symbols indicate intermediate-depth earthquakes with magnitudes between 2.6 and 4.1 as reported by Ramdani (1998). The open circles are the locations for the great 1954 Spanish deep-focus earthquake, and intermediate-depth earthquake of 1941 reported by GR. Lines indicate shorelines and the African–Eurasian plate boundary; the dashed line shows the direction of African–Eurasian (af–eu) plate convergence.

In the western Alboran Sea, between 1964 and 2000 the ISC located 16 earthquakes with depths exceeding 60 km; only one had a reported focal depth greater than 115 km. The largest was the earthquake of 7 August 1975 ($h_{pP-P} = 98$ km; $h_{EHB} = 91$ km; $m_{b(ISC)} = 5.1$), recorded by 198 stations; this is the only deep event in the entire Spain–Morocco region in the EHB catalog. However, from locations determined using local stations, Seber *et al.* (1996) and Buforn *et al.* (1997) plot numerous hypocenters with depths down to about 130 km. Morales *et al.* (1997) suggested that these hypocenters form a Wadati–Benioff zone that dips to the southeast, reaching a depth of 100 km; however, Buforn *et al.* (1997; 2004) present a cross section where the hypocenters beneath 90 km appear to form a distinctly vertical, north–south striking planar surface. Focal mechanisms determined using various methods are available for about 25 of these earthquakes (Grimison and Chen, 1986; Buforn *et al.*, 1991a; 2004; 1997; Buforn and Coca, 2002). Calvert *et al.* (2000) relocated selected earthquakes recorded by the Spanish and Moroccan regional networks and found a Wadati–Benioff zone that begins beneath the southern coast of Spain and plunges southward beneath the Alboran Sea. They also located some intermediate-depth activity beneath northern Morocco.

In the Gulf of Cadiz the ISC reports 7 intermediate-depth hypocenters with m_b of 4.0 or greater occurring between 1964 and 2000. Only one had a magnitude greater than 5.0 or a focal depth constrained by pP-P intervals; this occurred on 6 September 1969 with m_b of 5.5 and depth h_{pP-P} of 67 km; the EHB catalog assigned it a depth of 40 km. However, for an earthquake occurring on 27 December 1941 GR assigned magnitude M_S of 6.8 and a depth of 60 km; relocation for the EV Centennial catalog places its depth as 25 km. The depth extent of seismicity isn't well-constrained here because the reported hypocenters lie outside the boundaries of the Spanish and Moroccan local networks. Moreover, Seber *et al.* (1996b) show seismograms that indicate that mantle attenuation prevents some of the closer stations from recording pickable S arrivals. But, after rereading phases and performing relocations Seber seemed convinced that the depths are reliable. All Seber's hypocenters are at depths of 150 km or less except for one with a focal depth of 200 km. López-Casado *et al.* (2001) report that the deepest earthquakes here have depths of 180 km. Incidentally, the great shallow-focus Lisbon earthquake of 1 November 1755 probably occurred somewhere within the Gulf of Cadiz.

Calvert *et al.* (2000) review various models that aspire to explain the causes for the deep earthquakes in the Alboran Sea and Gulf of Cadiz. All the models presume that there was convergence between Africa and Europe during the last 30 million years. The models differ in that some presume that the seismic activity occurs in a detached or relic subducted slab; others assert that significant crustal thickening produced cold material that detached from the underside of the lithosphere and sank into the mantle (e.g., see Mezcua and Rueda, 1997). Using P and S phases from regional stations, Calvert *et al.* (2000) performed travel-time tomography and identified a high-velocity region in the mantle that extends to a depth of about 400 km and a localized zone of high velocity at 600 km that is approximately coincident with the 1954 focus. Gutscher *et al.* (2002) report that a marine seismic survey has imaged an active accretionary wedge west of the Straits of Gibraltar. Altogether, these data, the tomographic results, and the deep seismicity suggest that there is an east- or southeast-dipping subduction zone currently active in this region. However,

Stich *et al.* (2005) do not agree. Athough they use waveform modeling to confirm that some Gulf of Cadiz earthquakes have depths of about 60 km, they conclude that the focal mechanisms and scattered locations of the seismicity are inconsistent with on-going subduction.

Beneath the Atlas mountains, Hatzfeld and Frogneux (1981) and Ramdani (1998) report that the Spanish and Moroccan local networks locate several earthquakes per year with depths between 60 km and about 160 km; all have magnitudes of 3.9 or less. Hatzfeld and Frogneux reproduce seismograms with impulsive P and S arrivals for an earthquake located at 125 km depth; it lies within the boundaries of the Moroccan network and thus the depth should be accurate. Based on these data, Chen and Molnar (1983) proposed that this seismicity is characteristic of regions where the crust is very thick and the mantle is unusually cold. There is a gap in seismicity between depths of about 25–65 km; Chen and Molnar propose that crustal material within this gap fails by ductile flow whereas there is brittle failure (and earthquakes) above and below. Below the gap the earthquakes occur in the mantle, which remains brittle to higher temperature than the crust.

However, not everybody is convinced that the reported intermediate depths beneath the Atlas mountains are accurate. There were no such events among the selected, relocated earthquakes reported by Calvert *et al.* (2000). And, from an analysis of teleseismic P residual recorded by the regional network, Seber *et al.* (1996a) found no evidence for either thickened or high-velocity lithospheric material beneath the Atlas region.

10.21.2 Significant earthquakes

In the Spain–Morocco region I have found only five earthquakes with depths of 60 km or more and reported magnitudes of 5.0 or greater. Four of these are mentioned above (1941, 1954, 1969, and 1975); the fifth occurred in the Alboran Sea on 12 December 1988 ($m_{b(ISC)} = 5.1$; $h_{ISC} = 94$ km).

29 March 1954: $h_{EVC} = 627$ km; $M_{W(Chung\&Kanamori)} = 7.9$. It is arguable that no deep earthquake in the twentieth century is more peculiar than this event. It is remarkable for its large size, for its location far distant from any ordinary subduction zone, and for the near absence any subsequent seismic activity near its focus. It produced property damage in Granada and Malaga, Spain, and when it occurred it was the second largest deep-focus earthquake known (after the 17 January 1922 Peru earthquake). A half-century later Huang and Okal (1998) rank it as the fifth largest. In spite of its size it produced no recorded aftershocks. And as of 2004 there have been exactly three earthquakes reported nearby, all with epicenters within 30 km of the 1954 focus (30 January 1973, $m_{b(ISC)} = 3.9$; 8 March 1990, $M_{b(ISC)} = 4.5$;

31 July 1993, $m_{b(ISC)} = 3.9$). Frohlich (1998) calculated that this corresponds to a b value of 0.08–0.20, which is five to ten times lower than "ordinary" seismic activity.

Chung and Kanamori (1976) analyzed the body waves for the 1954 earthquake (the inset presents their mechanism) and identified a number of distinct subevents within the main rupture, which had a duration of about 40 seconds. Several other investigators have determined a focal mechanism (e.g., Hodgson and Cock, 1956) and there are published mechanisms for the 1973 and 1990 earthquakes as well (Udías *et al.*, 1976; Grimison and Chen, 1986; Buforn *et al.*, 1991b; 1997).

10.21.3 Spain and Morocco references

Buforn, E., M. Bezzeghould, A. Udías, and C. Pro, 2004. Intermediate and deep earthquakes in Spain, *Pure Appl. Geophys.*, **161**, doi:10.1007/s00024-003-2466-1, 623–646.

Buforn, E. and P. Coca, 2002. Seismic moment tensor for intermediate depth earthquakes at regional distances in southern Spain, *Tectonophysics*, **356**, 49–63, doi:10.1016/S0040-1951(02)00376-1.

Buforn, E., A. Udías, and R. Madariaga, 1991a. Intermediate and deep earthquakes in Spain, *Pure Appl. Geophys.*, **136**, 375–393.

Buforn, E., A. Udías, J. Mezcua, and R. Madariaga, 1991b. A deep earthquake under south Spain, 8 March 1990, *Bull. Seismol. Soc. Amer.*, **81**, 1403–1407.

Buforn, E., P. Coca, A. Udías, and C. Lasa, 1997. Source mechanism of intermediate and deep earthquakes in southern Spain, *J. Seismology*, **1**, doi:10.1023/A:1009754219459, 113–130.

Calvert, A., E. Sandvol, D. Seber, M. Barazangi, S. Roecker, T. Mourabit, F. Vidal, G. Alguacil, and N. Jabour, 2000. Geodynamic evolution of the lithosphere and upper mantle beneath the Alboran region of the western Mediterranean: constraints from travel time tomography, *J. Geophys. Res.*, **105**, doi:10.1029/2000JB900024, 10871–10898.

Chen, W.-P. and P. Molnar, 1983. Focal depths of intracontinental and intraplate earthquakes and their implications for the thermal and mechanical properties of the lithosphere, *J. Geophys. Res.*, **88**, 4183–4214.

•Chung, W.-Y. and H. Kanamori, 1976. Source process and tectonic implications of the Spanish deep-focus earthquake of March 29, 1954, *Phys. Earth Planet. Int.*, **13**, 85–96.

Frohlich, C., 1998. Does maximum earthquake size depend on focal depth? *Bull. Seismol. Soc. Amer.*, **88**, 329–336.

Grimison, N. L. and W.-P. Chen, 1986. The Azores–Gibraltar plate boundary: focal mechanisms, depths of earthquakes, and their tectonic implications, *J. Geophys. Res.*, **91**, 2029–2047.

Gutscher, M.-A., J. Malod, J.-P. Rehault, I. Contrucci, F. Klingelhoefer, L. Mendes-Victor, and W. Spakman, 2002. Evidence for active subduction beneath Gibraltar, *Geology*, **30**, doi:10.1130/0091-7613(2002)030, 1071–1074.

• Hatzfeld, D. and M. Frogneux, 1981. Intermediate depth seismicity in the western Mediterranean unrelated to subduction of oceanic lithosphere, *Nature*, **292**, doi:10.1038/292443a0, 443–445.

Hodgson, J. H. and J. I. Cock, 1956. Direction of faulting in the deep-focus Spanish earthquake of March 29, 1954, *Tellus*, **8**, 321–328.

Huang, W.-C. and E. A. Okal, 1998. Centroid moment tensor solutions for deep earthquakes predating the digital era: discussion and inferences, *Phys. Earth Planet. Int.*, **106**, doi:10.1016/S0031-9201(97)00111-8, 191–218.

López-Casado, C., C. Sanz de Galdeano, S. Molina-Palacios, and J. Henares-Romero, 2001. The structure of the Alboran Sea: an interpretation from seismological and geological data, *Tectonophysics*, **338**, doi:10.1016/S0040-1951(01)00059-2, 79–95.

Mezcua, J. and J. Rueda, 1997. Seismological evidence for a delamination process in the lithosphere under the Alboran Sea, *Geophys. J. Int.*, **129**, F1–F8.

Morales, J., I. Serrano, F. Vidal, and F. Torcal, 1997. The depth of the earthquake activity in the Central Betics (Southern Spain), *Geophys. Res. Lett.*, **24**, doi:10.1029/97GL03306, 3289–3292.

Ramdani, F., 1998. Geodynamic implications of intermediate-depth earthquakes and volcanism in the intraplate Atlas mountains (Morocco), *Phys. Earth Planet. Int.*, **108**, doi:10.1016/S0031-9201(98)00106/X, 245–260.

Seber, D., M. Barazangi, B. A. Tadili, M. Ramdani, A. Ibenbrahim, and D. Ben Sari, 1996a. Three-dimensional upper mantle structure beneath the intraplate Atlas and interplate Rif mountains in Morocco, *J. Geophys. Res.*, **101**, doi:10.1029/95JB03112, 3125–3138.

Seber, D., M. Barazangi, A. Ibenbrahim, and A. Demnati, 1996b. Geophysical evidence for lithospheric delamination beneath the Alboran Sea and Rif-Betec mountains, *Nature*, **379**, doi:10.1038/379785a0, 785–790.

Stich, D., F. L. Mancilla, and J. Morales, 2005. Crust–mantle coupling in the Gulf of Cadiz (SW Iberia), *Geophys. Res. Lett.*, **32**, L13306, doi:10.1029/2005GL023098.

Udías, A., A. L. Arroyo, and J. Mezcua, 1976. Seismotectonics of the Azores–Alboran region, *Tectonophysics*, **31**, doi:10.1016/0040-1951(76)90121-9, 259–289.

10.22 Italy and the Tyrrhenian Sea

10.22.1 Regional tectonics and deep seismicity

In the Mediterranean near Italy the African and Eurasian plates converge and collide; presently the rate is slow – about 0.5 cm/yr – and the location of the plate boundary is uncertain or distributed over a broad region. However, there is evidence for ongoing subduction – there are active volcanoes[18] in Italy and, in the Tyrrhenian Sea just west of Italy's "toe," there is a distinct zone of deep earthquakes extending down to a depth of almost 500 km (Figs. 10.47 and 10.48). The usual interpretation is that these earthquakes form a Wadati–Benioff zone dipping about 70° towards the west-northwest, although in Fig. 10.48 the dip angle appears to be somewhat

---

[18] One of these is Vesuvius, perhaps the most famous volcano in the world.

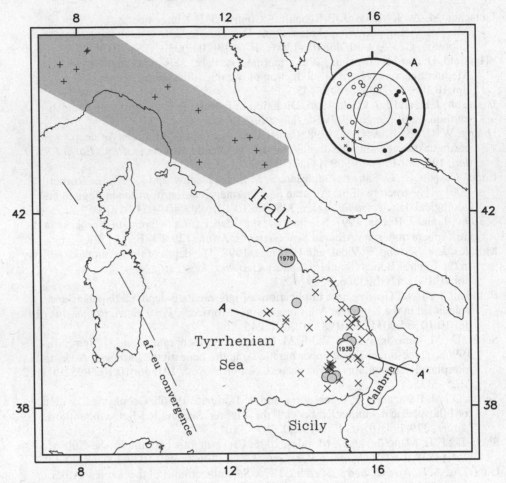

Fig. 10.47 Map of deep earthquake activity in Italy and the surrounding region. Filled circles, open circles, and crosses indicate earthquakes in the Harvard CMT and EHB catalogs. The open circle labeled "1938" indicates the GR location for the earthquake of 13 April 1938. The gray area in northern Italy is a region where existence of intermediate activity is uncertain; plus symbols are earthquakes with depths exceeding 60 km and magnitudes between 2.1 and 3.7 as reported by Selvaggi and Amato (1992) and Cattaneo *et al.* (1999). The solid line A–A' indicates the location and orientation of cross sections, and equal-area plots summarize T (solid circles), B (crosses), and P (open circles) axis orientations of selected Harvard CMT. Dashed lines show the direction of Africa–Eurasia (af–eu) plate convergence.

less. However, the lateral extent of this zone isn't much greater than its apparent thickness, and thus the strike-direction of the Wadati–Benioff zone is poorly defined.

In this region, intermediate- and deep-focus activity rates are low and the earthquakes are seldom very large. Between 1964 and 2000 the ISC located 218

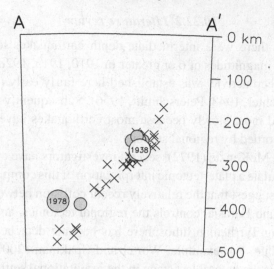

Fig. 10.48 Cross section A–A′ of intermediate- and deep-focus earthquakes in and near Italy. Earthquake symbols are as in Fig. 10.47.

intermediate- and deep-focus hypocenters; of these, only 10 had m_b of 5.0 or greater. Some older investigations report that earthquakes extended only to a depth of 350 km and state that there is a gap in activity between 50 and 200 km. However, more recent studies find nearly continuous activity that reaches a depth of at least 482 km, where a well-recorded earthquake with m_b of 5.1 occurred on 21 December 1996. Richter (1958) mentions an earthquake with magnitude 5.5 and a depth of 470 km that occurred on 17 February 1955. In the catalog prepared by the Istituto Nazionale di Geofisica in Rome there are hypocenters with depths exceeding 600 km (e.g., see Mele, 1998); however, all of these that the ISC locates have assigned depths shallower than 500 km. All catalogs agree that earthquakes are most numerous between about 250 km and 300 km depth.

From an evaluation of regional network data, Amato and Selvaggi (1991), Selvaggi and Amato (1992), and Cattaneo *et al.* (1999) report that there are hypocenters with focal depths exceeding 60 km occurring over a broad area in northern Italy (see gray area in Fig. 10.47). All are quite small, with magnitudes between 2.1 and 3.7, and all but two, with reported depths of 90 and 114 km, lie between 60 km and 80 km. The intermediate depths determined for all these earthquakes may be inaccurate, as the published information about S readings, etc., isn't adequate to confirm the reported depths, and because the network locates earthquakes with such depths only rarely. However, Amato *et al.* (1997), Carminati *et al.* (2005) and Chiarabba *et al.* (2005) accept the reported depths for these events, and conclude that they indicate there is ongoing subduction that dips westward at an angle of about 45° beneath the northern Appenines along the northeastern margin of Italy.

10.22.2 Literature review

Probably because there were intermediate-depth earthquakes located near Italy with GR-assigned magnitudes of 6 or greater in 1910, 1915, 1928, 1938 and 1941, the existence of deep activity was established here fairly early (DiFilippo, 1941; Gutenberg and Richter, 1948; Peterschmitt, 1956). Subsequently knowledge about the area developed more slowly because most earthquakes have low magnitudes and are poorly recorded by regional stations.

Beginning with McKenzie (1972), several investigators have evaluated the deep seismicity to formulate a plate-tectonic interpretation of this complex region. Generally these models suggest that the relatively recent collision between two continental plates, Africa and Eurasia, controls the regional tectonics, and some speculate that the subducting Tyrrhenian lithosphere has become detached or delaminated from its parent plate. For example, Wortel and Spakman (2000) state that here "slab detachment is the natural last stage in the gravitational settling of subducting lithosphere." However, the evidence published to date indicates that high-velocity, low-attenuation material extends continuously from the surface down to depths of at least 350 km (Caputo et al., 1972; Selvaggi and Chiarabba, 1995; Mele, 1998), and possibly much further (Lucente et al., 1999).

Hypocentral locations in the region became more reliable after about 1988 when there were improvements in the regional network (Selvaggi and Chiarabba, 1995). Before then the best information about Wadati–Benioff zone geometry came from carefully selected or relocated teleseismic data (e.g., Iannaccone et al., 1985; Anderson and Jackson, 1987; Giardini and Velona, 1991). There have been numerous compilations of focal mechanisms for the deep earthquakes in this region (McKenzie, 1972; Gasparini et al., 1982; 1985; Iannaccone et al., 1985; Anderson and Jackson, 1987; Caccamo et al., 1996; Frepoli et al., 1996; Bruno et al., 1999). This is one of the world's few regions where most mechanisms exhibit downdip compression.

10.22.3 Significant earthquakes

No individual intermediate- or deep-focus earthquakes near Italy have been extraordinarily damaging or the subject of numerous scientific investigations. The largest reported deep event occurred on 13 April 1938. GR assigned magnitude M_S of $6\frac{3}{4}$ and depth of 270 km, while Anderson and Jackson (1987) list m_B 7.1 and 290 km; in the EV Centennial catalog the relocated depth is only 35 km. DiFilippo (1941) published an investigation of two earthquakes on 16 March 1941 that GR placed at 100 km depth and assigned magnitudes of $6\frac{1}{2}$ and $5\frac{3}{4}$. The earthquake of 27 December 1978 ($M_{W(CMT)} = 5.9$; $h_{CMT} = 388$ km) is unusual in that its P axis is not aligned along the dip of the seismic zone and its reported location lies somewhat west of other earthquakes in the region. Thus it may be an "outboard" earthquake (see Section 5.1.4) that represents deformation separate from the main piece of subducted lithosphere.

10.22.4 Italy and Tyrrhenian references

Amato, A., C. Chiarabba, and G. Selvaggi, 1997. Crustal and deep seismicity in Italy (30 years after), *Annalli di Geofisica*, **40**, 981–993.

Amato, A. and G. Selvaggi, 1991. Terremoti crostali e sub-crostali nell-Appennino Settentrionale, *Studi Geol. Camerti*, **1**, 75–82.

• Anderson, H. and J. Jackson, 1987. The deep seismicity of the Tyrrhenian Sea, *Geophys. J. Roy. Astron. Soc.*, **91**, 613–637.

Bruno, G., I. Guerra, A. Moretti, and G. Neri, 1999. Space variations of stress along the Tyrrhenian Wadati–Benioff zone, *Pure Appl. Geophys.*, **156**, doi:10.1007/s000240050318, 667–688.

Caccamo, D., G. Neri, A. Sarao, and M. Wyss, 1996. Estimates of stress directions by inversion of earthquake fault-plane solutions in Sicily, *Geophys. J. Int.*, **125**, 857–868.

Caputo, M., G. F. Panza, and D. Postposchl, 1972. New evidences about the deep structure of the Lipari arc, *Tectonophysics*, **15**, doi:10.1016/0040-1951(72)90086-8, 219–231.

Carminati, E., A. M. Negredo, J. L. Valera, and C. Doglioni, 2005. Subduction-related intermediate-depth and deep seismicity in Italy: insights from thermal and rheological modeling, *Phys. Earth Planet. Int.*, **149**, doi:10.1016/j.pepi.2004.04.006, 65–79.

Cattaneo, M., P. Augliera, S. Parolai, and D. Spallarossa, 1999. Anomalously deep earthquakes in northwestern Italy, *J. Seismology*, **3**, doi:10.1023/A:1009899214734, 421–435.

Chiarabba, C., L. Jovane, and R. DiStefano, 2005. A new view of Italian seismicity using 20 years of instrumental recordings, *Tectonophysics*, **395**, doi:10.1016/j.tecto.2004.09.013, 251–268.

DiFilippo, D., 1941. Studio microsismico del terremoto del basso Tirreno del 16 Marzo 1941-XIX, *Soc. Sismol. Italiana, Boll.*, **39** (3–4), 3–25.

Frepoli, A., G. Selvaggi, C. Chiarabba, and A. Amato, 1996. State of stress in the southern Tyrrhenian subduction zone from fault-plane solutions, *Geophys. J. Int.*, **125**, 879–891.

Gasparini, C., G. Iannaccone, P. Scandone, and R. Scarpa, 1982. Seismotectonics of the Calabrian arc, *Tectonophysics*, **84**, doi:10.1016/0040-1951(82)90163-9, 267–286.

Gasparini, C., G. Iannaccone, and R. Scarpa, 1985. Fault-plane solutions and seismicity of the Italian Peninsula, *Tectonophysics*, **117**, doi:10.1016/0040-1951(85)90236-7, 59–78.

Giardini, D. and M. Velona, 1991. The deep seismicity of the Tyrrhenian Sea, *Terra Nova*, **3**, 57–64.

Gutenberg, B. and C. F. Richter, 1948. Deep-focus earthquakes in the Mediterranean region, *Geofisica pura e applicata*, **12**, 3–4.

Iannaccone, G., G. Scarcella, and R. Scarpa, 1985. Subduction zone geometry and stress patterns in the Tyrrhenian Sea, *Pure Appl. Geophys.*, **123**, 819–836.

Lucente, F. P., C. Chiarabba, G. B. Cimini, and D. Giardini, 1999. Tomographic constraints on the geodynamic evolution of the Italian region, *J. Geophys. Res.*, **104**, doi:10.1029/1999JB900147, 20307–20327.

• McKenzie, D., 1972. Active tectonics of the Mediterranean region, *Geophys. J. Roy. Astron. Soc.*, **30**, 109–185.

Mele, G., 1998. High-frequency wave propagation from mantle earthquakes in the Tyrrhenian Sea: new constraints for the geometry of the southern Tyrrhenian subduction zone, *Geophys. Res. Lett.*, **25**, doi:10.1029/98GL02175, 2877–2880.

Peterschmitt, E., 1956. Quelques données nouvelles sur les séismes profonds de la Mer Tyrrhénienne, *Annali di Geofisica (Roma)*, **9**, 305–334.

Richter, C. F., 1958. *Elementary Seismology*, San Francisco, W. H. Freeman, 768 pp.

Selvaggi, G. and A. Amato, 1992. Subcrustal earthquakes in the northern Apennines (Italy): evidence for a still active subduction? *Geophys. Res. Lett.*, **19**, 2127–2130.

Selvaggi, G. and C. Chiarabba, 1995. Seismicity and P-wave velocity image of the southern Tyrrhenian subduction zone, *Geophys. J. Int.*, **121**, 818–826.

Wortel, M. J. R. and W. Spakman, 2000. Subduction and slab detachment in the Mediterranean–Carpathian region, *Science*, **290**, doi:10.1126/science.290.5498.1910, 1910–1917.

10.23 Greece and Turkey

10.23.1 Regional tectonics and deep seismicity

One could argue that seismology began in Greece, as there are historical records of damaging Hellenic earthquakes extending back about 4000 years. This seismic activity occurs in response to convergence of about 1 cm/yr between the African and Eurasian plates. However, the regional tectonics are complex; there has been extension since the Tertiary in the Aegean and some reconstructions define a separate Aegean plate. There is ongoing subduction as there is a seismically active arc-shaped trench south of Greece and Crete and volcanic islands in the sea between Greece, Crete, and Turkey.[19] There are intermediate-depth earthquakes down to at least 160 km (Figs. 10.49 and 10.50) that form a Wadati–Benioff zone extending from the trench approximately northward under the Aegean Sea.

In the GR catalog there are about 30 intermediate-depth Hellenic earthquakes with magnitudes of 6.0 or greater occurring between 1908 and 1950. However, since 1977 the CMT catalog reports only one (23 May 1994, $M_{W(CMT)} = 6.1$, $h_{CMT} = 81$ km). Although some GR depths are questionable, intermediate-depth earthquakes clearly do occur here. The CMT catalog lists one earthquake (27 September 1983, $M_{W(CMT)} = 5.4$) at 170 km, and there are several earthquakes in the EHB catalog with depths of 150–170 km. The ISC catalog reports four earthquakes with depths of 350 km or more (including one at 709 km); however, none of these depths are supported by convincing evidence.

---

[19] One of these islands is Santorini, where a large and explosive volcanic eruption occurred about 1640 BC. This eruption, thought to be comparable in size and character to the 1883 Krakatoa eruption, destroyed the Minoan civilization and may be the origin of the Atlantis legend.

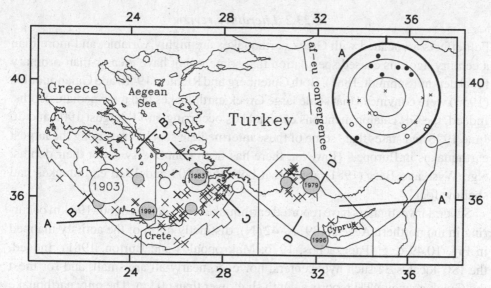

Fig. 10.49 Map of intermediate-depth earthquakes in Greece and Turkey. Filled circles, open circles, and crosses indicate selected earthquakes in the Harvard CMT, EV Centennial, and EHB catalogs (see description of selection process in the introduction to this chapter). Solid lines A–A' to D–D' indicate locations and orientations of cross sections, and equal-area plots summarize T (solid circles), B (crosses), and P (open circles) axis orientations of selected Harvard CMT for intermediate-depth earthquakes within the indicated regions. The dashed line shows the directions of African–Eurasian (af–eu) plate convergence.

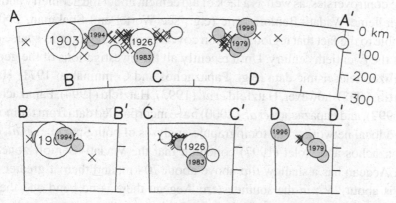

Fig. 10.50 Cross sections A–A' to D–D' of intermediate-depth earthquakes in Greece and Turkey. Earthquake symbols are as in Fig. 10.49.

There is an additional group of intermediate-depth earthquakes that occur further east, beneath Turkey and Cyprus. The GR catalog lists events with magnitudes of $6\frac{1}{4}$ and $6\frac{1}{2}$ in 1911, 1927, and 1941, and the largest recent earthquake occurred on 28 May 1979 ($h_{\text{CMT}} = 96$ km; $M_{\text{W(CMT)}} = 6.0$). Although GR reports a depth of 180 km for the 1911 earthquake, none of the depths in the EHB catalog exceed 126 km.

10.23.2 Literature review

Felt effects associated with Greek earthquakes are highly variable, and more than a century ago this caused speculation that some might have greater-than-ordinary focal depths (Schmidt, 1881). Both Gutenberg and Richter (1948) and Galanopoulos (1953) were convinced that some large Greek earthquakes had intermediate depths. Indeed, the GR catalog mentions earthquakes occurring on 11 August 1903 and 26 June 1926 and states that "some of these intermediate shocks are among the largest earthquakes in Europe." However, there has been controversy about their depths; e.g., Wyss and Baer (1981) argue that both the 1903 and 1926 earthquakes had shallow foci.

Several investigators have reported earthquakes with depths of 60–190 km occurring in the northern Aegean (39° N–42° N), distinctly north of the activity mapped in Fig. 10.49 (e.g., Papazachos, 1976; Makropoulos and Burton, 1981). Indeed, the ISC locates 34 such hypocenters; however, nearly all are small, and for most the Greek agency ATH reports a depth shallower than 60 km. The only earthquake with m_b exceeding 4.8 occurred on 4 March 1967 ($m_b = 5.9$). For this quake the intermediate focal depth is questionable; although the ISC fixed the depth at 60 km, the EV Centennial relocation found a depth of only 9 km. Furthermore, McKenzie (1972), who examined records to determine a focal mechanism, lists the depth as 33 km. Considering these uncertainties, it is possible that the reported intermediate focal depths are erroneous for all these northern Aegean earthquakes.

These controversies, as well as a lack of agreement about the geometry and extent of the Hellenic Wadati–Benioff zone (e.g., see Wortel and Spakman, 2000) are attributable to the fact that regional station coverage in this region was sparse during much of the twentieth century. Until recently all the investigations of the geometry have utilized teleseismic data (e.g., Papazachos and Comninakis, 1971; Hatzfeld and Martin, 1992); however, Hatzfeld *et al.* (1993), Hatzfeld (1994), Papazachos and Nolet (1997), and Papazachos *et al.* (2000) have incorporated data from temporarily-deployed local networks. The tomographic analyses of both Spakman *et al.* (1988) and Papazachos and Nolet (1997) conclude that the Wadati–Benioff zone in the western Aegean has a shallow dip above about 70 km, and then at greater depths the dip is about 25°; in the southeastern Aegean there is no bend and the dip is about 35°. Focal mechanisms generally exhibit downdip tension, and have been compiled by McKenzie (1972), Papazachos (1977), Kondopoulou *et al.* (1985), Beisser *et al.* (1990), and Benetatos *et al.* (2004).

For reasons that are unclear, Greek earthquakes and the seismologists who interpret them generate a surprising number of research papers that report periodicity in the intermediate-depth earthquake activity as well as relationships between intermediate seismicity, volcanism, and other geophysical phenomena (e.g., Comninakis

and Papazachos, 1980; Polimenakos, 1981; Papadopoulos, 1986, 1989; Papazachos, 1993; Nomikos and Vallianatos, 1997). Since these relationships haven't
been shown to occur in other regions and since the statistical evidence presented
for the Greek seismicity isn't very convincing, it seems likely that future studies
won't confirm their existence.

Concerning the intermediate earthquakes beneath southern Turkey and Cyprus
(Fig. 10.49), there is now consensus that at least some of the reported depths between
about 60 km and 120 km are genuine (e.g., Jackson and McKenzie, 1984). However,
there have been a variety of models proposed to explain the regional plate structure
here. This seismic activity isn't far from the presumed triple junction for the African,
Eurasian, and Arabian plates. But some reconstructions define a separate Turkish
plate to the north, and place the Africa–Arabia boundary either in this region (Wortel
and Spakman, 2000) or somewhat to the east (McKenzie, 1972).

10.23.3 Significant earthquakes

The reported intensity patterns for some of the historical earthquakes in Greece
and the surrounding islands suggest they may have had intermediate focal depths.
The EV Centennial catalog reports a depth of 80 km and a magnitude of 8.3 for
the earthquake of 11 August 1903, listing the original source as "unknown" but
referencing Båth and Duda (1979). Dunbar *et al.* (1992) report focal depths of 100
km for earthquakes occurring in Greece, Crete, or the Dodecanese Islands in 368
BC, 222 BC, 183 BC, AD 155, AD 365, AD 796, 1304, 1508, 1810, 1842, 1856,
1863, and 1867 (see section 3.6.2). The depths are subject to dispute; e.g., Wyss and
Baer (1981) present arguments that the earthquakes of 1810, 1856, 1863 and 1867
are shallow. However, Ambraseys and Adams (1998) evaluate seismograms and
intensity distributions which demonstrate clearly that the 26 June 1926 earthquake
had a depth of about 115 km, and thus it seems likely that at least some of the older
earthquakes were intermediate as well. Pilidou *et al.* (2004) argue that a 9 October
1996 Cyprus earthquake ($M_{\text{W(CMT)}} = 6.8$) had a depth of about 80 km, although
both Harvard and Arvidsson *et al.* (1998) obtain a shallow depth.

26 June 1926: $h_{\text{EVC}} = 102$ km; $m_{\text{B(Abe)}} = 7.7$. This shock was felt throughout
the eastern Mediterranean and caused some damage in Crete, Egypt, Turkey, and
Greece. The most severe intensities occurred on the island of Rhodes, where it
ruined or damaged 2000 of the 10,000 houses on the island.[20] Ambraseys and

---

[20] Considering the damage the 1926 quake caused, one wonders if the earthquake of about 227 BC that destroyed
the Colossus of Rhodes, one of the seven wonders of the ancient world, had an intermediate-depth focus.
Probably it was shallow, since Dunbar *et al.* (1992) do not list it as intermediate, and Galanopoulos (1960)
states that it was "possibly [accompanied by] a seismic sea wave." He also discusses tsunamis accompanying

Adams (1998) thoroughly review the literature reporting the intensity distribution for this earthquake and covering the dispute about its focal depth. I have been unable to find a focal mechanism.

10.23.4 Greece and Turkey references

Ambraseys, N. N. and R. D. Adams, 1998. The Rhodes earthquake of 26 June 1926, *J. Seismology*, **2**, doi:10.1023/A:1009706415417, 267–292.

Arvidsson, R., Z. Ben-Avraham, G. Ekström, and S. Wdowinski, 1998. Plate tectonic framework for the October 9, 1996, Cyprus earthquake, *Geophys. Res. Lett.*, **25**, doi:10.1029/98GL01547, 2241–2244.

Båth, M. and S. J. Duda, 1979. *Some Aspects of Global Seismicity*, Rept. 1–79, Uppsala Sweden, Seismological Institute.

Beisser, M., M. Wyss, and R. Kind, 1990. Inversion of source parameters for subcrustal earthquakes in the Hellenic arc, *Geophys. J. Int.*, **103**, 439–450.

Benetatos, C., A. Kiratzi, C. Papazachos, and G. Kirakaisis, 2004. Focal mechanisms of shallow and intermediate depth earthquakes along the Hellenic arc, *J. Geodynamics*, **37**, doi:10.1016/j.jog.2003.11.001, 253–296.

Comninakis, P. E. and B. C. Papazachos, 1980. Space and time distribution of the intermediate focal depth earthquakes in the Hellenic arc, *Tectonophysics*, **70**, doi:10.1016/0040-1951(80)90278-4, T35-T40.

Dunbar, P. K., P. A. Lockridge, and L. S. Whiteside, 1992. *Catalog of Significant Earthquakes 2150 BC– AD 1991*, Rept. SE 49, Boulder,CO., World Data Center A, U.S. Department of Commerce.

Galanopoulos, A., 1953. On the intermediate earthquakes in Greece, *Bull. Seismol. Soc. Amer.*, **43**, 159–178.

  1960. Tsunamis observed on the coasts of Greece from antiquity to present time, *Annali di Geofisica*, **13**, 369–386.

• Guidoboni, E., A. Comastri, and G. Traina, 1994. *Catalogue of Ancient Earthquakes in the Mediterranean Area Up to the 10th Century*, Rome, Italy, Istituto Nazionale di Geofisica, 504 pp.

Gutenberg, B. and C. F. Richter, 1948. Deep-focus earthquakes in the Mediterranean region, *Geofisica Pura e Applicata*, **12**, 3–4.

Hatzfeld, D., 1994. On the shape of the subducting slab beneath the Peloponnese, Greece, *Geophys. Res. Lett.*, **21**, doi:10.1029/93GL03079, 173–176.

• Hatzfeld, D. and C. Martin, 1992. Intermediate depth seismicity in the Aegean defined by teleseismic data, *Earth Planet. Sci. Lett.*, **113**, doi:10.1016/0012-821X(92)90224-J, 267–275.

Hatzfeld, D., M. Besnard, K. Makropoulos, N. Voulgaris, V. Kouskouna, P. Hatzidimitriou, D. Panagiotopoulos, G. Karakaisis, A. Deschamps, and H. Lyon-Caen, 1993. Subcrustal microearthquake seismicity and fault plane solutions beneath the Hellenic arc, *J. Geophys. Res.*, **98**, 9861–9870.

Jackson, J. and D. McKenzie, 1984. Active tectonics of the Alpine–Himalayan belt between western Turkey and Pakistan, *Geophys. J. Roy. Astron. Soc.*, **77**, 185–264.

the earthquakes of AD 365 and 1867, which both he and Dunbar indicate are intermediate, and suggests that slumping, rather than displacement of submarine blocks, may be responsible for these tsunamis. Galanopoulos also suggests that the 27 August 1886 earthquake was "probably intermediate." A comprehensive reference on ancient Mediterranean earthquakes is Guidoboni *et al.* (1994), who presents many original sources both in their original languages and in translation.

Kondopoulou, D. P., G. A. Papadopoulos, and S. B. Paulides, 1985. A study of the deep seismotectonics in the Hellenic arc, *Bol. Geofisica Teorica et Applicata*, **27**, 197–207.

Makropoulos, K. C. and P. W. Burton, 1981. A catalogue of seismicity in Greece and adjacent areas, *Geophys. J. Roy. Astron. Soc.*, **65**, 741–762.

McKenzie, D., 1972. Active tectonics of the Mediterranean region, *Geophys. J. Roy. Astron. Soc.*, **30**, 109–185.

Nomikos, K. and F. Vallianatos, 1997. Transient electric variations associated with large intermediate-depth earthquakes in South Aegean, *Tectonophysics*, **269**, doi:10.1016/S0040-1951(96)00157-6, 171–177.

Papadopoulos, G. A., 1986. Large intermediate depth shocks and volcanic eruptions in the Hellenic arc during 1800–1985, *Phys. Earth Planet. Int.*, **43**, 47–55.

1989. Forecasting large intermediate-depth earthquakes in the south Aegean, *Phys. Earth Planet. Int.*, **57**, 192–198.

Papazachos, B. C., 1976. Seismotectonics of the northern Aegean area, *Tectonophysics*, **33**, doi:10.1016/0040-1951(76)90057-3, 199–209.

1977. A lithospheric model to interpret focal properties of intermediate and shallow shocks in Central Greece, *Pure Appl. Geophys.*, **115**, 655–666.

1993. Long-term prediction of intermediate depth earthquakes in the southern Aegean region based on a time-predictable model, *Natural Hazards*, **7**, 211–218.

Papazachos, B. C. and P. E. Comninakis, 1971. Geophysical and tectonic features of the Aegean arc, *J. Geophys. Res.*, **76**, 8517–8533.

Papazachos, C. and G. Nolet, 1977. P and S deep velocity structure of the Hellenic area obtained by robust nonlinear inversion of travel times, *J. Geophys. Res.*, **102**, doi:10.1029/96JB03730, 8349–8367.

Papazachos, B. C., V. G. Karakostas, C. B. Papazachos, and E. M. Scordilis, 2000. The geometry of the Wadati–Benioff zone and lithospheric kinematics in the Hellenic arc, *Tectonophysics*, **319**, doi:10.1016/S0040-1951(99)00299-1, 275–300.

Pilidou, S., K. Preistley, J. Jackson, and A. Maggi, 2004. The 1996 Cyprus earthquake: a large, deep event in the Cyprean arc, *Geophys. J. Int.*, **158**, doi:10.1111/j.1365-246X.2004.02248.x, 85–97.

Polimenakos, L. C., 1993. Search for a seasonal trend in earthquake occurrence in the Hellenic arc, *Phys. Earth Planet. Int.*, **76**, 253–258.

Schmidt, J., 1881. *Studien über Vulkane und Erdbeben*, Leipzig.

Spakman, W., M. J. R. Wortel, and N.J. Vlaar, 1988. The Hellenic subduction zone: a tomographic image and its geodynamic implications, *Geophys. Res. Lett.*, **15**, 60–63.

Wortel, M. J. R. and W. Spakman, 2000. Subduction and slab detachment in the Mediterranean–Carpathian region, *Science*, **290**, doi:10.1126/science.290.5498.1910, 1910–1917.

Wyss, M. and M. Baer, 1981. Earthquake hazard in the Hellenic arc. In *Earthquake Prediction: an International Review*, eds. D. W. Simpson and P. G. Richards, Washington, DC, American Geophysical Union, 153–172.

10.24 Romania

10.24.1 Regional tectonics and deep seismicity

The Southern and Eastern Carpathian mountains form a tilted "V" with arms extending east–west and northwest–southeast across Romania; precisely where

I thank Lani Ocescu and Cezar Trifu for reviewing an earlier draft of this section.

these arms meet lies a region called Vrancea (see Fig. 1.7); beneath this region large intermediate-depth earthquakes occur and have been recorded for more than 600 years. Often these earthquakes are highly destructive, especially to the city of Bucharest, which lies 160 km south of their geographic focus.

The lithosphere in this region is undeniably continental and ages of volcanic centers indicate that active subduction ceased about 10 Ma. The larger earthquakes occur at depths from about 70 km to 180 km (see Fig. 1.9). Activity rates are low between depths of 40–70 km and 110–130 km; the shallower gap may indicate that the subducted slab is detached. Beneath about 70 km relocated hypocenters occupy a nearly vertical tabular region that strikes SW–NE with a width of 20–30 km and a thickness of 10–15 km. The ISC occasionally reports small-magnitude earthquakes with depths between 200 and 300 km.

10.24.2 Literature review

There is a truly enormous literature concerning the intermediate-depth earthquakes of Romania, more than for any other geographic area except Japan and possibly South America. Presumably this is because Romania has a long urban history and experiences about three earthquakes per century with magnitude 7 or greater (Oncescu *et al.*, 1999). Compared to other geographic regions, a disproportionate number of papers concern strong motion and microzonation. Much of the literature is not in English or appears in relatively obscure publications. General monographs on Romanian seismicity include Draghicénu (1896), Atanasiu (1961), Constantinescu and Enescu (1985), and Wenzel *et al.* (1999).

It is unclear to me when scientists first realized that the Romanian earthquakes weren't shallow. However, the GR catalog includes 12 intermediate-depth quakes occurring between 1908 and 1945, and states that they were:

probably from the same focus, reported destructive at Bucharest, and perceptible at surprisingly great distances. [They] resemble those of the Hindu Kush in location under a disturbed structure and in frequent repetition from nearly the same focus, which lies here at a depth between 100 and 150 km.

Santo (1970) characterized the Vrancea intermediate earthquakes as a "nest" and compared them to those in the Hindu Kush and Bucaramanga, Colombia. However, the Vrancea hypocenters are considerably less localized in space, and the largest events much larger than those in Bucaramanga. From an analysis of nine years of data recorded by the Romanian regional network, Trifu *et al.* (1990) and Trifu and Radulian (1991) concluded that the magnitude distribution differed somewhat from a Gutenberg–Richter relationship; however, this may not be particularly unusual for localized groups of earthquakes.

There have been numerous summaries of the locations, times, and source properties of Vrancea earthquakes. Studies presenting detailed maps and cross-sections of
Wadati–Benioff geometry include Oncescu (1984), Trifu (1990), Trifu *et al.* (1991)
and Oncescu and Bonjer (1997). Radu and Purcaru (1964), Oncescu (1987), and
Bala *et al.* (2003) compile focal mechanisms for Vrancea earthquakes, demonstrating their tendency to have near-vertical T axes but quite varied orientations for their
P axes. All the well-determined Harvard CMT have thrust or dip-slip mechanisms;
nearly all have a SW–NE trending steeply dipping nodal plane. Gusev *et al.* (2002)
evaluate the spectral and time-domain properties of sources for 16 Vrancea earthquakes with magnitudes ranging from 3.7 to 7.4. Enescu *et al.* (2005) present an
analysis of the temporal properties of 5630 intermediate-depth earthquakes occurring between 1972 and 2002.

Because Vrancea earthquakes occur regularly and are distributed in a vertically-
distributed zone beneath continental lithosphere occupied by a well-distributed
regional network, their travel times provide nearly ideal sources for tomographic
studies of regional structure. The inversions of both Koch (1985) and Fan and Wallace (1998) found a nearly vertical region of high-velocity mantle material that
coincided roughly with the observed locations of intermediate-depth seismicity.
Ismail-Zadeh *et al.* (2000) perform finite element modeling and conclude that the
geometry of the high-velocity region and distribution and mechanisms of hypocenters are consistent with a relic oceanic slab sinking into a viscous mantle. There
has been persistent interest in the regional structure here as various investigators
believe it may represent the terminal phase of subduction, representing break-off of
the slab as it detaches from the overlying lithosphere (Wortel and Spakman, 1992;
2000; Linzer, 1996; Wenzel *et al.*, 1998; Ismail-Zadeh *et al.*, 2005).

10.24.3 Significant earthquakes

Oncescu *et al.* (1999) claim that the historical record of Vrancea earthquakes
is complete between 1411 and the present for magnitudes exceeding 7.0. For
events prior to 1900 there seem to be few detailed descriptions in English;
however, quakes occurring in 1802, 1829, and 1838 caused severe damage and
undoubtedly had magnitudes of 7 or greater. The earthquake of 30 May 1990
($h_{CMT} = 74$ km; $M_{W(CMT)} = 7.0$) caused about ten fatalities and a significant
amount of damage. There are a number of publications describing it; however,
it was neither as large nor as damaging as the other events discussed below.

26 October 1802: $M_W = 7.9$. This may have been the largest Romanian earthquake in historic times (see discussion in Section 1.2) as it caused severe damage and was felt throughout Europe (see Figure 1.10 for Radu and Utale's, 1992,

intensity map). Nevertheless, Atanasiu (1961) and Constantinescu and Enescu (1985) barely mention it, and I have found no thorough modern discussion of its effects. However, Schuller (1882), Draghicénu (1896), and Ştefánescu (1902) each have a few pages of quotes from contemporary sources describing damage.

10 November 1940: $h_{EVC} = 122$ km; $m_{B(Abe)} = 7.3$. This earthquake killed about 1000 people and destroyed many buildings in Bucharest including the new thirteen-story Carlton Hotel, where 267 people died (Mândrescu and Radulian, 1999). The quake was the subject of a special volume (Romanian Academy, 1941) and had a profound influence on subsequent building codes in Romania. Both Radu and Purcaru (1964) and Oncescu and Bonjer (1997) (see inset) published focal mechanisms. Oncescu and Bonjer report that there were two shocks separated by about 6 s with the rupture propagating generally downward at 3.7 km/s.

4 March 1977: $h_{CMT} = 84$ km; $M_{W(CMT)} = 7.5$. Mândrescu and Radulian (1999) describe this earthquake as "the most damaging event that struck the Romanian territory in modern time." Fatality and damage estimates for this earthquake vary; however, it is clear that it killed approximately 1500 people and caused damage of at least $800,000,000 (e.g., Fig. 1.8). Müller *et al.* (1978), Hartzell (1979), Trifu (1980), Räkers and Müller (1982), and Silver and Jordan (1983) all investigated the source process; the analysis indicated that four subevents occurred over about 20 seconds and were confined to a region about 70 km in extent. Over the next two months there were about 140 aftershocks recorded (Fuchs *et al.*, 1979). This earthquake was the subject of several papers in a special section of *Tectonophysics* and two extensive special reports (Cornea and Radu, 1979; Bălan *et al.*, 1982). Fattal *et al.* (1977) describe the damage to buildings in some detail, and present numerous pictures.

30 August 1986: $h_{CMT} = 133$ km; $M_{W(CMT)} = 7.2$. Trifu and Oncescu (1987) relocated 79 aftershocks for this earthquake and found that the majority occupied a roughly planar zone between 125 and 148 km depth. Oncescu (1989) made a thorough study of the source process and concluded that it had a rather high dynamic stress drop of 962±571 bars. Monfret *et al.* (1990) investigated the source-time function and found it had little or no complexity and a duration of 5–10 s. Radulian and Trifu (1991) and Trifu and Radulian (1991) investigated *b* values and found that they decreased during a six-year period prior to the mainshock and then increased suddenly about two months afterward.

10.24.4 Romanian references

Atanasiu, I., 1961. *Cutremurele de pămînt din România (Earthquakes of Romania)*, Bucharest, Romania Romanian Academy Publishers.

Bala, A., M. Radulian, and E. Popescu, 2003. Earthquakes distribution and their focal mechanism in correlation with the active tectonic zones of Romania, *J. Geodynamics*, **36**, doi:10.1016/S0264-3707(03)00044-9, 129–145.

Bălan, Ş., V. Cristescu, and I. Cornea, eds., 1982, *Cutremurul de pămînt din România de la 4 martie 1977 (The March 4, 1977 Earthquake)*, Bucharest, Romania, Editura Academiei Republicii Socialiste România, 516 pp.

Constantinescu, L. and D. Enescu, 1985. *Cutremurele din Vrancea în Cadru Ştiinţific si Tehnologic (Vrancea Earthquakes Within Their Scientific and Technological Framework)*, Bucharest, Romania, Editura Academiei Republicii Socialiste România, 230 pp.

Cornea, I., and C. Radu, eds., 1979. *Seismological Studies on the March 4, 1977 Earthquake* (in Romanian), Bucharest-Magurele, Romania, Central Institute of Physics.

Draghicénu, M. M., 1896. *Les Tremblements de Terre de la Roumanie et des Pays Environnants*, Bucharest, Romania, Géologie Appliquée, L'Institut d'Arts Graphiques Carol Göbl.

Enescu, B., K. Ito, M. Radulian, E. Popescu, and O. Bazacliu, 2005. Multifractal and chaotic analysis of Vrancea (Romania) intermediate-depth earthquakes: investigation of the temporal distribution of events, *Pure Appl. Geophys.*, **162**, doi:10.1007/s00024-004-2599-x, 249–271.

Fan, G. and T. C. Wallace, 1998. Tomographic imaging of deep velocity structure beneath the eastern and southern Carpathians, Romania: implications for continental collision, *J. Geophys. Res.*, **103**, doi:10.1029/97JB01511, 2705–2723.

Fattal, S. G., E. Simiu, and G. Culver, 1977. *Observations on the Behavior of Buildings in the Romania Earthquake of March 4, 1977*, Washington, DC, U.S. National Bureau of Standard Special Publication 490, 168 pp.

Fuchs, K., K.-P. Bonjer, G. Bock, I. Cornea, C. Radu, D. Enescu, D. Jianu, A. Nourescu, G. Merkler, T. Moldoveanu, and G. Tudorache, 1979. The Romanian earthquake of March 4, 1977. II. Aftershocks and migration of seismic activity, *Tectonophysics*, **53**, 225–247, doi:10.1016/0040-1951(79)90068-4.

Gusev, A., M. Radulian, M. Rizescu, and G. F. Panza, 2002. Source scaling of intermediate-depth Vrancea earthquakes, *Geophys. J. Int.*, **151**, doi:10.1046/j.1365-246X.2002.01816.x, 879–889.

Hartzell, S., 1979. Analysis of the Bucharest strong ground motion record for the March 4, 1977 Romanian earthquake, *Bull. Seismol. Soc. Amer.*, **69**, 513–530.

Ismail-Zadeh, A. T., G. F. Panza, and B. M. Naimark, 2000. Stress in the descending relic slab beneath the Vrancea region, Romania, *Pure Appl. Geophys.*, **157**, 111–130.

Ismail-Zadeh, A., B. Mueller, and G. Schubert, 2005. Three-dimensional modeling of contemporary mantle flow and tectonic stress beneath the earthquake-prone southeastern Carpatians based on integrated analysis of seismic, heatflow, and gravity data, *Phys. Earth Planet. Int.*, **149**, doi:10.1016/j.pepi.2004.08.012, 81–98.

Koch, M., 1985. Nonlinear inversion of local seismic travel times for the simultaneous determination of the 3D-velocity structure and hypocenters – application to the seismic zone in Vrancea, *J. Geophys.*, **56**, 160–173.

Linzer, H.-G., 1996 Kinematics of retreating subduction along the Carpathian arc, Romania, *Geology*, **24**, doi:10.1130/0091-7613(1996)024, 167–170.

Mândrescu, N. and M. Radulian, 1999. Macroseismic field of the Romanian intermediate-depth earthquakes. In *Vrancea Earthquakes: Tectonics, Hazard and Risk Mitigation*, eds. F. Wenzel, D. Lungu, and O. Novak, The Netherlands, Kluwer, 374 pp., 163–174.

Monfret, T., A. Deschamps, and B. Romanowicz, 1990. The Romanian earthquake of August 30, 1986: a study based on GEOSCOPE very long-period and broadband data, *Pure Appl. Geophys.*, **133**, 367–379.

Müller, G., K.-P. Bonjer, H. Stöckl, and D. Enescu, 1978. The Romanian earthquake of March 4, 1977: I. Rupture process inferred from fault-plane solution and multiple-event analysis, *J. Geophys.*, **44**, 203–218.

Oncescu, M. C., 1984. Deep structure of the Vrancea region, Roumania, inferred from simultaneous inversion for hypocenters and 3-D velocity structure, *Annales Geophysicae*, **2**, 23–27.

    1987. On the stress tensor in the Vrancea region, *J. Geophys.*, **62**, 62–65.

    1989. Investigation of a high stress drop earthquake on August 30, 1986 in the Vrancea region, *Tectonophysics*, **163**, doi:10.1016/0040-1951(89)90116-9, 35–43.

• Oncescu, M. C. and K. P. Bonjer, 1997. A note on the depth recurrence and strain release of large Vrancea earthquakes, *Tectonophysics*, **272**, doi:10.1016/0040-1951(96)00263-6, 291–302.

Oncescu, M. C., V. I. Marza, M. Rizescu, and M. Popa, 1999. The Romanian earthquake catalogue between 984–1997. In *Vrancea Earthquakes: Tectonics, Hazard and Risk Mitigation*, eds. F. Wenzel, D. Lungu, and O. Novak, The Netherlands, Kluwer, 374 pp., 43–47.

Radu, C. and G. Purcaru, 1964. Considerations upon intermediate earthquake-generating stress systems in Vrancea, *Bull. Seismol. Soc. Amer.*, **54**, 79–85.

Radu, C. and A. Utale, 1992. The Vrancea (Romania) earthquake of October 26, 1802, *Proc. XXIII ESC General Assembly*, Prague, 110–113.

Radulian, M. and C.-I. Trifu, 1991. Would it have been possible to predict the August 30, 1986 Vrancea earthquake? *Bull. Seismol. Soc. Amer.*, **81**, 2498–2503.

Räkers, E. and G. Müller, 1982. The Romanian earthquake of March 4, 1977. III. Improved focal model and moment determination, *J. Geophys.*, **50**, 143–150.

Romanian Academy, 1941. *Comptes Rendus des Séances de l'Academie des Sciences de Roumanie – Numéro Consacré aux Recherches sur le Tremblement de Terre du 10 Novembre 1940 en Roumanie.*

Santo, T., 1970. Regional study on the characteristic seismicity of the world. Part VI. Colombia, Rumania and South Sandwich Islands, *Bull. Earthquake Res. Inst., Tokyo Univ.*, **48**, 1089–1105.

Schuller, G., 1882. Report on the earthquakes occurred in Wallachia in 1838 (in Romanian), *Bull. Soc. Geogr. Rom.*, **III** , Bucharest.

Silver, P. G. and T. H. Jordan, 1983. Total-moment spectra of fourteen large earthquakes, *J. Geophys. Res.*, **88**, 3273–3293.

Ştefănescu, G., 1902. Cutremurele de pămînt în România în timp de 1391 de ani, de la 455 până la 1874, *Analele Acad. Rom., Memorille Sect. Ştiinţific, (II)*, **XXIV**, 1–34, Bucharest.

Trifu, C.-I., 1980. The mechanism of March 4, 1977 Vrancea earthquake inferred from Rayleigh wave analysis, *Boll. Geofisica Teorica Applicata*, **22**, 311–320.

1990. Detailed configuration of intermediate seismicity in the Vrancea region, *Revista de Geofisica*, **46**, 33–40.

Trifu, C.-I. and M. C. Oncescu, 1987. Fault geometry of the August 30, 1986 Vrancea earthquake, *Annales Geophysicae*, **87/06B**, 727–729.

Trifu, C.-I. and M. Radulian, 1991. Frequency–magnitude distribution of earthquakes in Vrancea: relevance for a discrete model, *J. Geophys. Res.*, **96**, 4301–4311.

Trifu, C.-I., M. Radulian, and E. Popescu, 1990. Characteristics of the intermediate depth microseismicity in Vrancea region, *Revista de Geofisica*, **46**, 75–82.

•Wenzel, F., D. Lungu, and O. Novak, eds., 1999. *Vrancea Earthquakes: Tectonics, Hazard and Risk Mitigation*, The Netherlands, Kluwer, 374 pp.

Wenzel, F., U. Achauer, D. Enescu, E. Kissling, R. Russo, V. Mocanu, and G. Musacchio, 1998. Detailed look at final stage of plate break-off is target of study in Romania, *EOS, Trans. Amer. Geophys. Un.*, **79**, 589–600.

Wortel, M. J. R. and W. Spakman, 1992. Structure and dynamics of subducted lithosphere in the Mediterranean region, *Proc. Koninklijke Nederlandse Akad. v. Wetenschappen*, **95**, 325–347.

2000. Subduction and slab detachment in the Mediterranean-Carpathian region, *Science*, **290**, doi:10.1126/science.290.5498.1910, 1910–1917.

10.25 The Middle East

10.25.1 Regional tectonics and deep seismicity

Earthquakes in this region occur in response to convergence between the Arabian and Eurasian plates. However, as befits a zone where two continental plates collide, the deformation is distributed; this provokes some investigators to define various combinations of microplates. For example, McKenzie (1972) states that:

The principle problems [in this region] are the nature of the connection between the activity of Iran with that of Turkey, and the relative motions between the Arabian, Iranian, South Caspian and Eurasian plates. The boundaries of these plates are marked by fault systems rather than single faults, and the concept of plates in this region should be used with some caution since deformation is occurring at present over most of Iran.

I thank Muawia Barazangi and James Jackson for reviewing an earlier draft of this section.

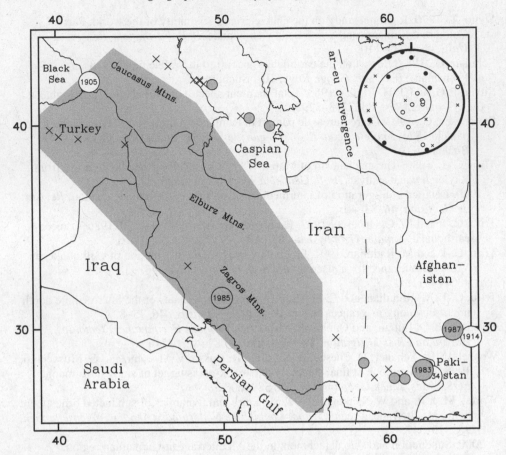

Fig. 10.51 Map of intermediate-depth earthquakes in the Middle East. Filled circles, open circles, and crosses indicate earthquakes in the Harvard CMT, EV Centennial, and EHB catalogs. The shaded area shows where several catalogs report intermediate-depth earthquakes but where recent analysis indicates all seismicity is shallow, including the 1905 and 1985 earthquakes (see text). The equal-area plot summarizes T (solid circles), B (crosses), and P (open circles) axis orientations of selected Harvard CMT for intermediate-depth earthquakes within this region. The dashed line shows the direction of convergence for Arabia–Eurasia (ar–eu) plates; other lines are political boundaries and shorelines.

The literature about seismicity in this region is confusing since various catalogs and numerous publications report deep activity extending from eastern Turkey along the Zagros mountains to the northern boundary of the Persion Gulf (Fig. 10.51). However, recent analysis indicates that these depths almost certainly are inaccurate, and intermediate-depth earthquakes occur only in two regions, beneath southeastern Iran/western Pakistan, and beneath the central Caspian Sea.

Within the Iran–Pakistan border region, known as the Makran, intermediate-depth earthquakes occur because of northward subduction along the coast of the Indian Ocean. Some of these earthquakes are relatively large or have been studied in detail and thus the reported intermediate focal depths are credible. For example, the CMT catalog reports a depth of 157 km and M_W of 6.1 for the earthquake of 10 August 1987. Laane and Chen (1989) modeled the $M_{W(CMT)}$ 6.7 earthquake of 18 April 1983 and found a depth of 65 km. Jackson and McKenzie (1984) constructed synthetic seismograms to demonstrate that the earthquake of 17 November 1972 had a focal depth of 65 km. And both Jackson (1980) and Kadinsky-Cade and Barazangi (1982) confirm a depth of about 105 km for the earthquake of 9 November 1970 in eastern Iran.

There are also some genuine intermediate-depth earthquakes in the central Caspian Sea. Both the EHB and CMT catalogs show activity here with focal depths between 60 km and about 80 km. The largest of these have M_W of 5.3. Priestley *et al.* (1994) and Jackson *et al.* (2002) reviewed these hypocenters in detail and conclude that they arise from "the onset of subduction of the South Caspian Basin beneath the central Caspian, a process that seems to occur aseismically at shallower levels."

Elsewhere within Fig. 10.51, all reported intermediate focal depths are probably spurious. There has been a lively debate about this. For example, Nowroozi (1971; 1972; 1976) asserted that there were intermediate earthquakes beneath the Zagros and west of the Caspian Sea, and Niazi and Basford (1968) reported hypocenters with depths of about 100–130 km in eastern Turkey and southern Iran. However, analysis of both teleseismic and temporary local network data convinced Ambraseys (1978), Asudeh (1983), Niazi *et al.* (1978), Berberian (1979), Ni and Barazangi (1986), Maggi *et al.* (2000), Talebian and Jackson (2004), and Tatar *et al.* (2004) that teleseismic locations in Iran were uniformly poor, and that all reported intermediate depths beneath the Zagros were erroneous. In eastern Turkey, Turkelli *et al.* (2003) operated a 29-station temporary array for a two-year period and found no earthquakes with focal depths exceeding 32 km.

Events reported as deep which are almost certainly shallow include the 27 March 1985 ($h_{CMT} = 84$ km; $M_{W(CMT)} = 5.2$) in the CMT catalog, which Maggi *et al.* (2000) modeled and found to have a focal depth of 10–15 km. Similarly, there have been various reports (e.g., Tskhakaya, 1962; etc.) of earthquakes with focal depths between 60 km and 150 km in the Caucasus and Elburz mountains adjacent to the Caspian Sea. However, Priestley *et al.* (1994) and Jackson *et al.* (2002) evaluate modern waveform data and conclude that activity here is shallow. Finally, in eastern Turkey the EV Centennial catalog follows Båth and Duda (1979) and lists the 21 October 1905 earthquake with a depth of 60 km and a magnitude of 7.5; however, it does not appear in the Abe or GR catalogs and there has been no well-located deep activity there subsequently.

10.25.2 Middle Eastern references

Ambraseys, N. N., 1978. The relocation of epicenters in Iran, *Geophys. J. Roy., Astron. Soc.*, **53**, 117–121.

Asudeh, I., ISC, 1983. Mislocation of earthquakes in Iran and geometric residuals, *Tectonophysics*, **95**, doi:10.1016/0040-1951(83)90259-7, 61–74.

Båth, M. and S. J. Duda, 1979. *Some Aspects of Global Seismicity*, Rept. 1–79, Uppsala, Sweden, Seismological Institute.

Berberian, M., 1979. Evaluation of the instrumental and relocated epicenters of Iranian earthquakes, *Geophys. J. Roy. Astron. Soc.*, **58**, 625–630.

Jackson, J., 1980. Errors in focal depths determination and the depth of seismicity in Iran and Turkey, *Geophys. J. Roy. Astron. Soc.*, **61**, 285–301.

Jackson, J. and D. McKenzie, 1984. Active tectonics of the Alpine–Himalayan belt between western Turkey and Pakistan, *Geophys. J. Roy. Astron. Soc.*, **77**, 185–264.

• Jackson, J., K. Priestley, M. Allen, and M. Berberian, 2002. Active tectonics of the South Caspian Basin, *Geophys. J. Int.*, **148**, doi:10.1046/j.1365-246X.2002.01588.x, 214.

Priestley, K., C. Baker, and J. Jackson, 1994. Implications of earthquake focal mechanism data for the active tectonics of the South Caspian Basin and surrounding regions, *Geophys. J. Int.*, **118**, 111–141.

Kadinsky-Cade, K. and M. Barazangi, 1982. Seismotectonics of southern Iran: the Oman line, *Tectonics*, **1**, 389–412.

Laane, J. L. and W.-P. Chen, 1989. The Makran earthquake of 1983 April 18: a possible analogue to the Puget Sound earthquake of 1965? *Geophys. J. Int.*, **98**, 1–9.

• Maggi, A., J. A. Jackson, K. Priestley, and C. Baker, 2000. A re-assessment of focal depth distributions in southern Iran, Tien Shan and northern India: do earthquakes really occur in the continental mantle? *Geophys. J. Int.*, **143**, doi:10.1046/j.1365-246X.2000.00254.x, 629–661.

McKenzie, D., 1972. Active tectonics of the Mediterranean region, *Geophys. J. Roy. Astron. Soc.*, **30**, 109–185.

Ni, J. and M. Barazangi, 1986. Seismotectonics of the Zagros continental collision zone and a comparison with the Himalayas, *J. Geophys. Res.*, **91**, 8205–8218.

Niazi, M. and J. R. Basford, 1968. Seismicity of the Iranian plateau and Hindu Kush region, *Bull. Seismol. Soc. Amer.*, **58**, 417–426.

Niazi, M., I. Asudeh, G. Ballard, J. Jackson, G. King, and D. McKenzie, 1978. The depth of seismicity in the Kermanshah region of the Zagros mountains, *Earth Planet. Sci. Lett.*, **40**, 270–274.

Nowroozi, A., 1971. Seismo-tectonics of the Persian plateau, eastern Turkey, Caucasus, and Hindu Kush regions, *Bull. Seismol. Soc. Amer.*, **61**, 317–341.

1972. Focal mechanisms of earthquakes in Persia, Turkey, West Pakistan and Afghanistan and plate tectonics of the Middle East, *Bull. Seismol. Soc. Amer.*, **62**, 823–850.

1976. Seismotectonic provinces of Iran, *Bull. Seismol. Soc. Amer.*, **66**, 1249–1276.

Priestley, K., C. Baker, and J. Jackson, 1994. Implications of earthquake focal mechanism data for the active tectonics of the South Caspian Basin and surrounding regions, *Geophys. J. Int*, **118**, 111–141.

Talebian, M. and J. Jackson, 2004. A reappraisal of earthquake focal mechanisms and active shortening in the Zagros mountains of Iran, *Geophys. J. Int.*, **156**, 506, doi:10.1111/j.1365–246X.2004.02092.x.

Tatar, M., D. Hatzfeld, and M. Ghafory-Ashtiany, 2004. Tectonics of the Central Zagros (Iran) deduced from microearthquake seismicity. *Geophys. J. Int.*, **156**, 255–266, doi:10.1111/j.1365–246X.2003.02145.x.

Tskhakaya, A. D., 1962. Depths of Caucasian earthquakes, *Bull. Acad. Sci. USSR, Geophys. Ser. (English trans.)*, **5**, 379–383.

Turkelli, N., E. Sandvol, E. Zor, R. Gok, T. Bekler, A. Al-Lazki, H. Karabulut, S. Kuleli, T. Eken, C. Gurbuz, S. Bayraktutan, D. Seber, and M. Barazangi, 2003. Seismogenic zones in eastern Turkey, *Geophys. Res. Lett.*, **30**, doi:10.1029/2003GL018023.

10.26 Pamir–Hindu Kush

10.26.1 Regional tectonics and deep seismicity

Just as the Tonga–Kermadec arc is the type example of subduction induced by the convergence of two oceanic plates, the collision of India and Eurasia is the corresponding example for the collision of two continental plates. This has produced the world's highest mountains (>8 km elevation) and the thickest crust (~70–80 km). The shallow seismicity in the collision zone is distributed across a broad region rather than concentrated along one or more well-defined faults. And, at intermediate depths the Wadati–Benioff zone doesn't consist of a single plane; rather, the seismicity reported in catalogs appears to form a diffuse mass (Fig. 10.52). However, if the better-recorded hypocenters are carefully relocated, they form distinct clusters and those with depths of 100–200 km occur within an S-shaped region separated by one or perhaps two gaps (Fig. 10.53). The westernmost concentration of activity occurs beneath the Hindu Kush while the easternmost group of hypocenters lies beneath the Pamirs.

Beneath 175 km the activity is most intense beneath the eastern edge of the Hindu Kush zone (Fig. 10.54). There are about 500 hypocenters in the EHB catalog within a cube having dimensions of about 100 km centered on 36.5° N, 70.7° E, 220 km depth; about four earthquakes each year have magnitudes of 4.8 or greater. Relocations indicate that the hypocenters in this "knot" or nest occur within distinct northern and southern groups separated by about 25 km. Since 1908 this nest has produced more than 15 intermediate-depth earthquakes with magnitudes of 7.0 or greater, as listed in the EV Centennial catalog. Among these are a few highly unusual events reported to have numerous aftershocks (see discussion below for the 14 March 1965 and 30 December 1983 earthquakes).

I thank Shamita Das and James Ni for reviewing an earlier draft of this section.

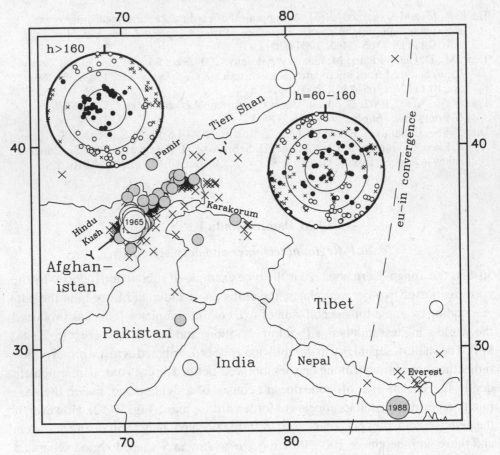

Fig. 10.52 Map of intermediate-depth earthquakes in the Pamir–Hindu Kush region. Filled circles, open circles, and crosses indicate selected earthquakes in the Harvard CMT, EV Centennial, and EHB catalogs (see description of selection process in the introduction to this chapter). The solid line Y–Y′ indicates location and orientation of cross sections, and equal-area plots summarize T (solid circles), B (crosses), and P (open circles) axis orientations of selected Harvard CMT with depths 60–160 km (right) and > 160 km (left). The dashed line shows the direction of Eurasia–India (eu–in) plate convergence.

There is also sporadic intermediate-depth activity reported in several regions outside the Pamir–Hindu Kush zones, especially beneath southernmost Tibet near Mt. Everest, the western Tien Shan, and the Karakorum. All but a handful of these earthquakes have reported depths of 100 km or less.

10.26.2 Literature review

The Pamir–Hindu Kush region has a long history in the deep earthquake literature. After H. H. Turner proposed that some earthquakes had deeper-than-normal

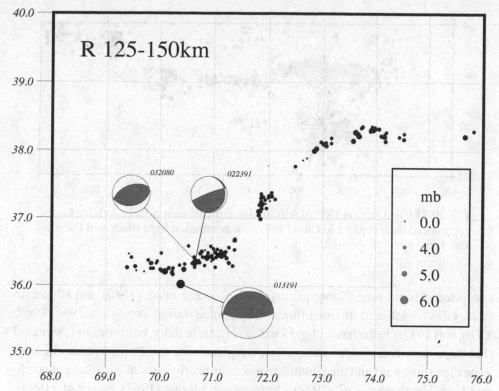

Fig. 10.53 Relocated Pamir–Hindu Kush hypocenters with depths of 125–150 km. Figure reproduced from Pegler and Das (1998) with permission from Blackwell Publishing, Ltd.

focal depths (see Section 3.3.2), Jeffreys (1923) presented data from the shallow-focus 18 February 1911 Pamir earthquake to refute this suggestion. However, after Wadati confirmed the existence of intermediate-depth earthquakes it soon became clear that some occurred beneath the Hindu Kush; there are about 70 in the GR catalog.

Santo (1969) first called attention to the very high rate of seismic activity at a depth of about 200 km beneath the Hindu Kush. He concluded that this "nest" was a cube with dimensions of about 30 km. He also stated that it occurred at the bottom of a "V-shaped pocket" of activity. Nowroozi (1971) published cross sections that showed intermediate-depth seismicity occurring in two thin zones, one striking E–W beneath the Hindu Kush, and the other oriented NE–SW beneath the Pamirs.

An open question concerns whether or not the Pamir and Hindu Kush sides of the "pocket" represent different subduction zones. Billington *et al.* (1977) inter-preted selected and relocated hypocenters and concluded that they formed a single,

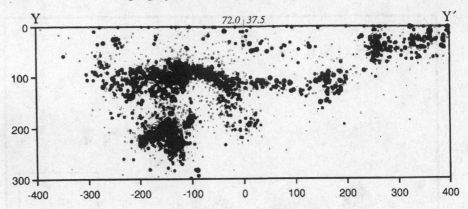

Fig. 10.54 Cross section Y-Y′ of relocated intermediate earthquake activity. Figure
reproduced from Pegler and Das (1998) with permission from Blackwell Publishing, Ltd.

contorted seismic zone. Subsequently, both Roecker *et al.* (1980) and Chatelain
et al. (1980) evaluated microearthquakes recorded during surveys in 1966, 1967,
1976, and 1977. On the basis of gaps in the seismic activity, both studies concluded
that there were two distinct zones of convergence; i.e., the Pamir Wadati–Benioff
zone dips southward and the Hindu Kush zone is nearly vertical with a slight northward dip. Hamburger *et al.* (1992), Burtman and Molnar (1993), Fan *et al.* (1994),
and Khan (2003) have reached similar conclusions.

The most comprehensive study of the Pamir–Hindu Kush seismicity is
that of Pegler and Das (1998), who used phases reported to the ISC and performed
JHD relocations for both shallow and intermediate earthquakes. In this paper and
in Pavlis and Das (2000) they present evidence that seismic activity occurs within a
single, steeply-dipping, S-shaped subducting slab. In the western section, beneath
the Hindu Kush, the slab plunges northward. Then, east of 71.6° E a tear in the
slab produces the gap observed between activity in the Hindu Kush and in the
Pamirs, where the slab has overturned and plunges southeastward. The hypothesis
of oblique motion is consistent with investigations such as Schoenecker *et al.* (1997)
that find that SKS and SKKS phases from intermediate-depth Hindu Kush and
Pamir earthquakes exhibit significant amounts of shear-wave splitting at teleseismic
distances.

The paucity of seismic stations and the unusually thick crust in this region (e.g,
see Zhao *et al.*, 1996) have raised concerns about the quality of reported locations.
Sverdrup *et al.* (1994) relocated earthquakes in northern Pakistan and obtained locations for some that differed by as much as 100 km from the ISC epicenters; however,
the discrepancies were 25 km or less for all earthquakes located with phases from

20 or more stations. Zhao and Helmberger (1991) and Zhu *et al.* (1997) performed waveform modeling on earthquakes in Tibet and the Hindu Kush, and concluded that most earthquakes with reported depths of 30–50 km actually had depths of 25 km or less. Langin *et al.* (2003) compared depths determined from teleseismic recordings to those determined by a temporary local network and concluded that "[intermediate] depths reported for events located solely with teleseismic data are unreliable." Nevertheless, Chen *et al.* (1981), Zhu and Helmberger (1996), and Chen and Yang (2004) demonstrated that some earthquakes in Tibet do have depths of 70–90 km, and also that some of the earthquakes well south of the main Pamir–Hindu Kush cluster have depths of 75–100 km. Beneath the Hindu Kush nest there is some activity reported with depths as great as almost 300 km. And Katok (1988) presents convincing evidence that one earthquake which occurred on 10 June 1971 ($m_b = 4.3$) had a focal depth of 380 km.

Possibly because large intermediate earthquakes are so common in this region, a number of investigators have published compilations of focal mechanisms (Ritsema, 1966; Ram and Yadav, 1984; Verma and Sekhar, 1985; Pegler and Das, 1998; Singh, 2000; Khan, 2003). In the Hindu Kush nest the mechanisms all have approximately vertical T axes; at shallower depths and in the central Pamir zone the mechanisms are highly variable (Fig. 10.52); beneath the Karakorum the only available mechanism is similar to those in the Hindu Kush nest (see Fan and Ni, 1989).

10.26.3 Significant earthquakes

During the twentieth century there were about 15 intermediate earthquakes with magnitudes of 7 or greater in this region. Except the earthquake of 29 July 1985 ($M_{W(CMT)} = 7.4$; $h_{CMT} = 98$ km), all of these had depths near 200 km and locations within the Hindu Kush nest at about 36.5° E–70.5° E. Several of the nest earthquakes generated special interest because they produced unusually well-recorded phases. Lynch (1938) made a thorough study of SKS and SKKS phases from the earthquake of 14 November 1937 ($m_{B(Abe)} = 7.1$; $h_{EVC} = 200$ km). Lehmann (1964) investigated arrival times at European stations from the earthquake of 4 March 1949 ($m_{B(Abe)} = 7.4$; $h_{EVC} = 230$ km). Yoshida (1989) inverted waveforms to determine the spatial and temporal properties of the rupture front for the earthquake of 30 December 1983 ($M_{W(CMT)} = 7.4$; $h_{CMT} = 212$ km).

14 March 1965: $h_{CMT-Hist} = 217$ km; $M_{W(CMT-Hist)} = 7.5$. Both Lukk (1968) and Pavlis and Hamburger (1991) concluded that this earthquake generated a well-defined sequence of at least 20 aftershocks. Such numerous aftershocks had never previously been reported for earthquakes with depths as great as 200 km; it is possible that they were observed here only because the earthquake was quite large and because there were regional stations that detected smaller-magnitude events. Nyffenegger and Frohlich (2000) were unable to demonstrate clearly that it possessed an aftershock sequence.

30 December 1983: $h_{CMT} = 212$ km; $M_{W(CMT)} = 7.4$. Pavlis and Hamburger (1991) concluded that this earthquake had about 80 aftershocks. Nyffenegger and Frohlich (2000) agreed, and identified more than 100; they also calculated a p value (exponent describing the decay of the sequence), for which they found a rather ordinary value of 0.83. Pavlis and Helmberger note that this earthquake, the 1965 earthquake, and most other Hindu Kush events with reported well-developed aftershock sequences, all seem to occur at the same place and have similar focal mechanisms.

20 August 1988: $h_{EV} = 61$ km; $h_{CMT} = 35$ km; $M_W = 6.9$. This earthquake is significant as it killed about 1000 people, caused severe damage along the Nepal–India border (Kumar, 1990; Nandy *et al.*, 1993), and had an epicenter close to that of the M_S 8.3 earthquake of 15 January 1934 that killed about 10,000. Whether it qualifies as deep is uncertain; Harvard obtained a depth of only 35 km and it had numerous aftershocks (Pandey and Nicolas, 1991), but both the ISC and the EV Centennial catalogs report a depth between 60 and 70 km supported by pP-P intervals. However, Holt *et al.* (1991), Zhou *et al.* (1995) and Chen and Kao (1996) all determined depths between 50 and 55 km from analysis of waveforms. Nandy *et al.* (1993) summarize focal mechanisms determined by various investigators; those determined from first motions show more strike-slip motion than that reported by Harvard (see inset).

10.26.4 Pamir–Hindu Kush references

Billington, S., B. L. Isacks, and M. Barazangi, 1977. Spatial distribution and focal mechanisms of mantle earthquakes in the Hindu Kush–Pamir region: a contorted Benioff zone, *Geology*, **5**, 699–704.

Burtman, V. S. and P. Molnar, 1993. Geological and geophysical evidence for deep subduction of continental crust beneath the Pamir, *Geol. Soc. Amer. Spec. Pap.* **281**, 1–75.

Chen, W.-P. and H. Kao, 1996. Seismotectonics of Asia: some recent progress. In *The Tectonic Evolution of Asia*, eds. A. Yin and T. M. Harrison, Cambridge, Cambridge University Press, 37–62.

Chen, W.-P. and Z. Yang, 2004. Earthquakes beneath the Himalayas and Tibet: evidence for strong lithospheric mantle, *Science*, **304**, doi:10.1126/science.1097324, 1949–1952.

Chen, W.-P., J. L. Nabelek, T. J. Fitch, and P. Molnar, 1981. An intermediate depth earthquake beneath Tibet: source characteristics of the event of September 14, 1976, *J. Geophys. Res.*, **86**, 2863–2876.

Chatelain, J. L., S. W. Roecker, D. Hatzfeld, and P. Molnar, 1980. Microearthquake seismicity and fault plane solutions in the Hindu Kush region and their tectonic implications, *J. Geophys. Res.*, **85**, 1365–1387.

Fan, G., and J. F. Ni, 1989. Source parameters of the February 13, 1980, Karakorum earthquake, *Bull. Seismol. Soc. Amer.*, **79**, 945–954.

Fan, G., J. F. Ni, and T. C. Wallace, 1994. Active tectonics of the Pamirs and Karakoram, *J. Geophys. Res.*, **99**, 7131–7160.

Hamburger, M. W., D. R. Sarewitz, T. L. Pavlis, and G. A. Popandopulo, 1992. Structural and seismic evidence for intracontinental subduction in the Peter the First range, central Asia, *Geol. Soc. Amer. Bull.*, **104**, 397–408.

Holt, W. E., J. F. Ni, T. C. Wallace, and A. J. Haines, 1991. The active tectonics of the eastern Himalayan syntaxis and surrounding regions, *J. Geophys. Res.*, **96**, 14595–14632.

Jeffreys, H., 1923. The Pamir earthquake of 1911 February 18, in relation to the depths of earthquake foci, *Month. Not. Roy. Astron. Soc., Geophys. Supp.*, **1**, 22–31.

Katok, A. P., 1988. On the deepest earthquake in the Pamir–Hindukush zone, *Izv., Earth Phys.*, **24**, 649–653.

Khan, P. K., 2003. Stress state, seismicity and subduction geometries of the descending lithosphere below the Hindukush and Pamir, *Gondwana Res.*, **6**, 867–877.

Kumar, B., 1990. Preliminary isoseismal map of Bihar–Nepal earthquake of August 21, 1988, *Bull. Indian Soc. Earthquake Tech.*, **27**, 59–63.

Langin, W. R., L. D. Brown, and E. A. Sandvol, 2003. Seismicity of central Tibet from project INDEPTH III seismic recordings, *Bull. Seismol. Soc. Amer.*, **93**, 2146–2159.

Lehmann, I., 1964. The Hindu Kush earthquake of March 4, 1949 as recorded in Europe, *Bull. Seismol. Soc. Amer.*, **54**, 1915–1925.

Lukk, A. A., 1968. The aftershock sequence of the Dzhurm deep-focus earthquake of 14 March 1965, *Izv., Earth Phys.*, **4**, 83–85.

Lynch, J., 1938. The earthquake of November 14, 1937, *Bull. Seismol. Soc. Amer.*, **28**, 177–189.

Nandy, D. R., A. K. Choudhury, C. Chakraborty, and P. L. Narula, 1993. *Bihar-Nepal Earthquake, August 20, 1988*, Geological Survey of India Special Publication 31, 104 pp.

Nowroozi, A. A., 1971. Seismotectonics of the Pakistan plateau, eastern Turkey, Caucasus, and Hindu Kush regions, *Bull. Seismol. Soc. Amer.*, **61**, 317–341.

Nyffenegger, P. and C. Frohlich, 2000. Aftershock occurrence rate decay properties for intermediate and deep earthquake sequences, *Geophys. Res. Lett.*, **27**, doi:10.1029/1998GL010371, 1215–1218.

Pandey, M. R. and M. Nicolas, 1991. The aftershock sequence of the Udayapur earthquake of August 20, 1988, *J. Geol. Soc. Nepal*, **7**, 19–29.

Pavlis, G. L. and S. Das, 2000. The Pamir–Hindu Kush seismic zone as a strain marker for flow in the upper mantle, *Tectonics*, **19**, doi:10.1029/1999TC900062, 103–115.

•Pavlis, G. L. and M. W. Hamburger, 1991. Aftershock sequences of intermediate-depth earthquakes in the Pamir–Hindu Kush seismic zone, *J. Geophys. Res.*, **96**, 18107–18117.

•Pegler, G. and S. Das, 1998. An enhanced image of the Pamir–Hindu Kush seismic zone from relocated earthquake hypocentres, *Geophys. J. Int.*, **134**, doi:10.1046/j.1365-246x.1998.00582x, 573–595.

Ram, A. and L. Yadav, 1984. Focal-mechanism solutions of earthquakes and tectonics of the Hindukush region, *Tectonophysics*, **104**, doi:10.1016/0040-1951(84)90103-3, 85–97.

Ritsema, A. R., 1966. The fault-plane solutions of earthquakes of the Hindu Kush centre, *Tectonophysics*, **3**, doi:10.1016/0040-1951(66)90017-5, 147–163.

Roecker, S. W., O. V. Soboleva, I. L. Nersesov, A. A. Lukk, D. Hatzfeld, J. L. Chatelain, and P. Molnar, 1980. Seismicity and fault plane solutions of intermediate depth earthquakes in the Pamir–Hindu Kush region, *J. Geophys. Res.*, **85**, 1358–1364.

Santo, T., 1969. Regional study on the characteristic seismicity of the world. Part I. Hindu Kush region, *Bull. Earthquake Res. Inst., Tokyo Univ.*, **47**, 1035–1048.

Schoenecker, S. C., R. M. Russo, and P. G. Silver, 1997. Source-side splitting of S waves from Hindu Kush–Pamir earthquakes, *Tectonophysics*, **279**, doi:10.1016/S0040-1951(97)00130-3, 149–159.

Singh, D. D., 2000. Seismotectonics of the Himalaya and its vicinity from centroid-moment tensor (CMT) solution of earthquakes, *J. Geodynamics*, **30**, doi:10.1016/S0264-3707(00)00007-7, 507–537.

Sverdrup, K. A., G. J. Schurter, and V. S. Cronin, 1994. Relocation analysis of earthquakes near Nanga Parbat-Haramosh massif, northwest Himalaya, Pakistan, *Geophys. Res. Lett.*, **21**, doi:10.1029/94GL01935, 2331–2334.

Verma, R. K. and C. C. Sekhar, 1985. Seismotectonics and focal mechanisms of earthquakes from Pamir–Hindukush regions, *Tectonophysics*, **112**, doi:10.1016/0040-1951(85)90184-2, 297–324.

Yoshida, S., 1989. Waveform inversion using ABIC for the rupture process of the 1983 Hindu Kush earthquake, *Phys. Earth Planet. Int.*, **56**, 389–405.

Zhao, L.-S., and D. V. Helmberger, 1991. Geophysical implications from relocations of Tibetan earthquakes; hot lithosphere, *Geophys. Res. Lett.*, **18**, 2205–2208.

Zhao, L.-S., M. K. Sen, P. Stoffa, and C. Frohlich, 1996. Application of very fast simulated annealing to the determination of the crustal structure beneath Tibet, *Geophys. J. Int.*, **125**, 355–370.

Zhou, R., F. Tajima, and P. L. Stoffa, 1995. Earthquake source parameter determination using genetic algorithms, *Geophys. Res. Lett.*, **22**, doi:10.1029/94GL03345, 517–520.

Zhu, L. and D. V. Helmberger, 1996. Intermediate depth earthquakes beneath the India–Tibet collision zone, *Geophys. Res. Lett.*, **23**, doi:10.1029/96GL00385, 435–438.

Zhu, L., D. V. Helmberger, C. K. Saikia, and B. B. Woods, 1997. Regional waveform calibration in the Pamir–Hindu Kush region, *J. Geophys. Res.*, **102**, doi:10.1029/97JB01855, 22799–22813.

10.27 Burma

10.27.1 Regional tectonics and deep seismicity

Burma lies east of India, roughly along the northern extension of the Andaman arc. Oblique motion of about 5 cm/yr between the Indian and Eurasian plates dominates the tectonics and seismic activity occurs over a broad zone several hundred km wide (Fig. 10.55). Towards the western edge of this activity there are thrust faults that apparently connect to an eastward-dipping Wadati–Benioff zone that extends to about 180 km depth (Fig. 10.56); at the eastern edge there is a transcurrent fault; thus some investigators call the region between the Burma plate or platelet.

Regardless of how one chooses to define the plate boundaries, the crust is every-where continental and thus the intermediate earthquakes here are sometimes com-pared to those in Romania and the Hindu Kush. A novel feature of the region is the partitioning of oblique motion between the Indian and Eurasian plates. Although the overall motion is NNE and nearly parallel to the strike of the Wadati–Benioff zone, the majority of the intermediate quakes have east-trending downdip T axes and along-strike P axes; the shallow earthquakes have predominantly strike-slip mechanisms. The ISC reports about a dozen hypocenters with focal depths exceed-ing 200 km. However, an inspection of the reported arrival times suggests that none have very reliable depths.

10.27.2 Literature review

There are intermediate-depth earthquakes beneath Burma in both the GR and Rothé catalogs. And both GR and Santo (1969) compare this seismicity to that beneath the Hindu Kush, but note that the Burma earthquakes seem less likely to repeat at a single focus.

A peculiar feature of the subsequent literature is the remarkable preponderance of papers organized around the interpretation of focal mechanisms (e.g. Fitch, 1970; Chandra, 1975; Chen and Molnar, 1990; Kumar and Rao, 1995; Satya-bala, 1998; 2003). One of the clearest discussions of regional tectonics is Le Dain *et al.* (1984), who augmented focal mechanism information with Landsat images. In earlier papers there is some disagreement about the orientation of the Wadati–Benioff zone. Chandra (1975) and Verma *et al.* (1976) state that it is V-shaped,

I thank James Ni for reviewing an earlier draft of this section.

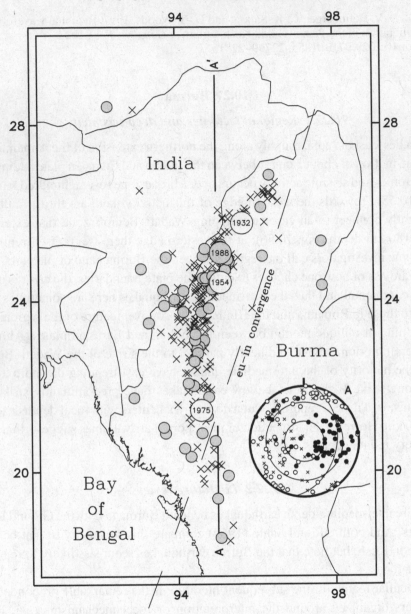

Fig. 10.55 Map of intermediate-depth earthquakes in Burma. Filled circles, open circles, and crosses indicate selected earthquakes in the Harvard CMT, EV Centennial, and EHB catalogs (see description of selection process in the introduction to this chapter). The solid line A–A' indicates the location and orientation of cross sections, and the equal-area plots summarize T (solid circles), B (crosses), and P (open circles) axis orientations of selected Harvard CMT. Other lines are political boundaries and shorelines; the dashed line shows the direction of Eurasian–Indian (eu–in) convergence.

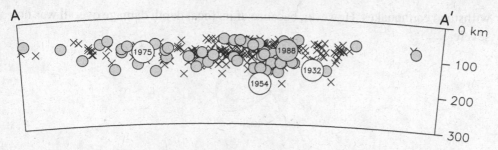

Fig. 10.56 Cross section A–A' of intermediate-depth earthquakes in Burma. Earthquake symbols are as in Fig.10.55 .

while Gupta and Bhatia (1986) disagree and describe it as "near vertical." However, Mukhopadhyay and Dasgupta (1988), Ni *et al.* (1989), and all subsequent investigators describe it as eastward dipping with a plunge of 30° –50° . Guzmán-Speziale and Ni (1996), Rai *et al.* (1996), and Dasgupta *et al.* (2003) plot cross sections of selected ISC hypocenters. Rai *et al.* also present travel time residuals for P phases from earthquakes at 124 and 146 km depth and conclude that the existence of negative residuals of 1–2 sec implies that the lithosphere must penetrate deeper than the deepest well-determined hypocenters, which occur at 160 km depth.

Kao and Rau (1999) infer that there should be a double seismic zone in Burma because it satisfies three conditions: oblique subduction, near the terminus of a subduction zone, and apparent lateral compression. However, the present location accuracy for Burma earthquakes is inadequate to test this.

10.27.3 Significant earthquakes

Only five intermediate earthquakes in this region have reported magnitudes of 7.0 or greater. These occurred on 9 February 1908 ($h_{Båth\&Duda} = 60\,km$, $M_{Båth\&Duda} = 7.3$), 14 August 1932 ($h_{EVC} = 144\,km$, $m_{B(Abe)} = 7.0$), 21 March 1954 (see below), 8 July 1975 ($h_{CMT-Hist} = 96\,km$, $M_{W(CMT-Hist)} = 7.0$) and 6 August 1988 ($h_{CMT} = 101\,km$, $M_{W(CMT)} = 7.3$).

21 March 1954: $h_{EVC} = 186\,km$; $m_{B(Abe)} = 7.4$. Rothé describes this as one of most violent earthquakes felt in India since 1898. Tandon and Mukherjee (1956) used first motions to determine a focal mechanism (see inset) and published an intensity map, noting that the radius of the macroseismic area was about 870 km. Its intensity reached MMI VI within a several-hundred-km region along the Burma–India border, where it reportedly damaged buildings designed to withstand earthquakes. However, because of its focal depth damage overall was not severe.

10.27.4 Burma references

Chandra, U., 1975. Seismicity, earthquake mechanisms and tectonics of Burma, 20° N–28° N, *Geophys. J. Roy. Astron. Soc.*, **40**, 367–381.

Chen, W. P. and P. Molnar, 1990. Source parameters of earthquakes and intraplate deformation beneath the Shillong plateau and the northern Indoburman ranges, *J. Geophys. Res.*, **95**, 12527–12552.

Dasgupta, S., M. Mukhopadhay, A. Bhattacharya, and T. K. Jana, 2003. The geometry of the Burmese–Andaman subducting lithosphere, *J. Seismology*, **7**, doi:10.1023/A:1023520105384, 155–174.

Fitch, T. J., 1970. Earthquake mechanisms in the Himalayan, Burmese, and Andaman regions and continental tectonics in central Asia, *J. Geophys. Res.*, **75**, 2699–2709.

Gupta, H. K. and S. C. Bhatia, 1986. Seismicity in the vicinity of the India–Burma border: evidence for a sinking lithosphere, *J. Geodynamics*, **5**, doi:10.1016/0264-3707(86)90016-5, 375–381.

• Guzmán-Speziale, M. and J. Ni, 1996. Seismicity and active tectonics of the western Sunda arc. In *The Tectonic Evolution of Asia*, eds. by A. Yin and T. M. Harrison, Cambridge, Cambridge University Press, 63–84.

Kao, H. and R.-J. Rau, 1999. Detailed structures of the subducted Philippine Sea plate beneath northeast Taiwan: a new type of double seismic zone, *J. Geophys. Res.*, **104**, doi:10.1029/1998JB/900010, 1015–1033.

Kumar, M. R. and N. P. Rao, 1995. Significant trends related to the slab seismicity and tectonics in the Burmese arc from Harvard CMT solutions, *Phys. Earth Planet. Int.*, **90**, doi:10.1016/0031-9201(94)03012-8, 75–80.

Le Dain, A. Y., P. Tapponier, and P. Molnar, 1984. Active faulting and tectonics of Burma and surrounding regions, *J. Geophys. Res.*, **89**, 453–472.

Mukhopadhyay, M. and S. Dasgupta, 1988. Deep structure and tectonics of the Burmese arc: constraints from earthquake and gravity data, *Tectonophysics*, **149**, doi:10.1016/0040-1951(88)90180-1, 299–322.

Ni, J. F., M. Guzmán-Speziale, M. Bevis, W. E. Holt, T. C. Wallace, and W. R. Seager, 1989. Accretionary tectonics of Burma and the three dimensional geometry of the Burma subduction, *Geology*, **17**, 68–71.

Rai, S. S., D. Srinagesh, and P. V. S. S. R. Sarma, 1996. Morphology of the subducted Indian plate in the Indo-Burmese convergence zone, *Proc., Earth Planet. Sci., Indian Acad.*, **105**, 441–450.

Santo, T., 1969. Regional study on the characteristic seismicity of the world. Part II. From Burma down to Java, *Bull. Earthquake Res. Inst., Tokyo Univ.*, **47**, 1049–1061.

Satyabala, S. P., 1998. Subduction in the Indo-Burma region: is it still active? *Geophys. Res. Lett.*, **25**, doi:10.1029/98GL02256, 3189–3192.

2003. Oblique plate convergence in the Indo-Burma (Myanmar) subduction region, *Pure Appl. Geophys.*, **160**, doi:10.1007/s00024-003-2378-O, 1611–1650.

Tandon, A. N. and S. M. Mukherjee, 1956. The Manipur–Burma border earthquake of 22 March 1954, *Indian J. Meteorol. Geophys.*, **7**, 27–36.

Verma, R. K., M. Mukhopadhyay, and M. S. Ahluwalia, 1976. Earthquake mechanisms and tectonic features of northern Burma, *Tectonophysics*, **32**, doi:10.1016/0040-1951(76)90070-6, 387–199.

10.28 Other

10.28.1 Discussion

Do intermediate- or deep-focus earthquakes occur outside of the 27 terrestrial regions described earlier in this chapter? The following statements are true:

- There are only three among the ~5600 deep earthquakes in the Harvard CMT catalog 1977–2004 (Fig. 10.57). All three occur near the boundary of one of the 27 regions, and for all three the corresponding EHB location lies within the region and is 50 km or more distant from the CMT location.

- There are only 13 among the 38,757 deep earthquakes in the EHB catalog 1964–2004. Of these, the catalog designates 10 as solution type LEQ, indicating their depths are unreliable. Two of the remaining events occurred in eastern China in 1976 and were aftershocks of the Tangshan earthquake of 28 July 1976 which had magnitude M_S 7.8 and reportedly killed 242,000 people. However, Shedlock *et al.* (1987) reread regional phase arrivals and relocated hypocenters for this aftershock sequence and concluded that maximum depths were about 20 km. The third outlier occurred on 19 October 2002 on the Cocos Ridge and had a reported depth of 72 km and m_b of 4.6. The closest reporting station was at a distance of 5.8° and the vast majority of stations recording it were in the northern quadrant; the ISC fixed the depth at 10 km, and in this case the EHB depth is not convincing.[21]

I thus hypothesize that the answer is "No." Or, if such earthquakes exist they are small and/or rare.

Let's now consider the evidence that the answer is "Yes," i.e. that earthquakes with depths of 60 km or greater do occur outside of the 27 regions. First, in the GR, Rothé, Abe, and EV Centennial catalogs for the period 1897–1963 there are 65 such earthquakes reported, including 28 with focal depths of exactly 60 km and 37 deeper events. Of these 37, 16 occur near the boundaries of geographic regions, 17 occur

---

I thank Ray Willemann for reviewing an earlier draft of this section.

[21] This is in no way a criticism of the EHB catalog. Although the EHB group continues to revise and improve the EHB catalog they don't have sufficient resources to inspect each of the more than 100,000 events in the 2004 version of the catalog.

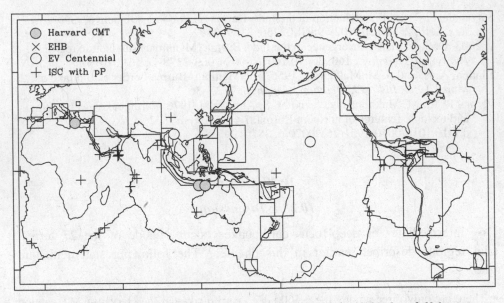

Fig. 10.57 Hypocenters in four global catalogs having focal depths of 60 km or more and reported epicenters outside of the 27 geographical regions, discussed in Chapter 10. The three mapped Harvard CMT hypocenters all lie near the boundary of one of the 27 geographical regions, and for all three the corresponding EHB hypocenters lie inside the region. For the EHB catalog this map shows three hypocenters designated as having reliable focal depths (i.e., EHB categories DEQ, HEQ, and FEQ). These all probably have inaccurate depths (see text). Similarly, none of the mapped hypocenters in the EV Centennial or the ISC pP catalog have convincing intermediate- or deep-focus depths.

close to oceanic ridge-transform systems, and the remaining four are on continents or in plate interiors. To construct the EV Centennial catalog, Engdahl and Villaseñor (2002) reviewed the GR, Rothé, Abe, and other catalogs, relocating many pre-1964 earthquakes with modern methods and selecting more reliable locations for others when several locations were available. Only five locations outside the 27 regions survived this process (Fig. 10.57) and all occurred prior to 1951. Of these five, three were on oceanic ridge-transforms, one is slightly east of the Burma subregion, and one is the 21 August 1951 Hawaii earthquake. The complete absence of similar events in the EHB and Harvard catalogs for 1964–2004, a period when the global seismic network was decidedly superior to that in the previous era, suggests that either the depths or epicenters are inaccurate for all or nearly all of these events.

Next, let's consider earthquakes in the ISC catalog. In the ISC catalog, depths determined from (pP-P) intervals are likely to be more accurate; there are 14,247 ISC earthquakes with depths of 60 km or more constrained by (pP-P) for the period 1964–2001. Only 31 of these earthquakes lie outside our 27 geographic regions, and of these 23 are situated along spreading centers or oceanic transforms. Three more occur beneath continents in western Australia, central Africa, and Asia; of these, the Australia event is the Calingiri earthquake of 10 March 1970, which produced a surface fracture 3 km long with vertical offsets as great as 30 cm (Gordon and Lewis, 1980). Two outliers occur in the Bangladesh–Bay of Bengal region; of these, the 12 June 1989 earthquake with M_W 5.8 produced damage (Karim, 1995) and the CMT catalog reports a depth of 15 km. Finally, three occur in the interior of oceanic plates, one beneath Hawaii, one in the southwest Pacific, and one in the Indian Ocean. Of these, the CMT catalog reports a depth of 5 km for the M_W 5.8 southwest Pacific earthquake of 23 May 1982.

The possibility that there are deep earthquakes beneath Hawaii is intriguing because there is vigorously active volcanism and because a local seismic network has monitored seismicity beneath Kilauea volcano for more than 40 years. The resulting catalog includes about 55 hypocenters with depths 60–95 km as well as huge numbers of hypocenters down to about 50 km depth. However, whether the greater depths are accurate is questionable. Klein *et al.* (1987) state that "none are located reliably beneath 60 km depth"; and Wolfe *et al.* (2003) recently applied cross-correlation methods to waveforms collected between 1988 and 1998 and reported none exceeding 60 km.

In summary, the evidence for a depth exceeding 60 km isn't compelling for any of the earthquakes discussed in this section. Indeed, after more than a century of monitoring, the complete absence of reliable deep locations outside of the 27 geographic regions suggests that deep earthquakes elsewhere must be both small and very rare, if they occur at all.[22] Although there is limited evidence from north Africa and central Asia that some intermediate-depth earthquakes aren't associated with subduction (Chen and Molnar, 1983; Chen, 1988; Chen and Yang, 2004), reports of such events are subject to debate, as multiple events, foreshocks, or noise may cause key phase arrivals to be misidentified and produce excessive focal depth as an artifact. Moreover, for deep-focus earthquakes the existence of very large events at all depths down to 650 km argues against the occurrence elsewhere of events too small to detect. Finally, while the mechanism of deep earthquakes remains in dispute, all plausible mechanisms require conditions (temperatures, metastable phases, pore fluids) that would persist only in lithosphere subducted within the last several tens of millions of years. This is significant since global plate reconstructions generally don't indicate the presence of such recently subducted lithosphere outside of the 27 regions discussed here.

---

[22] I have a wager with Scott Davis concerning whether any deep earthquakes whatsoever will occur outside of the 27 geographic regions. Prior to 2015, if there is one or more such well-located earthquake having a believable focal depth exceeding 60 km, I owe his family a fine dinner, prepared in the cuisine (possibly seafood) most characteristic of the region where the largest such event occurred.

10.28.2 References

Chen, W.-P., 1988. A brief update on the focal depths of intracontinental earthquakes and their correlations with heat flow and tectonic age, *Seismol. Res. Lett.*, **59**, 263–272.

Chen, W.-P., and P. Molnar, 1983. Focal depths of intracontinental and intraplate earthquakes and their implications for the thermal and mechanical properties of the lithosphere, *J. Geophys. Res.*, **88**, 4183–4214.

Chen, W.-P. and Z. Yang, 2004. Earthquakes beneath the Himalayas and Tibet: evidence for strong lithopheric mantle, *Science*, **304**, doi:10.1126/science.1097324, 1949–1952.

Engdahl, E. R. and A. Villaseñor, 2002. Global seismicity: 1900–1999. In *International Handbook of Earthquake and Engineering Seismology*, San Diego, CA, Academic Press for International Association of Seismology and Physics of the Earth's Interior, 665–690.

Gordon, F. R. and J. D. Lewis, 1980. The *Meckering and Calingiri Earthquakes, October 1968 and March 1970*, Perth, Australia, Geological Survey of Western Australia, Bull. 126.

Karim, N., 1995. Disasters in Bangladesh, *Natural Hazards*, **11**, 247–258.

Klein, F. W., R. Y. Koyanagi, J. S. Nakata, and W. R. Tanigawa, 1987. The seismicity of Kilauea's magma system. In *Volcanism in Hawaii*, eds. R. W. Decker, T. L. Wright, and P. H. Stauffer, Reston, VA, U.S. Geological Survey Professional Paper 1350, 1019–1185.

Shedlock, K. M., J. Baranowski, X. Weiwen, and H. X. Liang, 1987. The Tangshan aftershock sequence, *J. Geophys. Res.*, **92**, 2791–2803.

Wolfe, C. J., P. G. Okubo, and P. M. Shearer, 2003. Mantle fault zone beneath Kilauea Volcano, Hawaii, *Science*, **300**, doi:10.1126/science.1082205, 478–480.

10.29 Deep Moonquakes

10.29.1 Results from the Apollo missions

Between 1969 and 1972 the Apollo missions established a four-station seismic network on the Moon; between 1969 and 1977 it recorded more than 12,000 natural seismic events large enough to be cataloged (Nakamura *et al.*, 1982). Recently, Nakamura (2003), and Bulow *et al.* (2005) have reanalyzed these data utilizing modern computational methods. Nakamura reports that 14% of the catalog events were meteoroid impacts, 0.2% (28 events) were shallow moonquakes occurring within the Moon's crust and upper mantle; and 58% were deep moonquakes, occurring within the lunar mantle at depths of 700–1200 km (Fig. 10.58). The remainder were of uncertain origin, although many are likely to be deep moonquakes as well. The deep moonquake data provide essential information for constraining current models of the structure of the lunar interior (e.g., see Nakamura *et al.*, 1982; Khan *et al.*, 2000; Vinnik *et al.*, 2001; Nakamura, 2005, Khan and Mosegaard, 2001, 2002).

I thank Yosio Nakamura for reviewing an earlier draft of this section.

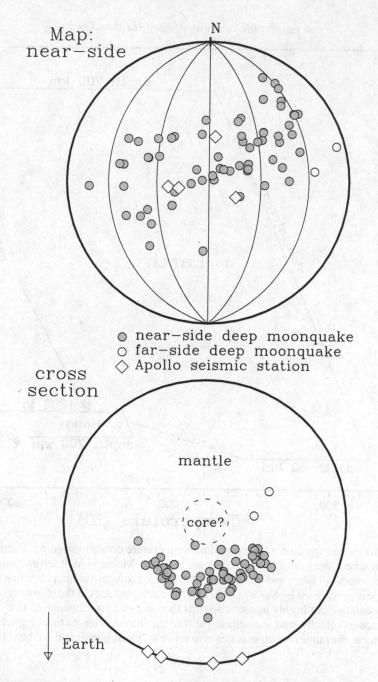

Map:
near-side

N

● near—side deep moonquake
○ far—side deep moonquake
◇ Apollo seismic station

cross
section

mantle

core?

Earth

Fig. 10.58 Locations of better-located deep moonquake source regions. (Top) Map of Moon's near side showing source region epicenters (circles) and Apollo seismic stations (diamonds). Note that known source regions concentrate in the Moon's middle latitudes. Locations are from Nakamura (2005). (Bottom) Equatorial cross section. Note that all but two better-located source regions are on the Moon's near side. The near-absence of far-side deep hypocenters may be real or may occur if scattering or attenuation in the deep lunar interior prevent far-side seismic signals from reaching the seismic stations.

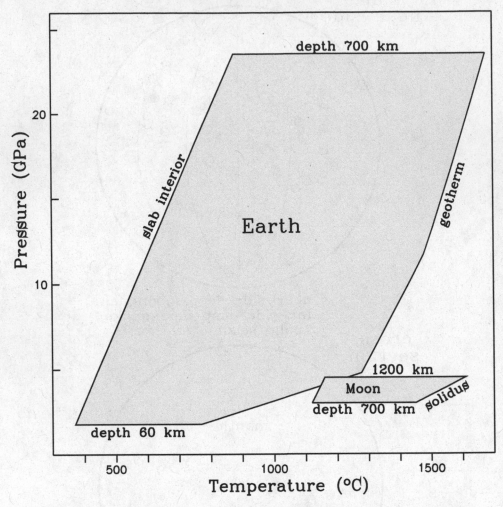

Fig. 10.59 Comparison of pressure and temperature conditions on the Earth and Moon where deep quakes occur. Pressure on the Moon is as inferred from the density model of Khan and Mosegaard (2001). For Earth, the temperature in normal mantle is as reported by Stacey (1992). Plotted temperatures in the slab interior are representative but highly approximate, as they depend on lithospheric thickness, subduction velocity, and slab dip angle (see the introduction to this chapter). On the Moon, the range of temperatures is as reported by Solomon and Toksoz (1973).

Mechanically, how similar are the conditions on the Earth and Moon where deep quakes occur? Estimates of pressure and temperature at depths of 700–1200 km in the Moon correspond roughly to those at depths of 100–150 km within normal mantle on the Earth (Fig. 10.59). However, the lunar temperatures are considerably warmer than within the cores of subducting slabs, where most deep earthquakes occur. It is unclear whether or not convection presently occurs within the lunar mantle (Schubert *et al.*, 2001); however, lunar surface geomorphology shows that plate tectonic processes such as subduction aren't active on the Moon. Instead, the occurrence times of deep moonquakes correlate strongly with the lunar monthly tidal cycle (Lammlein *et al.*, 1974; Toksoz *et al.*, 1977; Cheng and Toksoz, 1978). Tidal stresses are proportional to AM/R^3 (Melchior, 1983) where A is the surface area of the body, and R and M are the distance to and mass of the tide-inducing body. If we compare tidal stresses on the Earth and Moon the distance R is the same, but the ratio $(A_{Moon}M_{Earth})/(A_{Earth}M_{Moon})$ is 6.05, suggesting that monthly tidal shear stresses within the Earth should be about six times smaller than within the Moon, where they reach approximately 0.5 bars.

A peculiar feature of the deep moonquakes is that most appear to come from more than 160 known foci. While all but a few of these foci are on the Moon's near side (Fig. 10.58), Nakamura (2005) concluded that many unlocatable signals could originate from deep moonquakes on the far side. Thus it is unclear whether the near-absence of foci on the far side is real or whether deep moonquakes occur there but at locations for which much of the Apollo seismic network lies within a shadow zone. At each focus waveforms of different events are very similar; typically pairs have correlation coefficients of 0.50 or higher for a duration of a minute or more following P and S wave arrivals. This suggests that their focal regions are close together – perhaps within km. For some pairs the correlation coefficient is negative (Nakamura, 1978), and there are variations in the amplitude ratios of P waves and the polarizations of S waves (Koyama and Nakamura, 1980). All these observations suggest that tidally driven, repeated slip in different directions on a planar fault may generate these quakes.

Quantitative comparison of deep moonquakes and earthquakes is complicated because in some essential ways lunar seismograms are highly unlike terrestrial seismograms (Fig. 10.60). The lunar seismograms do not exhibit large-amplitude, impulsive phases that might indicate distinct changes in lunar structure; rather, the records are dominated by long trains of energy with slowly varying amplitude, probably caused by intense scattering within the lunar interior.[23] However, attenuation within the lunar crust and upper mantle appears to be extraordinarily low, with reported values for Q of P and S in the crust and upper mantle being between 3000 and 4800 (Nakamura and Koyama, 1982). For deep moonquakes the determination of focal depth depends on determining S–P intervals and the arrival times of P or S at at least three stations. The intense scattering prevents the identification of surface-reflected phases such as pP and sP. And, even for the better-recorded events, when different investigators determine arrival times for P and S there are

---

[23] Indeed, Nakamura (1976) pointed out that one can quite successfully model some lunar seismograms from shallow sources by solving the diffusion equation rather than the wave equation.

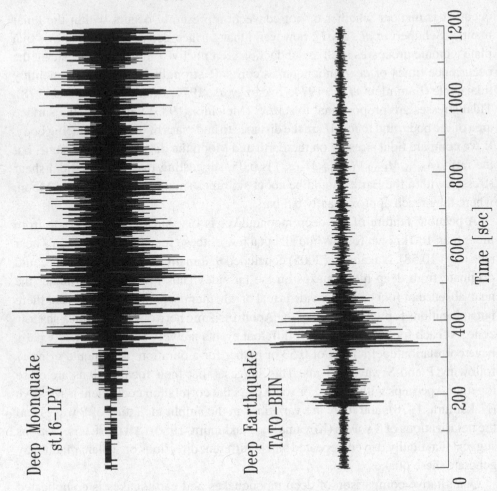

Fig. 10.60 Deep moonquake and deep earthquake seismograms, plotted with probable S arrivals aligned. (Top) Horizontal-component displacement seismogram for a deep moonquake occurring on 5 June 1973 at lunar station 16, at a source–station epicentral distance of approximately 1600 km. This was among the largest deep moonquakes occurring between 1969 and 1977 in the A1 source region, which had a focal depth estimated to be 870 km. The absence of an impulsive P phase and the long ringing coda are typical of moonquake seismograms. (Bottom) Horizontal-component displacement seismogram for a deep earthquake occurring in the Phillippines on 23 May 1998 and recorded at station TATO at a source–station epicentral distance of 1878 km. Harvard reported M_W of 6.0 and a focal depth of 629 km for this quake. Here the signal has been narrow-band filtered so that the instrument responses are similar for the two records shown.

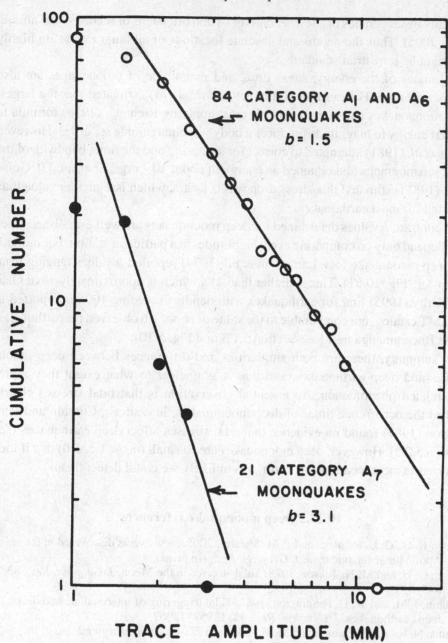

Fig. 10.61 Trace amplitudes and cumulative numbers of deep moonquakes from two foci. The b value is 1.5 for events in the most prolific deep moonquake focus A1/A6 (open circles); b is 3.1 for the A7 focus. Figure reproduced from Lammlein *et al.* (1974).

typically discrepancies of a few seconds or so and often tens of seconds (Nakamura, 1983; 2005). Thus the depths and absolute locations of all lunar events are highly uncertain by terrestrial standards.[24]

Estimates of the energy, stress drop, and magnitude of moonquakes are also highly uncertain. For example, Lammlein *et al.* (1974) estimated that the largest deep moonquakes had energies of 10^9 ergs; applying Richter's (1958) formula to convert energy to magnitude produces a body wave magnitude m_b of 1.3. However, Goins *et al.* (1981) attempted to correct for scattering and the finite bandwith of the lunar seismographs and obtained an energy of about 10^{13} ergs, or m_b of 3.0. Goins *et al.* (1981) estimated that stress drop was 0.1 bars, which is a smaller value than reported for most earthquakes.

In contrast, *b* values determined for deep moonquakes are well established since they depend only on comparing event amplitudes at a particular station. For individual deep moonquake foci, Lammlein *et al.* (1974) reported *b* values ranging from 1.5 to 3.1 (Fig. 10.61). This is higher than 1.0, which is approximately what Okal and Kirby (1995) find for earthquakes with depths exceeding 100 km reported in the CMT catalog, but comparable to the value of about 1.6 observed for earthquakes in the Bucaramanga nest (see Section 5.1.3 and Fig. 5.10).

In summary, there are both similarities and differences between deep moonquakes and deep earthquakes, and thus it is unclear to what extent they represent related phenomenon. An essential observation is that tidal stresses clearly control the occurrence times of deep moonquakes; in contrast, Curchin and Pennington (1987) found no evidence that tidal stresses affect deep earthquakes (see Section 5.2.5). However, deep moonquakes are so small ($m_b < 1.5$–3.0) that if they occurred in most locations on Earth, it is unlikely we could detect them.

10.29.2 Deep moonquake references

Bulow, R. C., C. L. Johnson, and P. M. Shearer, 2005. New events discovered in the Apollo lunar seismic data, *J. Geophys. Res.*, (in press).

Cheng, C. H. and M. N. Toksoz, 1978. Tidal stresses in the Moon, *J. Geophys. Res.*, **83**, 845–853.

Curchin, J. M. and W. D. Pennington, 1987. Tidal triggering of intermediate and deep focus earthquakes, *J. Geophys. Res.*, **92**, 13957–13967.

• Goins, N. R., A. M. Dainty, and M. N. Toksoz, 1981. Seismic energy release of the Moon, *J. Geophys. Res.*, **86**, 378–388.

Khan, A. and K. Mosegaard, 2001. New information on the deep lunar interior from an inversion of lunar free oscillation periods, *Geophys. Res. Lett.*, **28**, doi:10.129/2000GL102445, 1791–1794.

---

[24] However, in at least two ways the lunar data are superior to data collected on Earth. The amplitude of background noise on the Moon is about 1000 times smaller than on Earth in real time, so detection of smaller seismic events is possible. And the relative timing between lunar stations is superior because the data were transmitted by radio to a recording site on Earth, and thus the relative timing uncertainties are less than a millesecond.

2002. An inquiry into the lunar interior: a nonlinear inversion of the Apollo lunar seismic data, *J. Geophys. Res.*, **107**, 5036, doi:10.1029/2001JE001658.

Khan, A., K. Mosegaard, and K. L. Rasmussen, 2000. A new seismic velocity model for the Moon from a Monte Carlo inversion of the Apollo lunar seismic data, *Geophys. Res. Lett.*, **27**, 10.1029/1999GL008452, 1591–1594.

Koyama, J. and Y. Nakamura, 1980. Focal mechanism of deep moonquakes, *Proc. 11th Lunar Planet. Sci. Conf.*, 1855–1865.

•Lammlein, D. R., G. V. Latham, J. Dorman, Y. Nakamura, and M. Ewing, 1974. Lunar seismicity, structure, and tectonics, *Rev. Geophys. Space Phys.*, **12**, 1–21.

Melchior, P., 1983. *The Tides of the Planet Earth*, (2nd edition), 641 pp., Oxford, Pergamon Press.

Nakamura, Y., 1976. Seismic energy transmission in the lunar surface zone determined from signals generated by movement of lunar rovers, *Bull. Seismol. Soc. Amer.*, **66**, 593–606.

•1978. A1 moonquakes: source distribution and mechanism, *Proc. 9th Lunar Planet. Sci. Conf.*, 3589–3607.

1983. Seismic velocity structure of the lunar mantle, *J. Geophys. Res.*, **88**, 677–686.

2003. New identification of deep moonquakes in the Apollo lunar seismic data, *Phys. Earth Planet. Int.*, **139**, doi:10.1016/j.pepi.2003.07.017, 197–205.

•2005. Far-side deep moonquakes and deep interior of the Moon, *J. Geophys. Res.*, **110** 10.1029/2004JE002332, E01001.

Nakamura, Y. and J. Koyama, 1982. Seismic *Q* of the lunar upper mantle, *J. Geophys. Res.*, **87**, 4855–4861.

•Nakamura, Y., G. V. Latham, and H. J. Dorman, 1982. Apollo lunar seismic experiment – final summary, *J. Geophys. Res.*, **87** Supp., A117-A123.

Okal, E. A., and S. H. Kirby, 1995. Frequency–moment distribution of deep earthquakes: implications for the seismogenic zone at the bottom of slabs, *Phys. Earth Planet. Int.*, **92**, 1 doi:10.1016/0031-9201(95)03037-8, 69–187.

Richter, C. F., 1958. *Elementary Seismology*, San Francisco, W. H. Freeman.

Schubert, G., D. L. Turcotte, and P. Olson, 2001. *Mantle Convection in the Earth and Planets*, Cambridge, U.K., Cambridge University Press, 940 pp.

Solomon, S. C., and M. N. Toksoz, 1973. Internal constitution and evolution of the Moon, *Phys. Earth Planet. Int.*, **7**, 15–38.

Stacey, F. D., 1992. *Physics of the Earth* (3rd edition), Brisbane, Australia, Brookfield Press, 512 pp.

Toksoz, M. N., N. R. Goins, and C. H. Cheng, 1977. Moonquakes: mechanisms and relation to tidal stresses, *Science*, **196**, 979–981.

Vinnik, L., H. Chenet, J. Gagnepain-Beyneix, and P. Lognonné, 2001. First seismic receiver functions on the Moon, *Geophys. Res. Lett.*, **28**, doi:10.1029/2001GL012859, 3031–3134.

Earthquake index

Index of earthquakes mentioned in this book. Unless noted otherwise, all have been reported as intermediate- or deep-focus by at least one source. Dates in bold-faced type indicate significant events discussed in detail in Chapter 10. Entries followed by [f] indicate figures; entries followed by [t] indicate tables.

Index